DECOMMISSIONING FORECASTING AND OPERATING COST ESTIMATION

DECOMMISSIONING FORECASTING AND OPERATING COST ESTIMATION

Gulf of Mexico Well Trends, Structure Inventory and Forecast Models

MARK J. KAISER

Gulf Professional Publishing is an imprint of Elsevier
50 Hampshire Street, 5th Floor, Cambridge, MA 02139, United States
The Boulevard, Langford Lane, Kidlington, Oxford, OX5 1GB, United Kingdom

Library of Congress Cataloging-in-Publication Data
A catalog record for this book is available from the Library of Congress

British Library Cataloguing-in-Publication Data
A catalogue record for this book is available from the British Library

ISBN: 978-0-12-818113-3

For information on all Gulf Professional publications
visit our website at https://www.elsevier.com/books-and-journals

Publisher: Brian Romer
Senior Acquisition Editor: Katie Hammon
Editorial Project Manager: Aleksandra Packowska
Production Project Manager: Sruthi Satheesh
Cover Designer: Christian Bilbow

Typeset by SPi Global, India

CONTENTS

ACKNOWLEDGMENT

This work was prepared on behalf of the US Department of the Interior's Bureau of Ocean Energy Management and has not been technically reviewed by the BOEM. The opinions, findings, conclusions, or recommendations expressed in this book are those of the author and do not necessarily reflect the views of the BOEM. Funding for this research was provided through the US Department of the Interior's Bureau of Ocean Energy Management under BOEM contract M16PX00059.

ABBREVIATIONS AND UNITS

ABBREVIATIONS

API	American Petroleum Institute
°API	API gravity
BOEM	Bureau of Ocean Energy Management
BSEE	Bureau of Safety Environmental Enforcement
C	caisson
CAPEX	capital expenditures
CGOR	cumulative gas-oil ratio
DOI	Department of Interior
DVA	direct vertical access
FP	fixed platform
FPSO	floating production storage offload
GOR	gas-oil ratio
GR	gross revenue
GoM	Gulf of Mexico
LOE	lease operating expense
MD	measured depth
MMS	Minerals Management Service
MTLP	mini tension leg platform
MOPU	mobile offshore production unit
NR	net revenue
NTL	notice to lessee
OTC	Offshore Technology Conference
OPEX	operating expense
OOIP	original oil-in-place
OCS	outer continental shelf
PA	permanent abandonment
PV-10	present value at 10% discount
P	price
P2	probable reserves
P10	probability at 10% confidence
P50	probability at 50% quartile (median)
P90	probability at 90% confidence
Q	production
PHA	Production Handling Agreement
P1	proved reserves
PVT	pressure volume temperature
ROY	royalty payment
r	royalty rate
semi	semisubmersible
SPE	Society Petroleum Engineers
spar	surface piercing articulating riser
TA	temporary abandonment

TLP tension leg platform
TVD total vertical depth
US The United States
WAT wax appearance temperature
WP well protector

UNITS

ac acre
bbl barrel
boe barrel of oil equivalent
boepd barrel of oil equivalent per day
bopd barrel of oil per day
bpd barrel per day
B billion
Btu British thermal units
cf cubic feet
cfpd cubic feet per day
DWT dead weight ton
ft foot
L liter
m meter
MM million
M thousand
mg milligram
psi pounds per square inch
t tons
T trillion

BOX LIST

EXECUTIVE SUMMARY

OVERVIEW

The US Gulf of Mexico (GoM) is one of the oldest and most prolific offshore hydrocarbon basins in the world, producing over 52 billion barrels oil equivalent—20.7 Bbbl of oil and 187 Tcf of natural gas—from first production in 1947 through 2017.

Both oil (Fig. E.1) and natural gas (Fig. E.2) production in water depth less than 400 ft have been in decline for over two decades. Dwindling commercial prospects, sustained low oil and gas prices, smaller operators with smaller budgets, and the success of onshore shale development mean that drilling and installation activity in shallow water has been dramatically reduced in recent years and replaced with high levels of decommissioning (Fig. E.3). It is no longer possible for operators to continue "kicking the decommissioning can" down the road as fields turn marginal and production is no longer economic. Shelf production is not being replenished by drilling, and many industry observers believe the region is "fished out" and no longer prospective for future discoveries.

Fortunately, the deepwater GoM is also a highly prolific hydrocarbon basin, and as shelf production has declined, industry activity has moved into deeper water. Here, deepwater refers to water depth greater than 400 ft, but there is nothing special about this selection, and 600 ft, 1000 ft, or even 1500 ft cutoffs could be used without significantly changing the trends, inventory data, or model results that are presented herein. With participation of all the majors and several independents, deepwater oil production continues to increase Gulf-wide. Deepwater wells are expensive to drill and complete, and deepwater reservoirs are complex and difficult to develop, but high discovery rates and large deposits have kept operator interest and capital spending in the region high.

As with other large-scale development activities on public lands, offshore exploration and development in the US Outer Continental Shelf (OCS) requires environmental impact statements to be performed on a regular basis. The environmental impacts of leasing, development, and production are evaluated as part of the Bureau of Ocean Energy Management's (BOEM's) compliance with the National Environmental Policy Act that requires federal agencies to study the environmental impacts of their decision-making (Luther, 2005). One significant source of environmental impacts is the installation, operation, and decommissioning of offshore structures. To perform these evaluations, it is necessary to quantify how structure inventories are expected to change in the future with changes in business conditions and leasing programs.

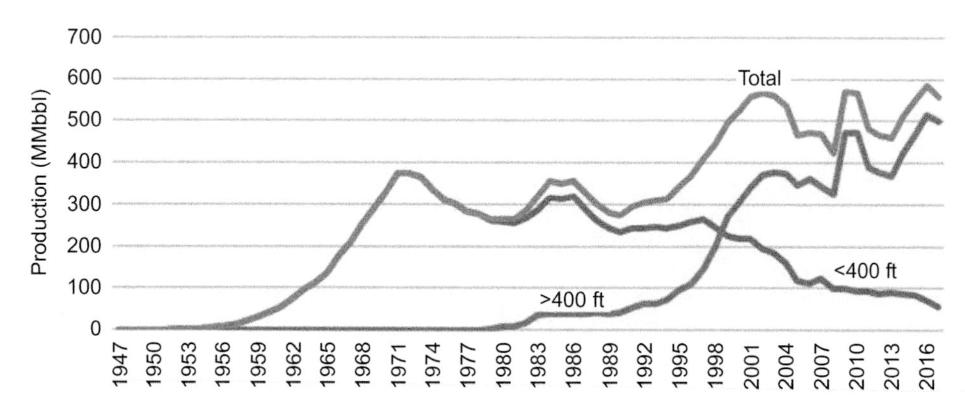

Fig. E.1 Oil production in the shallow-water and deepwater Gulf of Mexico. *(Source: BOEM 2018.)*

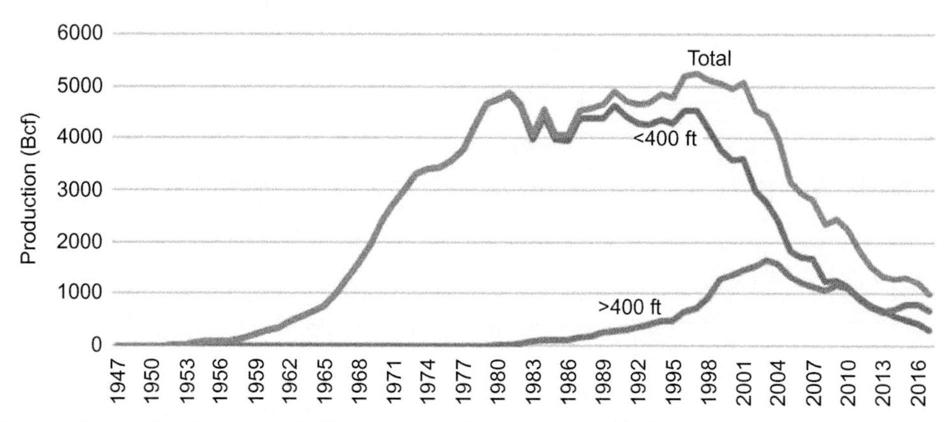

Fig. E.2 Gas production in the shallow-water and deepwater Gulf of Mexico. *(Source: BOEM 2018.)*

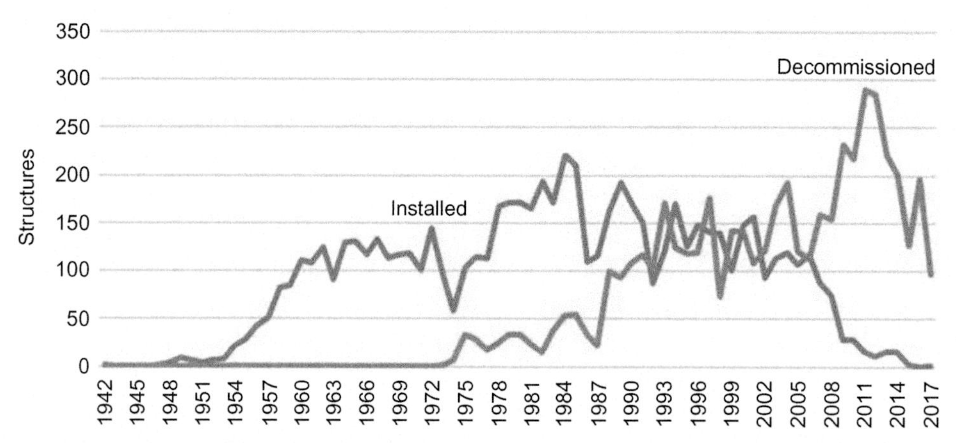

Fig. E.3 Structures installed and decommissioned in the shallow-water Gulf of Mexico. *(Source: BOEM 2018.)*

Infrastructure trends indicate a sharp decline in standing structures in shallow water and a relatively stable deepwater inventory (Figs. E.4 and E.5). Note the different scales in the two graphs. What are the expected trends in the two water depth classes in the future?

The primary purpose of this book is to construct decommissioning forecast models for the shallow-water and deepwater GoM based on economic and scientific considerations. Both water depth regions have played important roles in the development of the industry, but their future prospects are quite different. As shallow-water prospectivity declines and deepwater oil continues to produce at record levels, differences between the regions will become more pronounced.

The second objective of this book is to examine offshore operating costs in the GoM and bring together its many disparate branches. Offshore operations are expensive, but how expensive and how and why costs change over time have not been well documented and are rarely discussed in the literature.

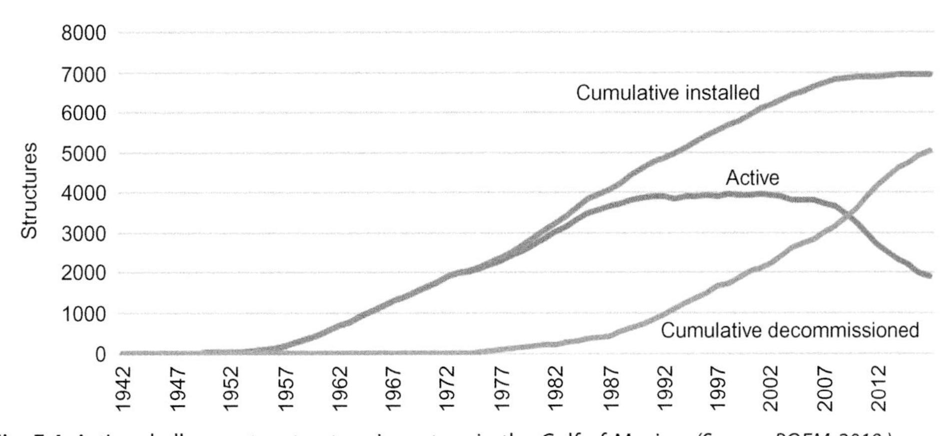

Fig. E.4 Active shallow-water structure inventory in the Gulf of Mexico. *(Source: BOEM 2018.)*

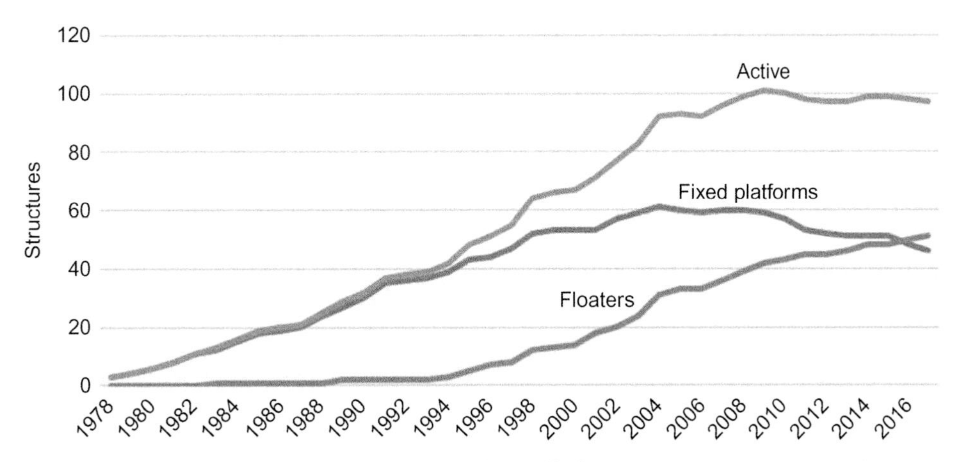

Fig. E.5 Active deepwater structure inventory in the Gulf of Mexico. *(Source: BOEM 2018.)*

ORGANIZATION

This text is organized in five parts. In Part 1, GoM production and activity statistics are reviewed circa 2017, structure classifications and economic limit factors are discussed, and reserves and resource estimates are quantified. In Part 2, well trends and structure inventories in the shallow-water and deepwater GoM are examined, and economic limit statistics are computed. In Part 3, the decommissioning forecast methodology is presented, and model results are described. Critical infrastructure issues and the benefits, costs, and risks of maintaining idle infrastructure are examined in Part 4, along with an update on the Louisiana and Texas Rigs-to-Reefs programs. The book concludes in Part 5 with a review of GoM operating cost data, factor models, and activity-based costing models.

OUTLINE

Part 1: Overview
Chapter 1 Production and Active Inventories
Chapter 2 Structure Classification
Chapter 3 Installation and Decommissioning Activity
Chapter 4 Economic Limit Factors
Chapter 5 Reserves and Resources

Part 2: Well Trends and Structure Inventory
Chapter 6 Well Trends
Chapter 7 Shallow-Water Structure Inventory
Chapter 8 Shallow-Water Economic Limit Statistics
Chapter 9 Deepwater Structure Inventory
Chapter 10 Deepwater Economic Limit Statistics

Part 3: Decommissioning Forecast
Chapter 11 Methodology and Parameterization
Chapter 12 Two Examples
Chapter 13 Shallow Water Decommissioning Forecast
Chapter 14 Deepwater Decommissioning Forecast

Part 4: Critical Infrastructure Issues
Chapter 15 Field Development Opportunities
Chapter 16 Costs, Benefits, and Risk
Chapter 17 Rigs-to-Reefs Programs

Part 5: Operating Cost Review
Chapter 18 Offshore Production Facilities
Chapter 19 Operating Cost Characteristics
Chapter 20 Financial Statements and Other Methods
Chapter 21 Field Examples
Chapter 22 Factor Models and Production Handling Agreements
Chapter 23 Activity-Based Costing

HIGHLIGHTS

In Part 1, background information on GoM production and activity statistics, economic limit factors, and reserves and resource estimates are presented. Chapter 1 begins with oil and gas production and active inventory trends broken out by water depth region.

Chapter 2 describes structure classifications and the role hub platforms play in field development. Structure type is based on the jacket and hull configuration and the manner in which the structure is fixed to the seabed. Production status describes whether the structure is producing, idle (formerly producing), or auxiliary (never or no longer producing). Manning status, number of wells, and major and minor structures are reviewed.

In Chapter 3, installation and decommissioning activity is described for the shallow-water and deepwater GoM circa 2017. There has been a dramatic decline in the number of shallow-water structures installed in recent years and record levels of decommissioning due to the maturity of the region, aging infrastructure, lack of prospects, and low oil and gas prices. In deepwater, active inventories are more stable, and about 100 structures have populated the region over the past decade.

In Chapter 4, the operating cost characteristics of offshore production and economic limit factors are described. When the net revenue from production falls below operating cost for a sustained period of time, there is no economic benefit derived, and operators will eventually shut in wells and idle structures. Economic limits are not publicly reported or directly observable, however, but using the revenue generated the last year a structure or well produces serves as a useful proxy of its value.

Reserves and resource assessments underlie all investment decisions made by oil and gas companies, and a clear understanding of geologic units and definitions is needed in evaluation. In Chapter 5, trend analysis of field counts, sizes, and distributions is presented followed by a description of the geology of the shallow-water and deepwater GoM.

In Part 2, GoM well and structure inventories are reviewed and the economic limit of production is evaluated. Wells are a primary evaluation unit, and in Chapter 6, trends in exploration and development drilling, well abandonments, producing and idle wells, and subsea drilling are summarized.

In Chapter 7, the primary features of producing, idle, and auxiliary structures in shallow water are examined. Production, cumulative production, and revenue are the most important features for producing structures. Idle age and idle age at decommissioning are the primary attributes for idle structures. Manning status and complex association are the most useful descriptors to characterize auxiliary structures.

In Chapter 8, the economic limits of shallow-water structures are quantified to gain insight into the conditions needed for commercial production at the end of a field's life. Decommissioned structures are treated as a statistical ensemble, and structure subclasses are defined and evaluated. Summary statistics for 3054 decommissioned shallow-water structures are tabulated by primary production, structure type, manned status, and water depth. The P50 adjusted gross revenue the last year of production is $1.2 million for gas structures and $630,000 for oil structures. Factor models are used to disentangle the impact of structure and operator attributes.

In Chapter 9, deepwater structures are organized into three subgroups to manage the discussion: fixed platforms in water depth 400–500 ft, fixed platforms and compliant towers in water depth > 500 ft, and floater inventories. Floater equipment capacity and capacity-to-reserves ratios are computed, followed by statistics for production, reserves, PV-10, and gross revenue. A snapshot of gross revenue data for individual structures circa 2017 are presented in the second half of the chapter.

In Chapter 10, the economic limits of deepwater decommissioned structures and subsea wells are examined. From 1977 to 2017, 23 deepwater structures were decommissioned and 486 wells were permanently abandoned in water depth greater than 400 ft. The average inflation-adjusted net revenue the last year of production of all deepwater structures was $13.4 million with a median value of $3.1 million. Deepwater dry tree wells exhibited an average economic limit of $3.8 million compared with $17.8 million for wet tree wells the last year of production, and gas wells generated $3.2 million at the end of their life compared with $5 million for oil wells.

In Part 3, the decommissioning forecast methodology and model results are described. Chapter 11 begins with a description of the forecast models and schedule approaches adopted for the producing, idle, and auxiliary structure classes. Different models are required to characterize the different types of structures that exist. For producing structures, decommissioning forecasts apply economic models based on decline curves and cash flow analysis, but for structures that are not producing or have never produced, different methods are needed. A user-defined schedule methodology is adopted for idle structure decommissioning, and statistical methods are implemented for auxiliary structures. The importance of the structure classes vary in shallow water and deepwater and require different emphasis in modeling.

In Chapter 12, two examples illustrate the decommissioning forecast methodology for producing structures. Chevron's Tick platform was evaluated circa 2012, and Anadarko's Horn Mountain spar was evaluated circa 2016. Decommissioning timing and reserves estimates under oil and gas price variation are used to illustrate model sensitivity.

In Chapter 13, the results from the three submodules for shallow-water producing, idle, and auxiliary structures are combined in a composite structure forecast, and sensitivity analysis is performed. Exponential and hyperbolic decline curves are used to bound the decommissioning time for producing structures, schedules are adopted for idle structures, and historic activity trends are employed for auxiliary structures. Transition probabilities are employed to capture structure transitions between classes. From 2017 to 2022, between 474 and 828 structures are expected to be decommissioned in the shallow-water GoM, and by 2027, between 704 and 1199 structures are expected to be decommissioned.

Historically, deepwater installation and decommissioning in the GoM have always been small with activity levels that can be counted on one hand most years and two hands in high-activity periods. Large capital expenditures and long time periods in planning and execution constrain investment and the number of structures that are installed, and because operators seek to maximize the value of their infrastructure, deepwater decommissioning activity is limited and often delayed as new opportunities for existing structures arise. These trends are not expected to change for the foreseeable future. In Chapter 14, model results indicate that between 27 and 51 deepwater structures will be decommissioned through 2031, and between 12 and 25 structure removals are expected from 2017 to 2022.

In Part 4, a review of critical infrastructure issues ties together several discussion threads and introduces new ideas in the conceptual framework. In Chapter 15, critical infrastructure is defined, and opportunities within and outside field developments are outlined. In Chapter 16, the potential benefits, costs, and risks associated with maintaining infrastructure beyond its useful life are described along with the role of government regulations in incentivizing operators and managing offshore activity. In Chapter 17, the Louisiana and Texas Rigs-to-Reefs programs are described along with recent changes in legislative activity.

The objective of Part 5 is to describe the elements that comprise offshore operating cost and how they change with the changing needs of production. Examples illustrate analytic techniques and help develop intuition. After an introductory chapter on offshore production facilities (Chapter 18), operating cost characteristics are covered (Chapter 19), followed by public disclosure requirements and computer methods (Chapter 20), field examples (Chapter 21), factor models and Production Handling Agreements (Chapter 22), and activity-based cost models (Chapter 23). These chapters can be read in any order but a linear progression is the most efficient.

In Chapter 18, the equipment and processes that comprise an offshore production facility is presented at a systems level. Facilities require labor to operate and maintain and consume chemicals, energy, and services that constitute production cost. Production processes are for the most part easy to understand, consisting mostly of phase separation using temperature and pressure changes that only require tanks, vessels, and piping to perform. Export pipeline specifications and ownership play an important role in dictating the degree of processing required.

Chapter 19 begins by reviewing cost categories and rules of thumb to illustrate cost ranges. Although elementary, this material sets the stage for more advanced discussion and ensures definitions and basic concepts are not neglected. The chapter concludes with a general description of operating cost factors that expands upon the discussion started in Chapter 4.

Chapter 20 summarizes the disclosure requirements of public companies and describes survey, software, and computer methods used in operating cost estimation. Operating costs for public oil and gas companies with the majority of their production and reserves in the GoM are examined. Most of these companies have now either left the region, sold their assets, filed for bankruptcy, or reemerged after bankruptcy as a private company illustrating the difficult and rapidly changing nature of marginal operations in the Gulf in recent years. A review of production and lifting cost metrics and the strengths and weaknesses of lease operating statements, survey, and computer methods are also outlined. The UK North Sea operating cost survey is the best publicly available data on offshore oil and gas production cost in the world, and results from the survey are highlighted.

According to US Securities and Exchange Commission regulations, if a field comprises more than 15% of an operator's total reserves in a given year, field data must be broken out separately and reported in financial statements. In Chapter 21, GoM fields reporting operating cost between 2010 and 2016 are presented. Although the sample is small, field data are granular and worth reviewing.

In Chapter 22, factor models and Production Handling Agreements are described. Factor models are popular in cost modeling because they are easy to implement in spreadsheets and allow for versatile and efficient implementation. Unfortunately, their reliability is often poor, and rarely is any attempt made to validate and/or calibrate models with empirical data. Factor models for lease operating cost, workovers, and gathering and transportation services are illustrated and their limitations are described. Production Handling Agreements are common in the deepwater GoM and govern the relationship between the owner of a production handling system and third-party producers who wish to use the facility for operations. Contract term negotiations and the manner in which risk enters agreements are illustrated.

The text concludes in Chapter 23 with a review of activity-based cost models for each of the main operating cost categories. Activity-based cost models provide a more accurate and auditable assessment of cost provided they are performed by experienced personnel with an understanding of their limitations. Activity-based cost models rely heavily on understanding the nature of commercial contracts, operational requirements, and regional markets, and require a different set of assumptions and expertise than the previous models discussed in Part 5.

DATA AND STATISTICS

The data and statistics employed in this book are straightforward to understand with minimal levels of reporting and interpretation uncertainty. That is, the data represent factual information that is relatively easy to interpret assuming the data are processed properly and definitions are understood. If confusion arises, it can usually be traced to incomplete information or misunderstanding operational issues, or both, or the exclusion of relevant factors in evaluation. In some cases, there may be problems with the data itself, and issues may arise in interpretation or unusual situations, but these are usually exceptional (one-off) events and remediated with additional work.

Operators report production, installation and removal activity, drilling, and wellbore abandonments in the Outer Continental Shelf of the United States electronically to the BOEM within a certain period of time after custody transfer of production and completion of operations. After review and quality control, which may take six months or more, the BOEM uploads the data to the Technical Information Management System database.

Wellbore data were assembled from the BOEM Borehole database (BOEM, 2017a). Structures were identified using the BOEM Platform Masters and Platform Structures databases (BOEM, 2017b,c). Field data were evaluated from the field directory and reserves history summaries (BOEM, 2017d,e).

Data for the installation and decommissioning trends (Part 1) and field data (Parts 1 and 4) were evaluated from February to April 2018. Data for the structure inventory and decommissioning forecast models (Parts 2 and 3) were evaluated from March to July 2017 and updated selectively. Data for the Louisiana and Texas reef programs were valid through mid-2018 (Part 4). Data from company financial statements (Part 5) were reviewed in 2016.

UNITS

English units are used following North American convention. Conversion to SI units is performed using Table E.1.

Table E.1 SI Metric Conversion Factors

ac × 4.047E-01 = ha
cf/bbl × 1.801E-01 = m^3/m^3
141.5/(131.5+°API) = g/cm^3
bbl × 1.589E-01 = m^3
ft × 3.048E-01 = m
ft^3 × 2.831E-02 = m^3
(°F − 32)/1.8 = °C
mi × 1.609E+00 = km
psi × 6.894E+00 = kPa

REFERENCES

Bureau of Ocean Energy Management, 2017a. BOEM Well Database.
Bureau of Ocean Energy Management, 2017b. BOEM Platform Master Database.
Bureau of Ocean Energy Management, 2017c. BOEM Structure Database.
Bureau of Ocean Energy Management Gulf of Mexico OCS Region, 2017d. OCS operations field directory. Quarterly Report, December 31, 2017.
Energy Management Gulf of Mexico OCS Region, 2017e. Reserves history for fields. Gulf of Mexico Outer Continental Shelf. December 31, 2016.
Luther, L., 2005. The National Environmental Policy Act: background and implementation. Congressional Research Service, Washington DC.

Mark J. Kaiser
Baton Rouge, Louisiana

February 2019

PART ONE

Overview

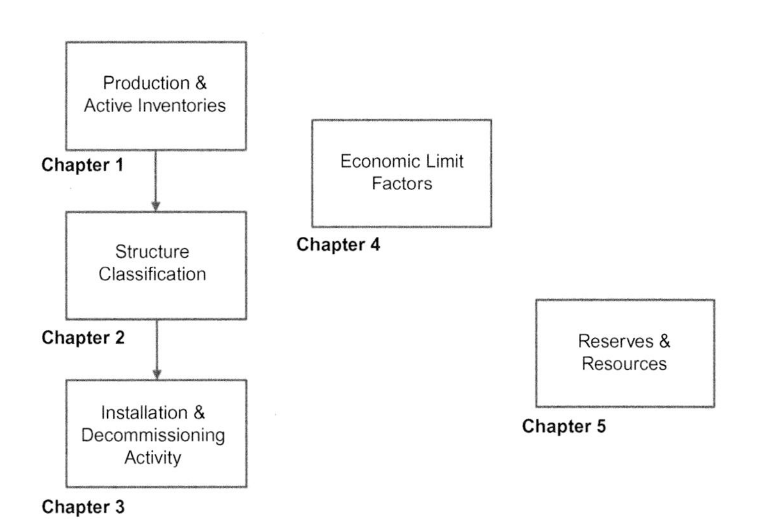

CHAPTER ONE

Production and Active Inventories

Contents

Abstract

Oil and gas production and active inventory trends in the shallow-water and deepwater Gulf of Mexico are described circa 2017, and structure terminology is introduced. The shallow-water Gulf has witnessed significant changes over the past decade and will continue to be subject to significant changes in the future. Shelf production is not being replenished by drilling, and most industry observers believe the shelf is fished out. Fortunately, the deepwater Gulf of Mexico is a highly prolific hydrocarbon basin, and oil production continues to increase Gulf-wide. Differences between the shallow-water and deepwater regions and future prospects are highlighted.

Decommissioning Forecasting and Operating Cost Estimation
https://doi.org/10.1016/B978-0-12-818113-3.00001-1

1.1 THE SETTING

The federal waters of the Gulf of Mexico (GoM) are administered by the Bureau of Ocean Energy Management (BOEM) and the Bureau of Safety and Environmental Enforcement (BSEE) and are described in terms of three administrative areas, referred to as the Western, Central, and Eastern planning areas.

1.1.1 Gulf of Mexico

The Gulf of Mexico basin formed approximately 160–200 million years ago during the Mesozoic Era with the breakup of Pangea and associated tectonic movements (Salvador, 1991; Galloway, 2008). Rifting probably continued through Early and Middle Jurassic time with the formation of stretched or transitional continental crust throughout the central part of the basin. There are two large carbonate platforms, the Yucatan Peninsula/Campeche Bank on the west and the Florida Platform on the east. Both platforms support wide and shallow submerged shelves (Fig. 1.1).

The deepest portion of the GoM is the Sigsbee[1] Deep, a flat region that lies in the western portion at a depth of 12,300 to 14,400 ft. Abyssal plains are thought to be derived

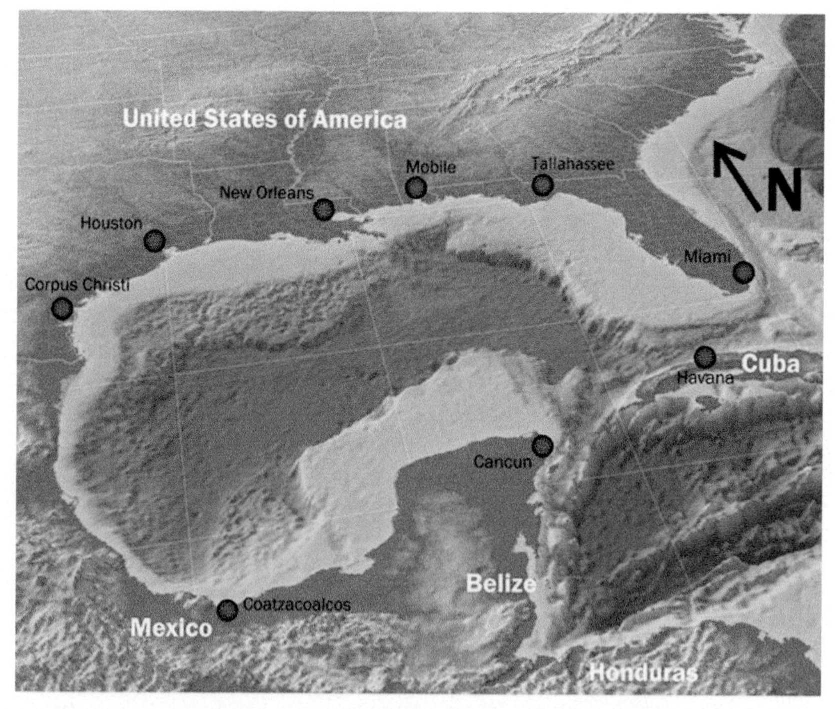

Fig. 1.1 Shelf and slope features of the Gulf of Mexico basin. *(Source: NOAA.)*

[1] Charles Dwight Sigsbee was the commanding officer of the steamer *George S. Blake* who discovered the feature in 1873–75 during mapping of the basin.

from sediment that accumulates in abyssal depressions. Seismic refraction and reflection measurements have revealed that the sedimentary layers in the GoM continue several miles beneath the seafloor.

1.1.2 Shelf vs. Slope

The continental shelf is a broad, relatively shallow submarine terrace of continental crust representing a continuation of the continental landmass and extending outward to the continental slope and rises where the deep ocean begins (Fig. 1.2). Continental shelves are part of the continental landmass during glacial periods and are inundated during interglacial periods. The width of the continental shelf around the United States varies from approximately 12 to 250 mi, depending on location.

The GoM continental shelf has an average gradient of about 1 in 500 and ranges from 600 to 1000 ft water depth, while the gradient of the continental slope increases to about 1 in 20 and extends up to about 9000 ft water depth (Garrison and Ellis, 2018).

The surface of the shelf is generally smooth with low-relief irregularities resulting from the presence of relict Pleistocene stream and shoreline deposits, scarps formed by active growth faults, and mounds produced by salt and shale diapirism. Seafloor topography on the slope and rise is complex and includes areas of possible landsliding and faulting, mud seeps, undulations, and rocky outcrops. Irregular rocky seafloor with sharp relief of tens of feet, fault scarps of 100–300 ft high, and areas of possible landslide contribute to the difficult conditions on the slope. Rugged topography on the flank of domes suggests failures in the geologically recent past. In water depths less than about 1500 ft, gas hydrates generally do not form gas hydrate hills, but large seep mounds may form around gas vents. Seep mounds may be steep or unstable.

1.1.3 State vs. Federal Waters

In the early years of offshore oil and gas production, California, Texas, Louisiana, and the federal government all claimed control over lands below the low-tide mark (Priest, 2008a). States challenged a 1945 proclamation made by President Truman that granted authority over the subsoil of the US continental shelf to the federal government and prompted the US Department of Justice to file suits against them. The legal

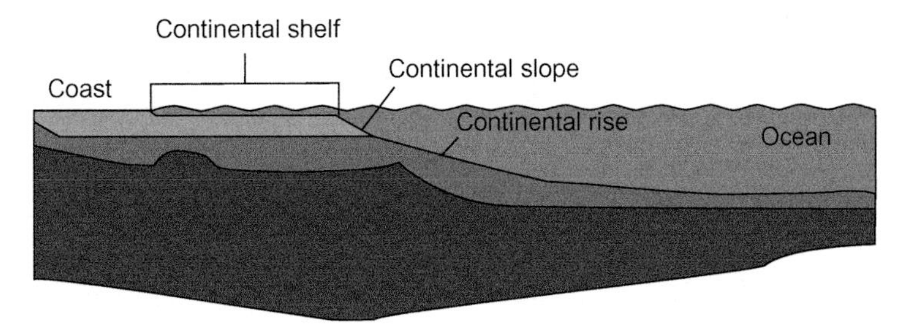

Fig. 1.2 Schematic representation of continental shelf, slope, rise, and abyssal plain. *(Source: ONR.)*

disputes ended when the US Supreme Court ruled in favor of the US government, against California in 1947 and against Louisiana and Texas in 1950 (National Commission, 2011).

In May 1953, Congress passed two pieces of legislation that put an end to the debate over federal and state jurisdiction. The Submerged Lands Act validated all state leases that had been awarded prior to the issuance of the Supreme Court's decision against California, Texas, and Louisiana and reserves to the states all land within 3 nautical miles (nm) of their shores (Box 1.1). For Texas and the west coast of Florida, the distance was 3 marine leagues (about 10 nm). The Outer Continental Shelf Lands Act (OCSLA) placed all offshore lands beyond the 3 nm limit under federal jurisdiction and gave the US Department of Interior the authority to issue leases.

1.1.4 Shallow Water vs. Deepwater

Shallow water and deepwater are relative terms, of course, and various thresholds have been used throughout the years to distinguish deepwater (Box 2.1). The late Griff Lee defined deepwater as "that depth just beyond the deepest existing platform...gradually increasing with age and the progress of the offshore industry." Today, the shelf is synonymous with shallow water and the slope and beyond with deepwater (Fig. 1.3), but this was not always the case.

In this book, shallow water is assumed to correspond to water depths less than or equal to 400 ft and deepwater to water depths >400 ft. In 1936, Louisiana sold its first offshore lease, and one year later Creole was the first oil field discovered in the GoM (Box 1.2). Production in federal waters began in 1947, where the first well drilled out of sight of land began producing 600 bopd 12 miles from the Louisiana coast. In 1978, Shell installed

BOX 1.1 Why Are (Most) State Waters Defined by a 3 mile Limit?

In Roman law, the sea was considered open to all persons and therefore incapable of appropriation and ownership (Percy, 1925). This doctrine held until the rise of the maritime nations of the Middle Ages (Venice, Denmark, Sweden, France, England, Spain, and Portugal) all asserted proprietorship and claims of the oceans of various extent and credibility (Bartley, 1953).

As the more grandiose claims faded away, a new doctrine for the measurement of a nation's dominion in the coastal sea arose, first stated in 1702 by Bynkershock (Bartley, 1953), but it may have arisen as early as 1609 by the Dutch legal philosopher Hugo Grotius who formulated the doctrine of "freedom of the seas" in support of his country's growing maritime influence and reach. The basis of the proposition was the effectiveness of occupation and control.

The distance that a cannon shot could reach and thereby control was considered the delineation in which the sea might be appropriated. Italian jurist Galiani proposed a range for cannon of 3 mi, a position subsequently adopted by other writers of international law. British common law after 1776 adopted the concept of the 3 mi limit into a doctrine of Crown ownership, which passed through to the current era.

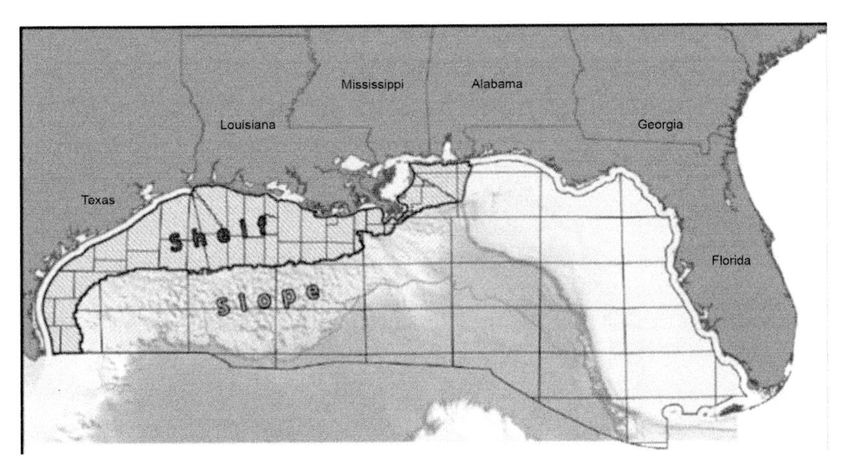

Fig. 1.3 Shelf and slope in the Central and Western Gulf of Mexico. *(Source: BOEM.)*

fixed platforms for the Bourbon project in 423 ft water depth and Cognac in 1023 ft water depth, and is generally considered the start of deepwater development in the GoM. The region between 400 and 1000 ft water depth has not been especially productive, and only 50 or so platforms have been installed throughout the region circa 2017. Using 1000 ft or even 1500 ft as the deepwater threshold will have no material impact on our discussion or statistics presented. In recent years, a new category ultradeepwater has been used to refer variously to water depth > 5000 ft, > 2000 m, and > 10,000 ft.

1.1.5 Sigsbee Escarpment

The Sigsbee Escarpment is a major geomorphological feature of the GoM seafloor, basically an underwater mountain about 2000 ft in height at its tallest, and extending for several hundred miles across the central and eastern GoM (Fig. 1.4). The escarpment and

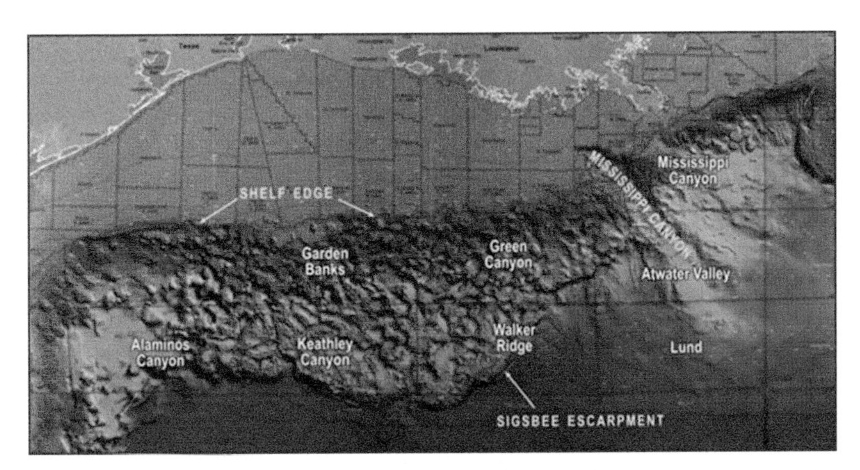

Fig. 1.4 Bathymetry of the northern Gulf of Mexico and Sigsbee Escarpment. *(Source: NOAA.)*

associated canyons were caused by deformation of underlying salt deposits and erosion during periods of low sea level in the geologic past.

Salt is a significant geologic feature in the GoM, and the salt canopy extends for hundreds of kilometers with thickness locally of as much as 3 to 6 mi. As the salt deforms, it creates faulting, scarps, and other potential geohazards that need to be understood and quantified to determine if development in the area is of acceptable risk. Below the slope break, the escarpment dips at steep angles (5 to 25 degrees), and near the base of the escarpment, major slump activity is observed out to several miles, clear evidence of numerous past failures that would have devastating consequences for any infrastructure in the debris flow.

1.1.6 Outer Continental Shelf Lands Act

In the GoM, leases for exploration and production on the outer continental shelf (OCS) are obtained via auction and the payment of a bonus bid and are subject to rent during the primary term and royalty during production. Section 8 of OCSLA stipulated that GoM tracts be auctioned by competitive, sealed bidding based on a cash bonus bid with a fixed royalty on oil and gas production paid to the government of not less than 12½ % (Priest, 2008b). Lease areas could not exceed 5760 ac (3 × 3 mi block size), and primary periods were specified for at least 5 years (Dupre, 2015).

From 1954 to 1982, areas for leasing were nominated by oil and gas companies, and lease terms were for 5 years, and the royalty rate was 1/6 (16.67%). In 1982, the Reagan administration announced a new area-wide leasing program that made available all unleased and available blocks in a planned area or program area. Under area-wide leasing, primary terms were extended to 8–10 years, and royalty rates on deepwater tracts >400 m were lowered to 1/8 (12.5%). Various modifications to these terms and conditions have been made over the years with changing economic and political conditions (US DOI 2001). Historically, area-wide sales in the central and western GoM occurred in March and August each year, but starting with lease sale 249 in 2017, the Western and Central planning areas are now combined, and BOEM holds two combined sales each year.

1.1.7 Protraction Areas

Gulf of Mexico planning areas are subdivided in named protraction areas, and each protraction area is divided into 3 × 3 mi (5760 ac) lease blocks, unless clipped by an existing marine boundary or edge of a planning area. The large mostly rectangular interior protraction areas reside on the deepwater slope and rise (Fig. 1.3) and are defined as 2 degrees east-west by 1 degree north-south or about 120 mi by 60 mi in areal extent. Approximately 800 lease blocks are contained within each deepwater protraction area.

1.2 PRODUCTION

From 1947 to 2017, the GoM produced 20.7 billion barrels (Bbbl) of oil and 187 trillion cubic feet (Tcf) of natural gas. More than half of total oil production (12.2 Bbbl) and about 80% of the Gulf's natural gas production (160 Tcf) have been produced in shallow water <400 ft and have been in decline since the mid-1990s (Figs. 1.5 and 1.6). In water depths >400 ft, oil production continues to increase, marked by significant fluctuations due to hurricane activity from 2004 to 2008 and the Macondo oil spill in 2010. The deepwater GoM is largely an "oily" province, and gas production in the region to date has not been enough to outpace the shallow-water decline.

In 2017, the GoM produced about 604 million barrels (MMbbl) of oil and 1 Tcf of natural gas, about 10% of domestic oil and gas production, generating around $34 billion in gross revenues. About 57 MMbbl oil and 318 Bcf natural gas were produced in water depth <400 ft, representing about 13% of the total offshore oil production and one-third of the total offshore gas production in the Gulf.

The mid-1950s through the early 1970s was a time of numerous discoveries and intense development activity, but by the late 1980s, discoveries on the shelf became less prolific, and in recent decades, exploration has been dramatically curtailed. Shelf production is not being replenished by drilling, and many industry observers believe the region is "fished out" and no longer prospective for future discoveries. Dwindling commercial prospects, sustained low oil and gas prices, smaller operators with smaller budgets, and the success of onshore shale development means that drilling and installation activity in shallow water has been dramatically reduced in recent years.

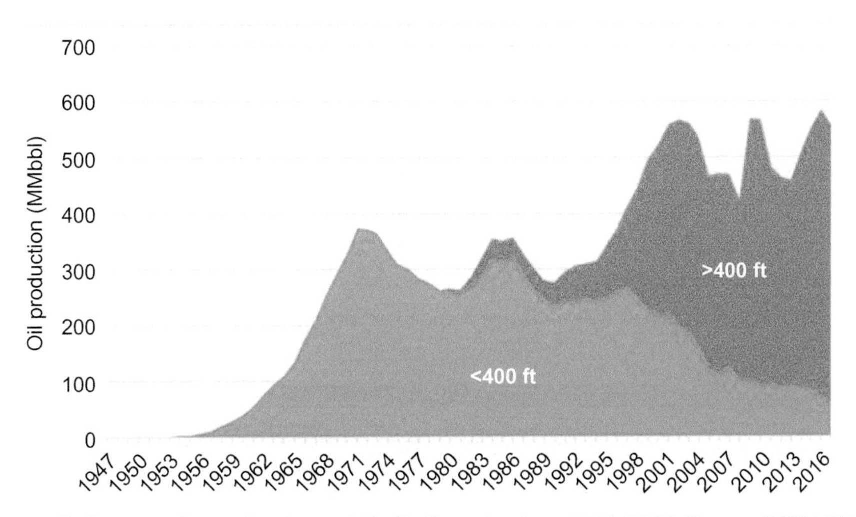

Fig. 1.5 Gulf of Mexico Outer Continental Shelf oil production, 1947–2017. *(Source: BOEM 2018.)*

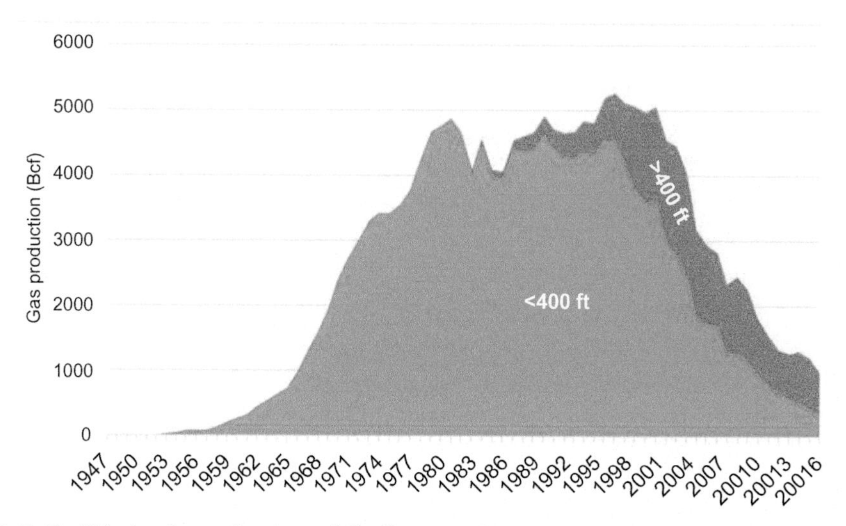

Fig. 1.6 Gulf of Mexico Outer Continental Shelf gas production, 1947–2017. *(Source: BOEM 2018.)*

Fortunately, the deepwater GoM is also a highly prolific hydrocarbon basin, and as shelf production has declined, industry activity has moved into deeper water with notable success. Circa 2017, the deepest structure in the GoM is Shell's Stones floating production storage and offloading vessel (FPSO) development in 9560 ft water depth. The most recent phase of deepwater exploration began in 2000–04 as companies pursued the Lower Tertiary trend that includes the stratigraphic interval between late Paleocene and early Eocene time. The Lower Tertiary Wilcox trend has been an important petroleum resource in the Gulf Coast states since the 1930s, producing primarily gas (Owen, 1975), but only in the early 2000s did penetrations of Lower Tertiary sands in the ultradeepwater GoM confirm the extension of the play there. Oil in place in Lower Tertiary reservoirs is huge, but reservoir conditions are difficult, and only a fraction of volumes will be recovered using primary depletion.

Deepwater produced 87% of the oil and condensate production in the Gulf and about two-thirds of the natural gas production in 2017, and over 80% of the remaining 3.5 Bbbl of oil and 7.3 Tcf of natural gas reserves are estimated to reside in deepwater (Burgess et al., 2018). Eighty-five percent of the 74 Bboe undiscovered technically recoverable resources are also estimated to reside in deepwater (US DOI, 2017) underlying the regions importance in the future. Currently, GoM oil production is near its all-time peak and is expected to plateau in the next few years, while GoM gas production has fallen steeply from its historic peak and is expected to continue to decline.

Table 1.1 Active Gulf of Mexico Inventory Circa 2017

Water Depth (ft)	Installed	Removed	Active
<400	6933	5024	1909
>400	120	23	97
Total	7053	5048	2005

Source: BOEM 2018.

1.3 ACTIVE INVENTORY

A total of 7053 structures have been installed in the GoM, and 5048 structures have been decommissioned through 2017, leaving an active (or "standing") inventory of 7053 − 5048 = 2005 structures circa 2017.

The vast majority of GoM structures reside in water depth <400 ft, and circa 2017, there were 1909 shallow-water structures and 97 deepwater structures (Table 1.1). Almost all installation and removal activity is in shallow water.

About 80% of shallow-water structures reside in water depth < 150 ft, and over two-thirds of all structures reside in <100 ft water depth (Table 1.2). Caissons and well protectors number about half of fixed platforms, and fixed platforms represent about two-thirds of the active inventory circa 2017.

In water depth > 400 ft, there have been 65 fixed platforms and 55 floaters installed through 2017, and 18 fixed platforms and five floaters have been decommissioned.

1.4 STOCK CHANGES

Active inventory represents the number of standing structures at a point in time and is determined simply as the difference between the cumulative number of structures installed up to that point and the cumulative number of structures decommissioned.

A bathtub analogy differentiates flow and stock variables (Fig. 1.7). Installations add to the inventory of active structures, while decommissioning removes structures from stock, and the relative difference between these two flows over a period of time (normally taken

Table 1.2 Active Shallow-Water Structures by Type and Water Depth Circa 2017

Water Depth (ft)	Caisson/WP	Fixed Platform	Total (%)
≤100	563	724	1287 (67%)
101–150	17	181	198 (10%)
151–200	10	152	162 (8%)
201–400	11	251	262 (14%)
Total	601	1308	1909

Source: BOEM 2018.

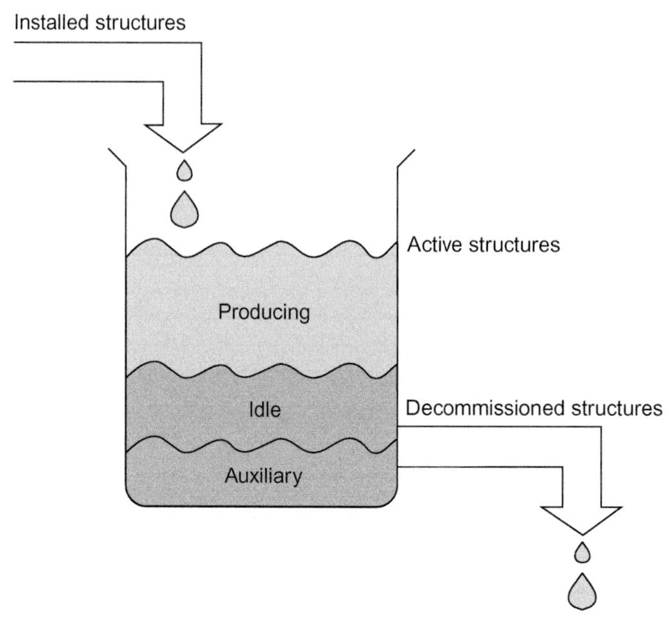

Fig. 1.7 Bathtub analogy for active inventory.

as one calendar year) determines the change in inventory over the period. When the number of installations during the year exceeds the number of decommissioned structures, active inventory will grow and vice versa; when decommissioning activity exceeds installation activity, standing inventories will decline.

At the beginning of offshore development, installations dominate, but as the region matures and fields deplete, decommissioning will increase and the number of active structures will decline. New discoveries will still be made, but the contribution of new structures relative to decommissioning activity will be small. The GoM shelf is in the decline phase.

To characterize the decline in standing structures, the change in the active inventory relative to its size is used to describe the inventory decline rate. Since 2009, the decline rate of the active inventory in the GoM has ranged between 5% and 9% per year (Table 1.3).

1.5 TRENDS

1.5.1 Shallow Water

The best single summary graph describing infrastructure trends in shallow water is shown in Fig. 1.8, where the cumulative number of installed and decommissioned structures is plotted along with their difference. In Fig. 1.9, the shallow-water active inventory is

Table 1.3 Active Inventory and Stock Changes, 2008–17

Year	Installed	Decommissioned	Active	Decline Rate (%)
2008	78	154	3755	
2009	32	234	3553	5.4
2010	30	220	3363	5.3
2011	18	294	3087	8.2
2012	11	286	2812	8.9
2013	17	223	2606	7.3
2014	21	203	2424	7.0
2015	4	128	2300	5.1
2016	2	200	2102	8.6
2017	2	108	1996	5.0

Source: BOEM 2018.

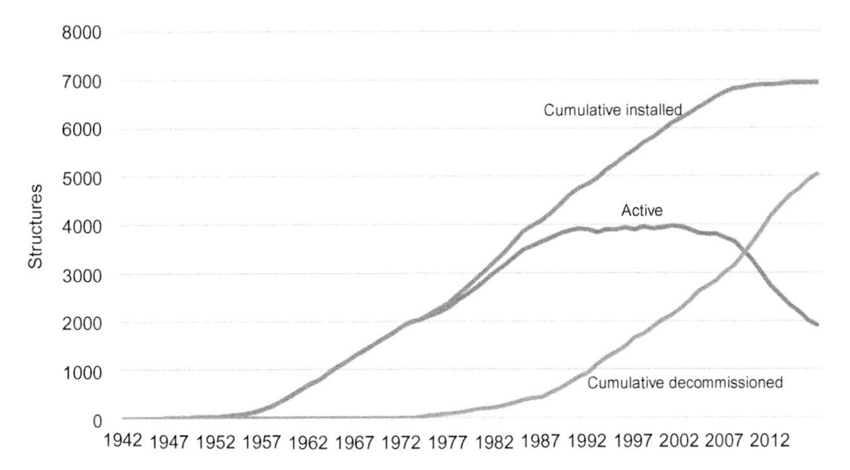

Fig. 1.8 Active structures in water depth <400 ft, 1942–2017. *(Source: BOEM 2018.)*

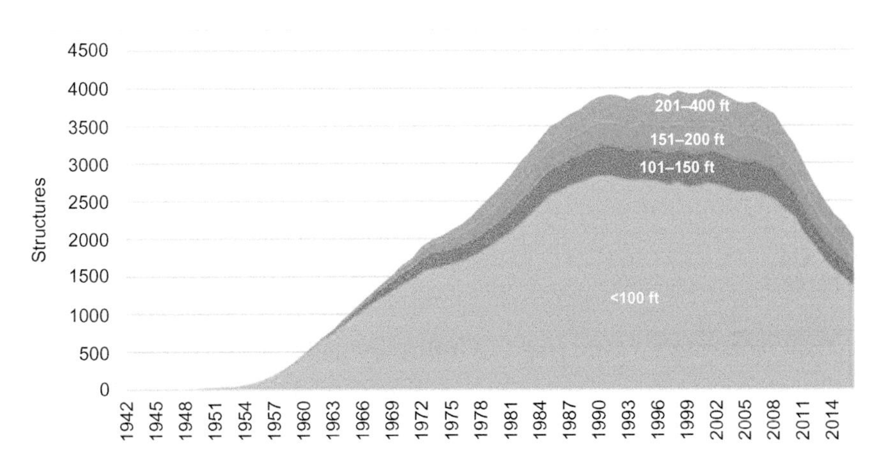

Fig. 1.9 Active shallow-water structures by water depth category circa 2017. *(Source: BOEM 2018.)*

broken out by water depth category. Most of the active shallow-water inventory resides in water depth < 100 ft

In 1973, the first structure in the GoM was decommissioned, and by the mid-1980s, removal rates regularly exceeded 100 structures per year and began to attract the attention of both state and federal regulators. Rigs-to-Reefs programs were developed during this time to help maintain the ecosystems established by the structures after decommissioning. For nearly 20 years beginning in the late 1980s, shallow-water inventories were essentially flat and ranged between 3800 and 4000 structures as installation and removal rates held each other in check and maintained quasi-equilibrium.

Before the mid-1980s, decommissioning was a relatively minor affair with activity levels averaging less than 30 structures per year. From the mid-1980s to 2006, cumulative removals approximately track installations, and the number of active structures held steady. Since 2008, installation rates have slowed down considerably, while decommissioning has accelerated, causing the active inventory to drop in a fashion that mimics its rise in the 1970s.

From 1987 to 2006, there were on average 134 shallow-water structures installed per year and 122 structures removed. From 2007 to 2017, there were on average 26 structures installed per year and 198 structures decommissioned. Over the past decade, decommissioning activity has been particularly intense with over 40% of all activity in shallow water taking place and at rates almost 10 times greater than installations.

1.5.2 Deepwater

In 1978, the first structure in water depth > 400 ft was installed in the GoM, and circa 2017, a total of 120 structures have been installed—65 deepwater fixed platforms, three compliant towers, and 52 floating structures. To date, a total of 18 deepwater fixed platforms and five floating structures have been decommissioned.

In 2002, deepwater fixed platform inventories appear to have peaked as decommissioning activity exceeded new installations for the first time (Fig. 1.10). Deepwater floater inventories continue to rise at a slow but steady pace as installation activity continues to dominate (Fig. 1.11). Deepwater structures maintain high residual value, and operators seek every opportunity to deploy them after their anchor fields have been exhausted.

1.6 OIL VS. GAS STRUCTURES

Structures are classified as "primarily oil" (or "oil") or "primarily gas" (or "gas") based on their cumulative gas-oil ratio (CGOR), defined as the total cumulative gas production measured in cubic feet to cumulative oil production measured in barrels. Typically, a CGOR threshold >10,000 cf/bbl is used to classify a well as primarily gas, while

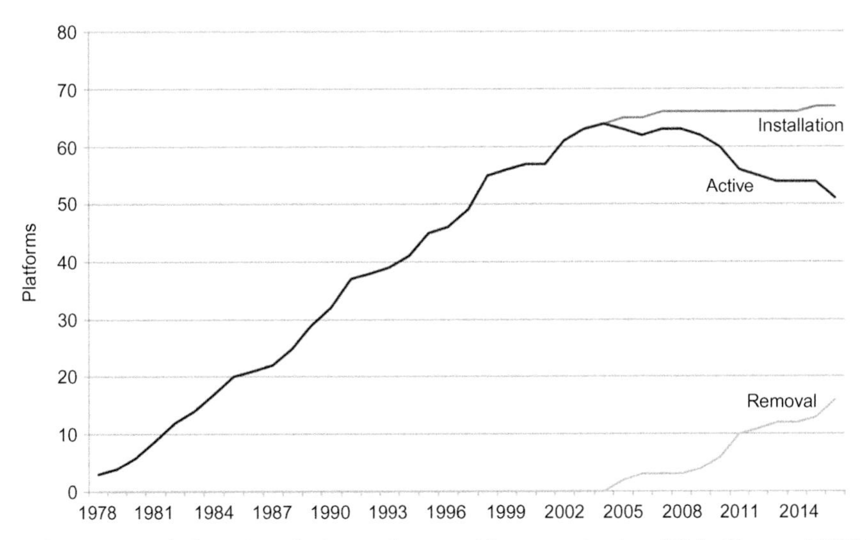

Fig. 1.10 Deepwater platform installation and removal in water depth >400 ft. *(Source: BOEM 2018.)*

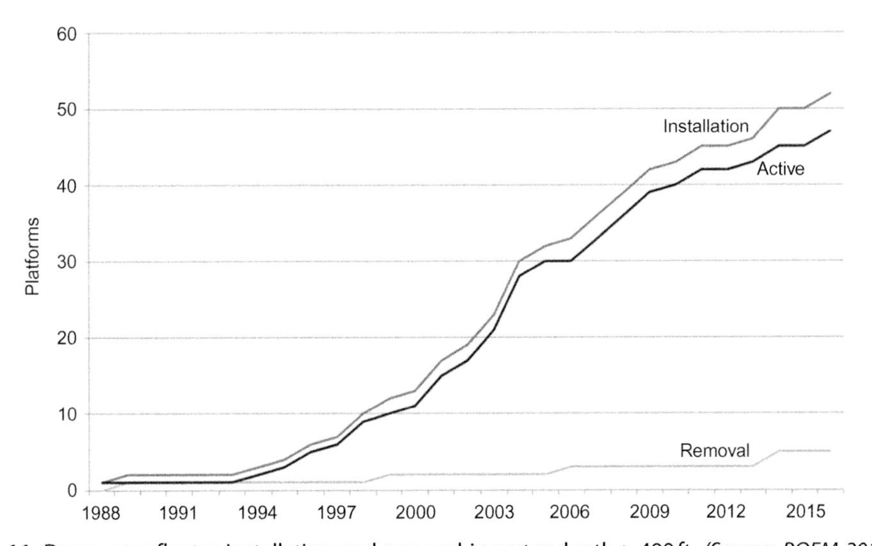

Fig. 1.11 Deepwater floater installation and removal in water depth >400 ft. *(Source: BOEM 2018.)*

CGOR <10,000 cf/bbl identifies wells as primarily oil, and this same classification can be used for structures. There is no universally agreed-upon value of the threshold, however, and cutoffs smaller than 5000 cf/bbl or values >10,000 cf/bbl may be employed. Note that lowering the threshold will reduce the oil structure count and increase the number of structures classified as gas producers.

1.7 PRODUCTION STATUS

1.7.1 Classification

Structures are classified according to their production status at a specific point in time (Fig. 1.12), whether they are currently producing, that is, structure produced during the year of evaluation (producing); currently not producing, that is, structure did not produce any hydrocarbons during the year of evaluation but previously produced (idle); or structures that have no (current) production links to wells but serve in a support role (auxiliary). Installation and decommissioning represent the beginning and end of a structure's life, but between these two end states there are multiple pathways a structure may take.

Producing, idle, and auxiliary classifications are not BOEM-assigned attributes but are inferred from publicly available data. Idle structures that plug and abandon all their wellbores and are repurposed for another function are reclassified as auxiliary even though they previously produced, but these structures are not distinguished in public data

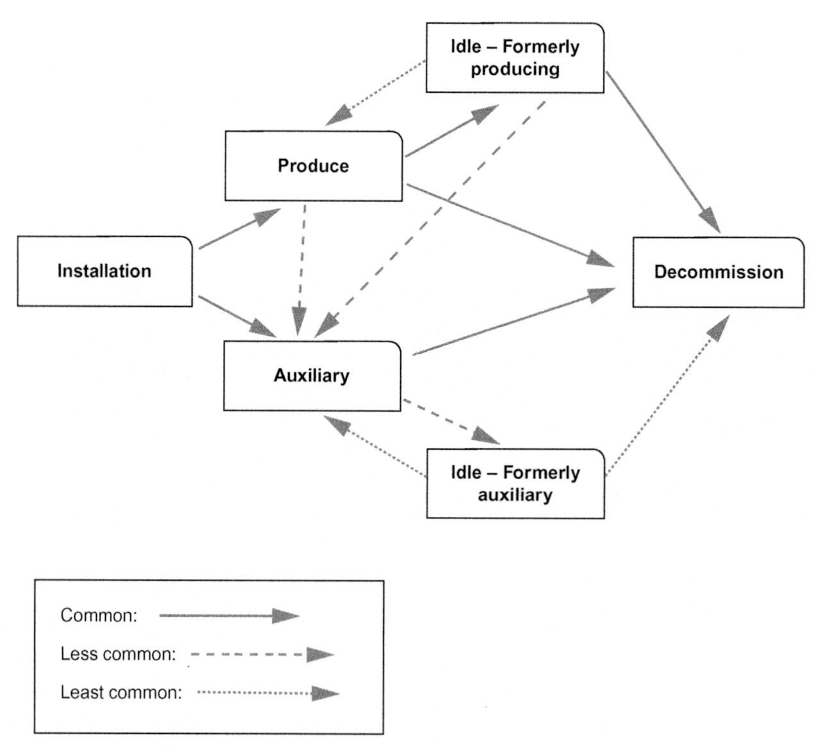

Fig. 1.12 Structure classifications and transition pathways over time.

introducing error in structure counts. Auxiliary structures that no longer serve a useful purpose but have not been decommissioned are considered a separate category, idle-formerly auxiliary. No attempt was made to identify members of this class.

1.7.2 Shallow Water

At the end of 2017, there were 1909 standing structures in water depth <400 ft in the Gulf of Mexico—887 producing structures, 660 idle structures, and 361 auxiliary structures (Fig. 1.13). Of the 887 producing structures, 564 were primarily oil producers, and 323 were primarily gas producers.

Idle structures are broken out into two idle age categories, 1–3 years idle and > 3 years idle, with caissons and well protectors consolidated into one category for simplicity (Table 1.4). Well protectors number about one-third of the number of caissons, and caissons and well protectors number about one-third of the total number of structures.

Fixed platforms are the majority structure type across all categories. A little less than half of all structures are producing, and about a quarter of all structures have been idle >3 years circa 2017. Auxiliary structures comprise about 20% of the total inventory.

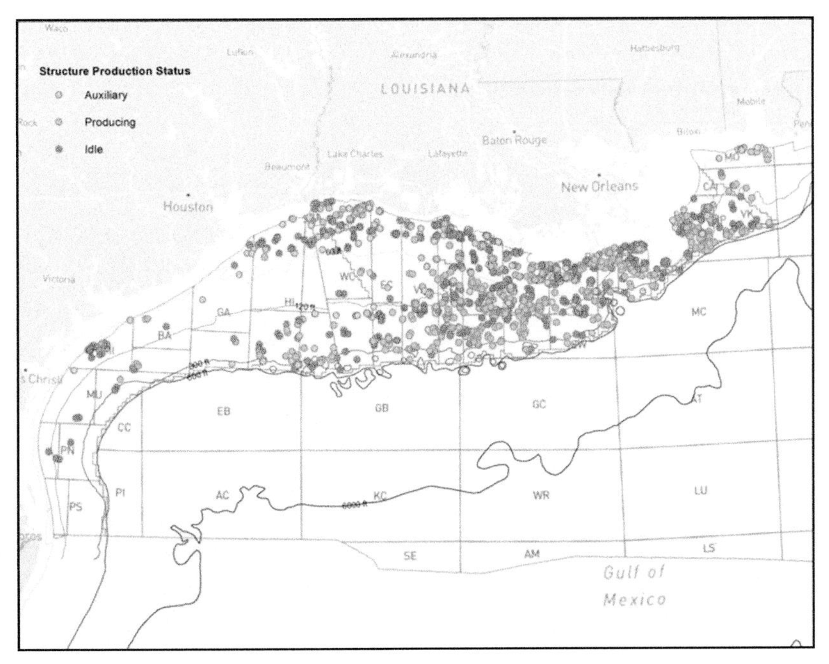

Fig. 1.13 Producing, idle, and auxiliary structures in shallow water circa 2017. *(Source: BOEM 2018.)*

Table 1.4 Shallow-Water Structures by Type and Production Status Circa 2017

	Caisson/WP	Fixed Platform	Total (%)
Idle (1–3 years)	85	134	219 (11%)
Idle (> 3 years)	212	229	441 (23%)
Auxiliary	43	318	361 (19%)
Producing	257	630	887 (46%)
Total	597	1311	1908

Source: BOEM 2018.

1.7.3 Deepwater

The inventory of structures in water depth > 400 ft circa 2017 consisted of 46 fixed platforms, three compliant towers, and 48 floaters (Tables 1.5 and 1.6). Of the 97 active structures, 86 were producing in 2017, six were idle, and five served as auxiliary structures.

Most deepwater fixed platforms reside along the edge of the continental slope near the base of the shallow-water planning areas in water depths up to 1350 ft (Fig. 1.14). About half of fixed platforms reside within 400–500 ft water depth, and the remaining half populate the 500–1000 ft water depth region. Six fixed platforms have been installed in water depth > 1000 ft, and all of these installations were producing circa 2017.

Most floaters reside in the Mississippi Canyon and Green Canyon protraction areas and range in water depths from 1500 to 9560 ft (Fig. 1.15). Most deepwater planning areas are sparsely populated in structures.

Table 1.5 Fixed Platform and Compliant Tower Inventory in the Gulf of Mexico Circa 2017

Complex	Operator	Block	Name	Water Depth (ft)	Install Year
70	Manta Ray Gathering	EC373		443	1998
113	W&T Energy VI	VK823	Virgo	1130	1999
147	W&T Offshore	EW910		557	1998
1052	Chevron U.S.A.	HI582	Cyrus	440	2002
1076	Renaissance Offshore	ST317		460	2002
1165	Tarpon Oper. & Dev.	WC661		484	2002
1279	W&T Offshore	ST316		447	2003
1320	ATP Oil & Gas	GB142		542	2003
1482	Ankor Energy	MC21	Simba	667	2005
1500	Fieldwood Energy	ST308	Tarantula	484	2004
2027	Bennu Oil & Gas	HI589		477	2007
2606	Walter Oil & Gas	EW834	Coelacanth	1186	2015
10178	Fieldwood SD Offshore	EB160	Cerveza	935	1981

Table 1.5 Fixed Platform and Compliant Tower Inventory in the Gulf of Mexico Circa 2017—cont'd

Complex	Operator	Block	Name	Water Depth (ft)	Install Year
10192	W&T Offshore	HI389		410	1981
10212	Fieldwood SD Offshore	EB159	Ligera	924	1982
10242	Fieldwood SD Offshore	EB110	Tequila	660	1984
10297	Fieldwood SD Offshore	EB165	Snapper	863	1985
22172	Energy XXI GOM	SP93		446	1978
22178	EnVen Energy Ventures	MC194	Cognac	1023	1978
22224	Fieldwood Energy	MC311	Bourbon	428	1978
22662	Fieldwood Energy	SP89		422	1982
22685	Energy XXI GOM	SP49		400	1981
22840	Exxon Mobil	MC280	Lena	1000	1983
23051	Taylor Energy	MC20		475	1984
23151	Energy XXI GOM	SP93		450	1985
23212	Manta Ray Gathering	SS332		438	1985
23277	Shell Oil	GC19	Boxer	750	1988
23353	Poseidon Oil Pipeline	SM205		457	1989
23503	Whistler Energy II	GC18	Boxer	750	1986
23552	Fieldwood Energy Offshore	GC65	Bullwinkle	1353	1988
23567–1	MC Offshore Petroleum	GC52	Marquette	604	1989
23567–2	MC Offshore Petroleum	GC52	Marquette	604	1989
23760	Chevron U.S.A.	GB189	Tick	720	1991
23788	Triton Gathering	GB191	Pimento	721	1993
23800	Fieldwood Energy	EW826		483	1988
23848	Arena Offshore LP	SP83		467	1990
23875	Eni US	MC365	Corrla	619	1992
23883	Stone Energy	MC109	Amberjack	1100	1991
23893	Energy XXI GOM	MC397	Alabaster	476	1991
23925	McMoRan Oil & Gas	EW947		477	1990
24129	EnVen Energy Ventures	EW873	Lobster	775	1994
24130	Stone Energy	VK989	Pompano	1290	1994
24201	Flextrend Development	VK817	Phar Lap	673	1995
27032	Flextrend Development	GB72	Spect. Bid	541	1995
27056	Shell Offshore	GB128	Enchilada	705	1997
33039	Hess	GB260	Baldpate	1648	1998
70012	Chevron U.S.A.	VK786	Petronius	1754	2000
70016	Fieldwood Energy	VK780	Spirit	722	1998
80015	Fieldwood Energy	SS354		464	1997
90014	Shell Offshore	GB172	Salsa	693	1998
90028	Energy Res. Tech.	EC381		446	1997

Note: Lena, Baldpate, and Petronius are compliant towers.
Source: BOEM 2017.

Table 1.6 Floating Structure Inventory in the Gulf of Mexico Circa 2017

Complex	Operator	Block	Name	Type	Depth (ft)	Install Year
67	Chevron U.S.A.	GC205	Genesis	SPAR	2590	1998
183	Exxon Mobil	AC25	Hoover	SPAR	4825	2000
235	Anadarko	VK915	Marlin	TLP	3236	1999
251	Eni US	GC254	Allegheny	MTLP	3294	1999
420	Shell Offshore	GC158	Brutus	TLP	2900	2001
811	EnVen Energy	EW1003	Prince	TLP	1500	2001
821	Anadarko	EB602	Nansen	SPAR	3675	2001
822	Anadarko	EB643	Boomvang	SPAR	3650	2002
876	Anadarko	MC127	Horn Mountain	SPAR	5400	2002
1001	BP E&P	MC474	Na Kika	SEMI	6340	2003
1035	Anadarko	GC645	Holsten	SPAR	4340	2004
1088	W&T Energy VI	MC243	Matterhorn	MTLP	2850	2003
1090	Murphy E&P	MC582	Medusa	SPAR	2223	2003
1101	BP E&P	MC778	Thunder Horse	SEMI	6200	2005
1175	Eni US	MC773	Devils Tower	SPAR	5610	2004
1215	BP E&P	GC782	Mad Dog	SPAR	4420	2004
1218	ConocoPhillips	GB783	Magnolia	TLP	4670	2004
1223	BP E&P	GC787	Atlantis	SEMI	7050	2007
1288	Anadarko	GB668	Gunnison	SPAR	3150	2003
1290	Murphy E&P	GC338	Front Runner	SPAR	3330	2004
1323	Anadarko	GC608	Marco Polo	TLP	4300	2004
1665	Anadarko	GC680	Constitution	SPAR	4970	2005
1766	Anadarko	MC920	Independ. Hub	SEMI	8000	2007
1799	BHP Billiton Pet.	GC613	Neptune	MTLP	4250	2007
1819	Chevron U.S.A.	GC641	Tahiti	SPAR	4000	2008
1899	BHP Billiton Pet.	GC653	Shenzi	TLP	4375	2008
1930	Chevron U.S.A.	MC650	Blind Faith	SEMI	6480	2008
2008	Shell Offshore	AC857	Perdido	SPAR	7835	2009
2045	Murphy E&P	MC736	Thunder Hawk	SEMI	6050	2009
2089	Bennu Oil & Gas	MC941	Mirage/Titan	SEMI	4050	2010
2133	Energy Res. Tech.	GC237	Helix	MOPU	2200	2009
2229	Petrobras America	WR249	Cascade/Chin.	FPSO	8300	2011
2385	Shell Offshore	MC807	Olympus	TLP	3028	2013
2424	LLOG Exp. Off.	MC547	Who Dat	SEMI	3280	2011
2440	Chevron U.S.A.	WR718	Jack-St Malo	SEMI	6950	2014
2498	Hess	MC724	Gulfstar	SPAR	4600	2014
2503	Shell Offshore	WR551	Stones	FPSO	9560	2016
2513	LLOG Exp. Off.	MC254	Delta House	SEMI	4400	2014
2576	Anadarko	KC875	Lucius	SPAR	7000	2014
2597	Anadarko	GC860	Heidelberg	SPAR	5300	2016
23583	MC Offshore	GC184	Jolliet	TLP	1760	1989
24080	Shell Offshore	GB426	Auger	TLP	2860	1994
24199	Shell Offshore	MC807	Mars	TLP	2933	1996
24229	Shell Offshore	VK956	Ram Powell	TLP	3216	1997
24235	Noble Energy	VK826	Neptune	SPAR	1930	1996
70004	Shell Offshore	MC809	Ursa	TLP	3970	1998
70020	Eni US	EW921	Morpeth	MTLP	1700	1998

Source: BOEM 2017.

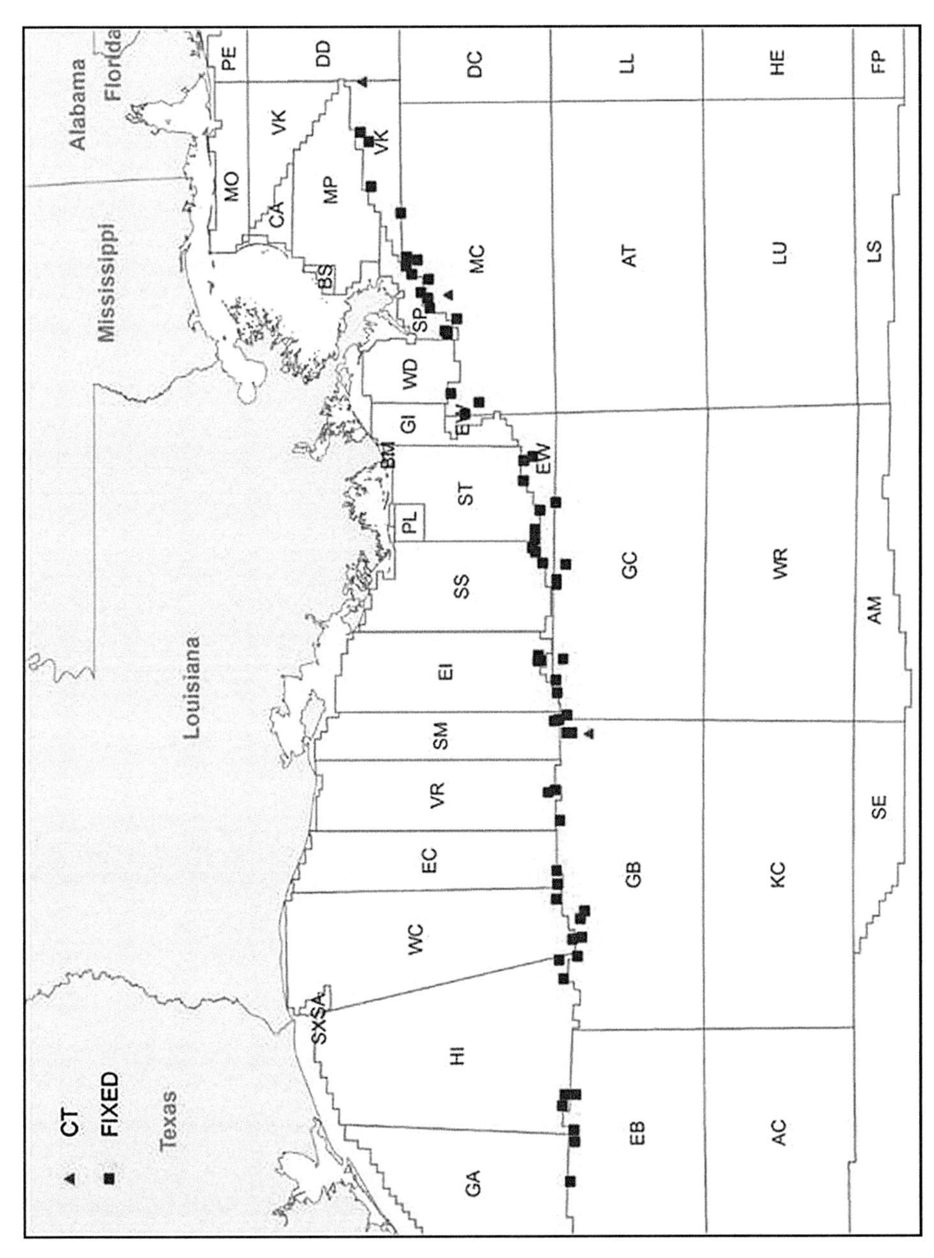

Fig. 1.14 Deepwater fixed platforms and compliant towers circa 2017. *(Source: BOEM 2017.)*

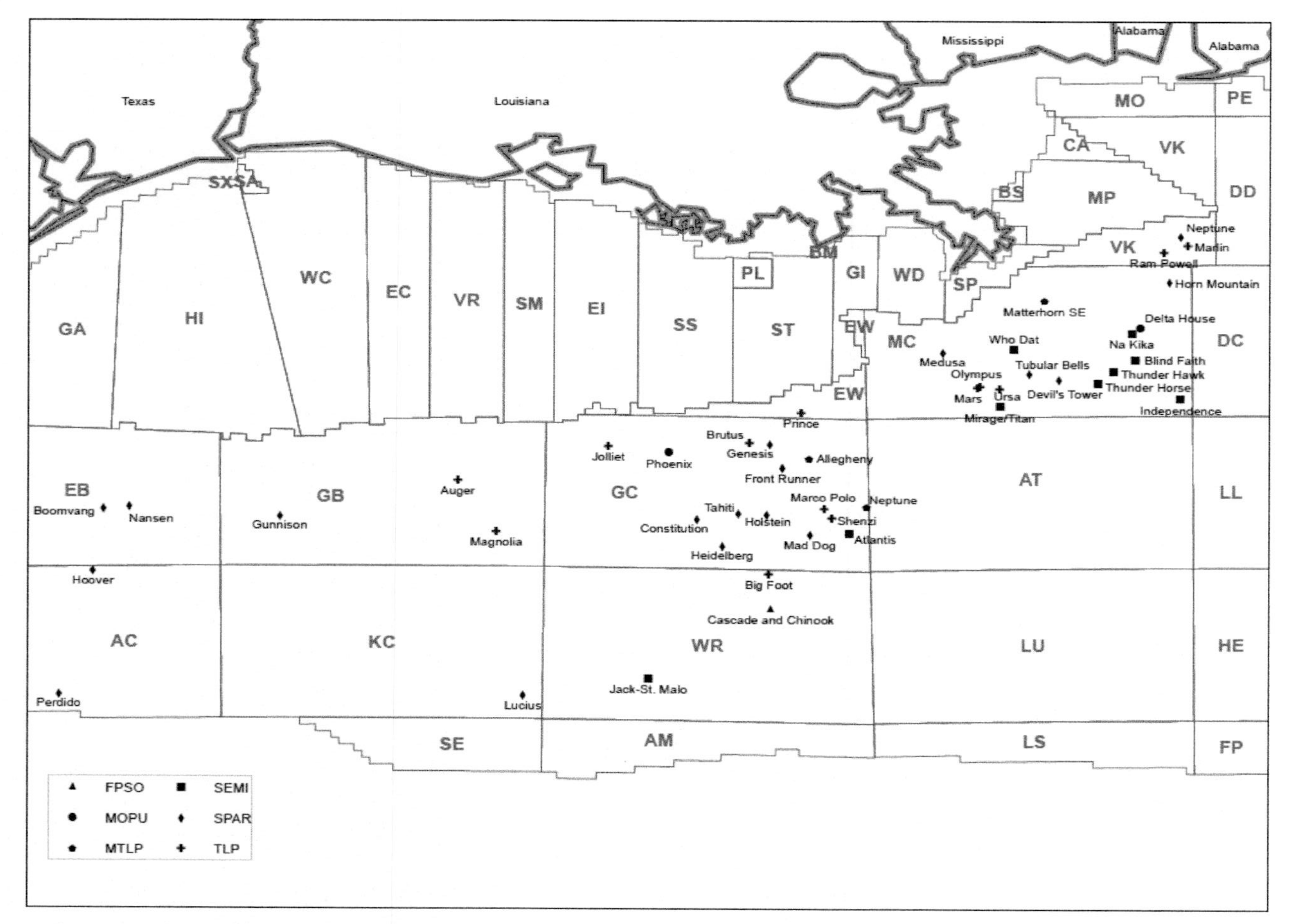

Fig. 1.15 Floaters in the Gulf of Mexico circa 2017. *(Source: BOEM 2017.)*

BOX 1.2 Creole Field—The Gulf's First Offshore Development

The State of Louisiana sold its first offshore lease in 1936, and one year later from a platform in 15 ft water depth and about a mile from the shore of Cameron Parish, Creole was the first oil field discovered in the GoM (Wasson, 1948). Nineteen directional wells were drilled and 10 completed as producers from one platform. A copper wire and cardboard model was constructed to represent the wells and fault planes of the field (Fig. 1.16). Horizontal and vertical scales are equal, and the observer is looking northeast.

Ten wells were drilled straight down to 1200 ft and then deflected toward the objective. If production was not found, the hole was plugged back far enough to start a new hole to test some other traps along its fault (Fig. 1.17). In this manner, all 10 wells were produced. Total depth is measured along the wellbore; vertical depth is straight down. Three Miocene sands between 5100 and 6900 ft were productive on closures on the upthrow side of faults, and a structure map on top of the Gulf sand is shown in Fig. 1.18.

Oil from the three sands at Creole was dark green ranging between 32.7° and 33.9° API with low sulfur and low vapor pressure. Oil was run by pipeline to a receiving station 17 mi northeast where it was stored before shipping by barge through the Intracoastal Waterway to a Texas refinery.

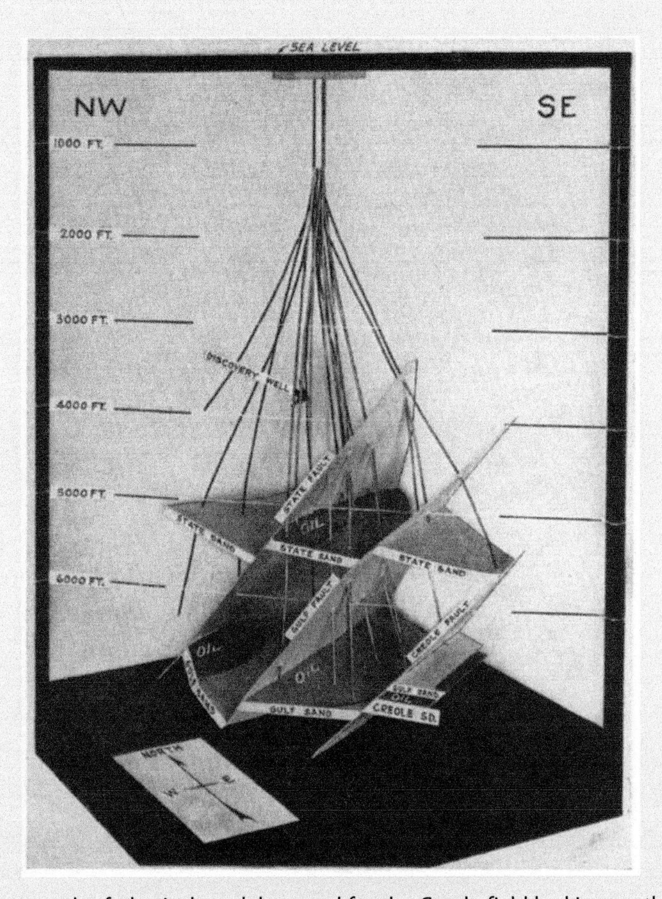

Fig. 1.16 Photograph of physical model created for the Creole field looking northeast. *(Source: Wasson, T., 1948. Creole field, Gulf of Mexico, coast of Louisiana. In: Structure of Typical American Oil Fields: SP 14, American Assoc. Pet. Geol. 3, 281–298.)*

Continued

BOX 1.2 Creole Field—The Gulf's First Offshore Development—cont'd

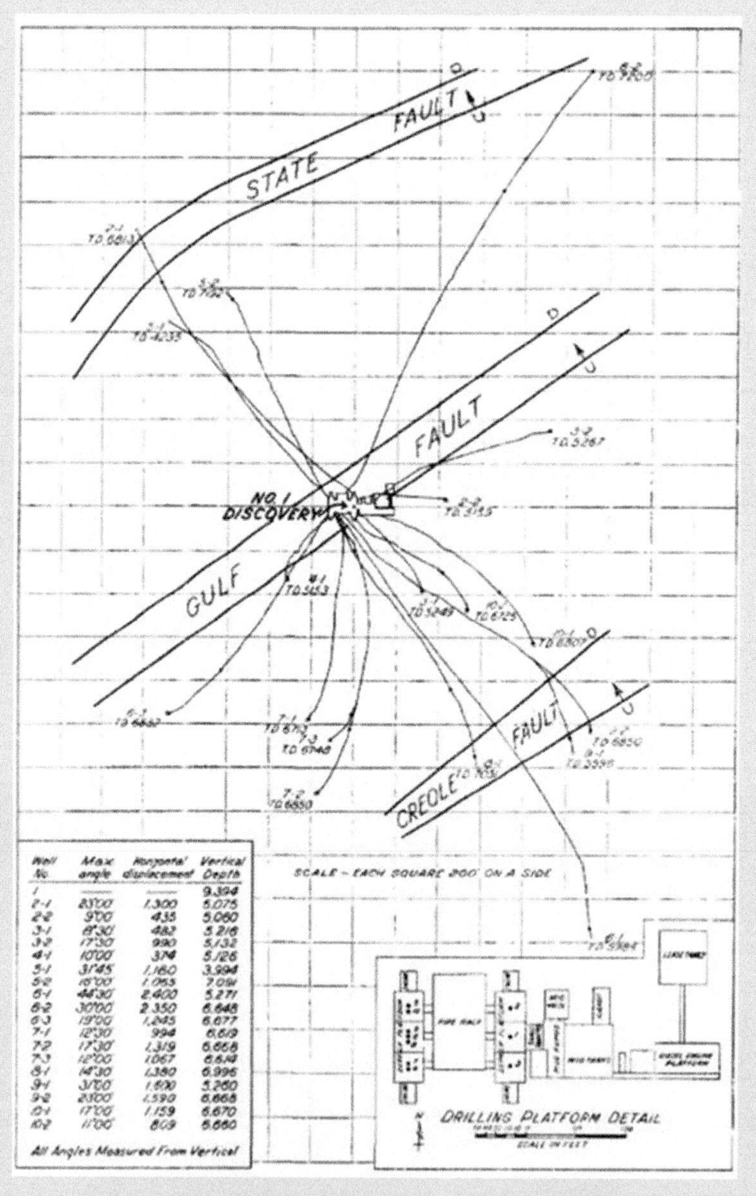

Fig. 1.17 Looking down on the drilling platform showing traces of deflected wellbores and total depth as measured along hole. *(Source: Wasson, T., 1948. Creole field, Gulf of Mexico, coast of Louisiana. In: Structure of Typical American Oil Fields: SP 14, American Assoc. Pet. Geol. 3, 281–298.)*

Continued

BOX 1.2 Creole Field—The Gulf's First Offshore Development—cont'd

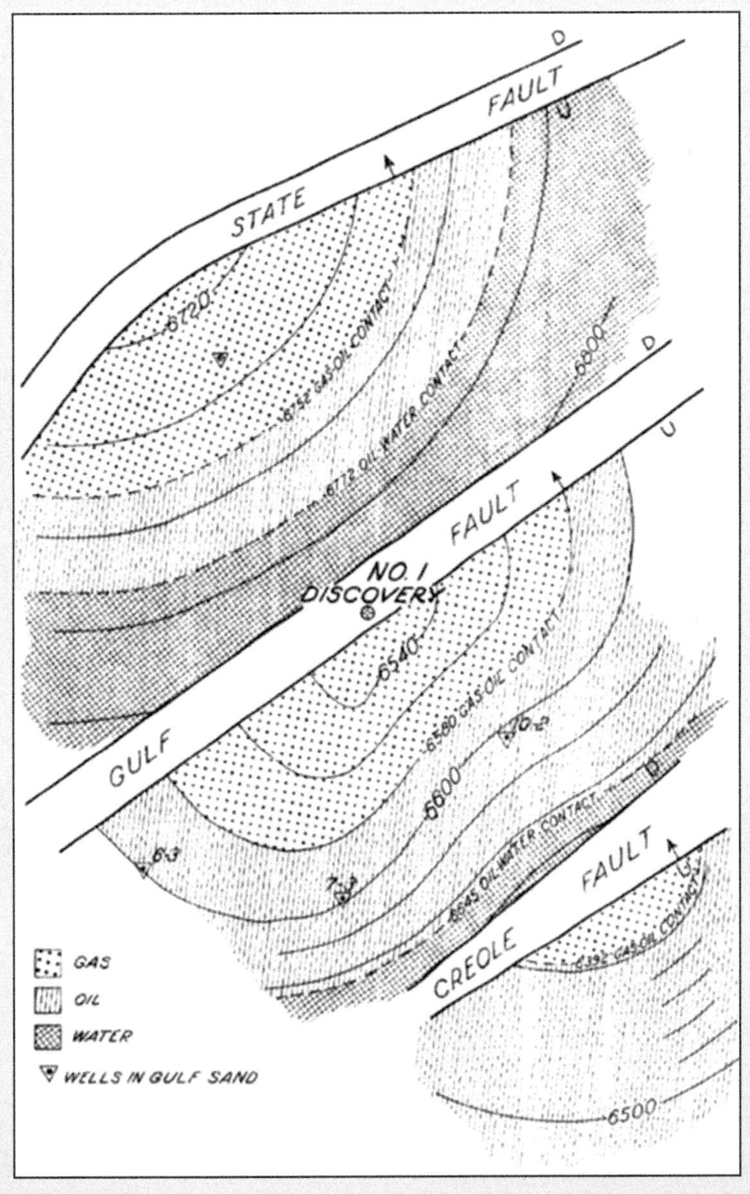

Fig. 1.18 Structure map on top of Gulf sand and traps on the upthrown of the three faults. *(Source: Wasson, T., 1948. Creole field, Gulf of Mexico, coast of Louisiana. In: Structure of Typical American Oil Fields: SP 14, American Assoc. Pet. Geol. 3, 281–298.)*

REFERENCES

Bartley, E.R., 1953. The Tidelands Oil Controversy: A Legal and Historical Analysis. University of Texas Press, Austin, TX.

Burgess, G.L., Kazanis, E.G., Cross, K.K., 2018. Estimated Oil and Gas Reserves, Gulf of Mexico OCS Region, December 31, 2016. U.S. Department of Interior, Bureau of Ocean Energy Management, Gulf of Mexico OCS Region, New Orleans, LA OCS Report BOEM 2018-034.

Dupre, D.E., 2015. What makes the United States offshore leasing system so special? A primer on the outer continental shelf oil and gas lease. LSU J. Energy Law Resources 4 (1), 37–58.

Galloway, W.E., 2008. Depositional evolution of the Gulf of Mexico sedimentary basin. In: Miall, A.D. (Ed.), The Sedimentary Basins of the United States and Canada: Sedimentary Basins of the World. Elsevier, Burlington, MA. pp. 505–549.

Garrison, T.S., Ellis, R., 2018. Essentials of Oceanography, 8th ed. Cengage Learning, Boston, MA.

National Commission on the BP Deepwater Horizon Oil Spill and Offshore Drilling, 2011. The History of Offshore Oil and Gas in the United States, Washington, DC.

Owen, E.W., 1975. Trek of the Oil Finders: A History of Exploration for Petroleum. AAPG, Tulsa, OK.

Percy, T.F., 1925. Justinian and the freedom of the sea. Am. J. Int. Law 19, 716–721.

Priest, T., 2008a. Claiming the coastal sea: the battle for the tidelands 1937–1953. In: Austin, D., Priest, T., Penney, L., Pratt, J., Pulsipher, A.G., Abel, J., Taylor, J. (Eds.), History of the offshore oil and gas industry in southern Louisiana. In: vol. 1. U.S. Department of Interior, Minerals Management Service, Gulf of Mexico OCS Region, New Orleans, LA, pp. 67–92, OCS Study MMS 2008-042.

Priest, T., 2008b. Auctioning the ocean: the creation of the federal offshore leasing program, 1954–1962. In: Austin, D., Priest, T., Penney, L., Pratt, J., Pulsipher, A.G., Abel, J., Taylor, J. (Eds.), History of the offshore oil and gas industry in southern Louisiana. In: vol. 1. U.S. Department of Interior, Minerals Management Service, Gulf of Mexico OCS Region, New Orleans, LA, pp. 93–116, OCS Study MMS 2004-049.

Salvador, A., 1991. Origin and development of the Gulf of Mexico basin. In: Salvador, A. (Ed.), The Gulf of Mexico Basin: The Geology of North America. Geological Society of America, Boulder, CO, pp. 389–444.

U.S. Dept. of the Interior. Minerals Management Service, 2001. Oil and gas leasing procedures guidelines. U.S. Dept. of the Interior, Minerals Management Service, Gulf of Mexico OCS Region, New Orleans, LA. OCS Study MMS 2001-76.

U.S. Department of the Interior, Bureau of Ocean Energy Management Gulf of Mexico OCS Region, Office of Resource Evaluation, 2017. Assessment of technically and economically recoverable hydrocarbon resources of the Gulf of Mexico Outer Continental Shelf as of January 1, 2014. OCS Report BOEM 2017-005.

Wasson, T., 1948. Creole field, Gulf of Mexico, coast of Louisiana. In: Structure of Typical American Oil Fields: SP 14, American Assoc. Pet. Geol. 3, 281–298.

CHAPTER TWO

Structure Classification

Contents

Abstract

There are many ways to classify offshore structures, and the most common categories are structure type and production status. Fixed platforms and floaters are the two primary structure types, and several subclasses are used to distinguish the jacket type and hull and the manner floating structures are fixed to the seabed. Production status classifies structures as producing, idle (formerly producing), and auxiliary (supporting role). Manning status, complex association, and number of wells are also useful attributes. The chapter concludes with a description of the role hub platforms play in development.

2.1 STRUCTURE TYPE

2.1.1 Shallow Water

Shallow-water structures in the GoM are distinguished according to their jacket type and number of pieces of equipment topsides. A caisson is a large-diameter cylindrical shell or tapered steel pipe that supports a small deck and a few pieces of equipment (Fig. 2.1). Structures with less than six pieces of equipment are classified as minor, and almost all caissons are minor (Fig. 2.2). Caissons are commonly used for small reservoirs that require one or two producing wells, and their production is delivered to another facility through

Decommissioning Forecasting and Operating Cost Estimation
https://doi.org/10.1016/B978-0-12-818113-3.00002-3

Fig. 2.1 Caisson, well protector, and fixed platform structures in the Gulf of Mexico. *(Source: BOEM.)*

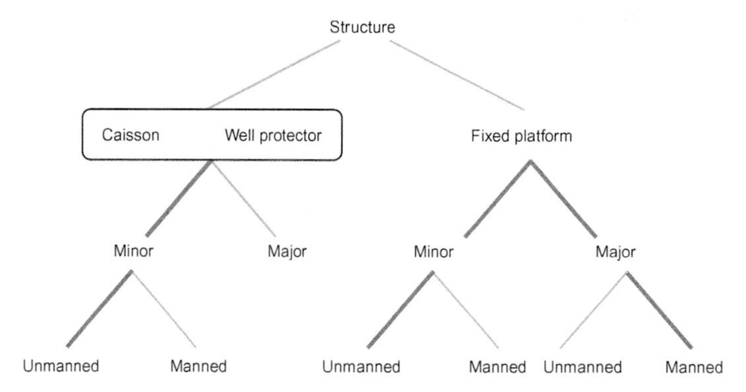

Fig. 2.2 Structure classification with dominant categories denoted with thick lines.

a flow line for processing or, if high-pressure gas, may be exported direct to a sales line or shore. In the GoM, caissons have been used in water depth up to about 150ft.

Jackets consist of three or more legs and the bracing system that connects the legs and are by far the most popular structure type in the Gulf. Jackets are secured to the seafloor with piles and are referred to variously as well protectors and fixed platforms (Gerwick, 2007). Well protectors usually have three or four legs, and most are minor structures, while fixed platforms have four or more legs and may be a major or minor structure. Major fixed platforms resemble the jacket structure of well protectors but are larger and more robust with facilities for drilling, production, and workover operations. Fixed platforms come in a wide variety of configurations and are secured to the seafloor with piles and skirt piles designed for the loads, seafloor terrain, and environmental conditions at location (Chakrabarti, 2005). Most well protectors are unmanned, and fixed platforms may be manned or unmanned.

Fixed platforms perform a variety of functions, and for structures with processing capacity, both oil and gas export lines will typically exit the structure, but this was not always the case, and at the beginning of the pipeline build-out in the GoM barges were used, and two-phase pipelines were common meaning only one export pipeline was employed (e.g., Massad and Pela, 1956; Lipari, 1962; Swift, 1966). By the early 1970s, most hydrocarbon streams from major processing platforms were separated and transported individually (e.g., Frankenberg and Allred, 1969; Hicks et al., 1971). Minor fixed platforms that do not perform processing or partial processing will usually have bulk and service lines similar to caissons and well protectors, and in cases where a structure was installed to support equipment capacity or quarters, no pipeline may be required. For pipeline junction platforms, oil and/or gas export lines will cross (enter and depart) the structure.

2.1.2 Deepwater

Deepwater development is of a completely different character than in shallow water. Deepwater is cold and has high hydrostatic pressure, both of which are drivers for solids deposition during production, especially the formation of hydrates. Reservoirs may be in unproduced geology in new or existing basins/plays, and deeper reservoirs are characterized by higher pressure/temperature and challenging reservoir-fluid properties and impurities (Thurmond et al., 2004; Chaudhury and Whooley, 2014; Reid and Dekker, 2014). Development factors can be driven by location (e.g., seabed terrain), water depth (Box 2.1), remoteness from existing infrastructure, and environmental conditions. Frequently, combinations of these factors are represented in a single project, contributing to high project complexity, high cost, and long periods between investment sanction and first production (Dekker and Reid, 2014).

BOX 2.1 Deepwater Thresholds

Water depth thresholds reflect oceanographic and biological processes, geologic conditions, jurisdictional issues, structure design considerations, and other factors.

Depth Zone. Oceanographers recognize three major depth zones in the ocean—the surface zone; a layer below the surface zone in which temperature, salinity, and density experience significant changes with increasing depth; and the deep zone that reaches to the seabed (Garrison and Ellis, 2017). The surface zone typically extends to depth of 100 to 500 m and is also called the mixed layer because winds, waves, and temperature changes cause extensive mixing within it and ocean parameters change seasonally. Below the surface zone lies a layer in which ocean-wave properties experience significant change with increasing depth, referred to as the thermocline (for temperature changes), halocline (for salinity changes), and pycnocline (for density changes). Each zone has different acoustic and light transmission properties, and the boundaries may give reflection from sonic transmission. Understanding the form of these changes is important for oceanographic modeling and for vessels that operate covertly (submarines) and rely upon accurate measurement of sound waves (Payne, 2010).

Photic Zone. The photic zone or euphotic zone is the depth of the water in a lake or ocean where most photosynthesis occurs (Skinner and Murck, 2011). Formally, it is defined where the amount of sunlight is such that the rate of carbon dioxide uptake (or the rate of oxygen production) is equal to the rate of carbon dioxide production (or the rate of oxygen consumption). The photic zone extends from the surface down to a depth where light intensity falls to 1% of

Continued

BOX 2.1 Deepwater Thresholds—cont'd

that at the surface and its thickness depends on the extent of light attenuation in the water column. Typical euphotic depths vary from only a few centimeters in highly turbid lakes to around 200 m in the open ocean. It also varies with seasonal changes in turbidity. Plant life is restricted to the upper 200 m of the ocean because in this zone, sufficient light energy is available for photosynthesis. Below the photic zone lie the aphotic (twilight) and abyssal (midnight) water depth zones.

Exclusive Economic Zone. Under the United Nations Convention on the Law of the Seas (UNCLOS), coastal states have jurisdiction over 200 nm of continental shelf adjacent to their coastlines, and UNCLOS Article 76 gives countries extended claims to up to 350 nm based on a complex formula (Cavnar, 2009). The term "continental shelf" as applied by UNCLOS is a legal phrase that refers to what scientists broadly call the continental margin but mixes technical and legal terminology in a confusing and often meaningless amalgamation.

Wave Base. Wave forces are usually the dominant design criterion affecting offshore structures (Gerwick, 2007). A wave is a traveling disturbance of the sea state. The disturbance travels, but the water particles within the wave move in a nearly elliptical orbit with little net forward motion. A deepwater wave is one for which the seafloor has no effect. Normally, waves do not exceed 400 m wavelength even under storm conditions, although in the Pacific Ocean wavelengths as long as 600 m have been measured. Since waves only generate orbital motion to a depth of about half a wavelength, referred to as the wave base, this has led to the adoption of 200 or 300 m water depth as being considered "deepwater" by design engineers and many regulatory agencies. In this context, deepwater refers to the water depth where surface waves do not impact the seafloor. At 300 m water depth, surface waves from the strongest recorded storm are not expected to move seafloor sediment on the continental shelf.

Depositional Models. Deepwater in a geologic context refers to an environment in which the reservoir sand is deposited by gravity-flow processes. Reservoirs deposited by gravity-flow processes, such as turbidities, typically consist of sheetlike to lenticular sands with rapid lateral facies changes and numerous interbedded shale breaks. The volume and producibility of turbidity reservoirs is markedly different from tabular shelf sands or growth-fault sands that have better lateral continuity and fewer shale breaks.

Deepwater facilities come in a variety of types and configurations that reflect the multiple trade-offs in cost and risk involved in development strategies (Ronalds and Lim, 2001; Ronalds, 2002; Cullick et al., 2007; D'Souza and Aggarwal, 2013; Reid et al., 2013; D'Souza et al., 2016). Deepwater facilities can host dry tree wells, direct vertical access wells, and/or subsea wells, and may process and/or transport production from third-party facilities. Deepwater structures may be owned and operated by the well owners or a third party. Whenever subsea wells are involved in development, integrated flow assurance is a key factor in design.

Fixed Platforms

Almost all of the deepwater fixed platforms in the GoM were installed to drill and complete wells from platform rigs with full oil and gas processing capacity. Deepwater fixed platforms host export lines and may serve as a pipeline junction for other export lines running across the structure and as a source of gas-lift or chemical injection to assist nearby production. If subsea well production is processed at the facility, the structure will host umbilical and electrohydraulic control systems, storage tanks, and related equipment. Deepwater fixed platforms can readily accommodate subsea tiebacks because of their structural stability and strength, and when production ceases, they remain prime candidates to transition into a support role for regional development.

Several deepwater fields are produced from fixed platforms, and many (but not all) are named including Bourbon (year of installation, 1978), Cognac (1978), Cerveza (1981), Ligera (1982), Tequila (1984), Snapper (1985), Boxer (1988), Bullwinkle (1988), Marquette (1989), Tick (1991), Amberjack (1991), Alabaster (1991), Corral (1992), Pimento (1993), Lobster (1994), Pompano (1994), Phar Lap Shallo (1995), Spectacular Bid (1995), Enchilada (1997), Spirit (1998), Salsa (1998), Virgo (1999), Cyrus (2002), Tarantula (2004), and Simba (2005). In 2015, Walter Oil & Gas installed the latest fixed platform in 1186 ft at the Coelacanth field in Ewing Bank block 834 (EW 834) (Fig. 2.3).

Cognac was the first structure in >1000 ft water depth (Fig. 2.4), and Bullwinkle is the deepest fixed platform in the GoM at 1350 ft water depth (Fig. 2.5).

Fig. 2.3 Sailaway of the Coelacanth platform offshore Texas on 15 October 2015. *(Source: Walter Oil & Gas.)*

Fig. 2.4 Cognac platform in Mississippi Canyon 194. *(Source: Shell, BOEM.)*

Fig. 2.5 Bullwinkle platform in Green Canyon 65. *(Source: Shell, BOEM.)*

Example. Mississippi Canyon 194 (Cognac) field

Cognac accumulations are trapped on a faulted, northwest plunging nose between 5000 and 11,000 ft covering lease blocks in Mississippi Canyon 194, 195, 150, and 151. A fixed platform was installed in 1978 in MC 194 in 1025 ft water depth with two rigs, one of which was later removed and replaced with

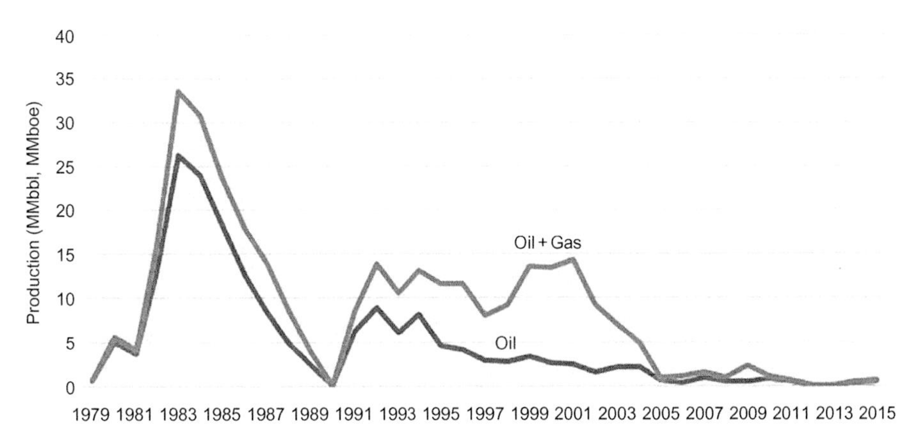

Fig. 2.6 Mississippi Canyon 194 (Cognac) field oil and gas production, 1979–2015. *(Source: BOEM 2017.)*

production equipment. The initial development plan included 72 drainage points and seven injection wells. From 1978 to 1981, a total of 61 wells were drilled and cased and 38 wells were completed. About half of the wells were high-angle (> 50°) extended-reach wells. First production occurred in 1979, and peak production was achieved in 1983 (Fig. 2.6). Cumulative production through September 2017 was 182 MMbbl of oil and 762 Bcf of natural gas. Circa 2015, production was 500,000 bbl oil and 600 MMcf gas, and remaining reserves are estimated at 3.1 MMbbl of oil and 5.4 Bcf gas.

Compliant Towers

There are only three compliant towers (CTs) in the GoM, and as their name suggests, the structures are compliant in the sense that they do not attempt to resist all environmental lateral forces through their piling alone. Instead, the structures are designed to permit limited movement with the waves. Thus, the structure requires less steel (strength) in construction, and their foundation bases do not expand with increasing water depth like fixed platform designs.

CTs have a constant cross section throughout the water column much like radio transmission towers on land. Lena is a guyed tower fixed with guy wires to the seabed, while Baldpate utilizes axial tubes as flex elements (articulated tower), and Petronius utilizes flex legs.

Lena was installed in 1983 and is near the end of its 35-year life, generating about $18 million in gross revenue in 2016. Baldpate and Petronius are of more recent vintage being installed in 1998 and 2000 and are still prolific producers, generating $153 and $188 million, respectively, in 2016.

Floaters

Floating structures are categorized into four main classes (Fig. 2.7): floating production storage and offloading vessel (FPSO), semisubmersible (semi), surface-piercing

Fig. 2.7 Main floater classes in the Gulf of Mexico. *(Source: Chevron, Shell, BOEM.)*

articulated riser (spar), and tension leg platform (TLP). Spars are the most common floater type circa 2017 followed by TLP/MTLPs and semisubmersibles (Table 2.1). Floaters span a wide range of production capacities from 40 to 300 Mboe per day and water depths from 1500 ft to the current maximum of 9560 ft. The two FPSO developments Cascade/Chinook and Stones are the deepest in the GoM.

Within each floater class, configurations have evolved into subclasses, and occasionally, new hybrid classes arise. Spars currently come in three varieties—classic spar, truss spar, and cell spar (Sablok and Barras, 2009). A smaller version of the TLP is referred to as

Table 2.1 Gulf of Mexico Floaters Installed Circa 2017

Class	Number
FPSO	2
MOPU	1
MTLP	4
Semi	11
Spar	18
TLP	12
Total	48

Source: BOEM 2018.

a mini TLP (MTLP) and classified by design as SeaStar MTLP and Moses MTLP (D'Souza and Aggarwal, 2013), while a larger version called an extended TLP (ETLP) is utilized at Chevron's Big Foot field. Mobile offshore production units (MOPUs) such as the MinDoc employed at the Telemark/Mirage/Titan fields represent a hybrid structure class (Bennett 2013).

TLPs can accommodate large topside payload, but water depth has been limited to about 5000 ft due to its heave restraining tendons (Fig. 2.8). Spars are not water depth limited because its heave response is achieved by its hull configuration, but topside payload capacity is restrained that limits the topside areal footprint and requires vertical

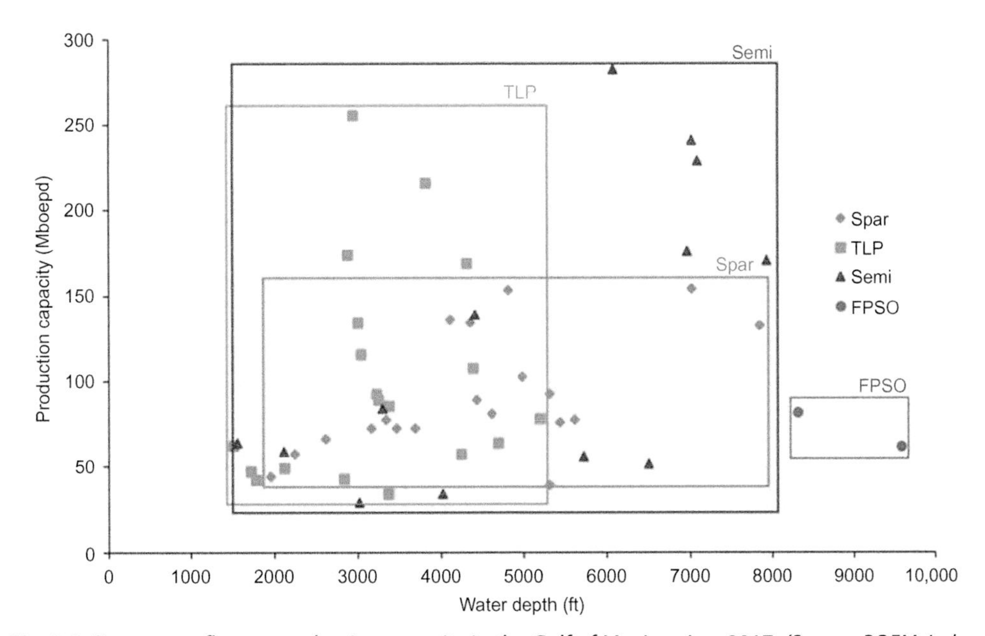

Fig. 2.8 Deepwater floater production capacity in the Gulf of Mexico circa 2017. *(Source: BOEM, Industry publications.)*

Table 2.2 Gulf of Mexico Host Facility Trends Since 2010

Field	Host Platform	Water Depth (ft)	Dry/Wet Tree	FEED Year	Sanction Year
Olympus	TLP	3100	Dry		2010
Big Foot	TLP	5200	Dry		2010
Jack-St Malo	Semi	7000	Wet		2010
Who Dat	Semi	3200	Wet		2010
Tubular Bell	Spar	4300	Wet		2011
Lucius	Spar	7100	Wet		2011
Delta House	Semi	4500	Wet		2012
Heidelberg	Spar	5300	Wet		2013
Stones	FPSO	9500	Wet		2013
Hadrian South	Semi	7200	Wet	2013	
Hadrian North	Semi	7000	Wet	2014	
Stampede	TLP	3500	Wet		2014
Hopkins	Semi	4000	Wet	2015	
Buckskin	Semi	7000	Wet	2015	
Appomattox	Semi	7200	Wet		2015
Shenandoah	Semi	5800	Wet	2016	
Mag Dog 2	Semi	4400	Wet		2017
Vito	Semi	4000	Wet	2017	
North Platte	Semi	4400	Wet	2018	

Source: BOEM 2018.

stacking of decks with offshore heavy lifts. Spars require mobilizing heavy lift cranes and accommodation vessels during installation, whereas TLPs, semis, and FPSOs allow quayside (dry-dock) integration and precommissioning.

Semis have been used in a wide range of water depths and production capacities with both dry tree and wet tree solutions and show the largest envelope of application. Before 2010, semis were considered unsuitable in water depth under 4000 ft due to its heave response in extreme seas and strength/fatigue performance with steel catenary risers, but with the acceptance and application of steel lazy risers, shallow waters are no longer limiting (D'Souza et al., 2016). Semis have a long history of redeployment and repurposing as opposed to TLPs or spars that are normally new builds.

Since 2010, semisubmersibles and wet tree systems have dominated development concepts and appear to be the most versatile combination (Table 2.2). Over 80% of sanctioned and front-end engineering design (FEED) developments are wet tree hosts, and every project sanctioned since 2010 has been a wet tree host. Projects in FEED will either move on to sanctioning with the concept shown, will be developed as a tieback, or shelved until a later date (Box 2.2).

BOX 2.2 Grow As You Know

Both Stones and Cascade/Chinook FPSO developments adopted "grow as you know" strategies that account for their small production capacity. For fields with a high level of reservoir uncertainty, operators pursue one or more phases of development to limit initial capital expenditures. Stones and Cascade/Chinook are Wilcox plays with the potential for large reserves, but operators don't know if they can produce the hydrocarbon resources economically. Hence, they have sought to minimize their financial risk until they learn more about the reservoirs and field development. Deepwater facilities can be leased to avoid large capital expenditures, and FPSOs have the added benefit of eliminating the capital cost of an oil export pipeline. To further minimize investment, fewer wells may also be drilled initially.

2.2 MANNED STRUCTURES

A manned platform is defined as having personnel normally present 24h a day, while on an unmanned platform or an 8h manned facility, personnel are not normally present 24h a day and must be transported off the structure to shore or a manned platform at night or be housed at a tender vessel.

All manned platforms have a helideck where the helicopter lands, which are usually located at the top of the structure and at maximum distance from the drilling rig, if present, and production equipment. Small structures can accommodate small helicopters, and larger platforms can accommodate both large and small helicopters. A few large structures have two heliports.

Production crew on a manned facility in the GoM can range from 4 to 100 or more. In shallow water, manned platforms typically serve as a hub from which staff manage and supervise regional operations, and shuttling to unmanned facilities via helicopter or crew boat is common. In deepwater, although the structure may serve as a regional hub for production, operating crews do not normally shuttle between facilities. All deepwater structures are manned 24h, while only about a third of the shallow-water inventory is manned.

In shallow water, 811 manned structures have been installed and 208 have been removed through 2017, leaving an active inventory of 603 manned platforms in water depth < 400 ft (Fig. 2.9). Most shallow-water manned structures are fixed platforms, and the few caissons and well protectors identified as manned are attached to manned complexes. According to BOEM structure naming convention, if a complex is manned, then all the structures in the complex are classified as manned.

About a third of manned platforms are auxiliary structures. Auxiliary structures do not have any wells boarding the structure. Auxiliary manned structures are used in field

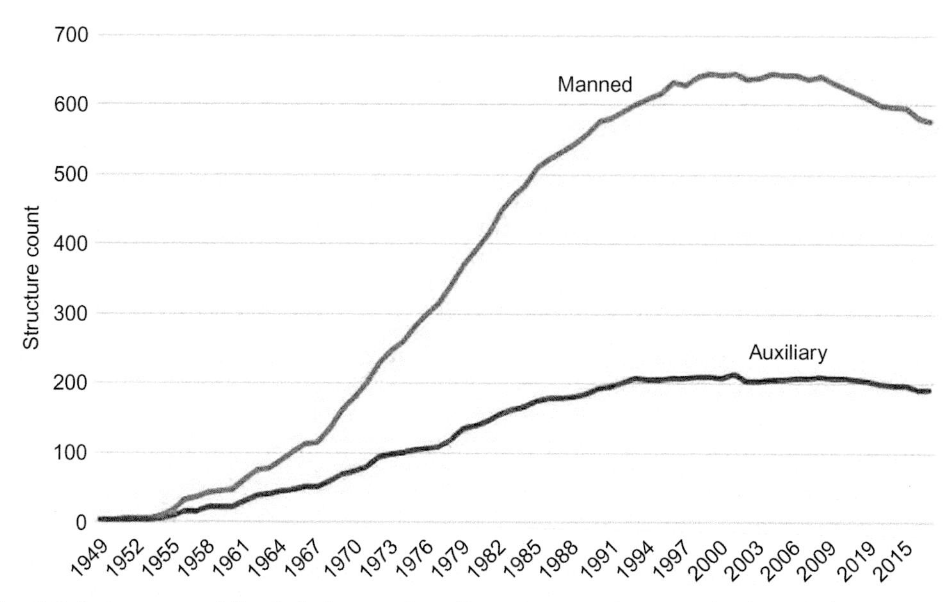

Fig. 2.9 Active manned fixed platform and auxiliary structures in water depth <400ft, 1949–2017. *(Source: BOEM 2018.)*

operations and pipeline transmission services and have been a very stable structure class over the past 30 years (Fig. 2.9).

The number of manned platforms in shallow water peaked at about the same time as the total number of active structures but at one-third the level, and whereas decommissioning has depleted shallow-water inventories by about half from its peak (from 3974 structures in 2001 to 1908 structures circa 2017), the number of manned platforms—being some of the most important structures anchored to the largest reservoirs with the most processing capacity and interconnections across multiple pipeline networks—has not been significantly impacted declining by only a few dozen structures from its peak. Shallow-water manned structures supporting deepwater facilities and pipelines have been immune from the high pace of decommissioning that pervades the shallow water because they are not tied to declining shallow-water fields.

2.3 MULTI-STRUCTURE COMPLEXES

A group of structures is referred to as a complex if they are physically connected by a walkway or bridge structure. A k-complex is defined to be the collection of k connected platforms, so, for example, a 2-complex is composed of two connected structures, a 3-complex is composed of three connected structures, and so on. The bridge that

Table 2.3 Active Structure Counts per Complex Excluding Compliant Towers and Floaters Circa 2017

Structures per Complex	Caisson	Fixed Platform	Mixed Platform	Well Protector	Structure Total
1	347	862	0	139	1348
2	13	86	31	1	262
3	2	38	11		153
4		12	8	1	84
5		8	1		45
6		3	1		24
7		1	1		14
9			1		9
15		1	0		15
Total	362	1011	54	141	1954

Note that three compliant towers and 48 floaters are excluded from the first row.
Source: BOEM 2018.

connects two offshore structures is called a catwalk and supports pipelines, pedestrian movement, and material handling. Connecting structures are usually fixed platforms, but in some cases minor structures may be used. A complex is really not much different than any group of structures used in field development, except that they are connected.

Multistructure complexes are entirely shallow-water phenomena since before the technology and installation capabilities were available to allow structures to be constructed vertically, they were built out horizontally to isolate and separate drilling operations from production and quarters and other functions. In many early developments from the 1950s to 1960s, after drilling platforms were installed, a central processing facility was used to process and export production from the producing platforms.

In 2017, there were 1348 single-structure complexes in the GoM, excluding three compliant towers and 48 floaters (Table 2.3). A total of 262 structures are members of two-structure complexes, 153 structures are associated with a 3-complex, 84 structures are associated with a 4-complex, and so on. When counting the total number of structures in a complex, only caisson-caisson (C-C), fixed platform-fixed platform (FP-FP), and well protector-well protector (WP-WP) types are represented in the caisson, fixed platform, and well protector columns in Table 2.3. Mixed platforms include any combination of C-WP, C-FP, or WP-FP.

There are 17 complexes with five or more structures per complex. The largest active complex in the GoM circa 2017 has 15 structures. Historically, three of the largest complexes in the GoM were the sulfur mining facilities operated by Freeport Sulfur when the Frasch process was competitive with oil and gas recovery operations (Fig. 2.10), but all these facilities ceased operations in the mid-1990s (see Box 21.1). About two-thirds of structures in multi-structure complexes are fixed platforms.

Fig. 2.10 Grand Isle sulfur mine in its heyday. *(Source: LDWF.)*

2.4 PRODUCTION STATUS

Structures transition between various states during their lifetime. Producing platforms will become idle when their reservoirs are depleted or transition into decommissioned status. A few idle structures may be repurposed to serve an auxiliary role if they are well positioned and can be used in field operations.

2.4.1 Producing Structures

Structures where wells first board and that have produced during the last year are classified as producing at the time of the assessment. Producing structures include minor structures such as caissons and well protectors as well as fixed platforms. Almost all caissons and well protectors are producing or formerly producing structures.

Platforms with a drilling or workover rig present or the capacity to accommodate a rig are either producing or formerly producing, and if conductors can be seen, this is direct evidence the structure is either producing or formerly producing. Rigs are often removed after drilling operations are complete, however, so the lack of a rig does not imply the structure does not hold wells. Also, conductors may be run through platform legs in some cases and removed in others and may not always be (visually) observed in photographs depending on the angle of the shot.

2.4.2 Auxiliary Structures

Auxiliary structures are used in support of operations but do not have any wells—or no longer have any wells—that board the structure. On auxiliary structures, no trees or conductors or active well slots will be found. Auxiliary structures are installed new or may be converted from formerly producing structures.

Auxiliary platforms have been used for many years for processing or export operations, to expand capacity at existing structures and for pumping or compressor stations, storage, quarters, pipeline junctions, and meters. Sometimes, small platforms are built adjacent to larger platforms to increase available space or to permit carrying heavier equipment loads on the principal platform. In other cases, large platforms may be installed next to small structures if development requires additional wells.

During the early years of offshore development in the GoM, it was common to separate drilling operations from production and quarters, and many auxiliary structures are associated with a complex. In later decades, as improved technology and construction capability allowed multiple functions to be combined, there was less need for auxiliary structures, and smaller numbers were installed.

One way to transform a formerly producing structure into an auxiliary structure is to plug and abandon all the wellbores and to run production from another field/platform across the facility. Another way is to recertify the structure as a transport and fuel depot hub. Many of the most recent installations of auxiliary structures have been in support of transportation companies to link deepwater production to shelf infrastructure.

2.4.3 Idle Structures

A third class of structure that is important to distinguish is formerly producing structures that have not produced for one year or more, referred to colloquially as "idle iron" or "idle—formerly producing" structures. By definition, all idle structures previously produced hydrocarbons, but at the time of evaluation, the structure did not produce during the previous 12 months. The reason for the cessation of production is not publicly reported, and because the utility of the structure is not (directly) observable, the conclusions that may be inferred about idle structure inventories are limited.

It is important to realize that idle structures are a natural feature of all offshore oil and gas developments and collect as part of the active inventory until decommissioned. An idle structure may return to production if any of its inactive or unplugged wells are brought back online, say through recompletion or a sidetrack operation, but the longer a structure is idle, the less likely it will return to production. Finally, if all the wells on an idle structure are permanently abandoned, it may be repurposed in a support function.

There are many reasons why a structure may stop producing, including reserves depletion, uneconomic production, well failure, scheduled workovers, weather damage, third-party problems (i.e., pipelines), recompletion delay, and investment review. Production cessation may be short or long term, and since BOEM databases do not identify utility, classifications that group idle structures within the same class will necessarily overestimate and bias structure counts.

Some idle structures may return to production, and others may be repurposed for another function, but the number of structures that make these transitions are believed to be relatively few because several conditions have to be satisfied simultaneously for the transition to occur. In some cases, operatorship and a review of active pipelines crossing the structure provide clues to make a meaningful inference (e.g., if the operator is a gas transmission company and all the wells on the structure have been permanently abandoned, the structure has likely been repurposed), but interpretations are subject to uncertainty and only applies for a subclass of structure. In deepwater, identification of structures that have made the transition from producing/idle to auxiliary is easier because the number of structures to examine is much smaller and pipeline interconnections are less complicated.

2.5 NUMBER OF WELLS

The number of wells required to develop a field and the number of wells that can be drilled from a structure are primary descriptors of a platform. Small reservoirs might contain two or three wells, while major fields will usually have many productive fault blocks and numerous pay sands and require dozens of wells to develop. Fields with large areal extent spread over several lease blocks will require several drilling platforms and numerous wells, while in deepwater, usually only one structure is deployed unless a field is unusually large or complex (e.g., Mars and Mad Dog).

Engineers determine how wide wells should be spaced without suffering any significant loss of reserves, and the depth and areal extent of the reservoirs will determine the number of structures required in development. One well may be used to access several vertically separated reservoirs, and directional drilling from platforms is common.

Example. East Breaks 160 (Cerveza) field

The East Breaks intraslope basin is located 100 mi offshore and south of Galveston (Braithwaite et al., 1988). The basin lies downslope of the modern shelf/slope break at water depths of 300–1300 ft and extends 35 mi northeast-southwest (Fig. 2.11). Proximity to the shelf was the key factor that controlled the depositional style and structural history of the field that determined the number and placement of wells.

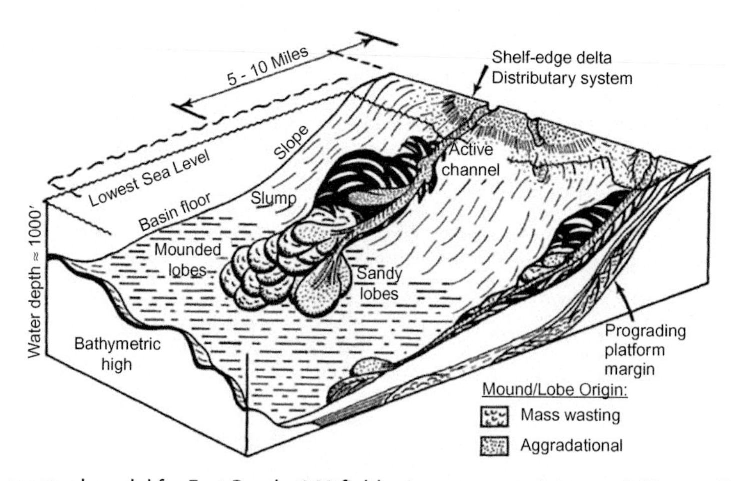

Fig. 2.11 Conceptual model for East Breaks 160 field primary reservoir interval. *(Source: Braithwaite, P., Armentrout, J.M., Beeman, C.E., Malecek, S.J., 1988. East Breaks block 160 field, offshore Texas: a model for deepwater deposition of sand. In: OTC 5696. Offshore Technology Conference, Houston, TX, May 2–5.)*

The East Breaks blocks 160–161 structure map and stratigraphic cross section are depicted in Figs. 2.12–2.13 for five exploration wells (Schanck et al., 1988). The wells at each end were drilled straight down, while wells 160#1, 160#2, and 161#3 were all drilled directionally. Well 160#2 penetrated five reservoir sands. A spider diagram shows the bottom-hole locations of development wells drilled from a platform in 935 ft water depth in the northeast corner of block EB 160 (Fig. 2.14).

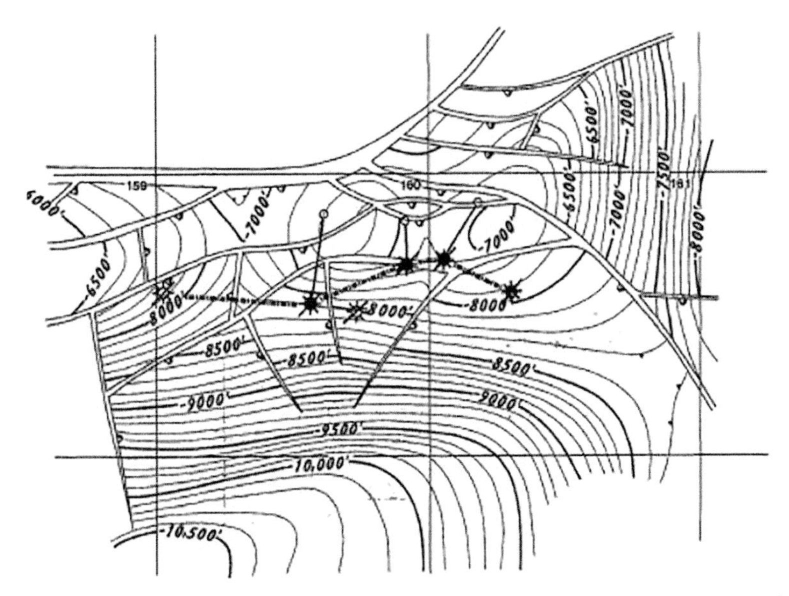

Fig. 2.12 East Breaks blocks 160–161 structure map with line of cross section. *(Source: Schanck, J.W., Cobb, C.C., Ivey, M.L., 1988. East Breaks 160 field on the offshore Texas shelf edge: a model for deepwater exploration and development. In: OTC 5697. Offshore Technology Conference, Houston, TX, May 2–5.)*

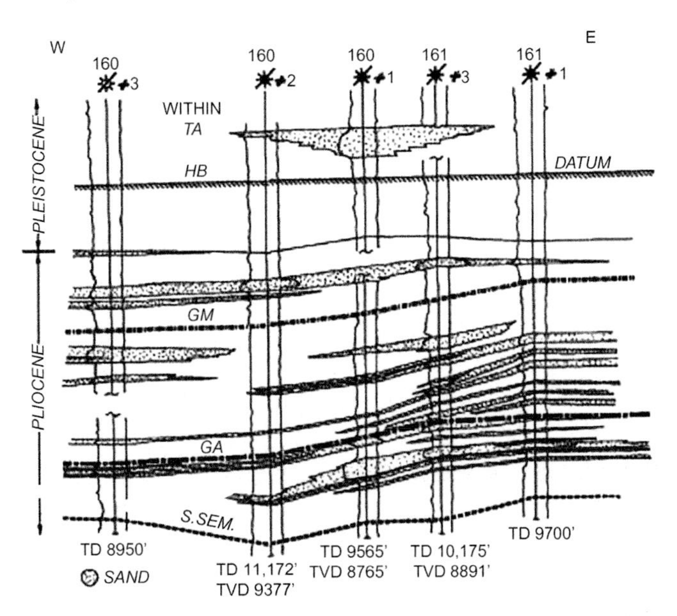

Fig. 2.13 East Breaks blocks 160–161 stratigraphic cross section. *(Source: Schanck, J.W., Cobb, C.C., Ivey, M.L., 1988. East Breaks 160 field on the offshore Texas shelf edge: a model for deepwater exploration and development. In: OTC 5697. Offshore Technology Conference, Houston, TX, May 2–5.)*

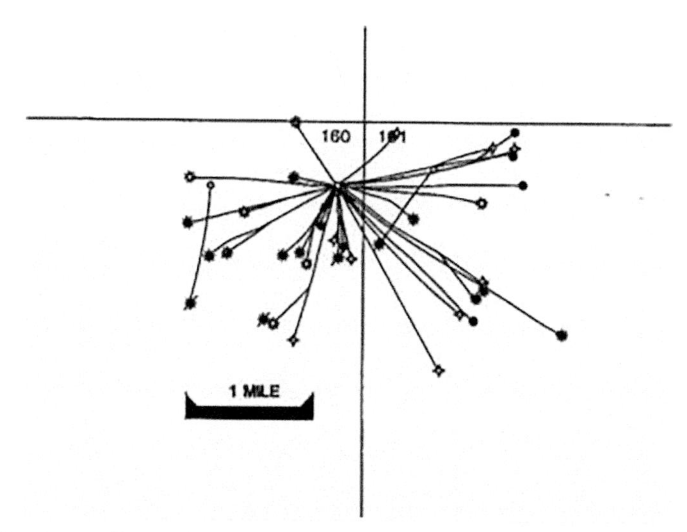

Fig. 2.14 East Breaks 160 (Cerveza) structure spider diagram. *(Source: Schanck, J.W., Cobb, C.C., Ivey, M. L., 1988. East Breaks 160 field on the offshore Texas shelf edge: a model for deepwater exploration and development. In: OTC 5697. Offshore Technology Conference, Houston, TX, May 2–5.)*

Through September 2017, the East Breaks 160 (Cerveza) structure had produced 14.8 MMbbl oil and 96.6 Bcf gas, while the adjacent East Breaks 159 (Ligera) structure in 924 ft water depth had produced 14.5 MMbbl oil and 181.1 Bcf gas.

Example. East Cameron 270 field

Block 270 of the East Cameron 270 field is a 2500 ac tract located about 65 mi from the shoreline in 170 ft water depth. The discovery well was drilled in 1970, and defined gas pays occur at 6470, 7680, 7970, and from 8310 to 8710 ft (Holland et al., 1975). Two 8-pile, 18-slot platforms were assigned to the block (Fig. 2.15), and several platforms on adjacent leases by other operators produce from the same field.

The A platform on EC 270 was set in August 1971, and 13 wells were drilled from the platform resulting in nine single and three dual completions. The B platform was installed in August 1971, and 12 wells were drilled resulting in five single and seven dual completions. Production commenced in 1973 and terminated in 2013 after producing 2 MMbbl of oil and 693 Bcf of gas.

Adjacent lease blocks 254, 255, and 273 terminated in 2007, 1986, and 1989, respectively, after producing 0.9 MMbbl oil and 95 Bcf gas (EC 254), 0.1 MMbbl oil and 26 Bcf gas (EC 255), and 0.2 MMbbl oil and 109 Bcf gas (EC 273). Block 272 is still active, and circa 2017 has produced 50.2 MMbbl oil and 199 Bcf gas.

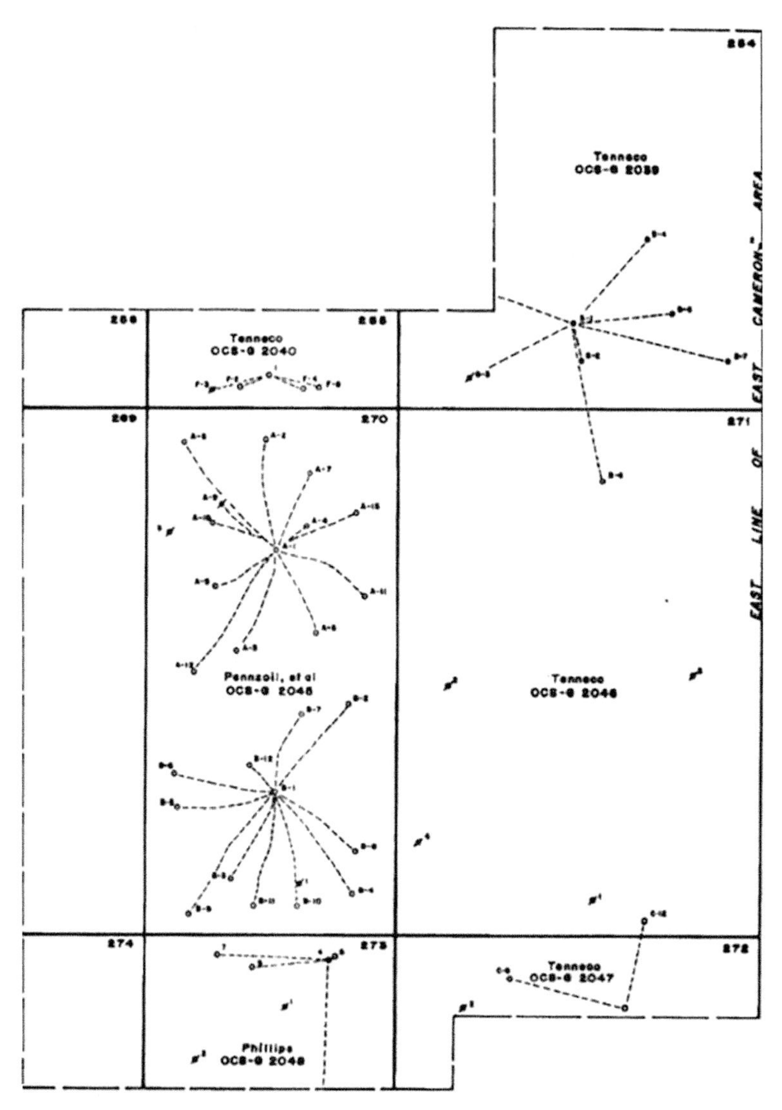

Fig. 2.15 East Cameron block 270 field development. *(Source: Holland, D., Sutley, C.E., Berlitz, K.E., Gilreath, J.A., 1975. East Cameron block 270, a Pleistocene field. Society of Petrophysicists and Well Log Analysts.)*

2.6 HUB PLATFORMS

Hub platforms are arguably the most important structure class in the GoM to maintain economic efficiency and commercial development. Unfortunately, there is no standard definition of a "hub" platform, and a variety of hubs can be defined based on

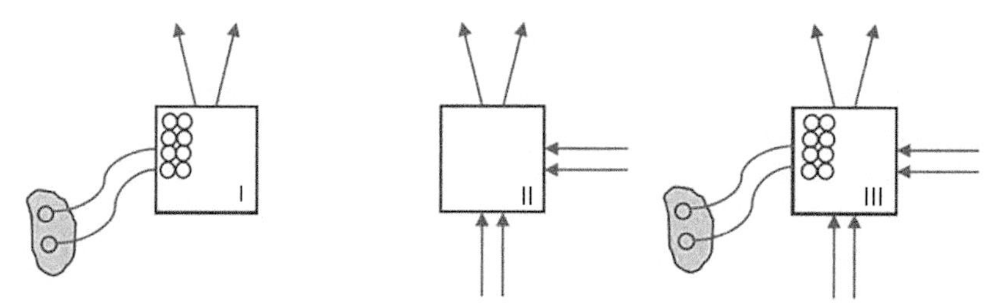

Fig. 2.16 Hub platform classes: field development (I), transportation (II), and combined services (III).

configuration/function and ownership. Generally, hub platforms are recognized as central points for gathering, redistribution, and transportation of oil and gas (Huff and Heijermans, 2003). Here, three types of hub platforms are distinguished, whether primarily serving in a field development role, primarily in transportation services, or for both development and transportation functions (Fig. 2.16).

2.6.1 Classification

Three hub classes are identified:

I. Structures that process production from one or more platforms or subsea wells
II. Structures that serve as a receiving station for processed production for export
III. Structures that process production and receive processed production for export

Historically, platforms were referred to as hubs when they acted as a central station to receive and process production from drilling platforms in a field (Fig. 2.17), and today,

Fig. 2.17 Central production facility at the Eugene Island 126 field circa 1956. *(Source: Massad, A.H., Pela, E.C., 1956. Facilities used in gathering and handling hydrocarbon production in the Eugene Island block 126 field Louisiana Gulf of Mexico. In: SPE 675-G. SPE AIME Annual Fall Meeting of AIME, Los Angeles, CA, Oct 14–17.)*

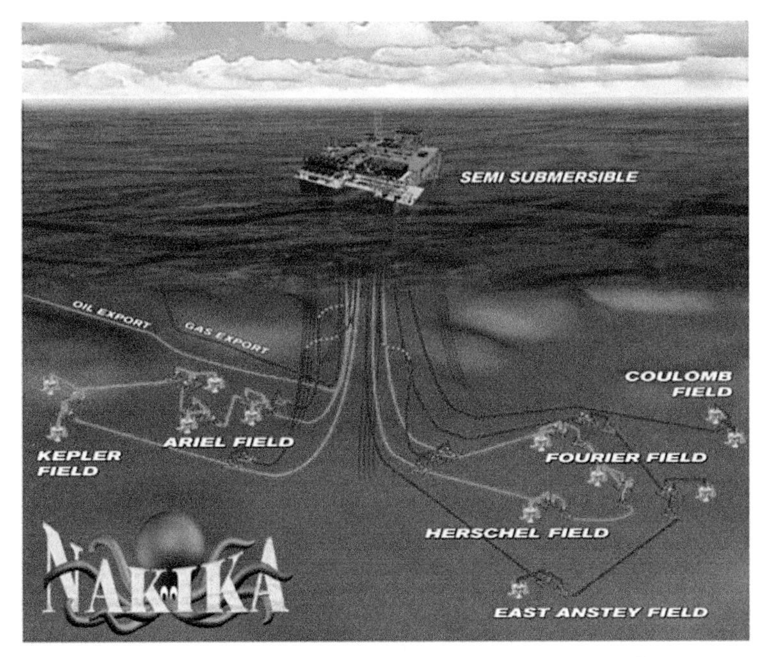

Fig. 2.18 Schematic of Na Kika host and subsea layout. *(Source: Keck, R.G., Colbert, J.R., Hardham, W.D., 2005. The application of flux-based sand-control guidelines to Na Kika deepwater fields. In: SPE 95294. SPE Annual Technical Conference and Exhibition, Dallas, TX, October 9–12.)*

this connotation still applies to facilities that develop multiple fields as in the Na Kika development (Fig. 2.18). In modern developments, the owners of the wells may be the same or different from the platform owner, and tiebacks may occur at the time the platform was installed or at a later date.

Example. Grand Isle 16 field

The Grand Isle 16 field was discovered in 1948 and developed with 20 producing platforms, one quarter platform, and one compressor platform in water depths from 40 to 65 ft (Frankenberg and Allred, 1969). Oil and gas flowed two phases to the centrally located "L" platform for separation (Fig. 2.19). Separator gas provided fuel supply to Freeport Sulfur's Caminada and Grand Isle mines. The remainder of the separator gas was either injected into reservoirs or compressed and transported via gas pipeline to a gas plant at Grand Isle. Circa 2016, cumulative production from the field had achieved 376 MMboe, and remaining reserves were estimated at 2.8 MMbbl oil and 6.5 Bcf natural gas.

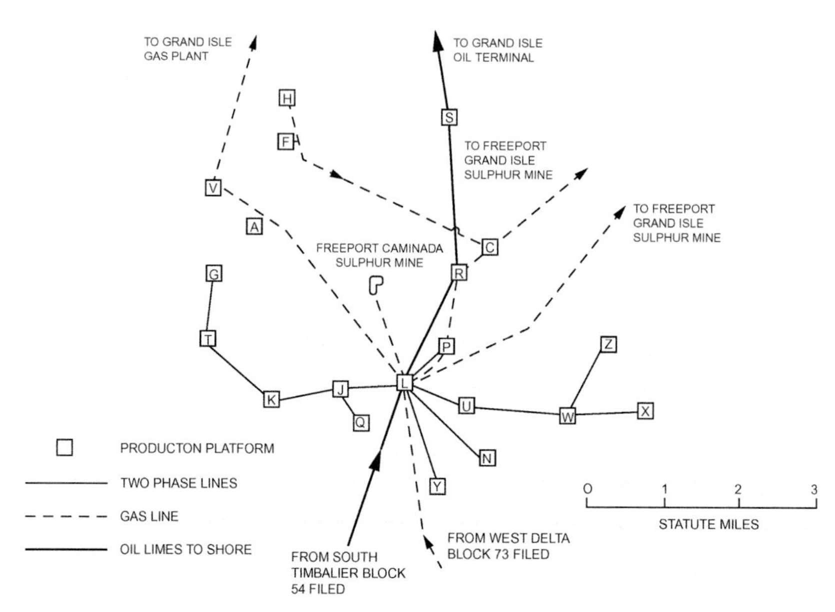

Fig. 2.19 Grand Isle block 16 field layout. *(Source: Frankenberg, W.G., Allred, J.H., 1969. Design, installation, and operation of a large offshore production complex. In: OTC 1082. Offshore Technology Conference, Houston, TX, May 18–21.)*

Structures may serve as a central point to gather and process production in field development or as host to tieback fields. Bullwinkle was already past its peak production when it began to accept tiebacks from the Rocky, Troika (Fig. 2.20), Aspen, and Angus fields. Falcon Nest is an example of a shallow-water platform built to accept deepwater subsea well tiebacks (Fig. 2.21), not a common strategy but one that has been deployed in several developments.

A structure with only dry tree wells may transition to hub status if wet wells are later tied back to the platform. Many of the deepwater fixed platforms in >400 ft water depth fall into this category, and several older floaters are still in operation because of new tiebacks. Bullwinkle and Auger are two early examples of structures installed for large anchor fields that later transitioned to hub platforms. If a structure's field production ceases and the structure is not used for subsea tiebacks, the structure may instead be used as a destination for export pipelines from the operator or third parties such as the Ship Shoal 332A platform.

Generally speaking, operators are better able to schedule and repurpose their own platforms than soliciting or commercializing production from third parties because of timing and logistic constraints and other issues. The deciding factors for operator-owned facilities and tiebacks are strategic and economic, while for third-party tiebacks, economics is usually the deciding factor. Planning, development, and negotiation between parties may take several years, and as long as the structure is producing, the owner(s) of the structure will maintain their bargaining power, but once the structure is near the end of its productive life, the balance of power will likely shift.

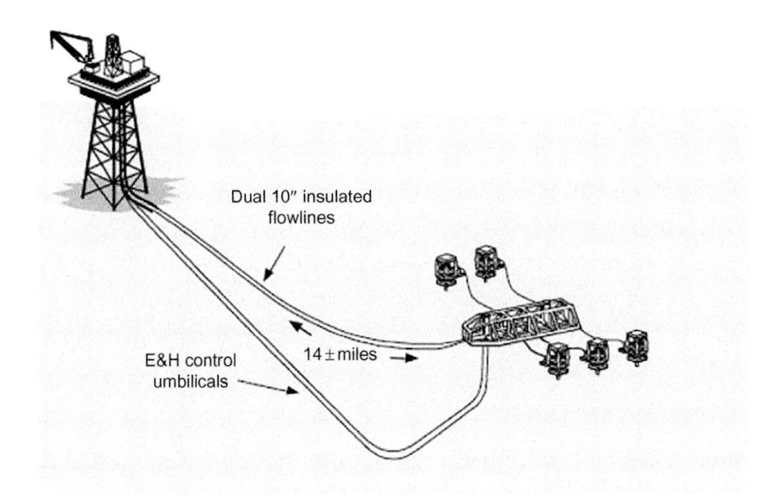

Fig. 2.20 Troika subsea system layout and tieback to host platform Bullwinkle. *(Source: Berger, R.K., McMullen, N.D., 2001. Lessons learned from Troika's flow assurance challenges. In: OTC 13074. Offshore Technology Conference, Houston, TX, Apr 30–May 3.)*

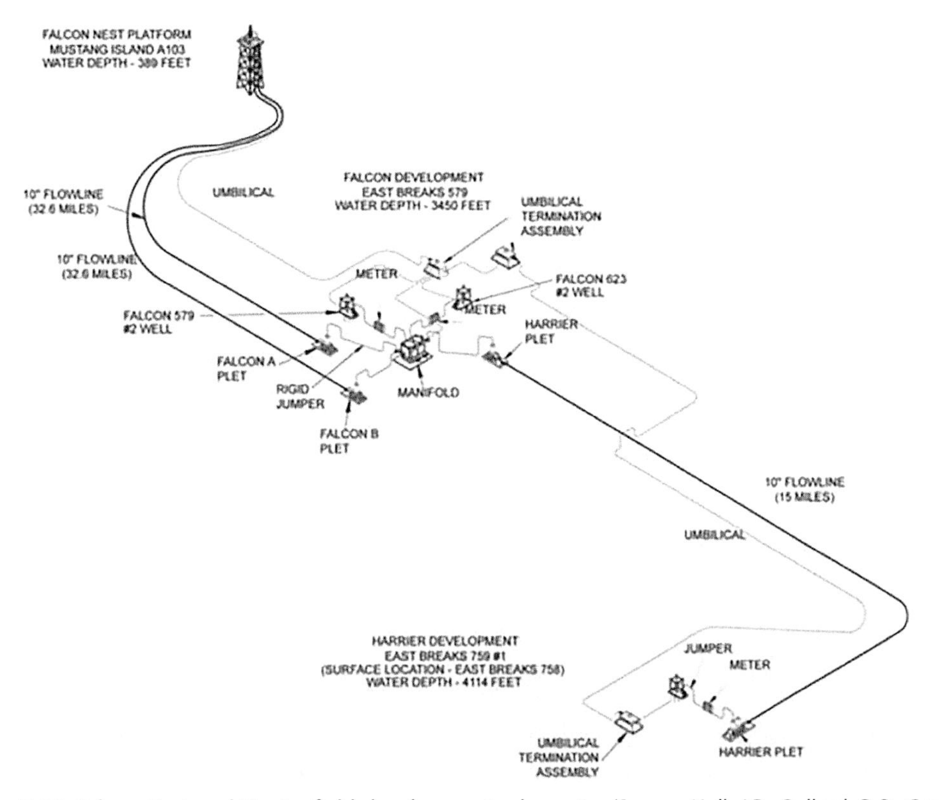

Fig. 2.21 Falcon Nest and Harrier field development schematic. *(Source: Hall, J.D., Bullard, D.B., Gray, W.M., Drew, M.C. 2004. Falcon Corridor: infrastructure with future marginal fields in mind. In: OTC 16527. Offshore Technology Conference, Houston, TX, May 3–6.)*

2.6.2 Process and Export Capacity

Hub platforms are described by their oil and gas processing capacity, number of interconnections, and oil and gas export volumes. Processing capacity refers to the equipment used to handle, separate, treat, heat, and cool raw hydrocarbons into pipeline quality oil and gas streams. Service and junction platforms do not have processing capacity and are primarily characterized by their pumping and compression capability, slug-catching facilities, metering, and dehydration services. For nonhub platforms, oil and gas processing capacity is a reasonably good indicator of oil and gas export capacity and pipeline diameter, but for hub platforms, export capacity usually greatly exceed processing capacity.

2.6.3 First Generation Hubs

Shell's Bullwinkle, Enchilada, and Auger developments were the first generation of deepwater hub platforms installed in the mid-1990s. The development formula was the same in each case. After anchor field production began to decline, nearby discoveries (mostly, but not always, from the owners of the platform) were tied back and processed as capacity became available. In many cases, expanding processing capacity at the host with subsea well tiebacks was considered more economic than installing new structures, would accelerate time to first production, and simultaneously extend the operating life of the facility that itself may create new future opportunities.

Example. Green Canyon 65 (Bullwinkle) platform

The Bullwinkle field was discovered in Green Canyon block 65 in 1983, and first production was in 1989. After peaking in 1992, subsea tiebacks at Rocky (1995), Troika (1996), Angus (1998), and Aspen (2002) were brought online, and production capacity at Bullwinkle was doubled to handle 200 Mbopd and 320 MMcfpd (Fig. 2.22). New export lines were also installed due to operator preferences on export destination. Through September 2017, cumulative production at the Bullwinkle field was 122 MMbbl oil and 208 Bcf gas, and facilities have handled about 284 MMbbl oil and 485 Bcf gas from tieback fields.

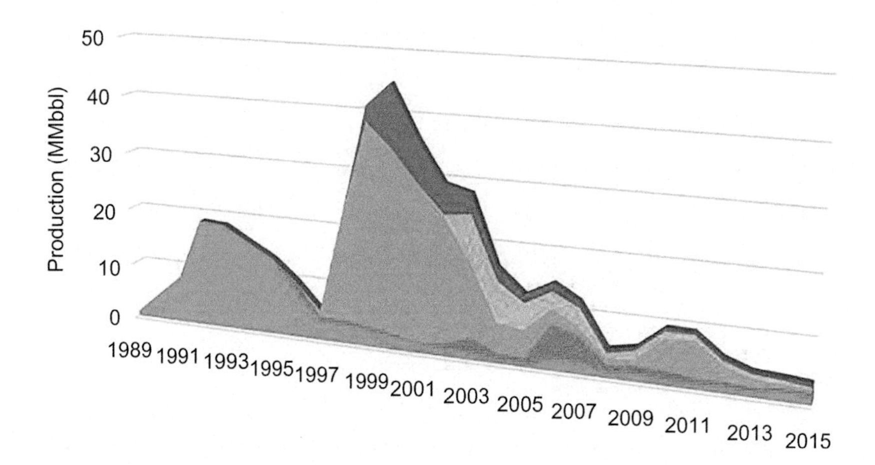

Fig. 2.22 Bullwinkle oil field production and subsea tiebacks Rocky, Troika, Angus, and Aspen. *(Source: BOEM 2017.)*

Example. Garden Banks 426 (Auger) TLP

The Auger field was discovered in 2860ft water depth in 1987, approved in 1990, and started production in 1994 from the first TLP to host a drilling rig with full processing facilities (Brock, 2000). The Auger field is located in Garden Banks blocks 426, 427, 470, and 471. By mid-2000, anchor field production was about half of equipment capacity, and remaining development and recompletion opportunities in the field were not adequate to offset production decline. In order to keep the facility full, a decision was made to transform Auger into an infrastructure hub serving as a subsea tieback host for nearby Shell fields and third-party production.

Production capacity was expanded in 1997 and again in 2000 to 100 Mbopd and 400 MMcfpd to serve as host for the Cardamom (GB 471), Macaroni (GB 602), Oregano (GB 559), Serrano (GB 516), and Llano (GB 387) tiebacks (Fig. 2.23). Macaroni was tied back in 1999, and in late 2001, the Serrano and Oregano fields were brought online. Northwest of the anchor field, Llano and Habanero were integrated into the system in 2002. Auger's cumulative production through 2017 was 267 MMbbl oil and 935 Bcf natural gas, and facilities have handled about 412 MMbbl oil and 1.3 Tcf natural gas in total (Table 2.4, Fig. 2.24).

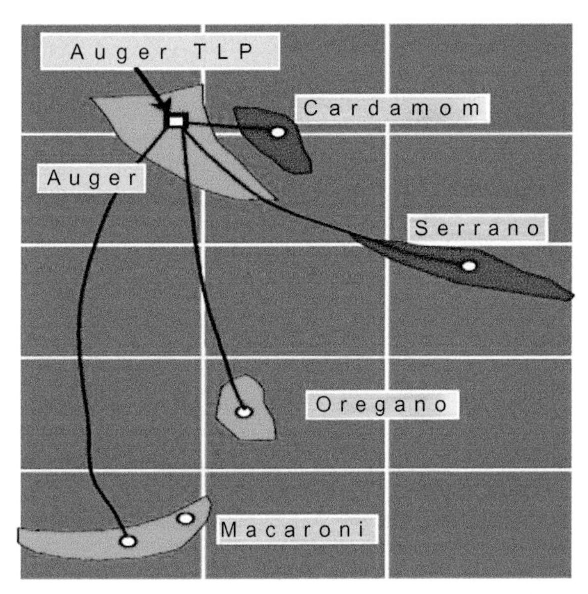

Fig. 2.23 Auger TLP subsea tiebacks Macaroni, Oregano, and Serrano. *(Source: Brock, V.A., 2000. Auger—an asset in transition. In: OTC 15111. Offshore Technology Conference, Houston, TX, May 5–8.)*

Table 2.4 Auger and Subsea Tieback Cumulative Production Circa September 2017

Name	Field	First Production	Lease Blocks	Oil (MMbbl)	Gas (Bcf)
Auger	GB 426	1994	426, 427, 470, 471	267	935
Macaroni	GB 602	1999	602	13.4	24.9
Oregano	GB 559	2001	559	33.1	49.5
Serrano	GB 516	2001	515, 516, 472	3.9	47
Llano	GB 387	2002	341, 385–387	94.7	223
Total				412	1279

Source: BOEM 2017.

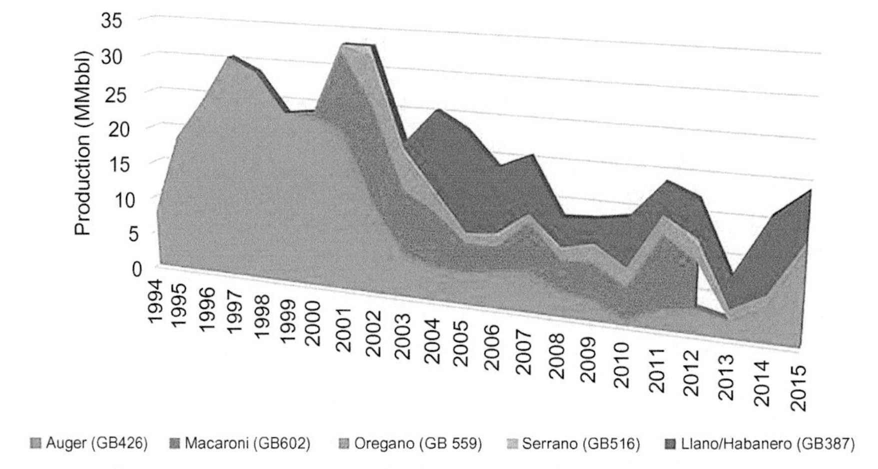

Fig. 2.24 Auger field production and subsea tiebacks Macaroni, Oregano, Serrano, and Llano. *(Source: BOEM 2017.)*

Auger oil and gas export lines provide access to multiple markets via existing shallow-water gathering systems (Fig. 2.25). The 12 in oil line, owned and operated by Shell Pipeline Company LP, is routed to GB 128 (Shell's Enchilada platform). At GB 128, Auger's oil ran to the Shell-operated EI 331A platform (before it was destroyed by Hurricane Ike in 2008) where it was delivered into multiple oil pipelines accessing different onshore market locations. One of the gas lines owned by Shell terminates at Enchilada and delivers into the Garden Banks Gas Pipeline System. The second gas line terminates at VR 397 and delivers to the ANR pipeline system (Kopp and Barry, 1994).

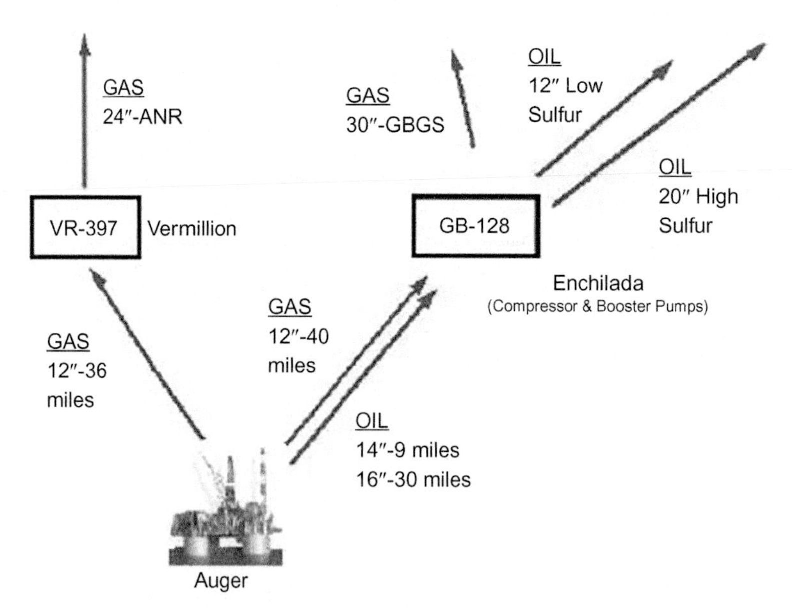

Fig. 2.25 Auger oil and gas export pipelines. *(Source: Brock, V.A., 2000. Auger – an asset in transition. In: OTC 15111. Offshore Technology Conference, Houston, TX, May 5–8.)*

2.6.4 Second Generation Hubs

Second-generation and later hubs were built with wider flexibility and equipment sizes, tied back to a greater number of subsea wells, and were designed with excess transportation capacity in addition to production processing. A greater variety of third-party operators also became interested in hub business models in the mid-2000 time period. Two examples illustrate the structure class.

Example. Garden Banks 72 (Spectacular Bid) platform

The GB 72 platform (aka Spectacular Bid) is located in Garden Banks block 72 in the western GoM in 514ft water depth (Fig. 2.26). The platform was designed and installed by a midstream company to use for off-lease processing and as a junction platform for its pipeline systems (Heijermans and Cozby, 2003). The platform originally processed production from four off-lease fields and served as the anchor portal for the deepwater Stingray gas pipeline and Poseidon oil pipeline systems. The Cameron Highway Oil Pipeline System (CHOPS) designed for the movement of the Atlantis, Mad Dog, and Holstein crude from the southern Green Canyon area also cross this platform to markets in Port Arthur and Texas City, Texas. Circa September 2017, cumulative production from the GB 72 field (covering blocks GB 28, GB 72, GB 73, and VR 408) was 13.5 MMbbl oil and 40 Bcf gas (Fig. 2.27).

Fig. 2.26 Garden Banks 72 (Spectacular Bid) platform. *(Source: Genesis Energy, BOEM 2017.)*

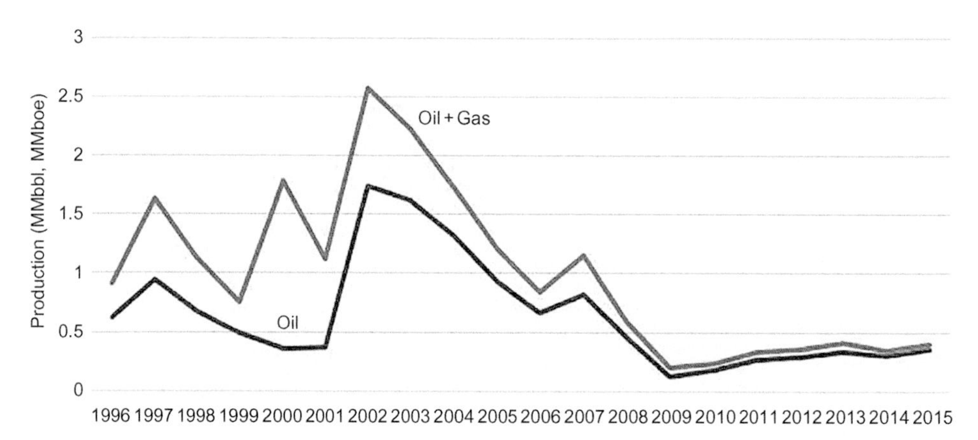

Fig. 2.27 Garden Banks 72 (Spectacular Bid) field oil and gas production profile, 1996–2015. *(Source: BOEM 2017.)*

Example. Ship Shoal 332A and 332B platforms

The SS 332A platform was installed in Ship Shoal block 332 in 438 ft water depth to develop a gas field (Fig. 2.28). El Paso Energy Partners (EPN) acquired the structure from Arco in the early 1990s after production ceased to support the Leviathan gathering system, the predecessor of the Manta Ray gathering system. Circa September 2017, cumulative production from the SS 332 field (covering SS 332 and SS 354) was 6.6 MMbbl oil and 105 Bcf gas (Fig. 2.29). In 1995, EPN constructed the Poseidon oil pipeline and the Allegheny oil pipeline in 1999, which utilized the SS 332A platform. A new platform SS 332B was connected adjacent to SS 332A to serve as transport hub for CHOPS and the interconnection between the Caesar oil pipeline and Cameron Highway.

Fig. 2.28 Ship Shoal 332A and 332B platform hub for the Cameron Highway Oil Pipeline System. *(Source: Genesis Energy, BOEM 2017.)*

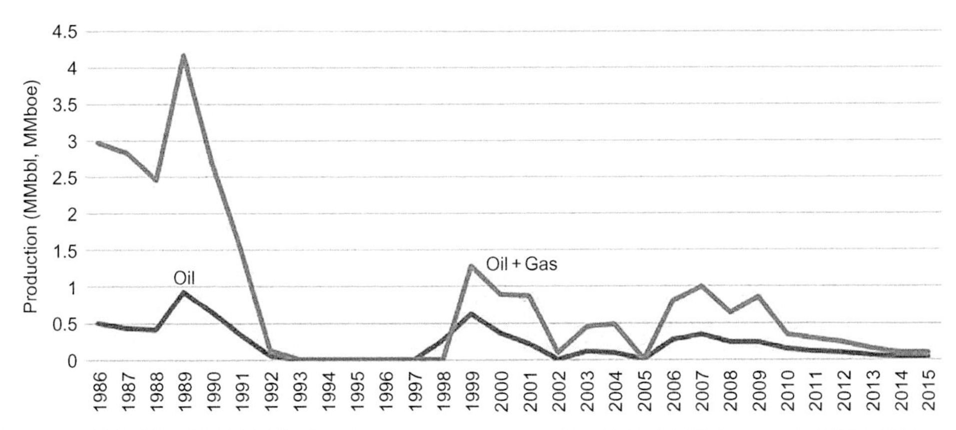

Fig. 2.29 Ship Shoal 332 field oil and gas production profile, 1986–2015. *(Source: BOEM 2017.)*

REFERENCES

Berger, R.K., McMullen, N.D., 2001. Lessons learned from Troika's flow assurance challenges. In: OTC 13074. Offshore Technology Conference, Houston, TX, Apr 30–May 3.

Braithwaite, P., Armentrout, J.M., Beeman, C.E., Malecek, S.J., 1988. East Breaks block 160 field, offshore Texas: a model for deepwater deposition of sand. In: OTC 5696. Offshore Technology Conference, Houston, TX, May 2–5.

Brock, V.A., 2000. Auger – an asset in transition. In: OTC 15111. Offshore Technology Conference, Houston, TX, May 5–8.

Cavnar, A., 2009. Accountability and the commission on the limits of the continental shelf: deciding who owns the ocean floor. Institute for International Law and Justice Emerging Scholars Paper 15.

Chakrabarti, S.K., 2005. Handbook of Offshore Engineering. Vol. I, II. Elsevier Ltd., Kidlington, Oxford, UK.

Chaudhury, G., Whooley, A., 2014. Art, science, and engineering of managing offshore field development economics and risks. In: OTC 25344. Offshore Technology Conference, Houston, TX, May 5–8.

Cullick, A.S., Cude, R., Tarman, M., 2007. Optimizing field development concepts for complex offshore production systems. In: SPE 108562. Offshore Europe Technology Conference, Aberdeen, Scotland, UK. 4–7 September.

D'Souza, R., Aggarwal, R., 2013. The tension leg platform technology – historical and recent developments. In: OTC 24512. Offshore Technology Conference, Rio de Janeiro, Brazil, October 29–31.

D'Souza, R., Basu, S., Khurana, S., 2016. Lessons learned from a comprehensive review of integration and installation phases of Gulf of Mexico floating production systems. In: OTC 27268. Offshore Technology Conference, Houston, TX, May 2–5.

Dekker, M., Reid, D., 2014. Deepwater development strategy. In: OTC 25135. Offshore Technology Conference, Houston, TX, May 5–8.

Frankenberg, W.G., Allred, J.H., 1969. Design, installation, and operation of a large offshore production complex. In: OTC 1082. Offshore Technology Conference, Houston, TX, May 18–21.

Garrison, T.S., Ellis, R., 2017. Essentials of Oceanography, 8th ed. Brooks Cole.

Gerwick, B.C., 2007. Construction of Marine and Offshore Structure, 3rd ed. CRC Press, Boca Raton, FL.

Hall, J.D., Bullard, D.B., Gray, W.M., Drew, M.C., 2004. Falcon Corridor: infrastructure with future marginal fields in mind. In: OTC 16527. Offshore Technology Conference, Houston, TX May 3–6.

Heijermans, B., Cozby, D., 2003. The commercial development of hub platforms and the value to E&P producers. In: OTC 15108. Offshore Technology Conference, Houston, TX, May 5–8.

Hicks, R.H., McDonald, W.M., Staffel, E.O., 1971. A prefabricated and automated offshore gas production and dehydration facility. In: OTC 1387. Offshore Technology Conference, Houston, TX, Apr 19–21.
Holland, D., Sutley, C.E., Berlitz, K.E., Gilreath, J.A., 1975. East Cameron Block 270, a Pleistocene field. Society of Petrophysicists and Well Log Analysts.
Holland, D.S., Nunan, W.E., Lammlein, D.R., Woodhams, R.L., 1980. Eugene Island Block 330 field, offshore Louisiana. In: Halbouty, M.T. (Ed.), Giant Oil and Gas Fields of the Decade: 1968-1979. American Association of Petroleum Geologists, Memoir 30, pp. 253–281.
Huff, S.S., Heijermans, B., 2003. Cost-effective design and development of hub platforms. In: OTC 15107. Offshore Technology Conference, Houston, TX, May 5–8.
Kopp, F., Barry, D.W., 1994. Design and installation of Auger pipelines. In: OTC 7619. Offshore Technology Conference, Houston, TX, May 2–5.
Lipari, C.A., 1962. An engineering challenge – development of south Louisiana's giant Timbalier Bay field. In: SPE 407. SPE Annual Fall Meeting. Los Angeles, CA, Oct 7–10.
Massad, A.H., Pela, E.C., 1956. Facilities used in gathering and handling hydrocarbon production in the Eugene Island block 126 field Louisiana Gulf of Mexico. In: SPE 675-G. SPE AIME Annual Fall Meeting of AIME, Los Angeles, CA, Oct 14–17.
Payne, C., 2010. Principles of Naval Weapon Systems, 2nd ed. United States Naval Institute, Annapolis, MD.
Reid, D., Dekker, M., 2014. Deepwater development non-technical risks – identification and management. In: SPE 170739. SPE Annual Technical Conference and Exhibition. Amsterdam, The Netherlands, Oct 27–29.
Reid, D., Dekker, M., Nunez, D., 2013. Deepwater development: wet tree or dry tree? In: OTC 24517. Offshore Technology Conference, Rio de Janeiro, Brazil, Oct 29–31.
Ronalds, B.F., 2002. Deepwater facility selection. In: OTC 14259. Offshore Technology Conference, Houston, TX, May 6–9.
Ronalds, B.F., Lim, E.F.H., 2001. Deepwater production with surface trees: trends in facilities and risers. In: SPE 68761. SPE Asia Pacific Oil and Gas Conference and Exhibition, Jakarta, Indonesia, April 17–19.
Sablok, A., Barras, S., 2009. The internationalization of the spar platform. In: OTC 20234. Offshore Technology Conference, Houston, TX, May 4–7.
Schanck, J.W., Cobb, C.C., Ivey, M.L., 1988. East Breaks 160 field on the offshore Texas shelf edge: a model for deepwater exploration and development. In: OTC 5697. Offshore Technology Conference. Houston. TX, May 2–5.
Skinner, B.J., Murck, B.W., 2011. The Blue Planet – An Introduction to Earth System Science, 3rd ed. John Wiley & Sons, Inc.
Swift, V.N., 1966. Production operations in the South Timbalier area of offshore Louisiana. In: SPE 1408. SPE AIME Symposium Offshore Technology and Operations, New Orleans, LA, May 23–24.
Thurmond, B.F., Walker, D.B.L., Banon, H.H., Luberski, A.B., Jones, M.W., Peyers, R.R., 2004. Challenges and decisions in developing multiple deepwater fields. In: OTC 16573. Offshore Technology Conference. Houston, TX, May 3–6.

CHAPTER THREE

Installation and Decommissioning Activity

Contents

Abstract

Structure installation and decommissioning activity reflect geologic prospectivity and production trends broadly and are useful indicators to track. The number of structures in the Gulf of Mexico in water depth <400 ft peaked in 2001 at 3972 and circa 2017 numbered 1908, while deepwater inventories numbered 97 circa 2017 and have not yet peaked. Record levels of shallow-water decommissioning in recent years are the result of the maturity of field production, sustained low oil and gas prices, and tougher regulatory conditions and oversight. Higher levels of deepwater decommissioning are expected in the years ahead unless alternative uses for structures are found. In this chapter, installation and decommissioning trends in the shallow-water and deepwater Gulf of Mexico is described.

3.1 CUMULATIVE ACTIVITY

A large number of oil and gas deposits spread throughout the GoM have required a large number of wells and structures to produce. A total of 7053 structures have been installed in the GoM through 2017 (Table 3.1), and 5048 structures have been decommissioned (Table 3.2), leaving an active inventory of 2005 structures circa 2017. Almost all of the installation and decommissioning activity has been in shallow water. Ninety-seven structures comprise the deepwater inventory circa 2017, 48 floaters, three compliant towers, and 46 fixed platforms. Shallow-water inventories have declined

Decommissioning Forecasting and Operating Cost Estimation
https://doi.org/10.1016/B978-0-12-818113-3.00003-5

Table 3.1 Installed Structures by Type and Water Depth, 1942–2017

Water Depth (ft)	C/WP	FP	Floater	All	Percent (%)
≤ 100	3010	2011	3	5024	71
101–150	182	553		735	10
151–200	98	452		550	8
201–400	40	584		624	9
> 400		65	55	120	2
All	3330	3665	58	7053	100

Source: BOEM 2018.

Table 3.2 Decommissioned Structures by Type and Water Depth, 1973–2017

Water Depth (ft)	C/WP	FP	Floater	All	Percent (%)
≤ 100	2447	1284	3	3734	74
101–150	165	372		537	10
151–200	88	300		388	8
201–400	33	333		366	7
> 400		19	4	23	1
All	2733	2308	7	5048	100

Source: BOEM 2018.

between 5% and 10% per year over the past decade, while deepwater inventories have remained relatively stable over the same period.

3.1.1 Shallow Water

Since all oil and gas fields have finite lives and the useful life of structures are usually tied to field production, the more structures installed in a region the more that will need to be removed, and at any point in time, the number of structures that have been decommissioned will be roughly proportional to the number installed. Regions with older structures are expected to have a higher decommissioning percentage compared with regions with structures of more recent vintage.

There have been three floating mobile offshore production units installed in shallow water on lease blocks West Cameron 44, South Timbalier 41, and South Timbalier 145 in 34, 68, and 92 ft water depth, respectively. Complex 1252 on WC 44 was installed in June 2003 and removed in July 2008, complex 1490 on ST 145 was installed in November 2004 and removed in May 2007, and complex 24205 on ST 41 was installed in November 1994 and removed in May 2003.

3.1.2 Deepwater

Only a small number of deepwater projects are sanctioned each year, and only a few structure installations occur annually. In total, 65 fixed platforms and three compliant towers have been installed in water depth >400 ft, and circa 2017, there were 46 fixed platforms and three compliant towers in the region. In total, 53 floating structures have also been installed in the GoM, and five floaters have been decommissioned circa 2017, leaving an activity inventory of 48 at year-end 2017.

Sixteen of the 18 decommissioned fixed platforms were primarily gas producers, and the average time to decommission after the last year of production was about 5 years with a range from 2 to 10 years (Table 3.3). Five floaters have been decommissioned through 2017: GC29/Llano (semi, 1989), Cooper (semi, 1999), Typhoon (MTLP, 2006), Red Hawk (spar, 2014), and Gomez (semi, 2014).

The first two deepwater structures decommissioned in the GoM were both semisubmersible developments. Placid Oil's GC52/Llano development in 1989 was decommissioned less than a year after installation because of well problems, and Newfield's Cooper development was decommissioned in 1999. Both semisubmersibles were towed away for decommissioning.

Table 3.3 Structures Decommissioned in the Gulf of Mexico in Water Depth > 400 ft

Operator	Complex	Type	Water Depth (ft)	First Production	Last Production	Decommissioned Year
Apache	85	FP	670	1998	2012	2016
Chevron U.S.A.	735	MTLP	2107	2001	2004	2006
W&T Offshore	1055	FP	472	2002	2008	2012
ATP Oil & Gas	1320	FP	542	2003	2012	2017
Anadarko Petroleum	1384	SPAR	5300	2004	2008	2014
ATP Oil & Gas	1771	SEMI	2975	2006	2013	2014
Fieldwood SD Off.	10242	FP	660	1985	2009	2017
Louisiana L&E	22274	FP	431	1981	2001	2005
Chevron U.S.A.	22372	FP	685	1980	2003	2010
Apache	22583	FP	651	1981	2005	2015
Apache	22705	FP	432	1989	2008	2016
Chevron U.S.A.	22846	FP	480	1984	2003	2011
EP Energy E&P	23004	FP	414	1986	2017	2011
W&T Offshore	23308	FP	415	1989	2008	2013
Placid Oil	23543	SEMI	1540	1988	1989	1989
Chevron U.S.A.	23581	FP	620	1990	2005	2011
Sojitz Energy Vent.	23859	FP	582	1990	2008	2010
Apache	23891	FP	531	1992	2011	2016
W&T Offshore	24021	FP	467	1992	2004	2006
Newfield Exp.	24079	SEMI	2097	1991	1999	1999
Louisiana L&E	24087	FP	420	1995	2001	2005
BP E&P	27014	FP	530	1995	2002	2009
W&T Offshore	28033	FP	430	1996	2008	2011

Source: BOEM 2018.

Chevron's Typhoon MTLP suffered catastrophic damage from Hurricane Rita in September 2005 and was decommissioned at a deepwater reef site at Eugene Island 367 in June 2006. After ATP Oil & Gas declared bankruptcy in 2013 and no buyers for its Gomez semisubmersible could be found, it was decommissioned in March 2014 and towed back to port at Ingleside, Texas. Anadarko decommissioned its Red Hawk spar in September 2014 at a deepwater reef site in Eugene Island 384.

The three semisubmersibles were each removed within a year of last production, Typhoon was decommissioned 2 years after last production, and Red Hawk was decommissioned 6 years after last production (Box 17.1). All the decommissioned semisubmersibles were towed back to shore, while the majority of the other deepwater structures—fixed platforms and floaters—were either reefed in place or towed to a deepwater reef site.

3.2 SHALLOW WATER TRENDS

3.2.1 Annual Activity

Decommissioning activity fluctuates year to year, and from 2007 to 2017 ranged between 108 and 295 structures annually (Fig. 3.1). Activity levels exceeded 200 removals per year from 2009 to 2014, peaking in 2011 at 295 structures decommissioned.

In 2016, there were 192 structures decommissioned (66 caissons and well protectors and 126 fixed platforms) and 697 well abandonments (585 permanent abandonments and 112 temporary abandonments) in water depth <400 ft. No structures were installed in 2016.

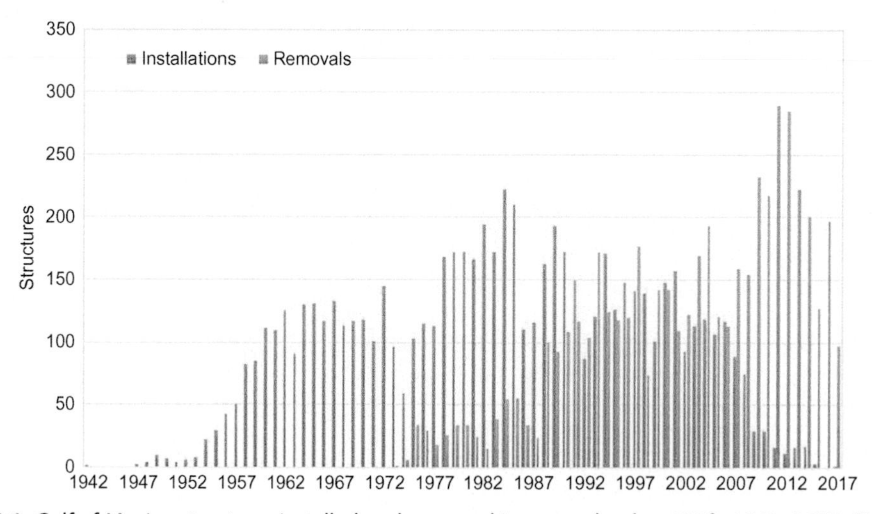

Fig. 3.1 Gulf of Mexico structures installed and removed in water depth < 400 ft, 1942–2017. *(Source: BOEM 2018.)*

In 2017, 108 shallow-water structures were decommissioned (40 caissons and well protectors and 68 fixed platforms), and 518 wells were abandoned (336 permanent abandonments and 182 temporary abandonments). One shallow-water structure was installed in 2017.

High levels of decommissioning activity in recent years is due in large part to the aging infrastructure and maturity of producing properties coupled with low oil and gas prices. Cleanup activity from the 2005 and 2008 hurricane seasons is nearly complete and has only contributed incrementally to decommissioning activity in recent years (Kaiser and Narra, 2017).

Notice to Lessee NTL 2010-G05 idle iron guidelines have also played a role in increased levels of decommissioning activity since operators are required to decommission structures that no longer serve a useful purpose within 5 years, unless special circumstances apply or a temporary exemption is granted.

3.2.2 Installation by Decade

There has been a dramatic decline in the number of shallow-water structures installed in recent years, and over the past decade, only 29 structures on average have been installed per year (Table 3.4). Caissons, well protectors, and fixed platforms all show similar installation trends across time, peaking in the 1977–86 period when about one-quarter of all installations occurred at an average rate of 170 structures/year, declining to 145 structures/year from 1987 to 1996 and 123 structures/year from 1997 to 2006. Four percent of shelf structures have been installed over the past decade, similar to the first decade of activity in the region.

Installation activity has fallen so dramatically so fast due in large part to the maturity of shelf production and the lack of new discoveries. Over 97% of proved reserves in shallow water have now been produced. New discoveries in mature areas are possible, but large discoveries are not likely unless new plays provide opportunities not previously considered. The shelf is gas-prone, and small offshore gas fields will have a difficult time competing with cheaper onshore shale plays when gas prices are in the $4/ Mcf range.

Oil wells on the shelf continue to produce, but new fields have been difficult to find. Large legacy fields such as those located at Eugene Island 330, West Delta 30, Grand Isle 43, Main Pass 41, Ship Shoal 208, West Delta 73, and Grand Isle 16 are probably the best opportunity for operators to seek additional (incremental) production. Small fields often have less upside potential compared with large fields, and reserves growth is smaller. Sidetrack drilling in depleted reservoirs is attractive mostly to niche players and those with large portfolios that can balance and manage the costs and risks involved. Structure installation in shallow water is not expected to add materially to shelf inventories in the near term unless there is a dramatic change in current conditions or new plays arise in the future.

Table 3.4 Shallow-Water Installation by Decade

Decade	C/WP	FP	Total	Percent (%)
1944–56	30	104	134	2
1957–66	607	425	1032	15
1967–76	441	659	1100	16
1977–86	775	924	1699	25
1987–96	708	738	1446	20
1997–2006	627	606	1233	18
2007–16	142	143	285	4
2017		1	1	
Total	3330	3600	6930	100

Source: BOEM 2018.

Table 3.5 Shallow-Water Decommissioning by Decade

Period	C/WP	FP	Total	Percent (%)
1973–76	65	5	70	1
1977–86	248	83	331	7
1987–96	636	443	1079	21
1997–2006	783	578	1361	27
2007–16	963	1121	2084	42
2017	40	68	108	2
Total	2735	2298	5033	100

Source: BOEM 2018.

3.2.3 Decommissioning by Decade

The first structure decommissioned in the GoM was in 1973, but it was not until the mid-1980s that activity levels began to pick up (Table 3.5). From 1987 to 1996, 108 structures per year were decommissioned on average, increasing to 136 structures per year from 1997 to 2006 and nearly doubling to 208 structures per year from 2007 to 2016 where over 40% of all decommissioning activity to date has occurred.

Historically, caissons and well protectors were removed in greater numbers than fixed platforms, but as their inventory declines, more fixed platforms are being removed than simple structures, a trend that will continue. More manned fixed platforms are also in queue for removal and will be decommissioned in the years ahead. Over the past decade, the number of structures decommissioned is almost an order of magnitude larger than installation activity.

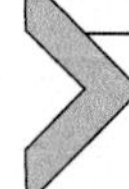

3.3 DEEPWATER TRENDS

3.3.1 Annual Activity

Only a handful of deepwater structures are installed or decommissioned most years (Fig. 3.2). Since 1978, the number of new deepwater structures can be counted on one hand every year except 1998 and 2004 when nine structures were installed, and

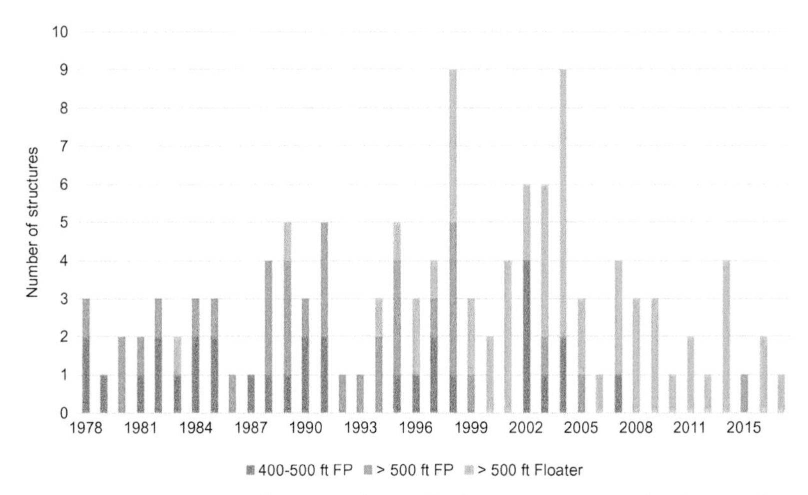

Fig. 3.2 Deepwater structure installation in the Gulf of Mexico in water depth > 400 ft, 1978–2017. *(Source: BOEM 2018.)*

in 2002 and 2003 when six structures were installed. Circa 2017, five deepwater floaters were sanctioned or under construction (Box 3.1). Significant capital expenditures and planning are required in deepwater development and execution. There are only a limited number of market participants with the requisite experience and capital to successfully operate in the deepwater space, and as these operators pursue investment opportunities worldwide, capital is allocated accordingly.

Drilling deepwater wells represent significant capital spending, and to "prove up" a development, several wells need to be drilled before a final investment decision is made. Deepwater drilling and completion costs about an order of magnitude greater than in shallow water. Over the past decade, for example, average drilling cost offshore has

BOX 3.1 Economic Lifetime

There is no one organizing principle to understand how the offshore sector is structured and how decisions are made, but an economic-risk framework is often useful for conceptual discussion since all significant capital investment decisions, that is, investments greater than a few hundred thousand dollars, are based upon economic and risk considerations. Conceptual models are most useful when only one or a limited number of factors influence the system. For complex systems, as the number of factors characterizing the system increases, conceptual models become much less useful since holding "all of the other factors" constant is neither feasible nor accurately reflects system behavior.

The economic lifetime of a producing asset is the point at which the net cash flow from operations turns negative. This is the time when the net income from production no longer exceeds the costs of production and a rational decision-maker will no longer wish to operate the venture because it costs more money than it generates. Production may continue beyond this point for strategic or other reasons but only by accepting financial losses.

Continued

BOX 3.1 Economic Lifetime—cont'd

There are two ways to extend the economic lifetime of an asset—reduce operating costs and increase revenue (Fig. 3.3). Offshore assets enter into economic decline when either income is falling or costs are rising, and for many mature developments, both are happening. A mature development must nonetheless continue to generate a positive net cash flow and compete against other projects for funds that become increasingly difficult because effort is usually rewarded with smaller and less certain benefits.

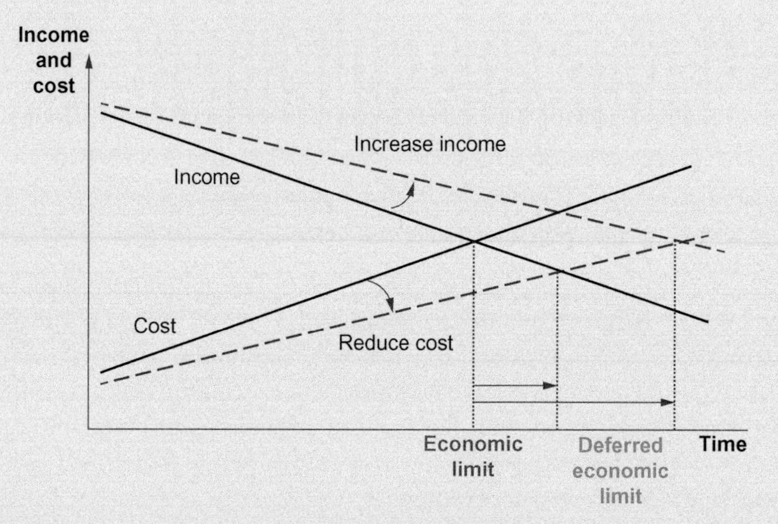

Fig. 3.3 Economic limit and means to defer economic lifetime.

Operating cost represents the major expenditures late in field life, and opportunities may exist to reduce operating expense. The number of staff assigned to a facility may be reduced, or maintenance budgets may be cut or delayed. In some cases, operations may be consolidated to achieve better transport and logistic rates or structures demanned and automated. Other means to reduce operating cost include outsourcing all or some operating and management functions, eliminating or reducing insurance coverage, and renegotiating service contracts. For marginal producers, end-of-life royalty relief may help to reduce cost to maintain profitable operations.

Income may increase by increasing production or by a rise in commodity prices. Increasing hydrocarbon production requires investment in workovers, infill drilling, sidetracking, or improved oil recovery techniques. Opportunities may be available to develop nearby fields through the infrastructure. Another way in which companies that own facilities but no longer have the hydrocarbons to fill them can continue to operate profitably is by renting equipment capacity and charging tariffs for the use of export routes, but the timing of these opportunities often presents obstacles.

Investment activity may occur at any time during the life cycle of the field and involve low-cost maintenance and workover activities such as changing production tubing, reperforating, and acidizing, to moderate cost activities such as sidetracking and equipment changes, to high-cost activity such as new well drilling and secondary recovery methods. Depending on the timing and type of activity, the uncertainty of potential impacts will vary from small to large, but generally speaking, the older the field, the greater the cost and risk of opportunities and the smaller the volumes that can be expected.

ranged from about $10 to $90 million per wellbore, compared with from $2 to $6 million per wellbore onshore, or on a per foot drilled basis from about $900 to $7500/ft offshore and from $300 to $800/ft onshore (API, 2011–2017).

Only the largest discoveries require a structure and once installed can be used to develop smaller fields in the region via subsea wells (Heijermans and Cozby, 2003; Reid and Dekker, 2014; Yoshioka et al., 2016). As new structures are installed in remote regions, they open up new developments with tieback fields.

3.3.2 Installation by Decade

In recent years, the number of deepwater platform installations has diminished significantly (Table 3.6), indicating that (large) reservoirs requiring stand-alone structures are not being discovered in the 400–1500 ft water depth region, and where finds are made, subsea tiebacks to existing infrastructure are the preferred development option. Installation activity for fixed platforms peaked from 1987 to 1996, while floater installations peaked from 1997 to 2006. Over the past decade, two floaters were installed on average every year, compared with three floaters per year from 1997 to 2006. Activity levels are small and will remain small for the foreseeable future.

3.3.3 Decommissioning by Decade

The first deepwater structure was decommissioned in 1989, and through 2009, only seven deepwater structures were decommissioned (Table 3.7). In every year since, however, one or more deepwater structures have been decommissioned. Through 2017, a total of 23 deepwater structures have been decommissioned, five floaters and 18 fixed platforms (Fig. 3.4).

Table 3.6 Deepwater Structure Installation by Decade

Period	Fixed Platform	Floater
1977–86	19	1
1987–96	26	5
1997–2006	18	29
2007–16	2	19
2017	0	1
Total	65	55

Source: BOEM 2018.

Table 3.7 Deepwater Structures Decommissioned in the Gulf of Mexico, 1989–2017

Depth (ft)	1989	1999	2005	2006	2009	2010	2011	2012	2013	2014	2015	2016	2017
400–500			2	1			3	1	1			1	
500–1500					1	2	1				1	2	2
> 1500	1	1		1						2			
Total	1	1	2	2	1	2	4	1	1	2	1	3	2

Source: BOEM 2018.

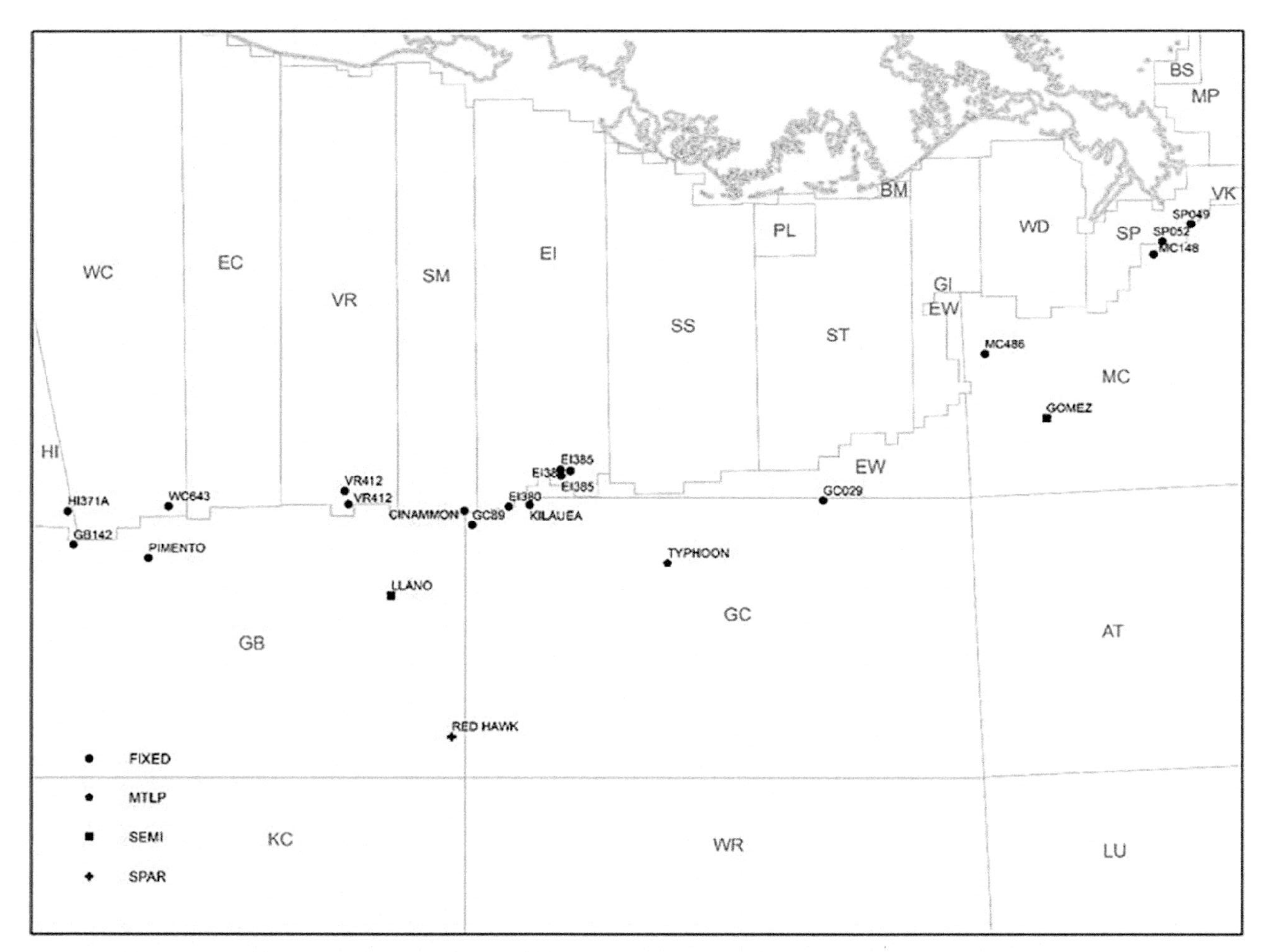

Fig. 3.4 Deepwater structures decommissioned in the Gulf of Mexico circa 2017. *(Source: BOEM 2018.)*

REFERENCES

American Petroleum Institute, 2011–2017. Joint Association Survey on Drilling Costs. American Petroleum Institute, Washington, DC.

Heijermans, B., Cozby, D., 2003. The commercial development of hub platforms and the value to E&P producers. In: OTC 15108. Offshore Technology Conference, Houston, TX, May 5–8.

Kaiser, M.J., Narra, S., 2017. Decommissioning of storm-destroyed assets nears completion. Offshore 77 (6), 34–35.

Reid, D., Dekker, M., 2014. Deepwater development non-technical risks – identification and management. In: SPE 170739. SPE Annual Technical Conference and Exhibition, Amsterdam, The Netherlands, Oct 27–29.

Yoshioka, R., Salem, A., Swerdon, S., Loper, C., 2016. Turning FPU's into hubs: opportunities and constraints. In: OTC 27167. Offshore Technology Conference, Houston, TX, May 2–5.

CHAPTER FOUR

Economic Limit Factors

Contents

Abstract

The economic limit of production refers to the production rate where it is no longer economic or profitable to produce. When the direct operating cost is greater than the net revenue of production, no economic value is being generated, and a rational operator will cease production if the condition persists for too long, that is, attempt to remedy the situation by spending capital to increase production or attempt to reduce cost, search for a buyer for the property, or consider other alternatives. Economic limits are not publicly reported or directly observable, but net revenue the last year of production for decommissioned structures and abandoned wells serves as useful proxies. The purpose of this chapter is to describe the operating cost characteristics of offshore production and the primary factors that impact economic limits. Flow assurance is the primary design issue for subsea wells.

Decommissioning Forecasting and Operating Cost Estimation
https://doi.org/10.1016/B978-0-12-818113-3.00004-7

4.1 OPERATING COST CHARACTERISTICS

The risk and cost in oil and gas operations is finding the resource and selecting the best method that maximizes the value to extract it. The cost of operations after concept selection is determined by the development type and complexity, production fluid, location, age, and design choices that are made. The majority of offshore field development cost occurs upfront in capital expenditures for the wells, platform, and export pipeline. If subsea wells are used in development, additional equipment, flow lines, risers, and umbilicals need to be installed, and operating cost is higher than for systems that employ only dry tree wells, for all else equal.

Wells are used to make contact with reservoir sands and serve as the conduit for the hydrocarbon fluids that flow to a facility for processing and export. Changes in temperature and pressure between the well and surface facilities are utilized for processing. Near the end of primary production, where the reservoir energy has been used up and equipment and wells are older, operating cost starts to increase. The volume of water produced will usually increase over time, increasing water cuts and the need to process unwanted fluids. As pressure declines and eventually dissipates, oil will no longer flow to the surface naturally and must be pumped using artificial lift and secondary production methods if economically viable.

Well interventions are performed throughout the life of a well to protect value (e.g., by repairing or preventing corrosion or scale and maintaining gas-lift systems) or to create value (e.g., shutting off water or adding gas lift to accelerate production) and are a primary means of increasing reserves and production. For wet tree wells, interventions are less frequent and recovery rates are smaller than dry tree wells. The primary objective of stimulation (aka workovers) is to restore impaired well/reservoir connectivity. The nature of impairment, treatment options, and posttreatment production issues often change over the life of the well, and the results of operations are uncertain and site-specific.

Production cost varies depending on the characteristics of the formation, location, method of recovery, cost and frequency of workover activities, and the combination of many other factors (Fig. 4.1). Operating cost depends on the type of production (heavy oil, condensate oil, wet gas, dry gas), quality of production (sweet, sour, corrosive), age of production (early, midlife, mature), operational requirements (chemical treatment and monitoring corrosion and scale), level of production, drive mechanism (solution gas, depletion drive, water aquifer), location (protected waters, shelf, deepwater), the number and type of wells (dry tree, wet tree), structure type (caisson, well protector, fixed platform, floater) and size (weight, equipment capacity, square footage), production characteristics (gas lift, water injection, gas injection), water production, distance to market, well servicing requirements (workovers, sand production, sand control), distance to port (boat and helicopter transportation), degree of automation (manned, unmanned),

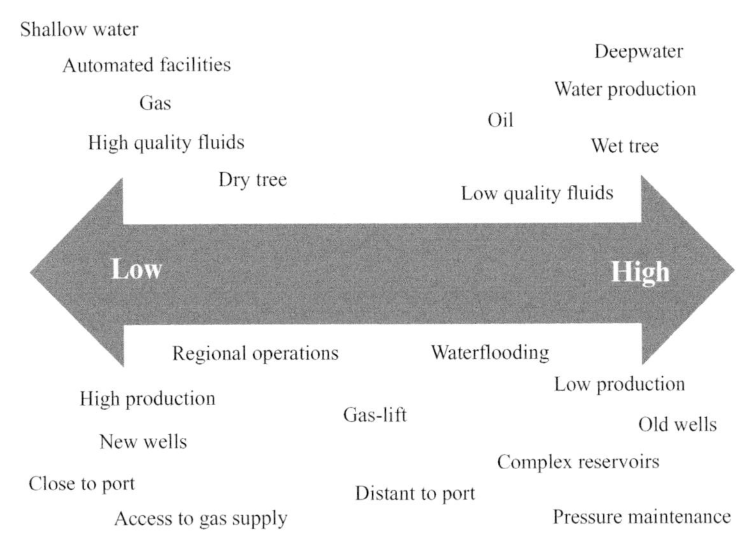

Fig. 4.1 Operating cost spectrum.

production crew size, structure function (producing, auxiliary), complex type (single, multiple), insurance requirements, and accounting methods used to allocate cost in regional operations.

Labor, transportation, well servicing, and contract services are usually the main expenses. Fuel, chemicals, insurance, gathering, and transmission are usually secondary. The contribution of each category to total cost is site- and time-specific. As properties age, they generally require more workovers and chemical cost may increase, but labor cost is mostly fixed unless significant reorganization occurs, while transportation cost and insurance will fluctuate with market conditions. Workovers are largely discretionary, and operators typically plan for a workover when they wish to accelerate or enhance production by first selecting high-producing wells in high-commodity-price environments. Insurance for large companies is discretionary (self-insurance), and mid- to small-size operators are the primary players in the market. Gathering and transmission fees are volume-based and are usually a small fraction of the total operating cost.

4.2 CASH FLOW MODEL

The economic limit is the production rate for a well, lease, structure, or other business unit at which the net revenue from the sale of production equals the cost of production:

$$\text{Net revenue} = \text{Production cost}$$

Net revenue NR is gross revenue GR less the royalty payment ROY, and lease operating expense LOE is synonymous with production cost or direct operating cost:

$$NR = GR - ROY = LOE$$

Gross revenue is estimated as the sum of the product of the (sales) price for oil and gas denoted by P^o and P^g and the quantity of oil and gas sold, Q^o and Q^g, respectively. Royalty in the US GoM is based on a fixed percentage of gross revenue, $ROY = r \cdot GR$, with r either 12.5%, 16.67%, or 18.75% depending on the water depth of the lease and time of the lease sale. For example, from 1983 to 2007, the royalty rate on leases offered in <200 m water depth was 16.67%, 18.75% from 2008 to 2016, and 12.5% from 2017 to the present.

Direct operating cost is defined to include all direct cost to maintain operations, property-specific fixed overhead charges, production, and property taxes but exclude depreciation, abandonment costs, and income tax (Gallun et al., 2001; Mian, 2002; Seba, 2003). Public companies in the United States are required to disclose lease operating expense according to Security and Exchange Commission (SEC) requirements using three categories: direct lease operating expenses, other lease operating expenses, and indirect lease operating expenses. In financial reports, companies combine all their properties on an aggregate or regional total (or by business segment), which does not allow for asset-specific information to be extracted except in special cases (see Chapter 21 for additional details).

If the lease operating expense, commodity price, and royalty rate are known or assumed for a given property, the cash flow relation can be solved for the economic production rate in terms of the primary (oil or gas) stream:

$$Q_{el} = \frac{LOE}{P(1-r)}.$$

Example. Economic limit calculation

If lease operating cost is $500,000 per year and future oil price is expected to be $50/bbl, the property needs to produce at least 12,000 bbl per year (33 bopd) to cover operating expense assuming a 16.67% royalty rate. For a smaller royalty rate, the denominator term will be larger, and therefore, the economic limit will be smaller. For example, at 12.5% royalty rate, the economic limit is 11,400 bbl per year (31 bopd). Conversely, if crude prices turn out to be higher than expected, the lease can produce to a lower level and still cover cost. At $80/bbl, if lease operating cost remains constant, the economic limit drops to 7500 bbl per year (21 bopd).

The economic limit Q_{el} represents when production is no longer profitable. Continued production at or below the economic limit creates no economic gain and would serve no economic purpose. Of course, operators may continue to produce for a period of time when production falls below Q_{el} if they believe prices will increase in the future allowing the property to return to profitable status or if adjustments can be made to reduce operating cost. Operators may be able to reduce their operating cost by delaying maintenance, while labor and logistics cost may be reduced in some cases by downsizing crew or sharing logistics (helicopter and marine vessel) services or renegotiating contracts.

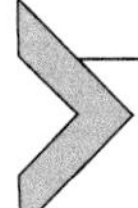

4.3 GENERAL CONSIDERATIONS

4.3.1 Reserves Application

Economic limits are used with decline curves to terminate production forecasts in reserves assessment and are used when modeling decommissioning time (Fig. 4.2). Wells are the basic unit of analysis in all production forecasts, and when consolidated at a lease, structure, or unit level, the economic limits correspond to the aggregation level applied.

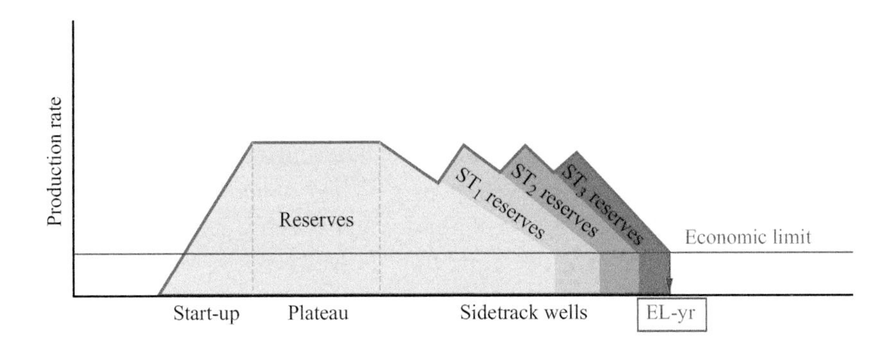

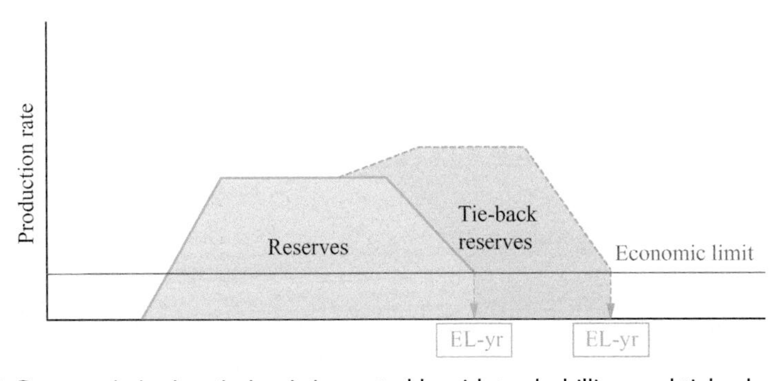

Fig. 4.2 Decommissioning timing is impacted by sidetrack drilling and tieback opportunities.

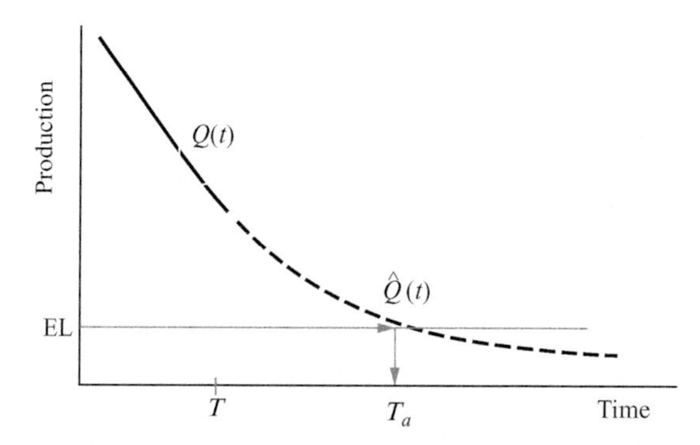

Fig. 4.3 Predicted abandonment time for a producing well using the economic limit cutoff.

At the time of evaluation T, well production is described by $Q(t)$, $t \leq T$, and assuming an economic limit EL, future commodity price P, and royalty rate r, well abandonment time T_a is determined using the economic limit EL as terminal criteria (Fig. 4.3):

Step 1. Forecast $\hat{Q}(t), t > T$, based on historic production data.

Step 2. Compute $NR(t) = \hat{Q}(t)P(1-r)$ for P and r assumptions.

Step 3. Determine $T_a = \min\{t \mid NR(t) = EL\}$, and compute remaining reserves.

4.3.2 Production Beyond Economic Limit Is Not Reserves

According to basic economic theory, a structure will produce hydrocarbons as long as its net revenue exceeds its direct operating expenses. When direct operating cost is greater than the net revenue of production, no economic value is being generated, and a rational operator will cease production if the condition persists for too long. Operators may attempt to remedy the situation by spending capital to increase production or reducing cost, search for a buyer for the property, or consider other alternatives. There are many reasons why an operator may continue to produce marginal properties even if it is not economic to do so (e.g., to hold the lease), but the production beyond the economic limit is not, by definition, reserves.

4.3.3 Strategic Factors Complicate Interpretation

Strategic factors impact economic limits and complicate interpretation. The operator may perform a workover in an attempt to restore production or drill new wells to increase production. The operator may produce at marginal rates to delay decommissioning, either to delay expenditures or to synchronize operations with other decommissioning activities to minimize costs and maximize efficiency (Price et al., 2016). The operator may continue to produce below the economic limit in anticipation of a future sales agreement. Historically, marginal properties have been sold as a package with other properties and unexplored acreage to avoid decommissioning altogether, but there are limits and risks to "kicking the decommissioning can" down the road.

Since the economic limit of a particular property or asset is not reported or known to outside observers, the length of time that an operator may produce below its economic limit is unobservable. Lease conditions explicitly forbid maintaining a lease if it is not producing in commercial quantities, but the interpretation of what is commercial varies by state, and in practice, landowners will have difficulty demonstrating such violation because of the lack of information on cost and the proprietary nature of operations. For the legal aspects of producing below the economic limit, see general treatments by Lowe (2003) and Maxwell et al. (2007).

When the value of an asset's reserves is less than its expected decommissioning cost, operators are unlikely to be able to divest the structure on a stand-alone basis and may have to pay the buyer or make other arrangements to assume responsibility for the property. A company may make a business decision to continue production below its economic limit to postpone incurring these expenses and to seek potential alternative uses of the facility. Many structures retain residual value beyond their productive life because of their (potential) ability to reduce (future) development cost, especially in deepwater, and operators may delay decommissioning unless holding cost (insurance and maintenance costs) and hurricane risk are considered to outweigh the potential benefits. Rarely can the trade-offs be quantified reliably. Regulations also play an important role in timing decisions.

4.3.4 Proxy for Commercial Operations

Economic limits serve as a proxy for rational decision-making, and as with all proxy measures, the correspondence is not perfect. An operator may shut in wells before the economic limit is reached if they decide for strategic reasons to exit the region or a hurricane severely damaged or destroyed facilities. Since production and revenue change over time and operating cost are dynamic and vary with a number of factors, a well may fall below its economic limit several times during its life cycle before returning to commercial production. Operating cost would have to exceed production revenue for a sustained period of time, perhaps one year or more, before an operator would consider shutting in a well. If an operator cannot achieve its economic threshold, operations will cease temporarily or permanently, and the structure idled, decommissioned, or divested.

As long as a well is not permanently abandoned, however, shut-in wells may be restarted and worked over at a later date or serve as a site for a sidetrack well (Brandt and Sarif, 2013). It is only when a well is permanently abandoned and a structure decommissioned that the revenue at the end of its life serves as a proxy for its economic limit. Frequently, only one or two producing wells hold offshore structures at the end of their life. Ultimately, the criterion operators employ when terminating production is unobservable, but that does not negate the application of the economic limit as a proxy for commercial operations.

4.4 FACTOR DESCRIPTION

Many factors impact the economic limit of offshore production (Fig. 4.4). Some of the factors are observable such as structure type and production, while other factors are unobservable or difficult to assess such as the impact of scale economies, asset concentration, and strategic decisions.

4.4.1 Structure Type

Simple structures like caissons and well protectors host a small number of wells and are expected to have a lower operating cost than fixed platforms, for all other things equal, and the average shallow-water fixed platform is expected to have a lower economic limit than the average deepwater structure. Platforms that are part of a multistructure complex are expected to achieve cost savings relative to isolated structures that are not part of a complex since multistructure platforms can be manned directly saving on logistics cost (personnel are within walking distance and not a boat ride or helicopter trip away) and economies similar to regional operations arise.

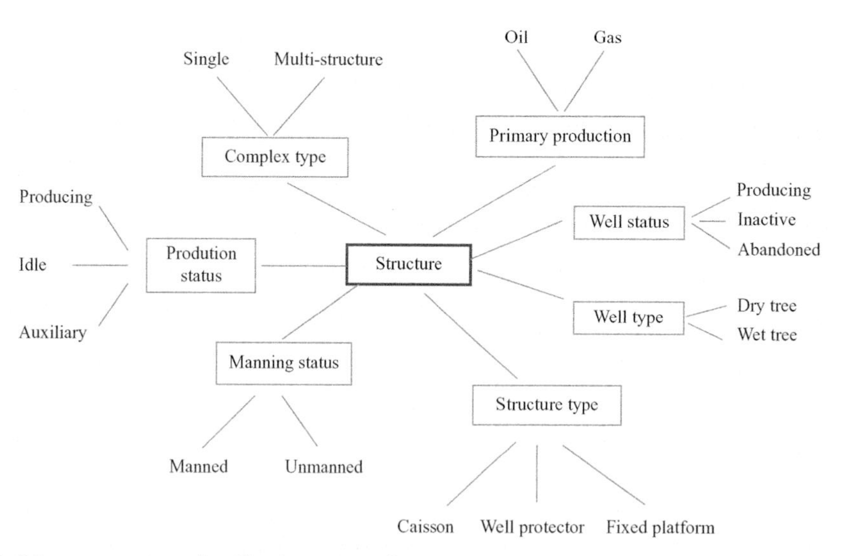

Fig. 4.4 Primary structure classification categories.

4.4.2 Water Depth

Water depth in the GoM is related to distance to shore and the type of structure required to resist environmental and operational loads. As water depth increases, the distance to shore increases approximately linearly (except for some locations around the mouth of the Mississippi River where the continental shelf is short), and as shore distance increases, so will the cost of transportation from onshore bases. The need for a robust platform and higher insurance and communication cost increases with water depth and contributes to higher operating cost. Differences in water depth should translate into differences in economic limits for all things equal.

4.4.3 Oil vs. Gas

Gas fields are generally cheaper to operate than crude oil (i.e., gas is less viscous and highly compressible and wants to come out of the ground), recover more resources (greater drainage area per well), and at higher rates (80% to 90% for gas vs. 20% to 40% for oil), and handling and treating are cheaper (gas streams typically only need to be dehydrated before export), and of course, wells producing gas provide a free source of energy to drive the equipment if adequate volumes are available. As properties mature, these cost advantages become less pronounced, and because oil is both more desirable and valuable as a commodity and is priced at a premium to natural gas (historically, about one-and-a-half times the price of gas on a heat-equivalent basis in the Gulf region), there is no obvious guidance for how product type influences economic limits.

4.4.4 Manned Status

Structures that are producing from one or two wells near the end of their life will generate similar levels of revenue, but the operating cost between structure types will differ, especially if a platform is manned or is part of a multistructure complex or floating structure. For manned platforms, direct operating cost will be greater than for unmanned platforms, and economic limits are expected to be greater for this structure class.

4.4.5 Operator

Independents typically produce from numerous marginal fields compared to large independents and majors which usually focus on fewer larger fields, and one would expect independents to be able to produce at lower rates, on average, than majors because of the nature and type of their asset portfolios and specialization, scale and regional economies, and organizational structure. Most majors do not operate properties once they reach a certain minimum level and regularly divest or decommission assets low on their decline curves.

Small operators seek scale and synergies to reduce operating cost and enhance downhole opportunities. One of the advantages of consolidating operations, as viewed by the large legacy property owners, is the ability to reduce logistics cost as they seek to squeeze out marginal oil and gas from mature properties. The sustainability of this business model over the years has been a mixed bag when viewed from the number of companies that stay in business, but value has been generated, and activities have added barrels making shallow-water GoM production decline less steep.

4.4.6 Well Type

Well type is an important factor if a platform handles wet tree (subsea) wells. Operating costs for subsea wells are much larger than for dry tree wells since platform rig access is not available, power and chemicals have to be supplied remotely, flow lines have to be maintained and regularly pigged, and the risks of decision-making increase. In the shallow-water GoM, the population of wet tree wells only number a few dozen (see Chapter 6), so the need to distinguish well type in shallow water does not arise, but in deepwater, subsea wells are quite common (about half the inventory of wells circa 2017) and need to be distinguished from dry tree wells in economic limit evaluations.

Subsea wells flow their production back to a host platform for processing and should have a higher economic limit than dry tree wells, for all things equal, since the wellbore is not accessible from a platform and will cost more to operate and maintain. Low flow rates present problems for all subsea wells and depending on the diameter of the pipe will present additional challenges. In gas wells, for example, low flows may not sweep produced water from the flow lines, requiring additional pigging frequency to prevent corrosion and creating a higher risk for hydrate formation or greater hydrate chemical inhibitor cost. One might expect that the greater the distance between well and host, the higher the economic limit and that elevation differences between the well and host may also be important.

4.4.7 Intervention Frequency

A well intervention is defined as any physical connection made to a completed well to alter production or address safety issues. Unlike dry tree wells, subsea wells are not accessible via a platform rig, and this lack of access has implications for performance and hydrocarbon recovery.

Subsea wells represent high-risk, high-cost assets compared with dry tree wells due to the complexities and difficulties of access and the costs associated with intervention. Since all wells require intervention to achieve their maximum recovery, and since under normal circumstances dry tree wells receive regular and planned interventions, one would expect subsea wells will not receive the same intervention frequency and subsequently have lower performance (Fig. 4.5).

Recovery factors for subsea wells are expected to be lower than dry tree wells, and economic limits are expected to be higher. Although the former hypothesis is not easily tested, the later claim can be evaluated statistically (see Chapter 8). The low frequency of interventions and greater back pressure that arises due to the distance to the host facility are the primary reasons for the reduced performance. Operators often forego production

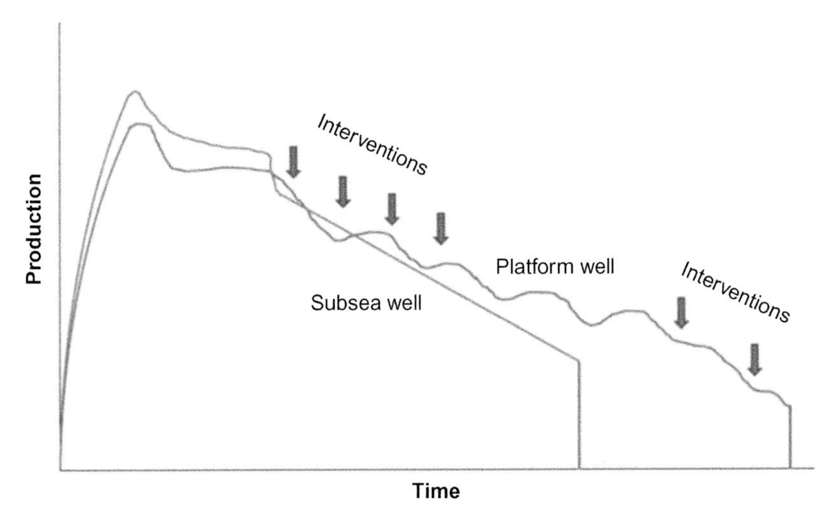

Fig. 4.5 Schematic of platform well production with regular interventions versus a subsea well without intervention.

improvement workovers on subsea wells because the combination of high cost and uncertain improvement fails to justify capital investment. Reservoirs with delayed pressure maintenance or nonproactive intervention programs will have recoveries less than equivalent wells with regular maintenance.

4.4.8 Other Factors

Age may play a role in structure maintenance requirements, but because maintenance costs are periodic, usually small, and often budget on a long-term basis, it is unlikely that a strong correlation exists between age and economic limit. For all things equal, older structures require greater maintenance cost than younger structures, and one might suspect that structures closer to cessation of production would have higher operating cost relative to structures more distant.

The manner in which oil is produced—via gas lift, waterflood, or natural flow—will impact operating expenditures. Gas lift is currently used on about one-third of producing oil wells in the shallow-water GoM, and access to gas lift allows operators to produce at lower production rates relative to oil wells without access to gas lift. Waterflooding has been employed on a number of shallow-water oil fields, while tertiary methods offshore (e.g., CO_2 flooding) are not common (or economic).

Water production increases at the end of the life cycle of oil and gas production, especially for black oil water-drive reservoirs, and because the treatment and disposal of produced water costs money, as water production increases so does operating expense. The cost to handle, process, and treat this waste stream may be a limiting factor in production.

Wells that water out are expected to reach their economic limit at a higher production level relative to wells that are not similarly burdened.

Companies with a geographic concentration of assets are more likely to achieve lower economic limits than a distributed collection of isolated assets because supply vessels and personnel can be utilized in a more effective and cost-efficient manner. Helicopter and boat charters in regional operations, for example, can be optimized to more efficient service operations, and labor may be more effectively utilized.

4.5 FLOW ASSURANCE

In subsea production systems, fluid from the reservoir travels through a number of system components, starting at the perforations at the completion and through the wellbore tubing and casing, past the subsea control valve, and into the tree, manifold, jumpers, flow line, riser and platform piping and equipment. The objective of flow assurance is to keep the flow path open for the life of the well.

4.5.1 Issues

The primary flow assurance issues for subsea systems are hydrate, wax, asphaltene, scale, and corrosion. Each of these components may occur at different places and at different times during the life cycle of production (Fig. 4.6), and therefore, anticipating their occurrence and designing systems to mitigate, reduce, or remediate are the key elements of flow assurance strategies.

The occurrence of hydrate, wax, asphaltene, scale, and corrosion may arise downhole in the production tubing of the well, at the wellhead or manifold on the seabed, in the connecting jumpers or flow lines, at the riser at the base of the host, and in the equipment

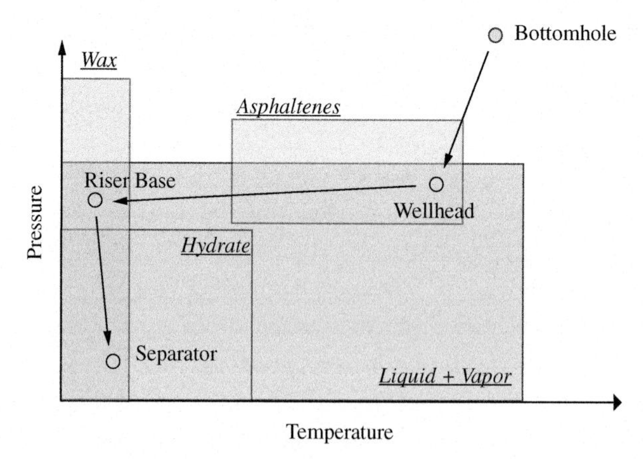

Fig. 4.6 Changes in pressure and temperature often create asphaltenes, wax, and hydrate production issues.

and piping topsides. They may occur early or late in life and during different operating states from normal production, shutdown, start-up, and remediation. Start-up/warm-up and shutdown/cooldown are transient states where flowing temperatures change and may enter regions of hydrate risk and wax appearance. Systems must be robust and flexible to handle dynamic and changing conditions.

Early in production, flow rates and temperatures are high, but later in life both will decline. If systems are not designed for these changing conditions, operational problems may occur. Early in production, fluid quality (viscosity, GOR, and water cut) is similar to the fluid samples obtained during well testing, but later in life, as reservoir pressures decline, fluid quality changes, which may cause operational problems if changes cannot be managed.

4.5.2 Subsea Production System Design

In subsea production system design, a whole-system modeling approach is used to define the hardware requirements and operating strategies (Kaczmarski and Lorimer, 2001; Arciero, 2017; Bomba et al., 2018). To optimize the system design, software models for each component are used to understand the trade-offs that result from balancing steady-state and transient operations with flow assurance management:

- Steady-state thermal and hydraulic
- Transient operations
  - Warm-up
  - Cooldown
  - Blowdown
  - Hot oil heating
- Flow assurance management and remediation
  - Hydrate prediction and inhibition
  - Wax deposition
  - Asphaltene deposition
  - Scale prediction
  - Internal corrosion
  - Erosion
- Corrosion

The flow assurance work process begins with the development of a thorough understanding of the fluid and reservoir properties (Joshi et al., 2017). A typical suite of analyses may include generation of hydrate stability curves, cloud-point and pour-point measurement, wax deposition patterns, asphaltene stability testing, and scale analysis based on water samples.

Results of fluid and reservoir analyses are combined with thermal-hydraulic modeling of the system to assess the flow assurance risks. The formation processes for the solids of concern (e.g., hydrates, wax, asphaltenes, and scale) are all driven by a combination of temperature and pressure, hence the need for accurate thermal-hydraulic modeling of system performance in steady-state operations and during transient operations such as start-up and shutdown. Corrosion, erosion, slugging, and emulsion formation are often included in the flow assurance analysis since they are also driven by a combination of temperature and pressure, flow rate, and flow regime.

The results of the flow assurance work process are expressed in the form of flow assurance strategies that combine elements of the system design, system operational strategy, and chemical treating requirements. Strategies vary greatly among projects because each development has a different set of driving factors of cost, project life, reliability targets, and environmental restraints. As a result, flow assurance must be integrated into the overall subsea systems engineering effort, not only to ensure that the analyses are performed at the correct time and desired accuracy, but also to integrate flow assurance with equipment selection and development of operational guidelines in a way that optimizes the final result for the project.

4.5.3 Hydrates

Hydrates are ice-like solids classed as clathrates, sometimes referred to as "dirty snow" (Fig. 4.7), which form when water and light hydrocarbons or other small compounds are present together at low temperatures and high pressures (Cochran, 2003). In deepwater development, ambient temperatures near the seabed are approximately 39°F (4°C), which is well within the hydrate formation region at typical operating pressures.

In hydrates, a gas molecule is trapped within a hydrogen-bonded cage of water molecules (Fig. 4.8). Many different gases are capable of forming hydrates provided

Fig. 4.7 Solid deposition such as hydrates (left) and wax (right) can result in a reduction or complete interruption of hydrocarbon flow. *(Source: Kaczmarski, A.A., Lorimer, S.E., 2001. Emergence of flow assurance as a technical discipline specific to deepwater: technical challenges and integration into subsea systems engineering. In: OTC 13123. Offshore Technology Conference. Houston, TX, Apr 30–May 3.)*

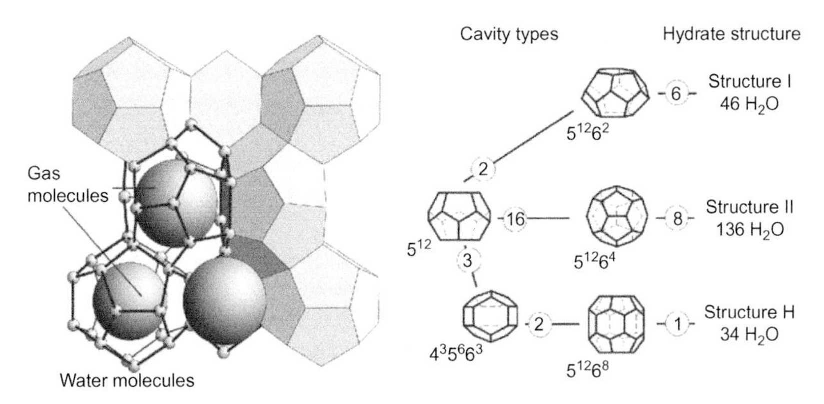

Fig. 4.8 Molecular representation of hydrate cages for different captive species. *(With permission from G Bohrmann).*

the molecules are small enough to fit within the cavity of the cage. High molecular weight gases are typically too large to form hydrates, but methane, ethane, propane, and butane, as well as N_2, CO_2, and H_2S, are small enough to fit inside. The crystalline structure of gas hydrate crystals depends on gas composition, pressure, and temperature.

Hydrates require certain conditions to form, namely,

- High pressure and low temperature,
- Gas component, and
- Water in sufficient amount.

Prevention of hydrates attempt to reduce or prevent one or more of these conditions via the following strategies:

- Water removal
- Insulation or heating to maintain temperature high enough to operate within the hydrate-free zone
- Operate at a sufficiently low pressure
- Use thermodynamic hydrate inhibitors (THI) such as methanol or monoethylene glycol (MEG)
- Use low-dosage hydrate inhibitors (LDHIs) such as kinetic hydrate inhibitors (KHIs) or antiagglomerants (AAs)

Thermodynamic inhibitors suppress the point at which hydrates form much like an anti-freeze for water ice, while KHIs interfere with the kinetics of hydrate formation, and AAs allow hydrates to form but only as a slush-like transportable material. KHIs bond with the hydrate cage and slow crystal growth. The more severe the hydrate problem, the more inhibitor is required, and production facilities can reach a limit rate of methanol treatment due to storage and injection constraints (Kopps et al., 2007). In recent years, LDHIs have been the most prominent advance in hydrate inhibitors.

For gas systems, the typical approach for hydrate control has been to rely on continuous dosing with glycol or methanol. For oil systems, insulated systems that maintain temperatures above the hydrate formation temperature during steady-state allow hydrates to be controlled with a minimum usage of glycol or methanol. During shutdowns and start-ups, the system drops below the hydrate temperature, and additional operational steps are required such as pressure relief (blowdown) or removal of wet fluids by pigging during shutdown and circulation of hot oil and methanol dosing during start-up.

Example. Mica hydrate control

The Mica field is located in 4350ft water depth and tied back about 29 mi to the Pompano platform in 1290ft water depth. The change in elevation is about 4400ft (Fig. 4.9). Three zones in the reservoir were initially produced using an oil flow line and a gas flow line connected to a four-slot manifold with pigging loop (Ballard, 2006). The oil flow line is an 8 × 12 in pipe-in-pipe insulated pipe, and the gas flow line is an 8 in uninsulated pipe.

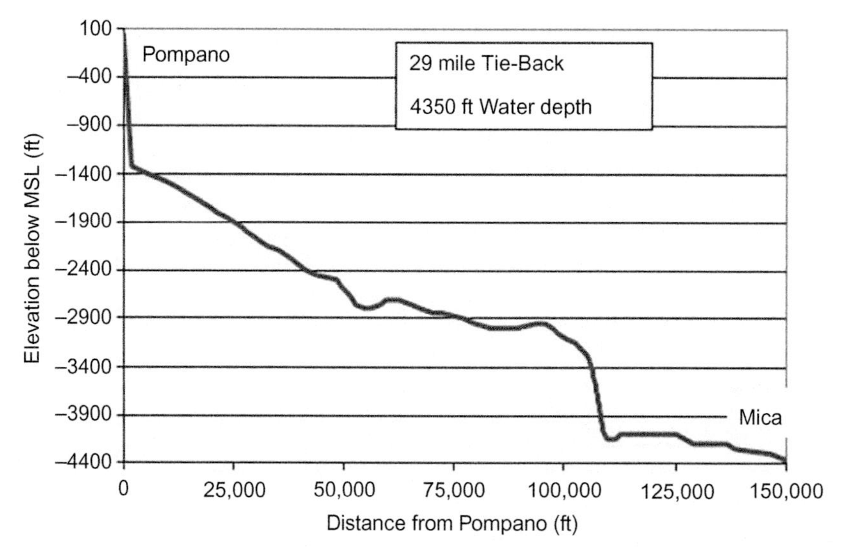

Fig. 4.9 Topography of seafloor between Mica and host Pompano. *(Source: Ballard, A.L., 2006. Flow-assurance lessons: the Mica tieback. In: OTC 18384. Offshore Technology Conference. Houston, TX, May 1–4.)*

The shallowest zone in the reservoir is a saturated gas and oil zone with a GOR of 1330 cf/bbl, gravity of 32° API, and a cloud point of 93–100°F. The middle zone is a near-critical fluid with a GOR >3000 cf/bbl, gravity of 37° API, and a cloud point of 90–94°F. The deepest zone is a dry gas with condensate-gas ratio of 5 bbl/MMcf and gravity of 44° API.

For the gas flow line, a highly insulated flow line would have caused the gas to arrive at very cold temperatures, and so, an uninsulated flow line was chosen to allow the seafloor to exchange heat with the pipeline to warm the fluids. Continuous methanol injection was applied at an expected dosage of 1 bbl of methanol per 2 bbl of water. The expected hydrate curve and temperature profile for the gas flow line are shown in Fig. 4.10. The hydrate region falls on the left of the hydrate curves. Thus, at a given pressure, as long as temperature is greater than the hydrate onset, no hydrates form in the flow line. As methanol is added, the hydrate curve shifts to the left with increasing methanol volumes, decreasing the operating envelope where hydrates may form.

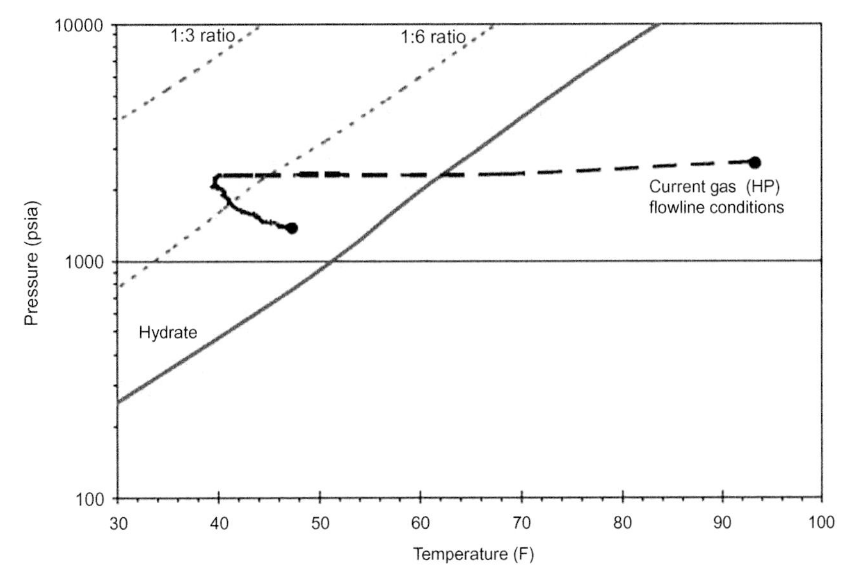

Fig. 4.10 Expected hydrate curve for gas flow line with temperature and pressure profile. *(Source: Ballard, A.L., 2006. Flow-assurance lessons: the Mica tieback. In: OTC 18384. Offshore Technology Conference. Houston, TX, May 1–4.)*

Scale, asphaltenes, and erosion were not expected to be issues, but with the long tieback distance and high cloud point temperatures of the oil, paraffin issues were expected in the two wells and oil flow line. It was not considered feasible to keep the fluids in the oil flow line above the wax appearance temperature during production, and pipe-in-pipe insulation was designed to spread out the deposited paraffin in the flow line as the oil cooled down and to perform regular pigging. Continuous methanol injection was incorporated at each tree upstream of the production choke and manifold. Blowdown was not considered an efficient strategy during shut-in due to the length and geometry of the flow line.

For a production rate of 15,000 bopd and water cut of 1%, the methanol usage in the oil flow line is estimated as

$$15{,}000\,\text{bopd} \times 0.01\frac{\text{bwpd}}{\text{bopd}} \times \frac{1\,\text{bbl MeOH}}{2\,\text{bbl H}_2\text{O}} = 75\,\text{bbl MeOH/d}$$

Methanol injection pumps were sized for 250 bpd, so as water cuts increase, the pumps will need to be replaced if production remains economic. At $2/gal, MeOH injection costs $6300/day at 1% water cut or $2.3 million per year. At 20% water cut, methanol costs increase to $46 million per year.

4.5.4 Waxes

Waxes are high molecular weight saturated organic mixtures of n-alkanes, i-alkanes, and cycloalkanes with carbon chain lengths typically ranging from C18 to C30 but can go as high as C75 or more (Speight, 1998; Riazi, 2005). Waxes precipitate from petroleum fluid as the temperature falls below the wax appearance temperature (WAT) and deposit to the contact surface of the well tubing, pipeline, or vessel (Fig. 4.7). If not under control, wax deposits may block the flow path completely.

The formation of wax crystals depends mostly on temperature change. Pressure and composition also affect wax formation but to a lesser extent than temperature change. The temperature at which crude oil develops a cloudy appearance due to its wax (paraffin) content precipitating out is called the WAT or cloud point. The pour point is defined as the lowest temperature crude oil flows. Both WAT and pour point are important design characteristics in subsea systems. Wax management is often considered with hydrate management strategies.

Most operators rely upon pigging for control of wax deposition supplemented with inhibitors for reduction of the deposition rate. As system offsets increase, however, the cost of a dual flowline system is substantial, and the time required for roundtrip pigging increases. Subsea pig launchers can be employed with a single line but incur the cost of intervention vessels. Wax deposition can be prevented, delayed, or minimized using paraffin inhibitors such as crystal modifiers or dispersants. The former are chemicals that interact with the growing wax crystallization, while dispersants prevent the wax nuclei from agglomerating on the pipe surface by disrupting its crystal growth.[1]

4.5.5 Asphaltenes

Asphaltenes are dark-colored, friable, and infusible hydrocarbon solids sometimes called the "cholesterol" of petroleum. Defined as the fraction of crude that precipitates upon the addition of an excess of n-alkane, the diversity of asphaltene production issues arises from the variety of oil types and production conditions. Asphaltenes are represented by the polynuclear aromatic layers with folded alkane chains, creating a solid structure known as a micelle. Some rings may be nonaromatic but many are fused and share at least one side. The tendency of asphaltenes to precipitate from a given crude is broadly related to the molecular weight, aromaticity, and polarity of the asphaltenes (Fig. 4.11).

Asphaltenes are common in heavy viscous crude and are usually controlled using inhibitors before destabilization and flocculation occurs (Jamaluddin et al., 2002; Tavakkoli et al., 2016). After destabilization, remediation is more difficult to control. Pressure and composition are the most significant factors in deposition mechanisms (Mullins et al., 2007). Asphaltene deposition tends to occur above the bubble point,

[1] Dispersants control wax by coating already formed paraffin crystals and causing the crystals to repel each other and metal surfaces that prevents solid deposits from forming and allows the crystals to be suspended in the crude.

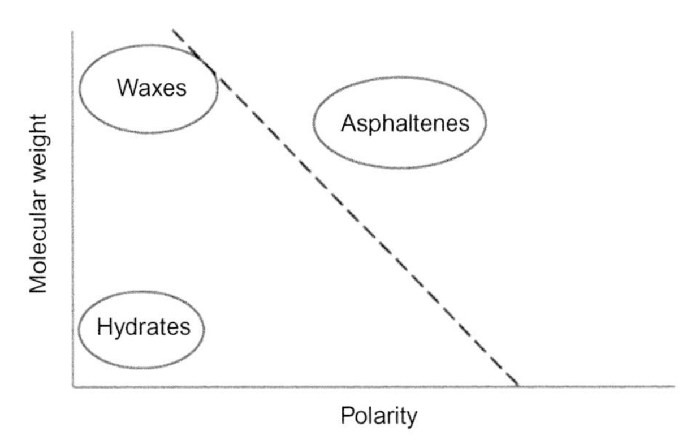

Fig. 4.11 Hydrocarbon molecular weight and polarity define the presence of asphaltenes.

concentrating the hazard to the wellbore and wellbore formation, and occasionally to the wellhead or flow line. Below the bubble point, asphaltene solids tend to redissolve, but the process can be slow and incomplete.

Chemical treatment is the most common and effective method for control, but much of the underlying chemistry remains uncertain. Prescreening criteria to evaluate the potential for asphaltene formation include the use of the de Boer plot and subsea safety valve analysis. The de Boer plot evaluates the loss of asphaltene solubility as a reservoir-fluid sample is depressurized and indicates regions that are likely, uncertain, and unlikely to precipitate asphaltenes. In the resin-to-asphaltene (RTA) ratio plot, statistical rules are used to evaluate the ratio required to keep the asphaltenes stable. For $RTA < 1.5$, asphaltenes are expected to be unstable during expansion of oil, $RTA > 2.5$ indicates the stable region, and $1.5 < RTA < 2.5$ is the transition zone.

Example. K2 flow assurance strategy

K2 and K2 North are located in the Green Canyon protraction area in about 4200ft water depth and are unitized across several blocks (Fig. 4.12). K2 is a deep subsalt play in the Mississippi Fan fold belt (Fig. 4.13) and the first phase development was comprised of eight subsea wells connected to the Marco Polo TLP 7 mi to the southeast in Green Canyon block 608 in 4300ft water depth (Fig. 4.14). The primary reservoirs are Miocene age turbidite sheet deposits located subsalt, and hydrocarbons are highly undersaturated, medium- to high-asphaltene oils with low to medium GORs (Sanford et al., 2006; Lim et al., 2008).

K2 and K2 North reservoirs had similar flow assurance issues and were designed in similar ways (Brimmer, 2006; Lim et al., 2008). Heat retention was determined to be the most effective mitigation solution for both hydrate and paraffin issues. Pipe-in-pipe flow lines and insulated risers with a chemical injection system for each well were selected in development with chemical injection used during start-ups as required. For hydrate mitigation, LDHI injected subsea between the surface-controlled subsurface safety valve (SCSSV) and the subsea tree is required during start-up only with methanol used as a backup system. At start-up after a shutdown, it is necessary to inject wax inhibitor to prevent wax formation while the flow rates are low and the temperature is below the WAT. Once the flow line temperature is greater than WAT, which is expected at flow rates >5000 bpd of produced fluid, the paraffin inhibitor is stopped.

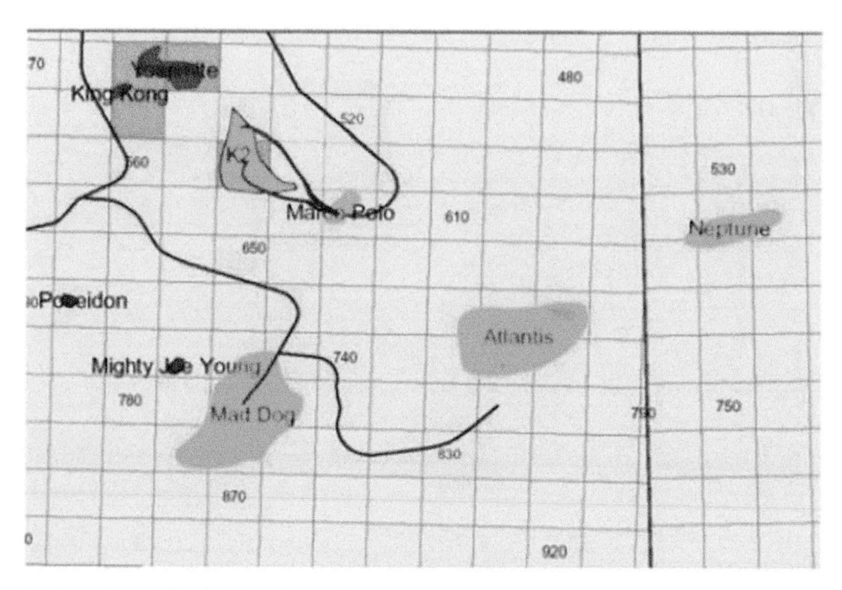

Fig. 4.12 Location of K2 field in Green Canyon protraction area. *(Source: Sanford, J., Woomer, J., Miller, J., Russell, C., 2006. The K2 project: a drilling engineer's perspective. In: OTC 18298. Offshore Technology Conference. Houston, TX, 1–4 May.)*

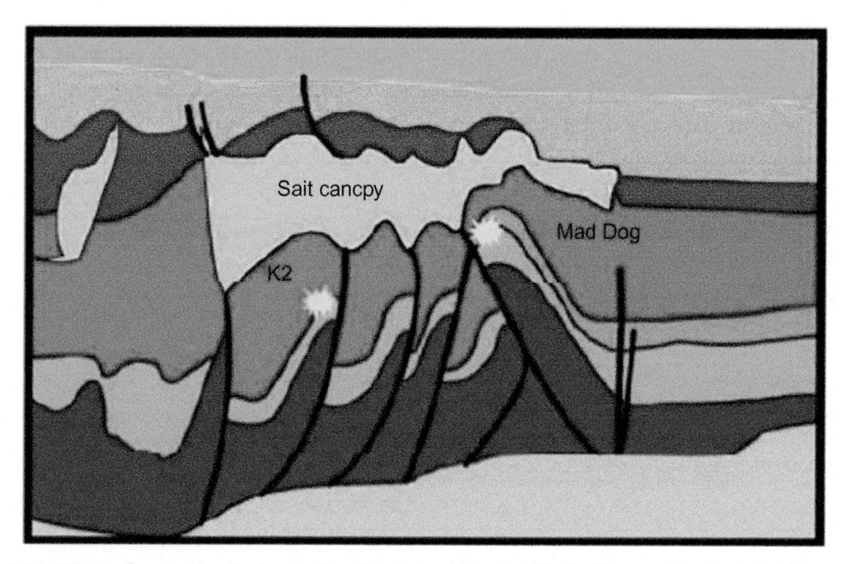

Fig. 4.13 K2 is a deep subsalt play in the Mississippi Fan fold belt. *(Source: Sanford, J., Woomer, J., Miller, J., Russell, C., 2006. The K2 project: a drilling engineer's perspective. In: OTC 18298. Offshore Technology Conference. Houston, TX, 1–4 May.)*

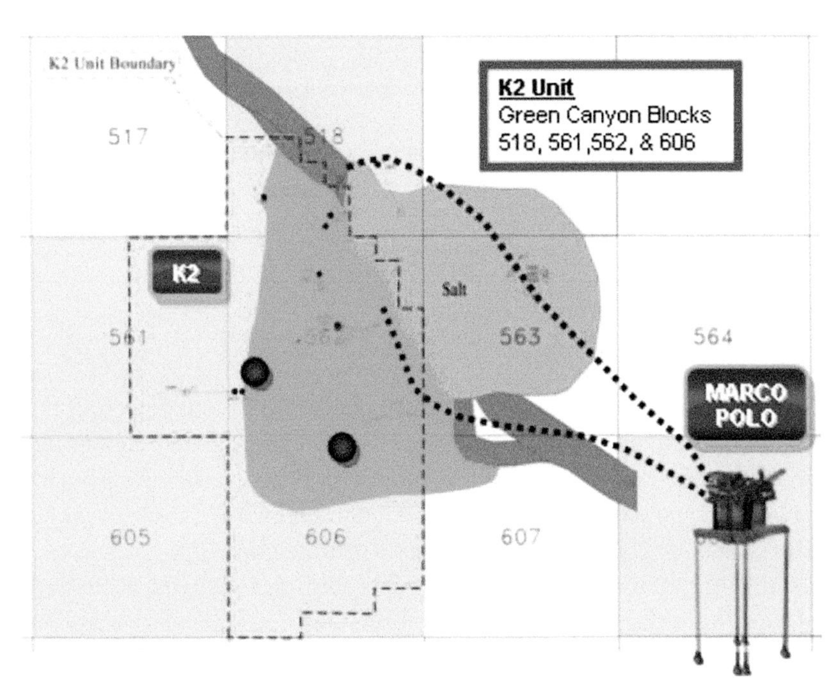

Fig. 4.14 K2 unit field development layout. *(Source: Dobbs, W., Browning, B., Killough, J., Kumar, A., 2011. Coupled surface/subsurface simulation of an offshore K2 field. In: SPE 145070. SPE Reservoir Characterization and Simulation Technical Conference and Exhibition, Abu Dhabi, UAE, Oct 9–11.)*

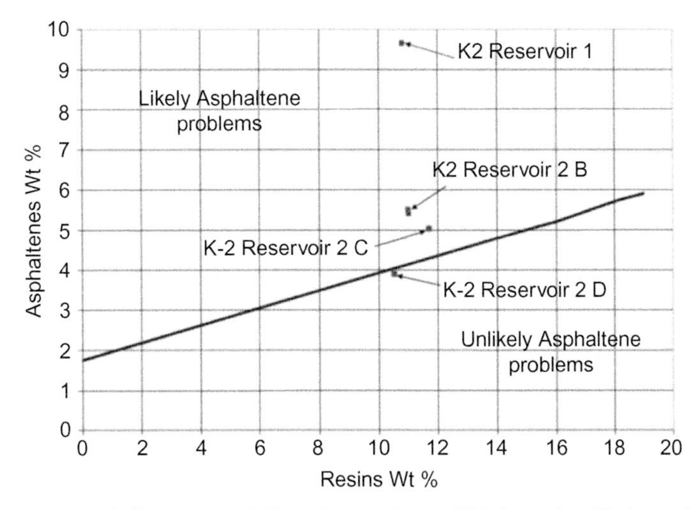

Fig. 4.15 Resin-to-asphaltene ratio delineating regions of likely and unlikely asphaltene problems. *(Source: Brimmer, A.R., 2006. Deepwater chemical injection systems: the balance between conservatism. In: OTC 18308. Offshore Technology Conference. Houston, TX, May 1–4.)*

Scale and corrosion issues were not expected to be severe and are mitigated using chemical injection. Slugging was not expected. Laboratory tests showed the propensity for solid formation and confirmed that a variety of reservoir fluids had an asphaltene deposition envelope that extended from the reservoir to the topside facilities (Fig. 4.15). Modeling indicated that asphaltene deposits occur around the bubble point where vaporization takes place and might stick to themselves or to the pipe wall causing a flow line restriction. Chemical injection to control asphaltenes is used when needed and varies with each well (Dobbs et al., 2011).

4.5.6 Inorganic Scale

Three water sources dominate the oil field environment—natural depletion resulting in formation water production, injection of seawater or aquifer water, and desulfated seawater (Jordan et al., 2001; Jordan and Feasey, 2008). Scale prediction programs can be used to predict the types, likelihood, and amount of mineral scales that can form. The key to managing scale control is to understand how the brine chemistry changes over the life cycle of the field for each individual well together with the ability to place scale control measures upstream of the point where scale formation is predicted to occur.

BOX 4.1 The Five Reservoir Fluids

There are five types of reservoir fluids often referred to as black oil, volatile oil, retrograde gas, wet gas, and dry gas (Fig. 4.16). The reservoir fluids are characterized by their producing GOR, API gravity, color, C6+ fraction, and viscosity (McCain, 1990). The C6+ fractions are those (liquid) hydrocarbon compounds with six or more carbon atoms also sometimes referred to as the heavy hydrocarbon liquid constituents.

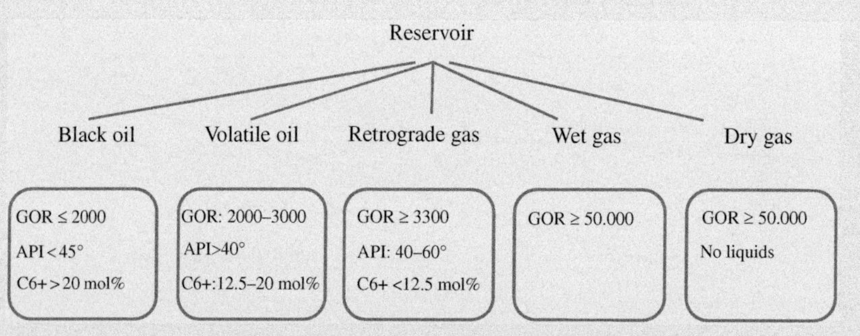

Fig. 4.16 The five reservoir fluids and their classification.

Black or heavy oils are characterized as having initial producing GORs <2000 cf/bbl and stock tank gravity below 45° API. Producing GORs increase during production after reservoir pressure falls below the bubble-point pressure of the oil. There is a large quantity of heavy hydrocarbons in black oils, and for low-gravity crude, sulfur content is high.

Continued

BOX 4.1 The Five Reservoir Fluids—cont'd

Volatile oils, also called high-shrinkage crude oils, release a large amount of gas in the reservoir when reservoir pressure falls below the bubble point of the fluid. The dividing line between black oils and volatile oils is somewhat arbitrary but usually identified with producing GOR between 2000 and 3300 cf/bbl. Oil gravity is usually 40° API or higher and increases during production after the bubble point is reached. Heptane plus content of crude ranges from 12.5% to 20 mol%.

Retrograde gases are so named because they exhibit a dew point, and when reservoir pressure falls below this point, liquid from the gas condenses in the reservoir. Unfortunately, this liquid in the reservoir does not flow and normally cannot be produced, and the result is that the composition of the reservoir fluid will change during production. The lower limit of the initial producing GOR for retrograde gas is approximately 3300–5000 cf/bbl, but the upper limit varies with the reservoir conditions and above 50,000 cf/bbl can be considered a wet gas. The surface gas for retrograde gases is rich in intermediates and processed to remove propane, butanes, pentanes, and heavier hydrocarbons. At high GOR, the quantity of retrograde liquid in the reservoir is very small. Stock tank liquid gravities vary between 40° and 60° API and increase as reservoir pressure falls below the dew-point pressure.

A wet gas exists solely as a gas in the reservoir throughout the reduction in reservoir pressure, and no liquid is formed in the reservoir. However, at the surface separator, some liquid, often referred to as condensate, will form. Condensate refers to hydrocarbons that are gaseous in the reservoir but liquid at the surface and are known in some circles as the "Champaign" of crude oil due to its clear light color. The "wet" gas designation does not mean the gas is wet with water, since all reservoir gas is normally saturated with water, but refers to the hydrocarbon liquids that condense out at the surface. Wet gases have high-producing GORs that remain constant during production, and the gravity of the stock tank liquid does not change noticeably during the life of the reservoir. Normally, a gas with a GOR $>$50,000 cf/bbl can be treated as a wet gas.

In a dry gas reservoir, no hydrocarbon liquid is formed in the reservoir or at the surface, although some liquid water usually condenses. The word dry indicates that the gas does not contain hydrocarbon liquids.

Engineers apply empirical pressure/volume/temperature (PVT) correlations as a function of commonly available field data (e.g., bubble-point pressure, GOR at bubble-point pressure, oil formation volume factor, oil compressibility, dead oil viscosity), in reservoir engineering studies (Dindoruk and Christman, 2002).

Example. Mississippi Canyon 696 (Blind Faith) reservoir fluids

The Blind Faith development is composed of four subsea wells from two oil reservoirs in Mississippi Canyon blocks 695 and 699 in approximately 7000 ft water depth (Subramanian et al., 2009). The wells are tied back about 5 mi via dual flow lines to block MC 650 in 6500 ft water depth. Two of the wells produce commingled fluid from multiple zones within the Pink reservoirs via stacked frac-pack completions. Two other wells produce from the deeper Peach reservoir via single frac-pack completions.

Continued

BOX 4.1 The Five Reservoir Fluids—cont'd

The Blind Faith reservoirs contain high-pressure, undersaturated oils (28°–33° API). The Pink and Peach fluids are quite different in terms of bubble point and GOR (Table 4.1). The fluids have a low paraffin content and a dead oil paraffin WAT from 80 to 100°F. Flow line insulation was designed to achieve 100°F arrival temperature topsides at turndown rates of 10,000 bpd, 0% water cut.

Table 4.1 Blind Faith Fluid Properties in the Pink and Peach Reservoir Sands

	Upper Pink	Lower Pink	Upper Peach	Lower Peach
Reservoir pressure (psia)	~13,500		~17,000	
Reservoir temperature (°F)	216–230		255–267	
Bubble point (psia)	4200–6500		1800–2000	
Density (°API)	28.3	29	28.5	33.3
Gas-oil ratio (scf/bbl)	500–1200		350–550	
Viscosity (cP) at 15 psia, 70°F	33.6	19.5	32	6.6
Wax content, C20+ (wt%)	2.59		2.17	2.51
Asphaltenes (wt%)	1.4	2.9	5.8	2.4

Source: Subramanian, S., Price, M.P., Pinnick, R.A., 2009. Flow assurance design and initial operating experience from Blind Faith in deepwater Gulf of Mexico. In: SPE 124535. SPE Annual Technical Conference and Exhibition. New Orleans, LA, Oct 4–7.

Asphaltenes in the Pink fluids range from 1.4% to 2.9 wt% and in the Peach fluids from 2.4% to 5.8 wt%. Asphaltenes are a solubility class of the crude oil defined as soluble in aromatic solvents and insoluble in paraffinic solvents. Asphaltenes may precipitate from the crude oil during depressurization and may deposit on the pipe walls and valves or accumulate in equipment. Mitigation techniques employed are asphaltene inhibitor (dispersant) injection.

REFERENCES

Arciero, B., 2017. Optimization of flow assurance applications through life-of-field: an integrated approach to production engineering and chemical management across Gulf of Mexico deepwater assets. In: OTC 27560. Offshore Technology Conference. Houston, TX, May 1–4.

Ballard, A.L., 2006. Flow-assurance lessons: the Mica tieback. In: OTC 18384. Offshore Technology Conference. Houston, TX, May 1–4.

Bohrmann, G., Torres, M., 2006. Gas Hydrates in Marine Sediment. In: Schulz, H, Zabel, M (Eds.), Marine Geochemistry. 2nd Ed., Springer-Verlag, pp. 482–507.

Bomba, J., Chin, D., Kak, A., Meng, W., 2018. Flow assurance engineering in deepwater offshore – past, present, and future. In: OTC 28704. Offshore Technology Conference. Houston, TX, Apr 30–May 3.

Brandt, H., Sarif, S.M., 2013. Life extension of offshore assets – balancing safety and project economics. In: SPE 165882. SPE Asia Pacific Oil and Gas Conference and Exhibition. Jakarta, Indonesia, 22–24 October.

Brimmer, A.R., 2006. Deepwater chemical injection systems: the balance between conservatism. In: OTC 18308. Offshore Technology Conference. Houston, TX, May 1–4.

Cochran, S., 2003. Hydrate control and remediation best practices in deepwater oil developments. In: OTC 15255. Offshore Technology Conference. Houston, TX, May 5–8.

Dindoruk, B., Christman, P.G., 2002. PVT properties and viscosity correlations for Gulf of Mexico oils. In: SPE 71633. SPE Annual Technical Conference and Exhibition. New Orleans, LA, Sep 30–Oct 3.

Dobbs, W., Browning, B., Killough, J., Kumar, A., 2011. Coupled surface/subsurface simulation of an offshore K2 field. In: SPE 145070. SPE Reservoir Characterization and Simulation Technical Conference and Exhibition, Abu Dhabi, UAE, Oct 9–11.

Gallun, R.A., Wright, C.J., Nichols, L.M., Stevenson, J.W., 2001. Fundamentals of Oil & Gas Accounting, 4th ed. PennWell, Tulsa, OK.

Jamaluddin, A.K.M., Nighswander, J., Joshi, N., Calder, D., Ross, B., 2002. Asphaltenes characterization: a key to deepwater development. In: SPE 77936. SPE Asia Pacific Oil and Gas Conference and Exhibition. Melbourne, Australia, Oct 8–10.

Jordan, M.M., Feasey, N.D., 2008. Meeting the flow assurance challenges of deepwater developments: from capex development to field start up. In: SPE 112472. SPE North Africa Technical Conference and Exhibition. Marrakech, Morocco, Mar 12–14.

Jordan, M.M., Sjuraether, K., Collins, I.R., Feasey, N.D., Emmons, D., 2001. Life cycle management of scale control within subsea fields and its impact on flow assurance, Gulf of Mexico and the North Sea basin. In: SPE 71557. SPE Technical Conference and Exhibition. New Orleans, LA, Sep 30–Oct 3.

Joshi, N.B., Li, Q., Champion, N., Kapadia, K., 2017. The relevance of chemistry in deepwater design and operations. In: OTC 27879. Offshore Technology Conference. Houston, TX, May 1–4.

Kaczmarski, A.A., Lorimer, S.E., 2001. Emergence of flow assurance as a technical discipline specific to deepwater: technical challenges and integration into subsea systems engineering. In: OTC 13123. Offshore Technology Conference. Houston, TX, Apr 30–May 3.

Kopps, R., Venkatesan, R., Creek, J., Montesi, A., 2007. Flow assurance challenges in deepwater gas developments. In: SPE 109670. SPE Asia Pacific Oil and Gas Conference and Exhibition. Jakarta, Indonesia, Oct 30–Nov 1.

Lim, F., Munoz, E., Browning, B., Joshi, N., Jackson, C., Smuk, S., 2008. Design and initial results of EOR and flow assurance laboratory fluid testing for K2 field development in the deepwater Gulf of Mexico. In: OTC 19624. Offshore Technology Conference. Houston, TX, May 5–8.

Lowe, J.S., 2003. Oil and Gas Law in a Nutshell, 4th ed. Thomson West, St. Paul, MN.

Maxwell, R.C., Martin, P.H., Kramer, B.M., 2007. The Law of Oil and Gas, 8th ed. Foundation Press, New York.

McCain Jr., W.D., 1990. The Properties of Petroleum Fluids, 2nd ed. PennWell, Tulsa, OK.

Mian, M.A., 2002. Project Economics and Decision Analysis, Vol. I: Deterministic Models. PennWell, Tulsa, OK.

Mullins, O.C., Sheu, E.Y., Hammami, A., Marshall, A.G., 2007. Asphaltene, Heavy Oils and Petroleomics. Springer Academic Press, New York.

Price, W.R., Ross, B., Vicknair, B., 2016. Integrated decommissioning – increasing efficiency. In: OTC 27152. Offshore Technology Conference. Houston, TX, 2–5 May.

Riazi, M.R., 2005. Characterization and Properties of Petroleum Fractions. ASME International, West Conshohocken, PA.

Sanford, J., Woomer, J., Miller, J., Russell, C., 2006. The K2 project: a drilling engineer's perspective. In: OTC 18298. Offshore Technology Conference. Houston, TX, 1–4 May.

Seba, R.D., 2003. Economics of Worldwide Petroleum Production. OGCI and Petroskills Publications, Tulsa, OK.

Speight, J.G., 1998. The Chemistry and Technology of Petroleum, 3rd ed. Marcel Dekker, Inc, New York.

Subramanian, S., Price, M.P., Pinnick, R.A., 2009. Flow assurance design and initial operating experience from Blind Faith in deepwater Gulf of Mexico. In: SPE 124535. SPE Annual Technical Conference and Exhibition. New Orleans, LA, Oct 4–7.

Tavakkoli, M., Boggara, M., Garcia-Bermudes, M., Vargas, F.M., 2016. Chapter 17: Asphaltene deposition: impact on oil production, experimental methods, and mitigation strategies. In: Riazi, M.R. (Ed.), Exploration and Production of Petroleum and Natural Gas. ASTM International Manual, Mayfield, PA.

CHAPTER FIVE

Reserves and Resources

Contents

Abstract

The US Gulf of Mexico is a world-class hydrocarbon basin, and industry has an impressive track record of advancing its capabilities in the region. From 1947 to 2017, the US Gulf of Mexico has produced over 52 billion barrels of oil equivalent (Bboe), 20.7 Bbbl of oil, and 187 Tcf of natural gas in federal waters. Remaining reserves are estimated at 4.8 Bboe, 80% of which are estimated to reside in deepwater, and undiscovered resources are estimated at 74 Bboe mean potential, 85% of which are estimated to reside in deepwater. In this chapter, the discovery and development activity of oil and gas fields in the US Gulf of Mexico are reviewed, and field discovery rates, sizes, distributions, and reserves statistics are summarized. The entire Gulf region was subject to rifting, salt deposition, and tectonic episodes that created the conditions for a large number of hydrocarbon deposits. Every decade or so, a new concept, play type, or technology has emerged, and most resource potential occurs in areas requiring technical advancement.

Decommissioning Forecasting and Operating Cost Estimation
https://doi.org/10.1016/B978-0-12-818113-3.00005-9

5.1 PROSPECTS, PLAYS, FIELDS AND RESERVES

During the first half of the 20th century, petroleum exploration focused on the prospect, a set of anomalous geologic criteria that define the location where an exploratory well is to be drilled to discover a hypothetical commercial accumulation of hydrocarbons (Levorson, 1967; Owen, 1975; Rose, 2001). By the mid-1930s, geologists recognized several different types of petroleum accumulations (e.g., anticlinal traps and stratigraphic traps), and in the 1950s and 1960s, stratigraphy and depositional systems were used to understand and predict the origin of sedimentary facies as they relate to the occurrence of reservoirs and seals and source beds. In the 1960s, revelations of plate tectonics and sedimentary basin analysis further advanced the field, which led to the exploration play concept in the early 1970s (Allen and Allen, 2005).

A play is a group of prospects and oil and gas fields all having similar geologic origins, basically a family of geologically similar traps. Fields comprising a play contain similar reservoir rocks that arose from similar depositional processes with similar histories of origin, emplacement, and preservation. Prospects and fields in a play have similar structural configurations and structural histories and generally form a lognormal distribution of reserves (Rose, 2001).

A field is an area consisting of a single reservoir or multiple reservoirs all grouped on or related to the same general geologic feature and/or stratigraphic trapping condition. A reservoir refers to an individual sealed hydrocarbon-bearing sand characterized by a single pressure system, sometimes called a block or pool to denote its compartmentalization. Fields refer to commercial accumulations and are defined according to a set of conventions/rules determined by regulatory agencies. There may be two or more reservoirs in a field separated vertically by impervious strata, laterally by local geologic barriers, or by both. The area may include one lease, a portion of a lease, or a group of leases with one or more wells that have been approved as producible. As a field is developed, its limits may expand.

Example. Garden Banks 426 (Auger) reservoirs

The Auger field was discovered in 1987 in 2847 ft water depth and is made up of four main pay horizons referred to as the Yellow "N," Blue "O," Green "Q," and Pink "S" sands ranging from 15,000 to 19,000 ft subsea and covering four blocks in Garden Banks 426, 427, 470, and 471 (Fig. 5.1).

The Auger field is made up of four main stacked pay horizons ranging in age from lower Pleistocene to lower Pliocene (Bilinski et al., 1992; Bourgeois, 1994). The two shallower pay sands contain condensate rich gas, and the deeper zones are volatile oils. Original ultimate recovery was estimated at 220 MMboe with a 2:1 oil/gas ratio, and circa September 2017, the Auger field had produced 267 MMbbl oil and 935 Bcf gas with a production resurgence in recent years (Fig. 5.2).

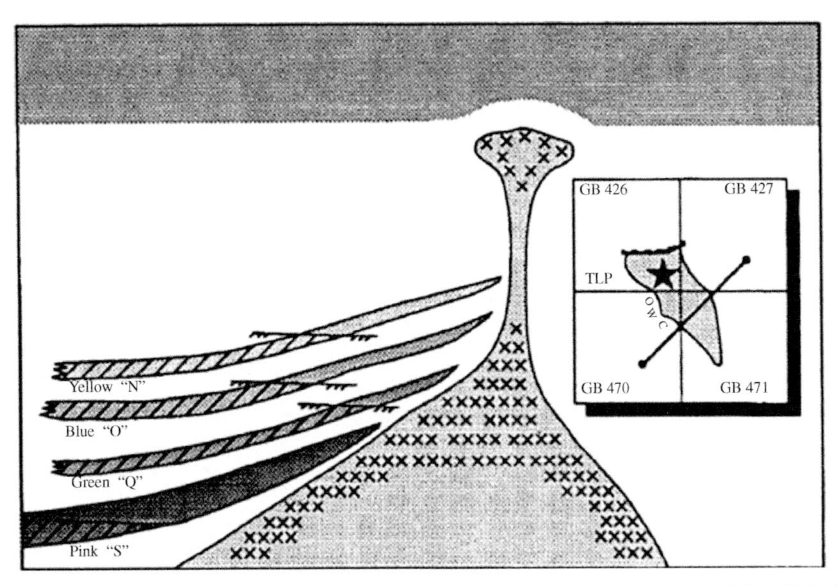

Fig. 5.1 Auger geologic structure and main reservoirs. *(Source: Bourgeois, T.M., 1994. Auger tension leg platform: conquering the deepwater Gulf of Mexico. In: SPE 28680. SPE International Petroleum Conference & Exhibition of Mexico. Veracruz, Mexico, Oct 10–13.)*

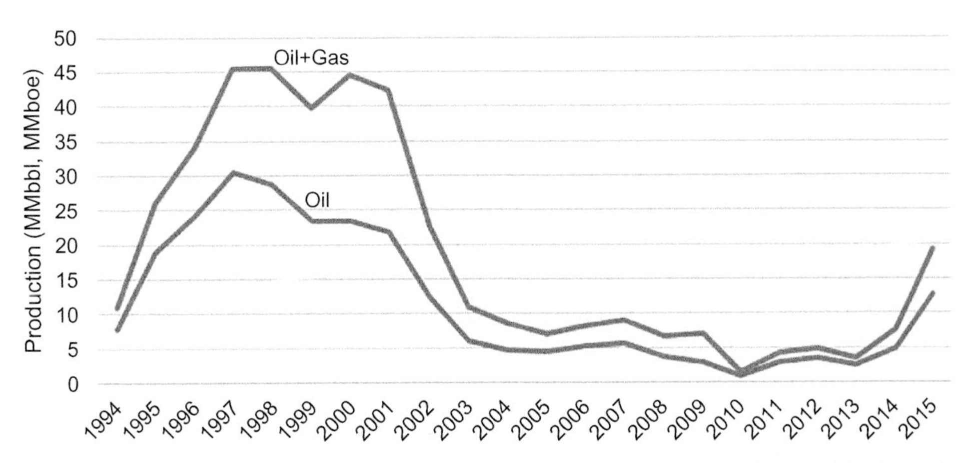

Fig. 5.2 Garden Bank 426 (Auger) field oil and gas production, 1994–2015. *(Source: BOEM 2017.)*

The BOEM Field Naming Committee determines leases capable of producing and associated with a specific geologic structure (BOEM, 1996). Fields are usually named after the area and block on which the discovery well is located, and when assigning leases to a field BOEM applies the following conventions:

- Structural lows are used to separate fields with structural trapping mechanisms.
- Faults are rarely used to separate fields and fields are never separated vertically.
- The structure or stratigraphic condition with pay having the largest areal extent on a lease determines the field expanse.
- Reservoirs that overlap areally are always combined into a single field.

Examples of BOEM field naming conventions are illustrated in Box 5.1.

Field size represents the projected final total production, which is cumulative production, plus proved reserves, plus projected future reserves additions. Reserves are created by capital, and many factors impact reserves estimates. BOEM performs reserves estimates for each field based on well logs, well data, seismic data, and production data (Burgess et al., 2018). Volumetric and performance methods are the primary techniques applied, and reserves estimates are reported deterministically using a single best estimate based on known geologic, engineering, and economic data.

Example. Stacked sands at South Timbalier 21

The South Timbalier 21 field was discovered in 1957 in 46ft water depth and consists of four South Timbalier blocks ST 21, 22, 27, and 28. South Timbalier block 21 is the largest producing block responsible for about two-thirds of field production. Through September 2017, the field produced 259 MMbbl and 426 Bcf gas, and remaining reserves are estimated at 1.5 MMbbl oil and 3.1 Bcf gas. There are six vertically stacked sands in the field, and the sand with the largest acreage is also the deepest and defines the areal extent of the field (Fig. 5.3).

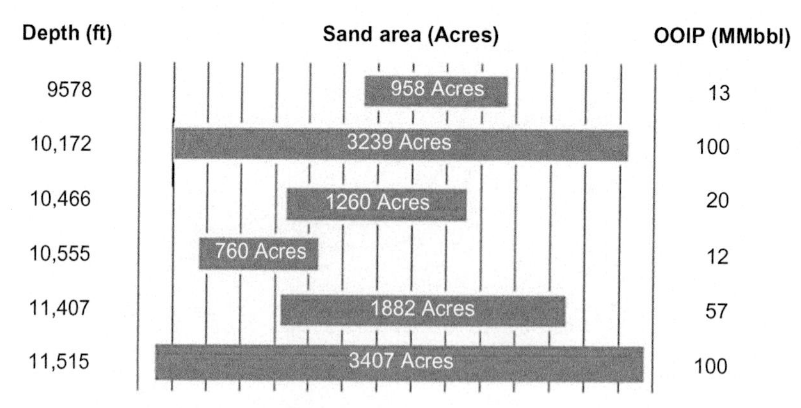

Fig. 5.3 Stacked sands at South Timbalier 21 field.

BOX 5.1 BOEM Field Naming Convention

Three examples of BOEM field naming conventions are illustrated. Additional examples can be found on the BOEM website.

Example. Salt dome with fault trap downdip on the domal structure

In Fig. 5.4, a piercement salt dome with traps against the salt and a fault trap down on the flank is depicted. Hydrocarbon discoveries are found in traps against the salt in blocks 2, 3, and 4, and the fault trap in block 13 also has a discovery. Blocks 2, 3, 4, and 13 would all be considered on the same structure and therefore in the same field.

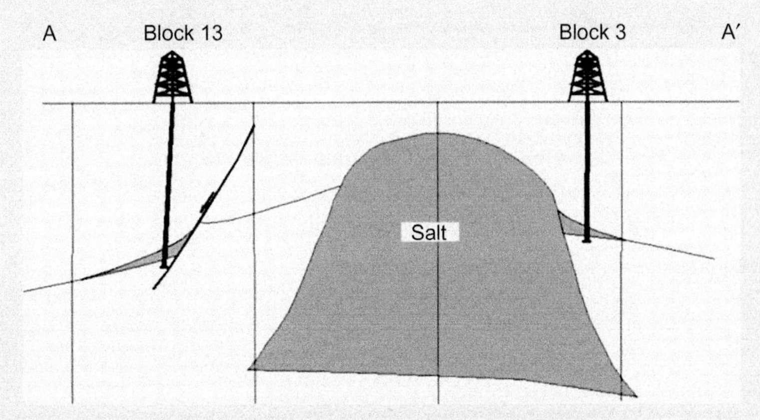

Fig. 5.4 Salt dome with fault trap downdip on the domal structure. *(Source: Bureau of Ocean Energy Management, 1996. Field Naming Handbook. New Orleans, LA.)*

Example. Two structural highs with a separating structural low

The structural low between the two anticlinal features in Fig. 5.5 is sufficient to designate two separate fields.

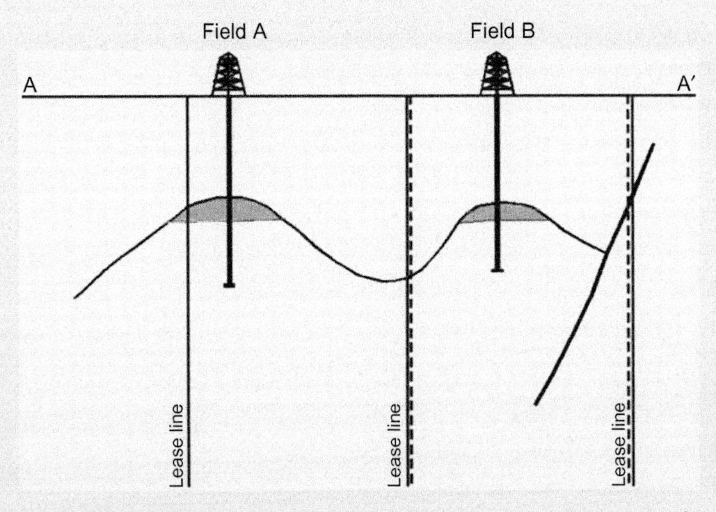

Fig. 5.5 Two structural highs with a separating structural low. *(Source: Bureau of Ocean Energy Management, 1996. Field Naming Handbook. New Orleans, LA.)*

Continued

BOX 5.1 BOEM Field Naming Convention—cont'd

Example. Series of traps against a large fault without separating structural lows

A long fault with a series of traps against the fault is shown in Fig. 5.6. In this case, there are no separating lows between the traps, and so, they are combined into a single field.

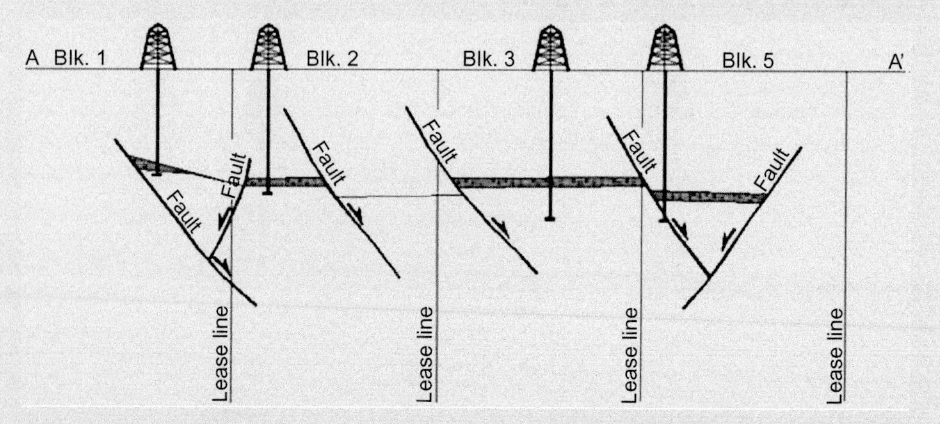

Fig. 5.6 Series of traps against a large fault without separating structural lows. *(Source: Bureau of Ocean Energy Management, 1996. Field Naming Handbook. New Orleans, LA.)*

Multiple structures may be used to develop one field, or multiple fields may be developed from one structure. In shallow water, platform construction and installation is cheap relative to well cost, and for fields with large areal extent spread over several lease blocks, multiple drilling platforms are common. In deepwater, fields are usually developed with one structure (D'Souza and Basu, 2011; Dekker and Reid, 2014). For some large areally extensive deepwater deposits such as Mars and Mad Dog, two structures may be used, but this is not common. After deepwater structures are installed, marginal reservoirs often become economic to produce when tied back to existing infrastructure.

5.2 GEOLOGIC TIME

The universe is estimated to be about 13.8 billion years old, and the Milky Way galaxy in which the Earth and its solar system inhabits is only a bit younger at about 12 billion years. The Earth was formed about 4.5 billion years ago, and the first life, probably bacteria followed by algae, appeared approximately 3 to 4 billion years ago and evolved in the oceans. There is relatively little organic matter preserved in Precambrian sedimentary rocks, and most are buried deep and are not good reservoir or source rocks for hydrocarbons. No significant deposits of oil and gas are known in Precambrian rocks.

Geologic timescales are used to classify the age of the rock that hydrocarbons inhabit, but there is usually no direct correspondence with the time hydrocarbons accumulated except as a lower bound (i.e., hydrocarbons are usually at least as old as the rock in which it is trapped). Large divisions of geologic time are called eras, and eras are subdivided into periods, periods into epochs, and epochs into ages (Table 5.1).

Within each period and epoch, reference is made to late, middle, and early times to describe specific intervals. For examples, the Late Pliocene runs from 3.6 to 1.8 million years before present, and the Early Pliocene runs from 5.3 to 3.6 million years before present. The Late Tertiary generally refers to the Eocene and Paleocene epochs. It is also common to group epochs when describing reservoir and depositional environments. The most common groupings are the Neogene, which groups the Miocene, Pliocene, and Pleistocene, and the Paleogene, which refers to the Eocene and Paleocene (or Late Tertiary).

At the start of the Paleozoic era 570 million years ago, known as the Cambrian explosion, a great abundance of diverse plants and animals is found in the fossil record. During the Ordovician period in the Paleozoic era, fish came into existence. Plants and animals adapted to life on land in the Silurian period. During the Mississippian and Pennsylvanian

Table 5.1 Geologic Timescale

Era	Period	Epoch	Age
Cenozoic	Quaternary	Holocene	0–10 ka
		Pleistocene	0.01–2 Ma
	Tertiary	Pliocene	2–5.3 Ma
		Miocene	5.3–38 Ma
		Eocene	38–55 Ma
		Paleocene	55–65 Ma
Mesozoic	Cretaceous		65–140 Ma
	Jurassic		140–200 Ma
	Triassic		200–250 Ma
Paleozoic	Permian		250–290 Ma
	Pennsylvanian		290–320 Ma
	Mississippian		320–365 Ma
	Deconian		365–405 Ma
	Silurian		405–425 Ma
	Ordovician		425–500 Ma
	Cambrian		500–570 Ma
Precambrian			0.57–4.5 Ga

Note: In geologic timescales, k denotes thousand, M denotes million, and G denote giga (billion).

periods, also known as the Carboniferous period, extensive land areas were covered by swamps and inland oceans that formed many of the world's coal deposits. Oil and gas was also being formed at this time but generally occurs in rock laid down at later time.

Example. Lower Tertiary trend

The most recent phase of deepwater exploration in the GoM began in the early years of the 21st century as companies moved east and further offshore pursuing the Lower Tertiary trend (Fig. 5.7). Confirmation of the Lower Tertiary began with discoveries in the Alaminos Canyon and Walker Ridge areas by Shell, Chevron, Unocal, Total, and others in the early 2000s. Large Lower Tertiary discoveries include Buckskin (KC 872), Great White (AC 857), Anchor (GC 807), Tiber (KC 012), St Malo (WR 678), and Kaskida (KC 292). Volumetric estimates of Lower Tertiary discoveries account for about 14% of total deepwater oil and gas recoverable volumes (Nixon et al., 2016).

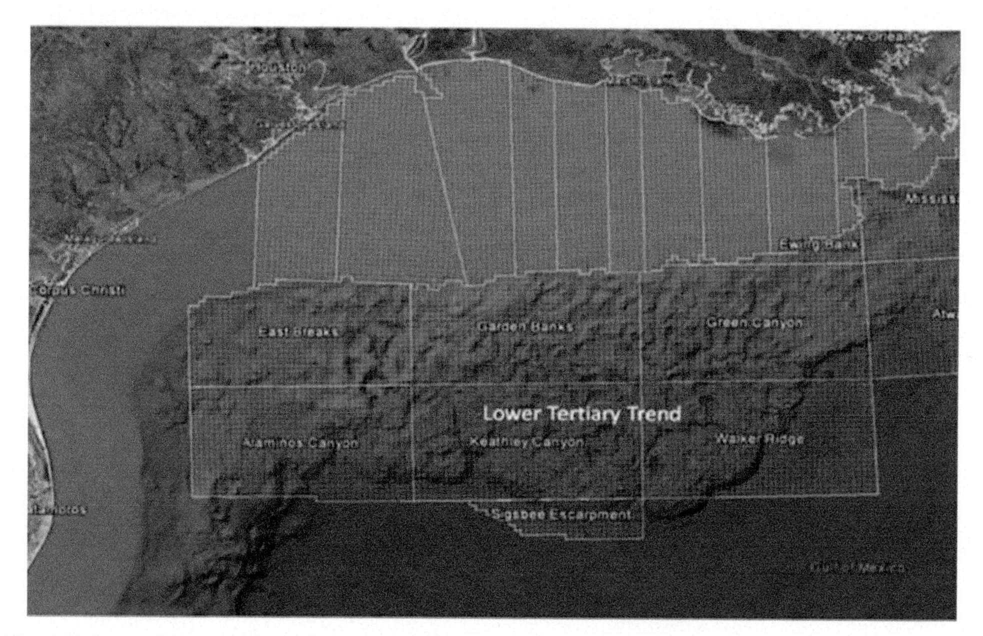

Fig. 5.7 Approximate outline of the Lower Tertiary trend in the Gulf of Mexico. *(Source: NOAA.)*

The Lower Tertiary trend (aka the Wilcox or Paleogene) refers to the trend of reservoirs that were formed between 65 and 38 million years ago. In contrast to the conventional deepwater Miocene reservoirs that are 5–24 million years old, the Lower Tertiary trend is buried deeper (25,000–35,000 ft), has higher pressures (17,000–24,000 psi), and has stronger rock formations. High temperature (220–270°F), low gas-oil ratio (170–250 cf/bbl), and low permeability contribute to the difficult reservoir characteristics (Oletu et al., 2013). The result of these conditions is that additional wells are required in development along with subsea pumps and boosting systems and regular intervention to maximize recovery rate (Close et al., 2008). Oil in place of Lower Tertiary reservoirs is huge, but only a fraction of volumes may be recovered using traditional primary depletion.

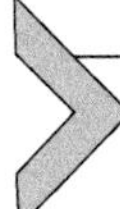

5.3 GULF OF MEXICO GEOLOGY

5.3.1 Formation

The Gulf of Mexico formed during the breakup of the megacontinent Pangea during the Triassic about 200 million years ago, when the North American plate pulled away from the South American and African plates, and the original basin was much larger than the basin of today (Fig. 5.8). Most of the geologic basin filled along the margins with sediments to form land leaving only a smaller portion of the original basin (Salvador, 1991). Rifting continued into the Middle Jurassic with the intermittent advance of a sea that resulted in extensive salt deposits. For succinct authoritative descriptions of the evolution of the GoM basin, Dribus et al. (2008), Galloway (2008), and Galloway (2009) are recommended.

In the northern Gulf, sedimentary materials from North America (particularly the Rocky Mountains) were deposited via ancient river systems throughout the Cretaceous and Tertiary periods. Traps associated with salt structures along the Gulf coast that are

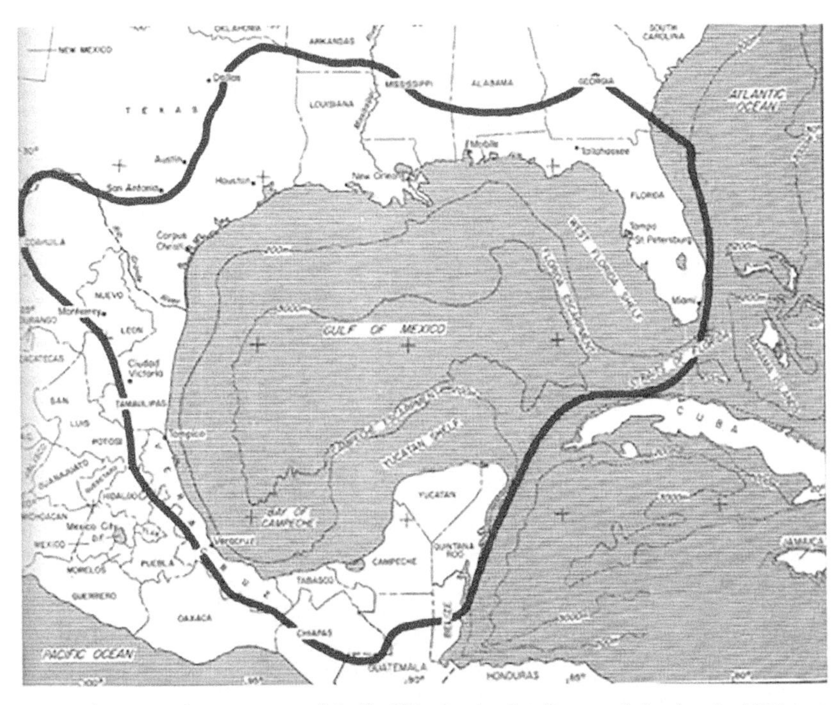

Fig. 5.8 Original geographic expanse of Gulf of Mexico basin. *(Source: Salvador, A., 1987. Late Triassic-Jurassic paleogeography and origin of Gulf of Mexico basin. AAPG Bull. 7 (4), 419–451. AAPG © 1987. Reprinted by permission of AAPG whose permission is required for further use.)*

well inland today originally evolved underwater within the early GoM basin. Turbidite systems were deposited in a series of minibasins on top of and in between allochthon salt bodies, providing reservoirs for many of the northern deepwater Gulf hydrocarbon systems. Allochthonous means out of place, and the allochthon salt is not in the position within the stratigraphic column where it was deposited. The majority of these reservoirs range in age from Miocene to Pleistocene in a seaward direction.

In the southern Gulf, massive layers of chemical carbonates were laid down on the Yucatan Shelf and the Florida Shelf during the Jurassic through the Cretaceous periods and provide Mexico's largest offshore hydrocarbon resources. In the Campeche subbasin off the Yucatan peninsula, the giant Cantarell discovery was made in 1976 and is a highly fractured dolomitized carbonate breccia deposited by the Chicxulub meteor impact when it struck the platform 65 million years ago.

5.3.2 Shallow Water (Modern Shelf)

The principal productive trends of South Louisiana and adjacent offshore shelf are those south of the Lower Cretaceous shelf edge. Stable, unfaulted, south-dipping beds lie north of the shelf edge with little conventional production. South of the shelf edge, salt structures and growth faults populate the region and provide multiple traps (Branson, 1986; Salvador, 1987). The productive trends strike roughly parallel to the coastline and become successively younger to the south and extend into Texas on the west.

Offshore Texas is characterized by a series of large, down-to-the-basin expansion fault systems that trend parallel to the Texas coastline (US DOI, 2017). The shallow sections of these fault systems have been thoroughly explored, and rollover anticlines on the downthrown sides of the faults have been prolific gas producers from Miocene reservoirs. The Louisiana shelf is characterized by a series of normal fault-related trends that generally become younger across the shelf, from the Miocene sediments in the inner shelf to the Pliocene and Pleistocene sediments on the middle and outer shelf (US DOI, 2017). The complexity and abundance of salt structures increase to the south, and near the shelf edge, significant tabular salt bodies form the Sigsbee Escarpment.

In the early days, explorers targeted shallow salt dome related traps, mostly Pliocene and Pleistocene reservoirs, similar to what was found onshore (Atwater, 1959). The Miocene fields offshore Louisiana produce from Upper Miocene sediments on the large salt domes located along the coastline. Examples include West Delta 30, West Bay, Grand Isle 16, Caillou Island, Timbalier Bay, and Bay Marchand. Pliocene development dates from the 1950s, and the trend is situated on the continental shelf with the Pliocene-Pleistocene trend boundary running eastward and close to the 100 ft water depth contour from West Cameron to South Timbalier.

Example. Eugene Island 126 field

The Eugene Island 126 field is a piercement-type salt-dome structure discovered in 1950 and typical of those found in the region with normal complex faulting characteristics (Fig. 5.9). The field is located approximately 25 mi offshore in 38–50 ft water depth in Eugene Island blocks 119, 120, 125, and 126. The salt reaches the surface of the seafloor, and the structure is mapped on top of the salt and on a reservoir sand below the Pliocene-Miocene contact (Atwater, 1959).

The salt stock is nearly circular and measures approximately 2 mi in diameter at a depth of 4000 ft (Fig. 5.10). The salt upwelling brought sands up to surface within 6000 ft before truncation, while on the south flank, beds terminate at about 10,000 ft. The major reserves of the field are found in a series of Pliocene sands on the northwest flank above the unconformities, and directional holes were drilled from multiple platforms located around the periphery of the salt (Massad and Pela, 1956).

Through September 2017, the field produced 141 MMbbl oil and 234 Bcf gas. In 2015, field production was 380,000 bbl oil and 565,000 boe or about 1041 bopd and 1548 boepd (Fig. 5.11). Remaining reserves are estimated at 4.9 MMbbl oil and 13.5 Bcf gas.

A snapshot of field development circa 1956 shows 64 wellbores drilled from 13 platforms and 53 producing wells (Fig. 5.12).

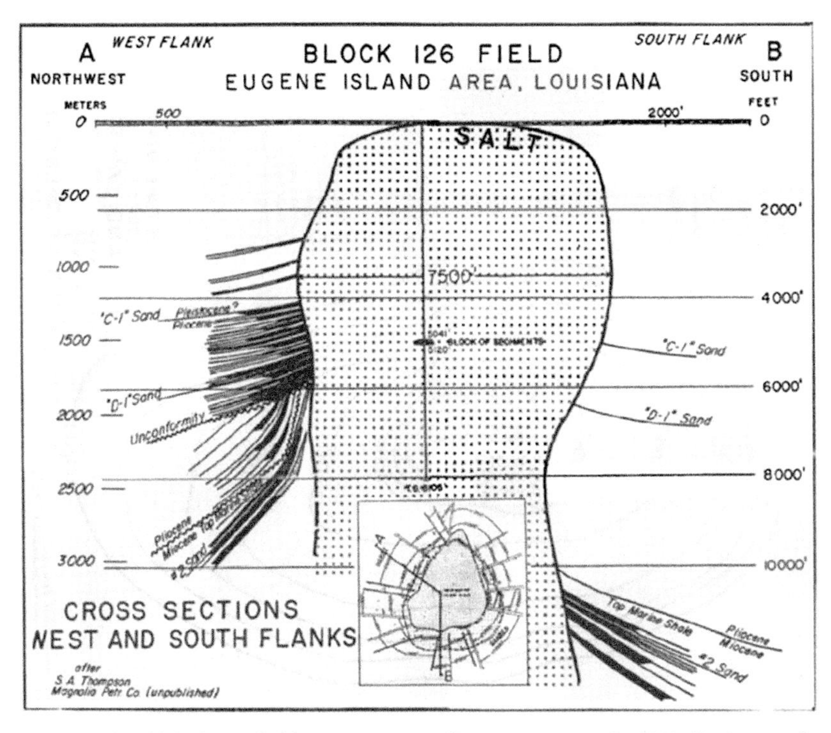

Fig. 5.9 Eugene Island block 126 field cross sections. *(Source: Atwater, G., 1959. Geology and petroleum development of the continental shelf of the Gulf of Mexico. Gulf Coast Assoc. Geol. Soc. Trans. 9, 131–145. Reprinted by permission of Gulf Coast Assoc. Geological Societies, 2018.)*

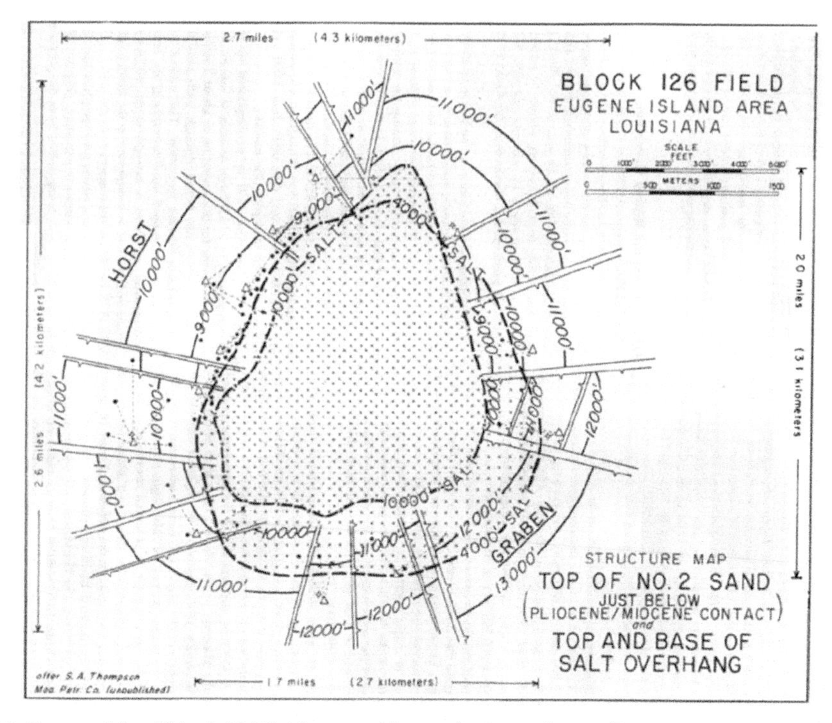

Fig. 5.10 Eugene Island block 126 field top and base of salt overhang. *(Source: Atwater, G., 1959. Geology and petroleum development of the continental shelf of the Gulf of Mexico. Gulf Coast Assoc. Geol. Soc. Trans. 9, 131–145. Reprinted by permission of Gulf Coast Assoc. Geological Societies, 2018.)*

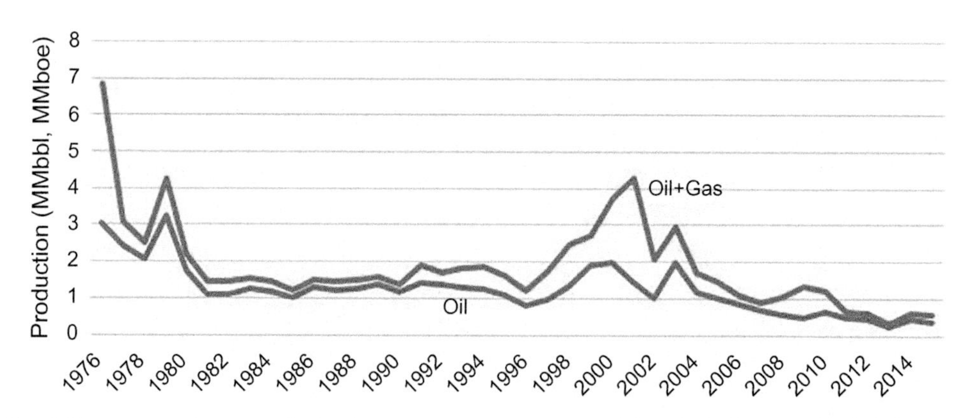

Fig. 5.11 Eugene Island 126 field oil and gas production plot, 1976–2015. *(Source: BOEM 2017.)*

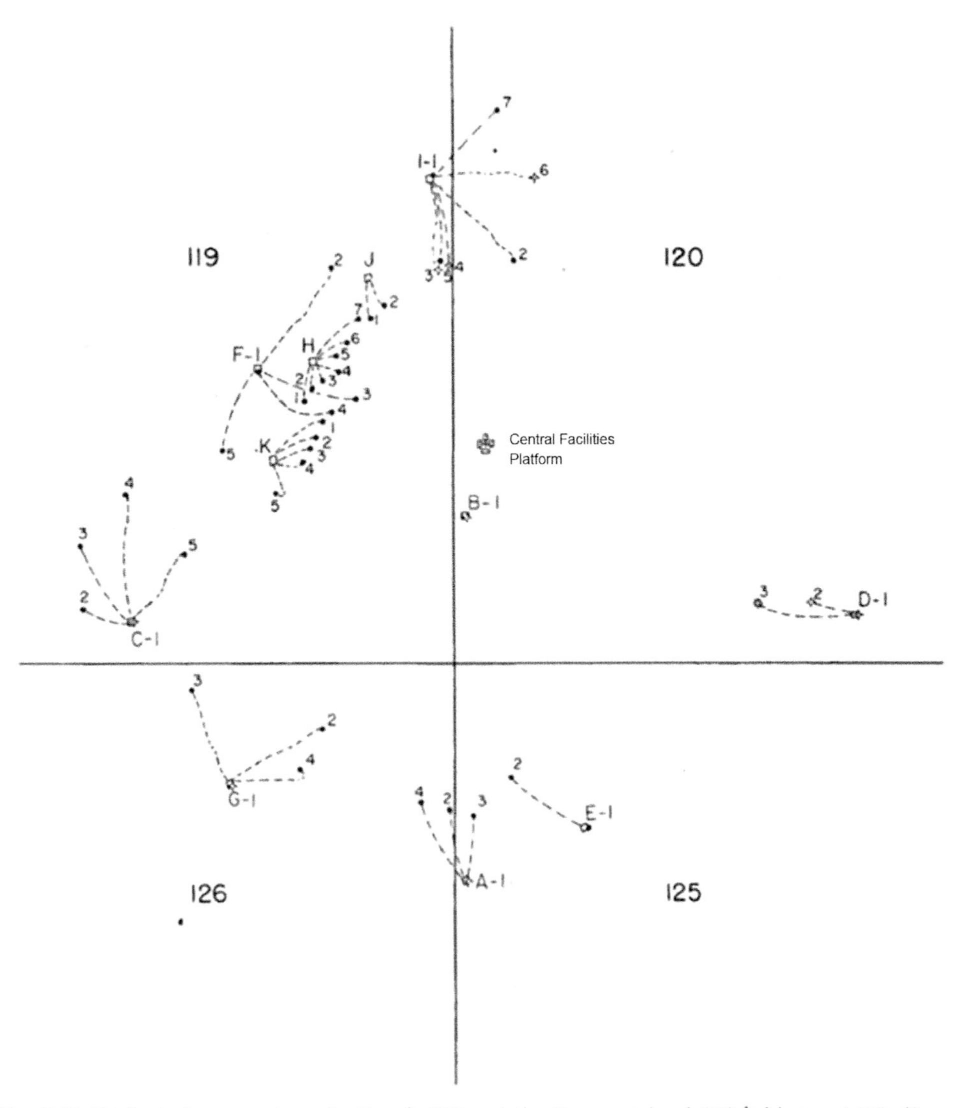

Fig. 5.12 Well platforms and production facilities at the Eugene Island 126 field circa 1956. *(Source: Massad, A.H., Pela, E.C., 1956. Facilities used in gathering and handling hydrocarbon production in the Eugene Island block 126 field Louisiana Gulf of Mexico. In: SPE 675-G. SPE AIME Annual Fall Meeting of AIME. Los Angeles, CA, Oct 14–17.)*

5.3.3 Deepwater (Modern Slope)

The slope occurs between the modern shelf edge and the abyssal plain and includes the Sigsbee Escarpment, the large compressional structures in front of the Sigsbee Escarpment, and the depositional limit of Louann Salt. Isolated salt bodies in the lower slope create complex traps for hydrocarbon accumulations (Nixon et al., 2016).

The mid-1970s saw a handful of pioneering companies move out from shallow waters to the deepwater Flextrend pursuing Pliocene and Upper Miocene plays. Shell's Cognac discovery in 1975 in Mississippi Canyon 194 opened this play. In the 1980s, minibasin and intraslope basin success during the decade and the first subsalt exploration were initiated. Shell's Bullwinkle (1983) and giant discoveries at Mars (1989), Ursa (1991), and Auger (1987) were made.

In the 1990s, exploration of the fold belt began, and BHP Billiton's 1995 Neptune discovery was the first in the Mississippi Fan fold belt. This play has been a prolific source of large discoveries such as the Mad Dog (1992) and Shenzi (2002) fields. The Perdido fold belt is Paleogene-aged reservoirs found in the western GoM trending southwest into Mexican waters, and more deeply buried formations were found in the central GoM at Cascade (2002) and Chinook (2003).

Deepwater exploration to date has focused on four major geologic provinces (Weimer et al., 2017): basins, subsalt, fold belt, and abyssal plain (Fig. 5.13). Basins refer to thick Neogene suprasalt sediments that overlie shallow allochthonous salt or welded to strata that once underlaid the welded-out basins. Subsalt prospects occur below allochthonous salt bodies and have considerable variation in salt thickness, styles and timing of formation, and boundaries marked by the edge of the Sigsbee Escarpment. Three

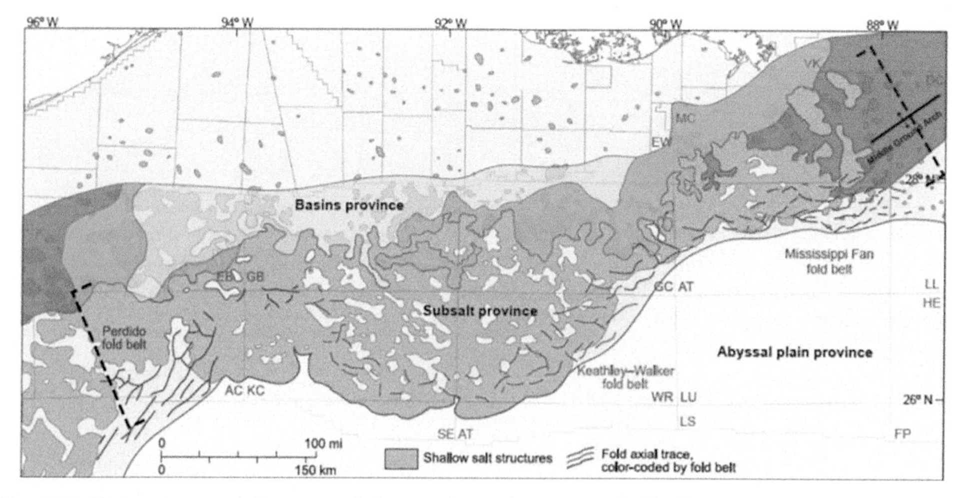

Fig. 5.13 Major structural features of the northern deepwater Gulf of Mexico. *(Source: Weimer, P., Bouroullec, R., Adson, J., Crossey, S.P.J., 2017. An overview of the petroleum systems of northern deep-water Gulf of Mexico. AAPG Bull. 101 (7), 941–993.)*

subregional fold belts have been identified and include the Perdido fold belt in the western GoM, the Mississippi Fan fold belt in the eastern GoM, and the Keathley-Walker fold belt near the Sigsbee Escarpment.

Example. Deepwater Gulf of Mexico reservoir characteristics

The Neogene reservoirs in the GoM from the Miocene, Pliocene, and Pleistocene epochs can be broadly characterized as overpressured, highly permeable, unconsolidated, and highly compacting containing black undersaturated oils of medium gravity and moderate GOR and often with some aquifer support (Lach and Longmuir, 2010). The vast majority of production in the deepwater GoM circa 2017 is from Neogene reservoirs. The P50 recovery factor for Neogene fields has been estimated at about 32% original oil in place (OOIP), with P90 and P10 recovery factors of 16% and 49%, respectively.

The Paleogene fields, except for those in the Alaminos Canyon, are deep, highly overpressured, and low to moderate permeability and cemented sandstones, containing black, highly undersaturated oils with moderate viscosity and low gravity and low GOR (Lach and Longmuir, 2010). The natural drive energy that has permitted good primary recoveries in the Neogene reservoirs is not expected in Paleogene reservoirs. For Paleogene fields with artificial lift, the P50 recovery factor is estimated at around 10%.

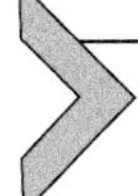

5.4 FIELD RESERVES

5.4.1 Data Source

GoM fields are described according to the location of the discovery well by lease block number and protraction area, field nickname (Box 5.2), discovery year, water depth, field class (developed producing, developed nonproducing, undeveloped, and justified for development), field type (oil and gas) using a GOR threshold of 9700 cf/bbl, original reserves (oil, gas, and boe), cumulative production (oil, gas, and boe), and (remaining) reserves (oil, gas, and boe) circa 2015 (BOEM, 2017a,b). For fields spanning state and federal waters, reserves are estimated for the federal portion only (Kaiser and Narra, 2018).

BOX 5.2 What's In a Name?

Development projects are typically named based on a common theme following the name of the prospect or reservoir by the geologist(s) who made the discovery. Mountains, rivers, fish, drinks, sharks, rock bands, scientists, gods and goddesses, movies, spouses, and even cartoon characters have been used. For example, the Perdido development is in the remote and isolated southwestern area of the GoM, about 6 mi north of the Mexican boundary. In Spanish, perdido means lost, which seems appropriate considering the location of the field. Three fields in the development are named after sharks and include the Great White, Silvertip, and Tobago. Na Kika is a Polynesian word for octopus, probably because the field development resembles the multiple arms of the arthropod (Box 15.1). The Auger Basin got its name from the tool theme Shell used to identify specific prospects, but its subsea tiebacks are named after spices: Cardamom, Macaroni, Serrano, and Oregano. Oil and gas from Auger are sent to the Enchilada platform.

Reserves can be expressed as recoverable volumes of oil or gas or of oil plus gas on an energy equivalent basis. The hydrocarbon liquids from a gas well are referred to as condensate and are added to oil production when reporting total liquid hydrocarbons, while the natural gas from an oil well is referred to as associated gas and combined with the gas from gas wells.

There is a six month delay in the time when oil and gas is produced and when it is uploaded on BOEM's web server, and because reserves estimation requires models and analysis in evaluation, there is another two year delay from the time production is recorded to when reserves are estimated. In this chapter, BOEM's 31 December 2015 reserves estimates are applied (Burgess et al., 2016).[1] Two water depth groups are adopted in analysis—shallow water (<400 ft) and deepwater (>400 ft)—instead of the seven water depth classes used in BOEM reports. And in place of the 25 deposit-size classes, two simplified categorizations are employed: <50, 50–100, 100–500, and >500 MMboe and < 1, 1–3, 3–5, 5–10, 10–25, 25–50, 50–100, 100–500, and >500 MMboe. Major fields are considered >50 MMboe, and giants refer to fields with >500 MMboe reserves.

5.4.2 Cumulative Production and Reserves

From 1947 through 2017, the US GoM has produced 20.7 Bbbl of oil and 187 Tcf of natural gas. More than half of total oil production (12.2 Bbbl) and about 80% of the Gulf's natural gas production (160 Tcf) have been produced in water depths <400 ft.

As of 31 December 2015, original reserves were estimated at 23.1 Bbbl of oil and 194 Tcf of gas from 1312 fields, and cumulative production totaled 19.7 Bbbl of oil and 186 Tcf of natural gas (Burgess et al., 2016). Remaining reserves (or simply reserves), the difference between original reserves and cumulative production, are estimated to be 3.5 Bbbl of oil and 7.3 Tcf of gas (combined 4779 MMboe) as of 31 December 2015 (Table 5.2).

Table 5.2 Gulf of Mexico Cumulative Production and Reserves by Water Depth Circa 2016

Water Depth (ft)	Number of Fields	Cumulative Production (MMboe)	Reserves (MMboe)
< 500	1082	41,331	899
500–999	54	1260	33
1000–1499	26	1388	98
1500–4999	97	6431	1993
5000–7499	35	1883	1403
≥ 7500	18	483	353
Total	1312	52,776	4779

Source: Burgess GL, Kazanis EG, Shepard NK. 2016. Estimated Oil and Gas Reserves, Gulf of Mexico OCS Region, December 31, 2015. U.S. Department of Interior, Bureau of Ocean Energy Management, OCS Report BOEM 2016-024. New Orleans, LA.

[1] BOEM's 2018 assessment estimates reserves through 31 December 2016, but does not differ materially from the 2016 report; for example, cumulative field counts in the 2018 assessment totaled 1315.

In water depth <500 ft, BOEM estimates shallow-water reserves at 899 MMboe, about 20% of the GoM total reserves of 4779 MMboe. In water depth >500 ft, BOEM estimates reserves at 3880 MMboe, about 80% of total reserves.

5.4.3 Field Counts and Reserves

In 2015, there were 1312 fields in the GoM: 1058 in shallow water and 254 in deepwater, 290 oil fields and 1022 gas fields, 960 fields in the central GoM and 352 fields in the western GoM. Eight out of 10 fields in shallow water are gas fields, while in deepwater, more than half of all fields are oil fields. In 2015, 496 of the 1312 fields were producing (Fig. 5.14).

Capital spending is required before reserves are booked and a field added to BOEM's list. Over time, fields become inactive and stop producing. Circa 2015, there were 496 producing fields, 334 in shallow water and 162 in deepwater (Fig. 5.15). The number of producing fields on the shelf peaked at around 700 for about a decade but has dropped rapidly thereafter. The number of producing deepwater fields has also declined as older fields outnumber new developments. As exploration progresses in a region or play, more fields are discovered, but they typically shift toward smaller sizes. Declining field counts in shallow water is the first indicator of the declining prospectivity in the region.

In shallow water, field discoveries peaked from the mid-1970s to 1990, both oil and gas (Fig. 5.16), while in deepwater, field counts peaked in 2001 at about one-fourth the level in shallow water (Fig. 5.17). In the 1980s, 316 oil and gas fields were discovered in shallow water, compared with 83 fields in deepwater from 2000 to 2009 (Table 5.3).

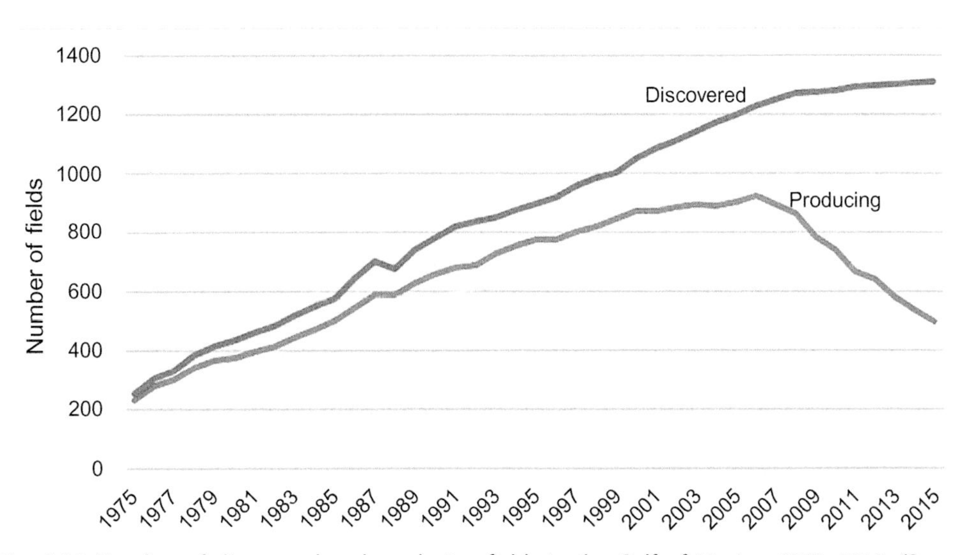

Fig. 5.14 Number of discovered and producing fields in the Gulf of Mexico, 1975–2015. *(Source: BOEM 2017.)*

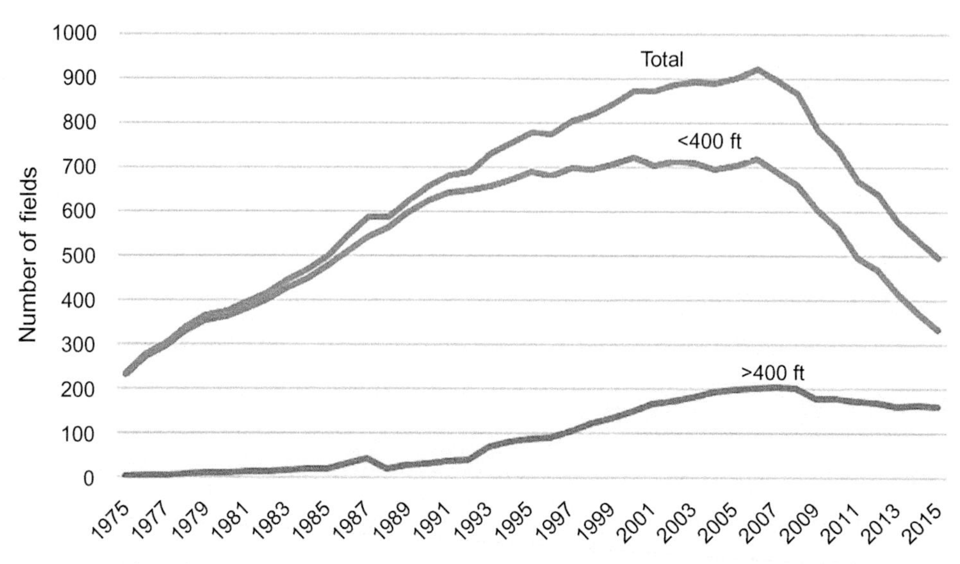

Fig. 5.15 Number of producing fields in the shallow-water and deepwater Gulf of Mexico, 1975–2015. *(Source: BOEM 2017.)*

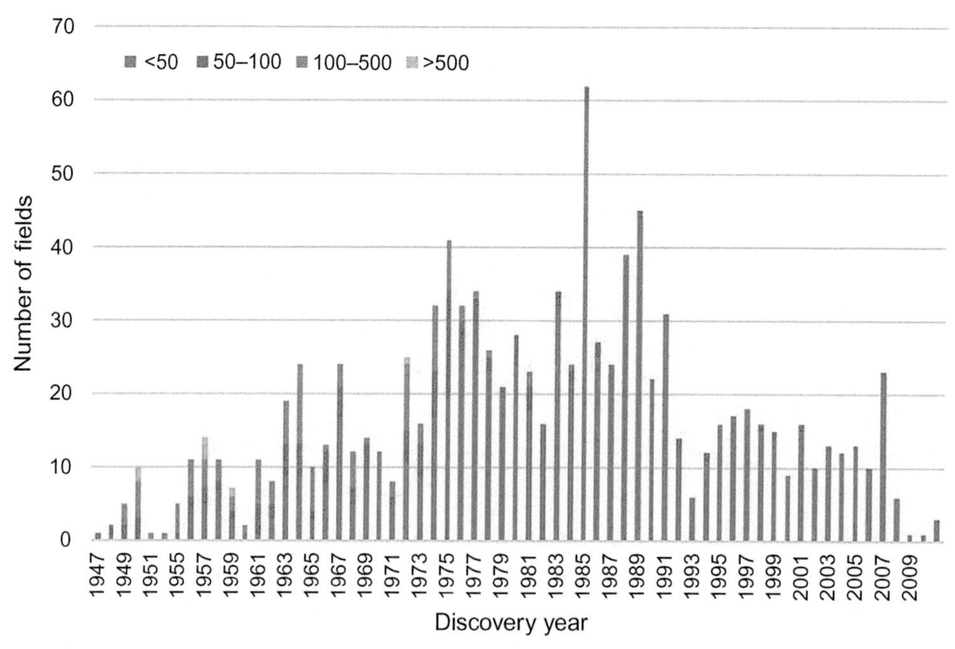

Fig. 5.16 Number of fields by discovery year and deposit size in the shallow-water Gulf of Mexico. *(Source: BOEM 2017.)*

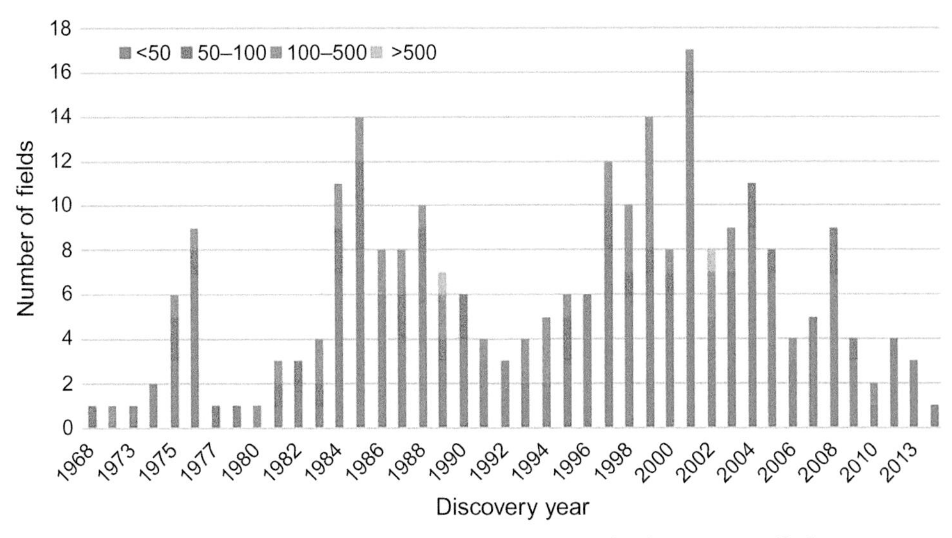

Fig. 5.17 Number of fields by discovery year and deposit size in the deepwater Gulf of Mexico. *(Source: BOEM 2017.)*

Table 5.3 Number of Fields by Type and Discovery Decade in Shallow-Water and Deepwater Gulf of Mexico

Water Depth	Decade	Oil	Gas	Field Count
Shallow	1940s	5	13	18
	1950s	21	30	52
	1960s	35	112	147
	1970s	46	217	263
	1980s	22	294	316
	1990s	14	140	154
	2000s	7	98	105
	2010s		3	3
	Subtotal	150	908	1058
Deep	1960s	2		2
	1970s	7	13	20
	1980s	36	33	69
	1990s	39	31	70
	2000s	47	36	83
	2010s	9	1	10
	Subtotal	140	114	254
Total		290	1022	1312

Source: BOEM 2017.

Seven giants >500 MMboe have been discovered in shallow water compared with just two in deepwater. Six of the seven shallow-water giants were discovered in the 1940s and 1950s, and all were producing circa 2018.

In shallow water, reserves volumes peaked in the 1960s and 1970s at about 12,300 MMboe each decade, from 147 fields in the 1960s to 263 fields in the 1970s (last column in Table 5.4). In the 1980s, 316 fields were discovered with 3870 MMboe reserves added. In the 1940s, the average field size was 228 MMboe, but by 2010s, it was two orders of magnitude smaller at 2 MMboe. Average field size by decade shows the trend: 152 (1950s), 83 (1960s), 47 (1970s), 12 (1980s), 5.8 (1990s), and 2.9 MMboe (2000s). In deepwater, average field size per decade shows a declining but less predictable trend, from 203 MMboe in the 1960s to 49 (1970s), 88 (1980s), 65 (1990s), 45 (2000s), and 29 MMboe (2010s).

Large fields discovered early continue to provide the vast majority of shallow-water production and remaining reserves (Table 5.5). Most of the remaining reserves in shallow water were discovered before 1980 (703/872, 81%). In contrast, in deepwater, large volumes of reserves are still being added, and the majority of remaining reserves are from 2000 and after (2144/3907, 55%). In shallow water, only 6 MMboe reserves have been discovered since 2010, half of which has been produced.

Table 5.4 Number of Fields by Original Reserves and Discovery Decade in the Gulf of Mexico

Water Depth	Discovery Decade	MMboe				Field Count	Reserves (MMboe)
		<50	50–100	100–500	>500		
Shallow	1940s	2	6	8	2	18	3872
	1950s	20	7	21	**4**	52	7891
	1960s	77	28	**42**		147	12,266
	1970s	185	**43**	34	1	263	12,345
	1980s	**302**	11	3		316	3870
	1990s	150	4			154	857
	2000s	104	1			105	309
	2010s	3				3	6
	Subtotal	843	100	108	7	1058	41,455
Deep	1960s		1	1		2	405
	1970s	14	4	2		20	971
	1980s	40	**13**	15	1	69	6087
	1990s	44	8	**18**		70	4566
	2000s	**67**	8	7	1	83	3779
	2010s	9		1		10	293
	Subtotal	174	34	44	2	254	16,101
Total		1017	134	152	9	1312	57,556

Source: BOEM 2017.

Table 5.5 Remaining Reserves by Discovery Year for the Gulf of Mexico Circa 2016

Discovery Decade	Shallow Water (MMboe)	Deepwater (MMboe)
1940s	60 (2%)	
1950s	172 (2%)	
1960s	266 (2%)	3 (1%)
1970s	205 (2%)	11 (1%)
1980s	90 (2%)	690 (11%)
1990s	49 (6%)	1059 (23%)
2000s	27(9%)	1892 (50%)
2010s	3 (50%)	252 (86%)
Total	872 (2%)	3907 (24%)

Note: The percentage of remaining reserves discovered by decade and in total is shown in parenthesis.
Source: BOEM 2017.

5.4.4 Creaming Curves

When cumulative reserves discovered are plotted by discovery year, the pattern is referred to as "creaming curves," and they provide useful information about the maturity of a region or play and future prospects (Rose, 2001). When the discovery curve is rising steeply, exploration is efficient and prospectivity is high because large reserves are being found quickly. When the discovery curve flattens out, incremental additions are small and exploration success rates decline, leading to declining prospectivity and higher risks associated with exploration.

The shallow-water GoM is primarily a gas province, and discoveries through the 1970s added reserves at a fast pace through the mid-1980s (Fig. 5.18). From 1947 to

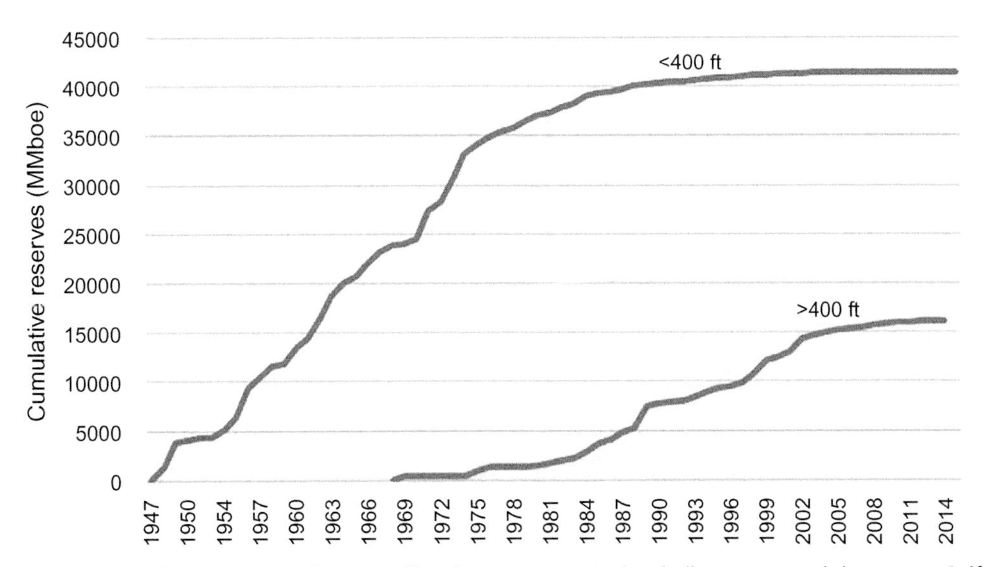

Fig. 5.18 Cumulative reserves discovered by discovery year in the shallow-water and deepwater Gulf of Mexico, 1947–2015. *(Source: BOEM 2017.)*

1975, about 35 Bboe reserves were added or 1.3 Bboe per year on average. Growth on the shelf continued through the 1980s, but discoveries were less frequent and smaller in size, and from 1975 to 1996, reserves additions averaged 286 MMboe per year and thereafter dropped off sharply. For the last two decades, annual additions in shallow water have been small or nonexistent.

The majority of fields discovered on the shelf are gas and gas-condensate fields and have yielded a faster and longer growth rate than oil discoveries, which by comparison has been essentially flat for the last half century (Fig. 5.19). Gas developments are also somewhat easier and quicker to develop than oil fields and accounts for some of the differences in development pace. The deepwater GoM is primarily an oil province, and reserves growth has been significantly slower than in shallow water (Fig. 5.20). It took about 30 years for the first 10 Bboe in deepwater to be booked, for example, compared with about a decade in shallow water, and unlike the incredible growth spurts in shallow water (e.g., from 1970 to 1974, almost 10 Bboe reserves were added), deepwater development requires more time in planning and execution.

Deepwater development requires field appraisal studies and large capital expenditures and takes much longer to develop on average than shallow-water fields. Terrain is more difficult, less infrastructure is available to provide support, and several options must be compared to select the optimal development. Deepwater field developments represent highly engineered systems, and there are many different options in design. Consequently, one would expect deepwater reserves to be added at a slower rate and production to exhibit a greater lag.

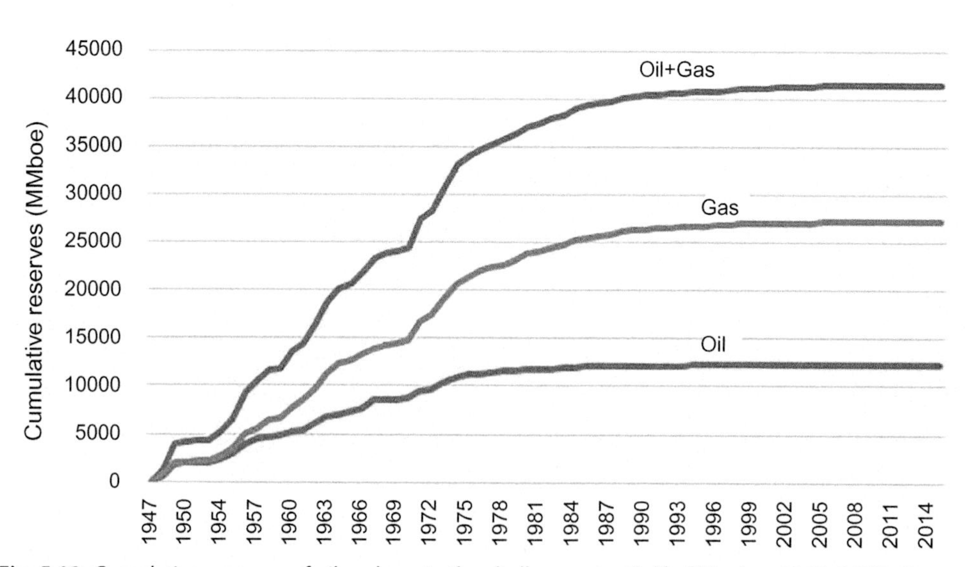

Fig. 5.19 Cumulative reserves of oil and gas in the shallow-water Gulf of Mexico, 1947–2015. *(Source: BOEM 2017.)*

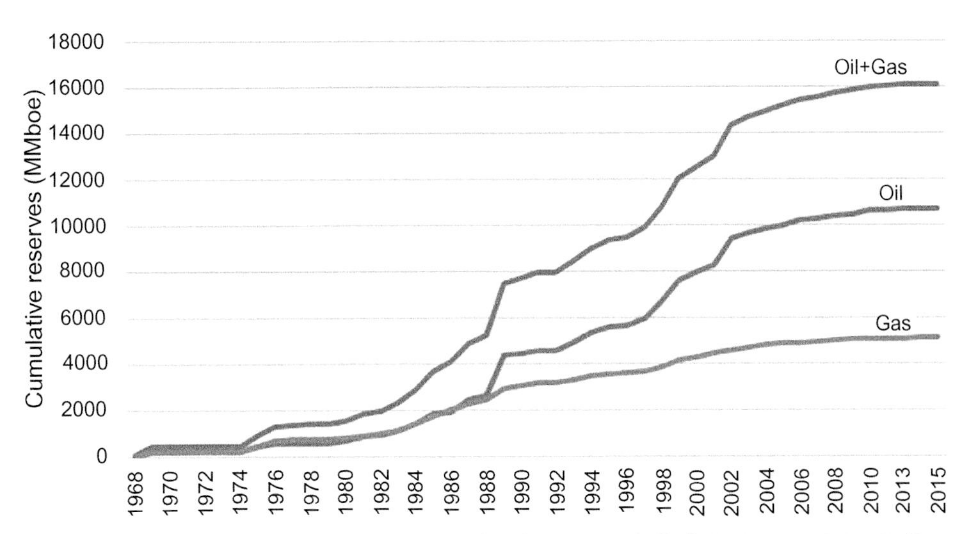

Fig. 5.20 Cumulative reserves of oil and gas in the deepwater Gulf of Mexico, 1947–2015. *(Source: BOEM 2017.)*

5.4.5 Reserves vs. Production

In exploration, there is a tendency for large fields to be found first followed by smaller fields during later times, and field counts generally peak later than reserves volumes as a greater number of smaller fields are developed. The build-out of infrastructure plays a role in this trend since once infrastructure is installed in a region, smaller fields are more likely to be developed as tiebacks than in earlier stages when such opportunities are not available or feasible. As technology matures and improves, multiple small fields may be consolidated into one project, which provides another development option not available in earlier years.

Onshore, if exploratory drilling is successful, the well is completed, facilities are installed, and the well may begin flowing within a few months. Offshore, a successful exploration well is the start of the evaluation process that often takes several years to first production. Delineation is required to gauge the size of the deposit and to ensure development is commercial. Production can also be delayed by many factors such as capital, development, infrastructure, regulatory, and/or technical constraints.

In shallow water, there is a 20-year lag between reserves booking and production for gas fields, while for oil fields, the gap started at about 20 years and has increased over time, indicating greater effort to extract remaining barrels (Fig. 5.21). In deepwater, the gap has been relatively constant at about 25 years (Fig. 5.22), and since most gas production is associated with oil fields, timing differences between deepwater oil and gas are not evident.

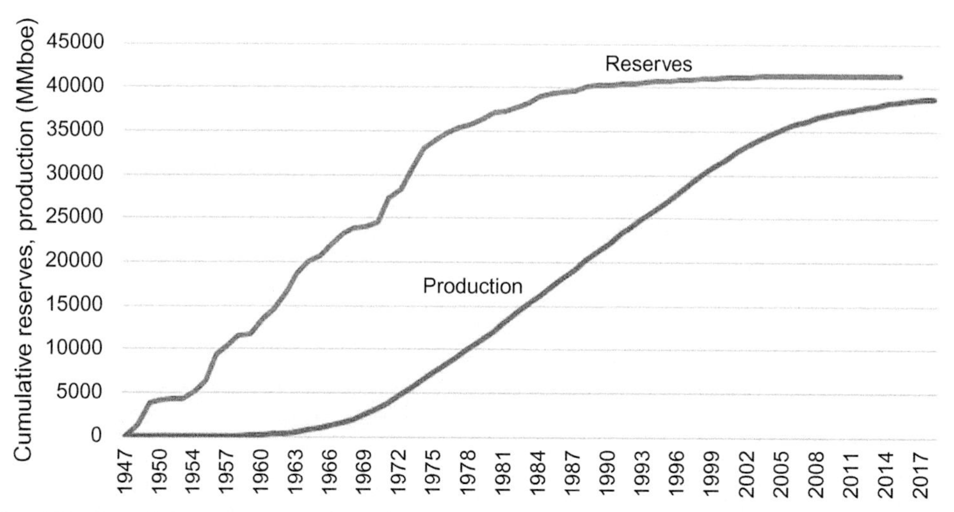

Fig. 5.21 Cumulative reserves and production in the shallow-water Gulf of Mexico, 1947–2017. *(Source: BOEM 2017.)*

5.4.6 Field-Size Distribution

Hydrocarbon deposits within a given trend, play, or basin typically follow lognormal distributions because the parameters that control field size are multiplicative independent random variables (e.g., reserves = field area × pay thickness × recovery factor), and whenever multiplicative relations are found, lognormal distributions generally follow.

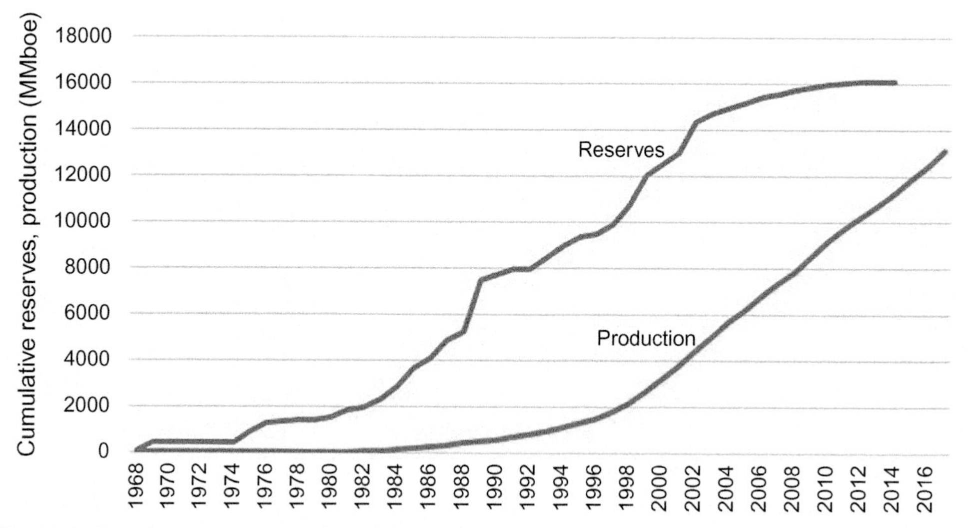

Fig. 5.22 Cumulative reserves and production in the deepwater Gulf of Mexico, 1947–2017. *(Source: BOEM 2017.)*

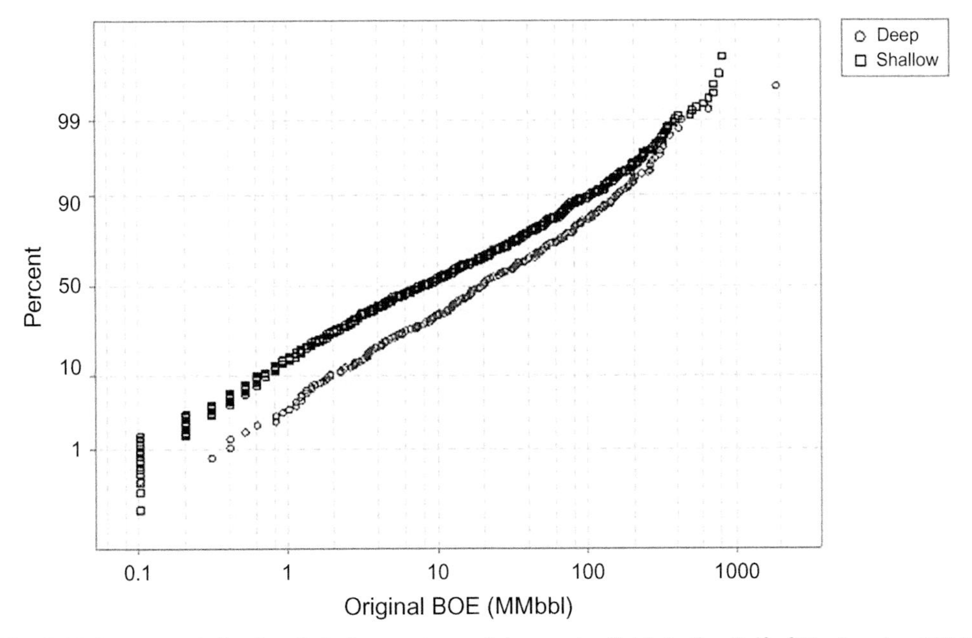

Fig. 5.23 Lognormal plot for all shallow-water and deepwater fields in the Gulf of Mexico circa 2015. *(Source: BOEM 2017.)*

On average, deepwater fields in the GoM are larger than shallow-water fields (Fig. 5.23), shallow-water oil fields are larger than shallow-water gas fields on a heat-equivalent basis (Fig. 5.24), and deepwater oil fields are larger than deepwater gas fields (Fig. 5.25). Volume comparisons based on heat equivalence are not without their limitations, of course, but serve as one (imperfect) means to compare liquid and gas hydrocarbon deposits.

In shallow water, median field size was 8.4 MMboe compared with 19.5 MMboe in deepwater, while the average field size in shallow water was 33 MMboe compared with 58 MMboe in deepwater (Table 5.6). The large difference between the mean and median statistics is due to the inclusion of many small fields developed in shallow water and the occurrence of a few giant deposits among the discoveries (i.e., statistical differences arising from lognormal distributions). P90 shallow-water fields are 99 MMboe compared with 166 MMboe in deepwater.

Central GoM oil and gas fields are larger than western GoM oil and gas fields, and the average gas field is smaller than the average oil field. In shallow water, the median gas field is 6.6 MMboe compared with 36 MMboe for the median oil field, about five times smaller on a heat-equivalent basis. In deepwater, the median gas field is 9.8 MMboe compared with 34 MMboe for the median oil field, about three times smaller. The mean shallow-water oil field is 124 MMboe compared with 24 MMboe for the mean gas field. In deepwater, the mean oil field is 101 MMboe compared with 22 MMboe for the mean gas field.

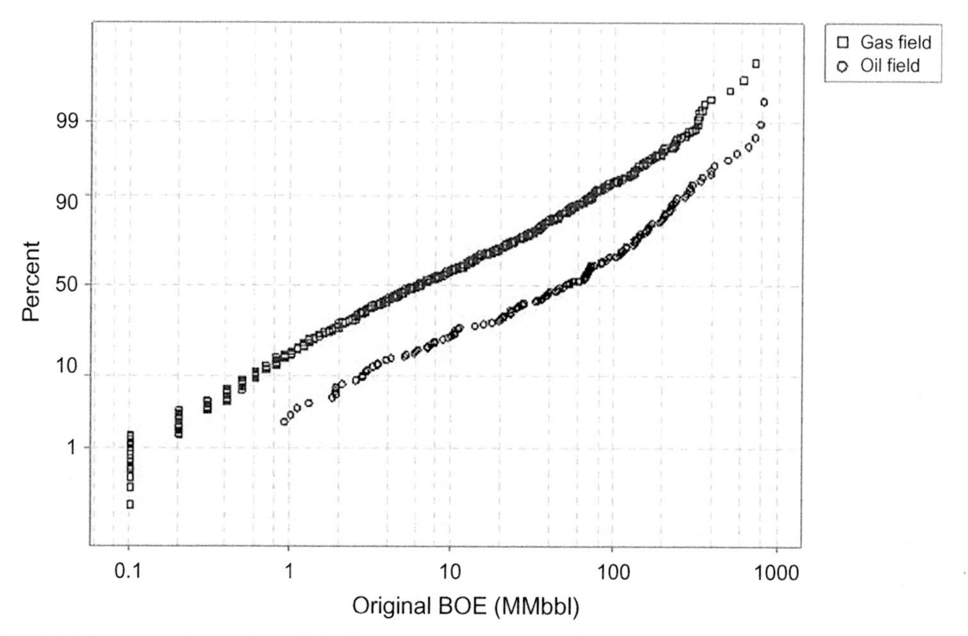

Fig. 5.24 Lognormal plot for all shallow-water gas and oil fields in the Gulf of Mexico circa 2015. *(Source: BOEM 2017.)*

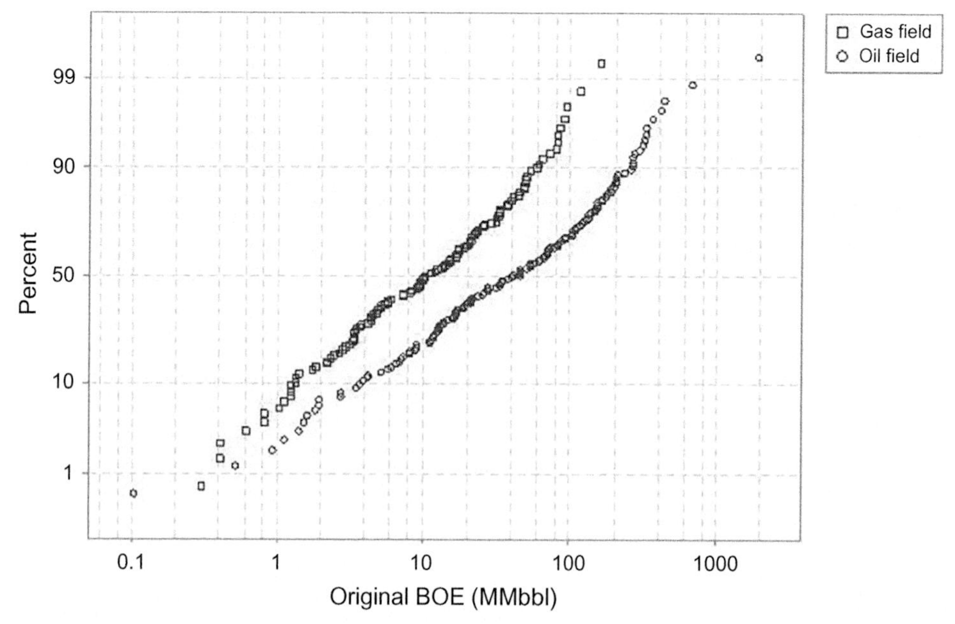

Fig. 5.25 Lognormal plot for all deepwater gas and oil fields in the Gulf of Mexico circa 2015. *(Source: BOEM 2017.)*

Table 5.6 Field-Size Statistics in the Gulf of Mexico Circa 2016

	Shallow			**Deepwater**		
Statistic	**Oil + Gas (MMboe)**	**Gas (MMboe)**	**Oil (MMboe)**	**Oil + Gas (MMboe)**	**Gas (MMboe)**	**Oil (MMboe)**
P99%	732	467	817	954	264	1615
P90%	99	69	362	166	60	286
P50%	8.4	6.6	36	19.5	9.8	34.3
P10%	0.72	0.63	3.5	2.9	1.6	4.1
P1%	0.1	0.1	0.5	0.4	0.4	0.7
MSw	33.2	23.5	124	58.3	22.4	101

Note: Swanson's mean (MSw) is calculated from the formula MSw = 0.3 (P90%) + 0.4 (P50%) + 0.3 (P10%).
Source: BOEM 2017.

5.4.7 Largest Fields

In Tables 5.7 and 5.8, the top 20 largest fields in shallow-water and deepwater GoM circa 2016 are depicted, and in Figs. 5.26 and 5.27, the top 50 fields are graphed with remaining reserves. Reserves growth will impact the rate of decline, and in many cases, the pie will grow larger; reserves downgrades also occur.

Table 5.7 Top 20 Shallow-Water Fields Based on Original Barrels of Oil Equivalent Circa 2016

					Reserves		
Rank	**Field Name**	**Discovery (Year)**	**Original Reserves (MMboe)**	**Cumulative Production (MMboe)**	**Oil (MMbbl)**	**Gas (Bcf)**	**BOE (MMboe)**
2	EI 330	1971	798.6	782.9	11.7	22.6	15.7
3	WD 30	1949	796.5	758.9	6.8	21.3	10.6
4	GI 43	1956	711.1	669.8	22.4	106.0	41.3
5	TS 000	1958	710.6	681.7	2.7	146.8	28.9
7	BM 002	1949	649.7	642.9	5.6	6.8	6.8
8	VR 14	1956	604.2	604.2	0	0	0
9	MP 41	1956	550.0	543.8	3.2	17.1	6.2
10	VR 39	1948	496.3	495.3	0.3	3.9	1.0
11	SS 208	1960	484.6	472.2	6.0	35.8	12.4
14	WD 73	1962	403.0	395.2	5.4	17.3	8.4
15	WI 238	1964	383.4	365.6	8.2	54.0	17.8
16	GI 16	1948	380.4	376.4	2.8	6.5	4.0
17	SP 61	1967	378.8	365.1	10.1	20.4	13.7
18	SP 89	1969	353.0	350.5	1.9	3.6	2.5
19	ST 172	1962	349.9	349.9	0	0	0
20	WC 180	1961	342.2	341.7	0	2.5	0.5
21	ST 21	1957	335.7	333.6	1.5	3.1	2.1
22	SS 169	1960	334.0	326.0	5.3	14.8	8.0
23	ST 176	1963	322.0	314.6	3.4	22.3	7.4
24	EI 292	1964	319.9	316.0	2.7	7.2	4.0

Note: Cumulative production through 2015.
Source: BOEM 2016.

Table 5.8 Top 20 Deepwater Fields Based on Original Barrels of Oil Equivalent Circa 2016

Rank	Field	Discovery (Year)	Original Reserves (MMboe)	Cumulative Production (MMboe)	Reserves		
					Oil (MMbbl)	Gas (Bcf)	BOE (MMboe)
1	Mars-Ursa	1989	1851.0	1504.5	260.8	481.6	346.5
6	Tahiti/Tonga	2002	660.5	265.0	363.7	178.8	395.5
12	Auger	1987	432.4	410.4	13.7	46.7	22.0
13	Atlantis	1998	411.6	279.2	113.7	105.0	132.4
25	N. Th. Horse	2000	319.5	227.8	78.3	75.1	91.7
26	Cognac	1975	319.3	315.2	3.1	5.4	4.1
30	King/Horn Mt.	1993	309.1	262.1	34.7	69.0	47.0
32	Shenzi	2002	303.2	254.4	45.6	18.1	48.8
35	Great White	2002	285.6	126.2	134.9	137.6	159.4
42	Salsa	1984	264.7	222.7	25.6	92.0	42.0
44	Holstein	1999	261.7	107.9	124.4	165.3	153.8
45	Troika	1994	260.3	240.5	13.4	36.0	19.8
46	Ram-Powell	1985	257.9	255.2	2.2	3.1	2.7
47	Thunder Horse	1999	251.2	124.7	111.4	84.8	126.5
55	Stampede	2006	229.0	0.0	214.3	82.7	229.0
63	Pompano	1981	206.8	188.6	15.1	17.1	18.2
68	Baldpate	1991	198.1	181.3	8.8	45.2	16.8
69	Petronius	1995	197.9	186.2	8.8	16.5	11.7
74	Genesis	1988	191.8	175.5	13.1	17.8	16.3
80	Mad Dog	1998	183.7	155.5	17.9	57.4	28.2

Note: Cumulative production through 2015.
Source: BOEM 2016.

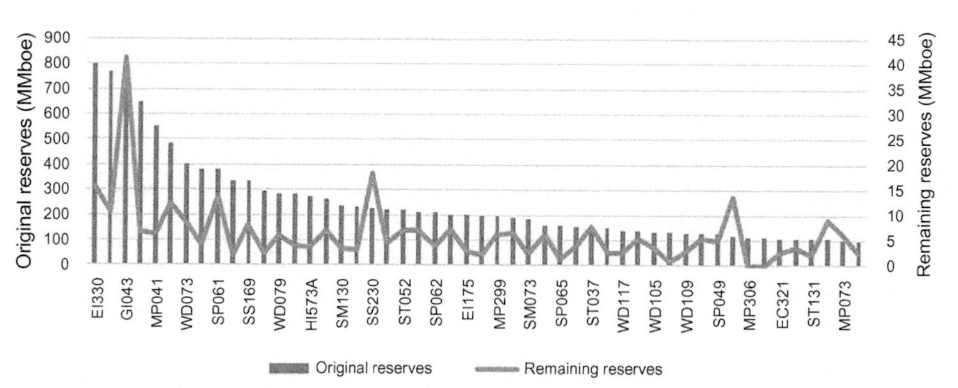

Fig. 5.26 Original reserves and remaining reserves in the top 50 shallow-water Gulf of Mexico fields. *(Source: BOEM 2017.)*

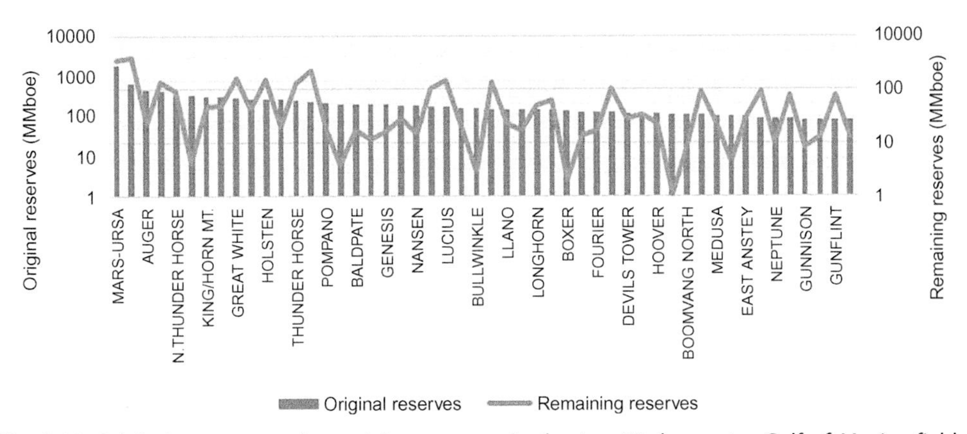

Fig. 5.27 Original reserves and remaining reserves in the top 50 deepwater Gulf of Mexico fields. *(Source: BOEM 2017.)*

Twenty of the top 25 fields and 36 of the top 50 fields in the GoM reside in shallow water, and almost all of these shallow-water fields have only a small percentage of their reserves remaining, collectively about 200 MMboe or about 20% of total shallow-water reserves. All of the shallow-water fields except Eugene Island 330 were discovered in the 1950s and 1960s and have produced for over 50 years. Most of the remaining reserves in shallow-water fields represent <5% of original reserves and for the top 50 fields average 2.7% (2.2% standard deviation) indicating their mature status.

The deepwater Mars-Ursa field is the largest deposit in the GoM at 1.5 Bbbl of oil and 2.1 Tcf of natural gas or 1.85 Bboe, followed by Tahiti/Tonga (661 MMboe), Auger (432 MMboe), and Atlantis (412 MMboe). The Eugene Island 330 field is the largest field in shallow water and the second largest field in the GoM at 799 MMboe, followed by shallow-water fields at West Delta 30 (797 MMboe), Grand Isle 43 (711 MMboe), and Tiger Shoal (711 MMboe). All of these fields are classified as giants since they have produced >500 MMboe, the traditional cutoff used to identify the largest fields.

5.4.8 Field-Size Distribution Shift

When a play is separated into earlier and later groups of field discoveries, the daughter distributions shift toward smaller field size during the life of the parent play. This tendency is illustrated by separating shallow-water and deepwater field discoveries into three time periods. In shallow water, the periods employed are 1940–69, 1970–99, and 2000–17 (Fig. 5.28). In deepwater, the periods employed are 1970–89, 1990–2009, and 2010–17 (Fig. 5.29). No distinction is made between the principle plays in either water depth category.

The overall reduction in field sizes in shallow water is significant (Table 5.9). Median field sizes drop from 51 MMboe (1940–69), to 6.6 MMboe (1970–99), to 1.2 MMboe (2000–17), and mean field sizes drop from 118 MMboe (1940–69), to 20.3 MMboe

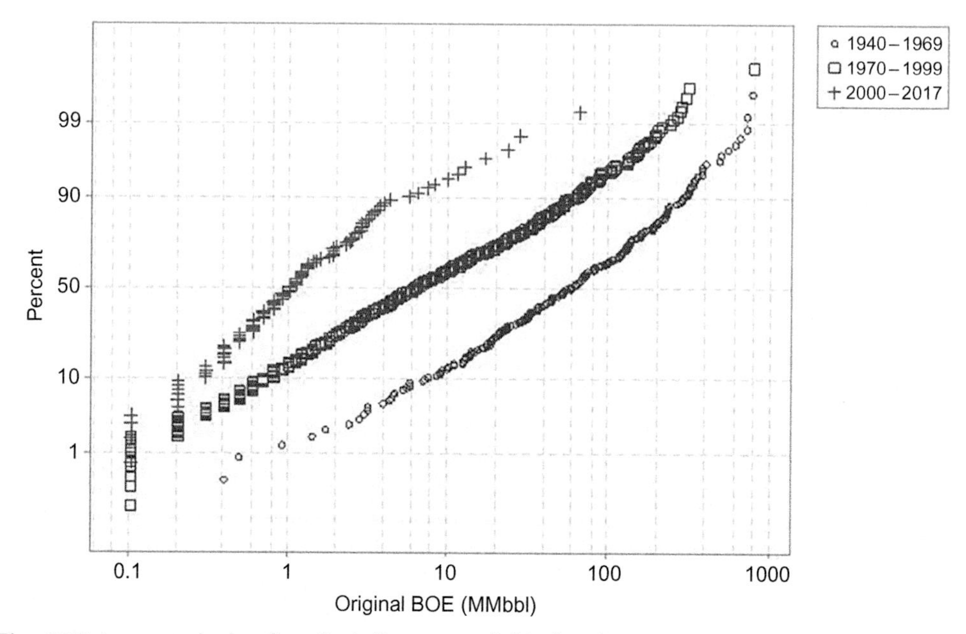

Fig. 5.28 Lognormal plot for all shallow-water fields by discovery period circa 2015. *(Source: BOEM 2017.)*

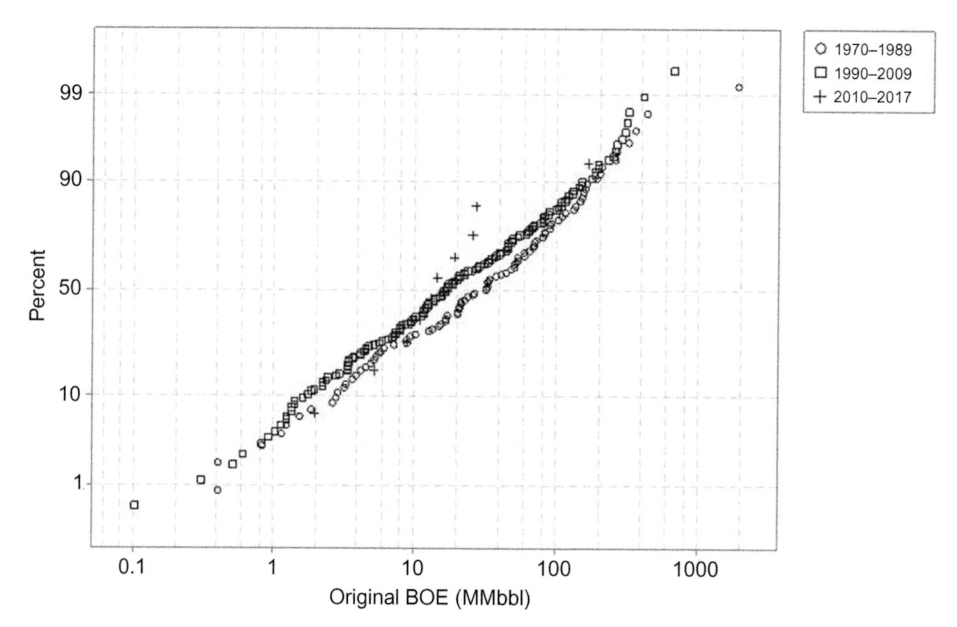

Fig. 5.29 Lognormal plot for all deepwater fields by discovery period circa 2015. *(Source: BOEM 2017.)*

Table 5.9 Shift of Field-Size Distribution in the Shallow-Water Gulf of Mexico (MMboe)

Statistic	1940–69	1970–99	2000–17
P99%	1416	346	20
P90%	318	58	5.6
P50%	51	6.6	1.2
P10%	8.1	0.8	0.2
P1%	1.8	0.13	0.07
MSw	118	20.3	2.2

Note: Swanson's mean (MSw) is calculated from the formula MSw = 0.3 (P90%) + 0.4 (P50%) + 0.3 (P10%).
Source: BOEM 2017.

(1970–99), to 2.2 MMboe (2000–17). Mean field declines are about an order of magnitude between each period. Commercial limits are truncated at about 100,000 boe. Large fields represented by P90% have declined from 318 MMboe (1940–69), to 58 MMboe (1970–99), to 5.6 MMboe (2000–17).

In deepwater, there is less distinction in field-size distributions over time due in part to the impact of different plays and development strategies that may confound the statistics (Table 5.10). At the low end, field reserves grew slightly larger from 2010 to 2017 compared with the previous two periods, but this might be due to sample bias since the length of the latest time period is less than half of these periods. Conversely, if the reserves size increase is statistically meaningful, it would indicate that marginal developments are now being economically produced in larger numbers. At the high end, declines are observed. Median field sizes dropped from 25 MMboe (1970–89), to 17 MMboe (1990–2009), to 14 MMboe (2010–17). Mean field sizes fell from 75 MMboe (1970–89), to 52 MMboe (1990–2009), to 26 MMboe (2010–17), not as dramatic as in shallow water but still significant. P90% field sizes dropped from 214 MMboe (1970–89), to 149 MMboe (1990–2009), to 65 MMboe (2010–17).

Table 5.10 Shift of Field-Size Distribution in the Deepwater Gulf of Mexico (MMboe)

Statistic	1968–89	1990–2009	2010–17
P99%	1223	874	219
P90%	214	149	65
P50%	25	17	14
P10%	3.0	2.0	3.2
P1%	0.5	0.3	0.9
MSw	75.2	52.4	26.3

Note: Swanson's mean (MSw) is calculated from the formula MSw = 0.3 (P90%) + 0.4 (P50%) + 0.3 (P10%).
Source: BOEM 2017.

5.5 RESERVES GROWTH

Reserves are created by capital, and therefore, it is not surprising that additional capital spending will typically lead to additional reserves or what is sometimes referred to as reserves growth (aka reserves appreciation or ultimate recovery appreciation). The topic is old and important, and the US Geologic Survey considers analysis of reserves growth as one of the most significant research problems in the field of resource assessment. Research has been conducted since the 1960s to better understand the phenomenon, beginning with work by Arrington (1960), Attanasi and Root (1994), and up through Cook (2013) and others. For a more complete review of the issues and literature, Morehouse (1997) and Cook (2013) are good starting points.

Many factors impact reserves estimates, and these same factors along with others also impact reserves growth calculations. Disentangling the various factors is difficult or impossible given the level of uncertainty and influence of unobservable and difficult-to-quantify factors such as capital spending, technology progress, improved knowledge, development strategies, and user preferences. These factors will always constrain the ability to predict ultimate recovery at the start of production. Only at the end of production are field reserves known with certainty.

Classification and estimation of reserves follow standard industry procedures (PRMS, 2007). Proved reserves (also called P1) are associated with a 90% level of confidence that the volumes specified will be produced and, therefore, represent a (very) high level of certainty that the quantities stated exist. Proved reserves are "much more likely than not" to be recovered.

Probable reserves (also called P2) are associated with a 50% confidence level of recovery, and therefore, the probability that the volumes of probable reserves will be achieved is associated with a flip of a coin. Probable reserves are "as likely as not" to be produced and lack for various reasons the certainty to be classified as proved. The capital necessary to develop and produce probable reserves has not been spent and, therefore, is neither as valuable or certain as proved developed reserves. The combination of proved and probable reserves is denoted as 2P (= P1 + P2) and is often used by management in their investment decision-making and by regulators (including BOEM) in their assessment of a field's ultimate recovery.

Using P1 or 2P at first production will therefore lead to reserves growth when measured at the end of the life of the field since it is "built into" the definition of the metric, obviously more for P1 than for 2P, depending on geologic conditions, capital expenditures, and other factors. Reserves estimates may also decline, and if 2P reserves are initially applied, this should happen about half the time. Researchers have attempted to explain and quantify these changes using models of various complexity but rarely (never) apply the most important factors in evaluation, namely, capital expenditures or their proxy wells.

Circa 2016, there were 815 fields that ceased production, 772 gas fields and 42 oil fields, 92 in deepwater and 722 in shallow water. Of the 815 fields, 381 were discovered before 1985, 332 were discovered between 1986 and 2000, and 101 were discovered after 2000. Using only fields that no longer produce, reserves growth is computed as the percentage difference between last year reserves and first producing year reserves estimate.

Gas fields realize larger reserves growth than oil fields, 91 versus 31% on average, with median values of 4 and −23%, respectively (Fig. 5.30). Shallow-water fields realized a larger increase in reserves compared with deepwater fields, with average values of 163 and 70% and median value of 6.4 and −16%, respectively (Fig. 5.31). Fields discovered earlier show a larger average growth in reserves compared with fields discovered at a later date, and median values reflect the same trend. For fields discovered before 1985, reserves growth was 143% on average (32% median), compared with 43% (−14% median) for fields discovered between 1986 and 2000 and 31% (−6% median) for fields discovered after 2000 (Fig. 5.32).

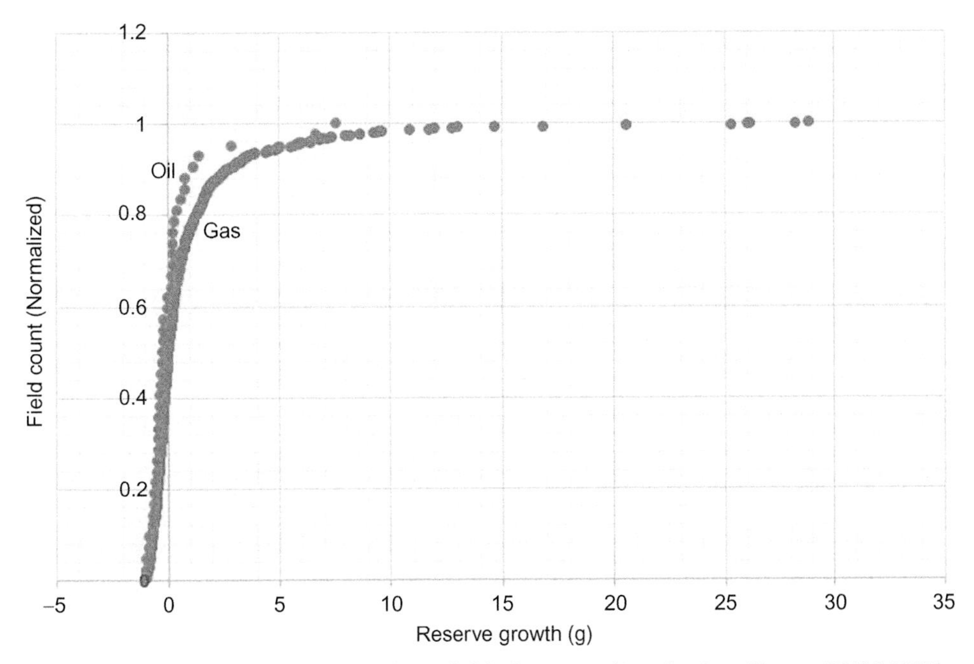

Fig. 5.30 Reserves growth for all oil and gas fields that ceased production. *(Source: BOEM 2017.)*

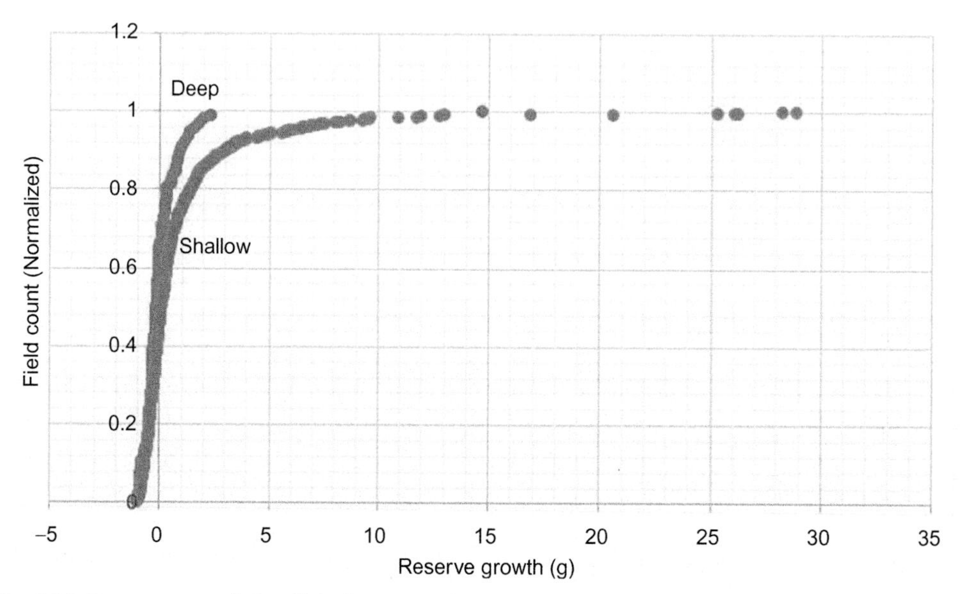

Fig. 5.31 Reserves growth for all shallow-water and deepwater fields that ceased production. *(Source: BOEM 2017.)*

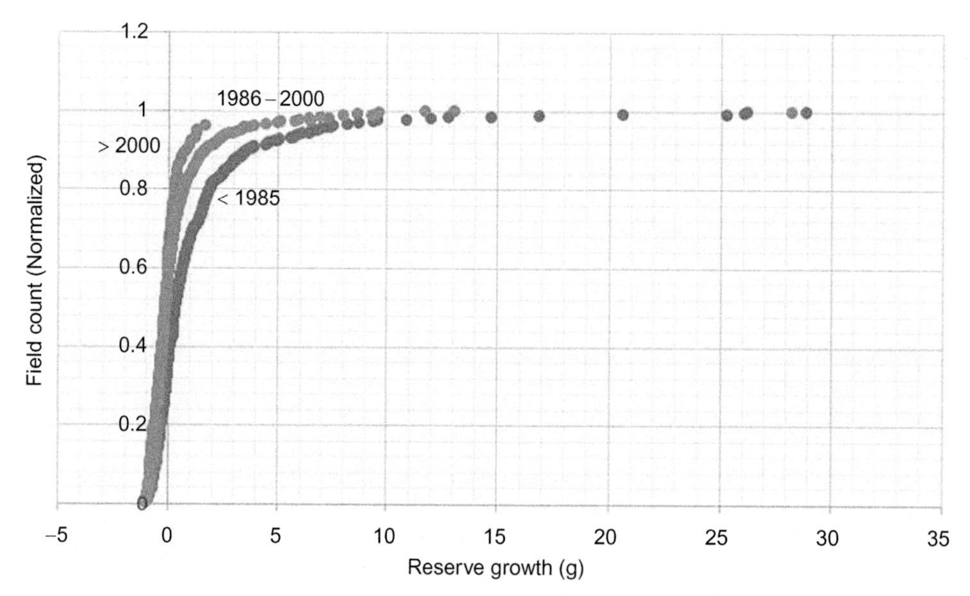

Fig. 5.32 Reserves growth for all oil and gas fields that ceased production. *(Source: BOEM 2017.)*

5.6 UNDISCOVERED RESOURCES

Every five years, BOEM geoscientists and engineers from the Alaska, GoM, and Pacific regions jointly develop a national assessment of undiscovered oil and gas resources on the 26 OCS planning areas of the United States. Each region creates a regional assessment of undiscovered technically and economically recoverable resources located outside of known oil and gas fields.

BOEM utilizes a play-based approach and models that simulate oil and gas accumulations based on user-defined geologic parameters and then aggregates those accumulations to estimate undiscovered hydrocarbon resources per play. Each reservoir in a BOEM-designated field is assigned to a distinctive play and is aggregated to the sand level, and each sand is aggregated to the pool level. Plays are aggregated into assessment units according to geography and rock age, and a probabilistic approach is applied with economic thresholds in evaluation.

Simulations are based on user-defined assumed oil and gas recovery factors, average acreage of a pool, and many other factors. These factors along with exploration probabilities are used to generate a distribution of various accumulation sizes expected to be found within a geologic play. The simulated accumulations are ranked and aggregated to estimate undiscovered technically recoverable resources (UTRR). Using the estimates of geologic resources, engineering and economic parameters are used to generate a value of resources that can be developed in a given price environment, referred to as undiscovered economically recoverable resources (UERR).

The values for both UTRR and UERR are aggregated from different plays within an area to obtain resource estimates for planning areas, OCS regions, and the entire US OCS. In the 2016 assessment, BOEM estimated a mean undiscovered resource that includes 90.6 Bbbl of oil and 328 Tcf of gas across the US OCS (US DOI, 2017). Of this total, about 49 Bbbl oil and 142 Tcf natural gas are located in the GoM OCS.

UTRR in the western GoM in <200 m (656 ft) water depth is estimated at 95% probability to be at least 0.51 Bbbl of oil and 15.8 Tcf of natural gas or 3.33 Bboe (Table 5.11).

Table 5.11 Undiscovered Technically Recoverable Resources in the Gulf of Mexico Circa 2015

	Oil (Bbbl)			Gas (Tcf)			BOE (Bboe)		
Planning Area	**95%**	**Mean**	**5%**	**95%**	**Mean**	**5%**	**95%**	**Mean**	**5%**
Western GoM (< 200 m)	0.51	0.75	0.98	15.8	24.9	35.7	3.3	5.2	7.3
Central GoM (< 200 m)	1.04	1.36	1.71	23.6	36.0	52.5	5.2	7.8	11.0
Total GoM	39.5	48.5	58.5	124	141.8	159.6	61.5	73.7	86.9

Source: US DOI, 2017.

The mean estimate of UTRR oil and gas in the western GoM is 5.2 Bboe, and the 5% probability upper estimate of UTRR is 7.28 Bboe. In the central GoM in <200 m (656 ft) water depth, the low, mean, and high estimates for UTRR resources are 5.2, 7.8, and 11.0 Bboe, respectively.

Gulf-wide, across all water depths and planning areas, including the eastern GoM and Straits of Florida, the low, mean, and high estimates for undiscovered technically and economically recoverable resources are 61.5, 73.7, and 86.9 Bboe, respectively.

REFERENCES

Allen, P.A., Allen, J.R., 2005. Basin Analysis. Blackwell Publishing, Malden, MA.

Arrington, J.R., 1960. Predicting the size of crude reserves is key to evaluating exploration programs. Oil Gas J. 58 (9), 130–134.

Attanasi, E.D., Root, D.H., 1994. The enigma of oil and gas field growth. AAPG Bull. 78 (3), 321–332.

Atwater, G., 1959. Geology and petroleum development of the continental shelf of the Gulf of Mexico. Gulf Coast Assoc. Geol. Soc. Trans. 9, 131–145.

Bilinski, P.W., Crabtree, P.T., Crump, J.G., McGee, D.T., Watson, R.C., 1992. Auger field subsurface overview: future site of world-record deepwater tension leg platform. In: SPE 24701. SPE Annual Technical Conference and Exhibition. Washington, DC, Oct 4–7.

Bourgeois, T.M., 1994. Auger tension leg platform: conquering the deepwater Gulf of Mexico. In: SPE 28680. SPE International Petroleum Conference and Exhibition of Mexico. Veracruz, Mexico, Oct. 10–13.

Branson, R.B., 1986. Productive trends and production history, South Louisiana and adjacent offshore. Gulf Coast Assoc. Geol. Soci. Trans. 36, 61–70.

Bureau of Ocean Energy Management, 1996. Field Naming Handbook. New Orleans, LA.

Bureau of Ocean Energy Management, 2017a. OCS operations field directory. Quart. Rep. December 31, 2017.

Bureau of Ocean Energy Management, 2017b. Reserves history for fields. Gulf of Mexico Outer Continental Shelf. December 31, 2016.

Burgess, G.L., Kazanis, E.G., Shepard, N.K., 2016. Estimated Oil and Gas Reserves, Gulf of Mexico OCS Region, December 31, 2015. U.S. Department of Interior, Bureau of Ocean Energy Management, New Orleans, LA OCS Report BOEM 2016-024.

Burgess, G.L., Kazanis, E.G., Cross, K.K., 2018. Estimated Oil and Gas Reserves, Gulf of Mexico OCS Region, December 31, 2016. U.S. Department of Interior, Bureau of Ocean Energy Management, Gulf of Mexico OCS Region, New Orleans, LA OCS Report BOEM 2018-034.

Close, F., McCavitt, B., Smith, B., 2008. Deepwater Gulf of Mexico development challenges overview. SPE 113011. In: SPE North Africa Technical Conference and Exhibition. Marrakech, Morocco. March 12–14.

Cook, T.A., 2013. Reserves growth of oil and gas fields – Investigation and applications. U.S. Geological Scientific Investigations Report 2013-5063, Denver, CO.

D'Souza, R., Basu, S., 2011. Field development planning and floating platform concept selection for global deepwater developments. In: OTC 21583. Offshore Technology Conference. Houston, TX, May 2–5.

Dekker, M., Reid, D., 2014. Deepwater development strategy. In: OTC 25135. Offshore Technology Conference. Houston, TX, May 5–8.

Dribus, J.R., Jackson, M.P.A., Kapoor, J., Smith, M.F., 2008. The prize beneath the salt. Oilfield Rev. Autumn, 4–17.

Galloway, W.E., 2008. Depositional evolution of the Gulf of Mexico sedimentary basin. In: Miall, A.D. (Ed.), The Sedimentary Basins of the United States and Canada: Sedimentary Basins of the World. Elsevier, pp. 505–549.

Galloway, W.E., 2009. Gulf of Mexico. GeoExPro 6 (3), 4–13.

Kaiser, M.J., Narra, S., 2018. A retrospective of oil and gas field development in the U.S. Outer Continental Shelf Gulf of Mexico, 1947–2017. Natural Resources Research, pp. 1–31.
Lach, J.R., Longmuir, G., 2010. Can IOR in deepwater Gulf of Mexico secure energy for America? In: OTC 20678. Offshore Technology Conference. Houston, TX, May 3–6.
Levorson, A.I., 1967. Geology of Petroleum. WH Freeman & Co, San Francisco, CA.
Massad, A.H., Pela, E.C., 1956. Facilities used in gathering and handling hydrocarbon production in the Eugene Island block 126 field Louisiana Gulf of Mexico. In: SPE 675-G. In SPE AIME Annual Fall Meeting of AIME. Los Angeles, CA, Oct. 14–17.
Morehouse, D.F., 1997. The Intricate Puzzle of Oil and Gas 'Reserves Growth'. Energy Information Administration Natural Gas Monthly, July.
Nixon, L., Kazanis, E., Alonso, S., 2016. Deepwater Gulf of Mexico December 31, 2014. U.S. Department of Interior, Bureau of Ocean Energy Management, Gulf of Mexico OCS Region, New Orleans, LA OCS Report BOEM 2016-057.
Oletu, J., Prasad, U., Ghadimipour, A., Sakowski, S., Vassilellis, K., Raham, B., Park, N., Li, L., 2013. Gulf of Mexico Wilcox play property trend review. In: OTC 24186. Offshore Technology Conference. Houston, TX, May 6–9.
Owen, E.W., 1975. Trek of the Oil Finders: A History of Exploration for Petroleum. AAPG, Tulsa, OK.
Petroleum Resources Management System, 2007. Society of Petroleum Engineers.
Rose, P.R., 2001. Risk Analysis and Management of Petroleum Exploration Ventures. American Association of Petroleum Geologists, Tulsa, OK.
Salvador, A., 1987. Late Triassic-Jurassic paleogeography and origin of Gulf of Mexico basin. AAPG Bull. 7 (4), 419–451.
Salvador, A., 1991. Origin and development of the Gulf of Mexico basin. In: Salvador, A. (Ed.), The Gulf of Mexico Basin: The Geology of North America. Geological Society of America, Boulder, CO, pp. 389–444.
U.S. Department of the Interior, Bureau of Ocean Energy Management Gulf of Mexico OCS Region, Office of Resource Evaluation, 2017. Assessment of technically and economically recoverable hydrocarbon resources of the Gulf of Mexico Outer Continental Shelf as of January 1, 2014. OCS Report BOEM 2017-005.
Weimer, P., Bouroullec, R., Adson, J., Crossey, S.P.J., 2017. An overview of the petroleum systems of northern deepwater Gulf of Mexico. AAPG Bull. 101 (7), 941–993.

PART TWO

Well Trends and Structure Inventory

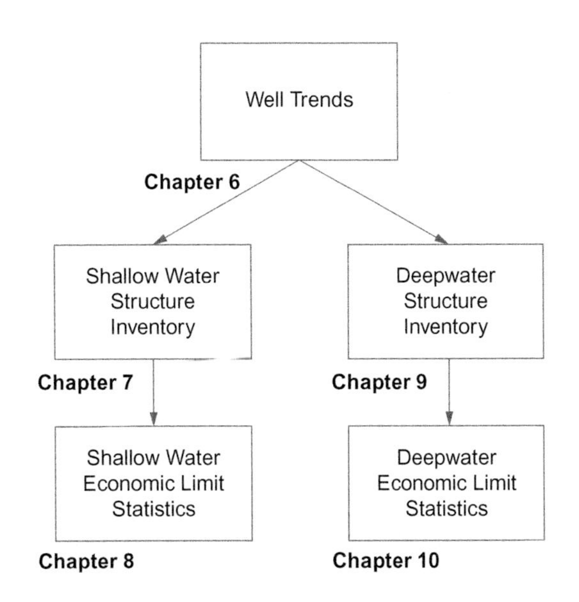

CHAPTER SIX

Well Trends

Contents

Abstract

Wells are a central feature in oil and gas development, and activity levels provide important information on trends in capital spending, the efficiency of recovery, and prospectivity. In this chapter, GoM well inventories and trends are described for exploration and development wells, abandoned wells, producing and idle wells, and subsea completions.

6.1 WELL TYPE

Exploration and development wells are used to find commercial accumulations of hydrocarbons and develop them. Exploration wells are drilled outside known reservoirs, and therefore, operations almost always take place from a mobile offshore drilling unit. Exploration is part of discretionary capital spending, and activity fluctuates with changes in oil and gas prices, discoveries, geologic prospectivity, company budgets, new technology, and other factors. Exploratory wells are not all successful, and not all development wells produce. For conventional targets, exploration success might range from 10% to 20%, with development well failures 10% or less (Box 6.1).

The number of wells required to develop a field and the number of wells that can be drilled from a structure are primary factors in development. Small reservoirs in shallow water might contain two or three wells and be developed with a simple structure, while major fields will usually have many productive fault blocks and numerous pay sands and require dozens of wells and multiple drilling platforms to develop. Reservoir sands that are deep and compact will require a smaller number of wells than a thin reservoir that is spread over a large areal extent. If there are faults and isolated reservoirs, more wells will be required to reach these locations. Engineers determine how wide wells should be

https://doi.org/10.1016/B978-0-12-818113-3.00006-0

BOX 6.1 Seven Decades and Still Going Strong

Several useful reviews, historic perspectives, and government reports are available on exploration and drilling activity in the GoM, ranging from corporate surveys and activity analysis to technology trends and interviews, beginning in the late 1940s through the present day. Interesting tidbits and sidebars abound, and a small sampling of this work spanning the decades includes McGee (1949), Atwater (1959), Owen (1975), Attanasi and Drew (1984), Beu (1988), Priest (2008), Francis (2007), and National Commission (2011).

spaced without suffering any significant loss of reserves, and the depth, areal extent, and geologic complexity of the reservoirs will determine the number of structures required.

Development wells offshore are usually drilled directionally into the target from a template or central drilling cluster using a mobile offshore drilling rig, a platform rig, or both. Sidetracks on development wells are drilled to access a different area of the reservoir when the original wellbore ceases to produce unless the original wellbore was unsuccessful or problems were encountered during drilling. In deepwater, it is common to transform exploration/delineation wells into development wells because of the high cost of drilling.

A sidetrack is a new section of wellbore drilled from an existing well. It is important to distinguish sidetracks from original wellbores in well counts, because although they represent new wells, they do not incur the full cost of drilling and require much less footage. In mature fields, sidetrack drilling can slow production decline if successful in adding new reserves, but this is a niche play and risky for old wells in depleted reservoirs, appealing mostly to small-size and midsize operators because of its specialized nature and limited capital requirements. At the beginning of field development and through midlife, sidetrack drilling is common, but as a field gets older, the potential for reserves additions is more limited increasing the risk of the investment. The uncertainty of outcome—the well may not pay off the cost to sidetrack—constrains opportunities to add reserves.

6.2 WELLS SPUD

A total of 52,964 wells have been drilled in the GoM through 2017: 46,126 wells (87%) in water depth <400 ft and 6838 wells (13%) in water depth >400 ft (Table 6.1). Well counts include all wellbores drilled; exploration and development, including sidetracks; and a few dozen geologic and stratigraphic wells.

The number of wells spud in shallow water peaked during the 1976–85 period, when >1000 wells were drilled annually, most years exceeding 1200 wells, and reaching a maximum of 1321 wells in 1984 (Fig. 6.1). In the mid-1990s and 2000, a resurgence in drilling occurred, but in subsequent years, the number of shallow-water spuds has dropped

Table 6.1 Exploration and Development Wells by Water Depth Including Sidetracks Circa 2017

Water Depth (ft)	Exploration	Development	Total
< 400	14,978	31,148	46,126
> 400	3792	3046	6838
Total	18,770	34,196	52,964

Source: BOEM 2018.

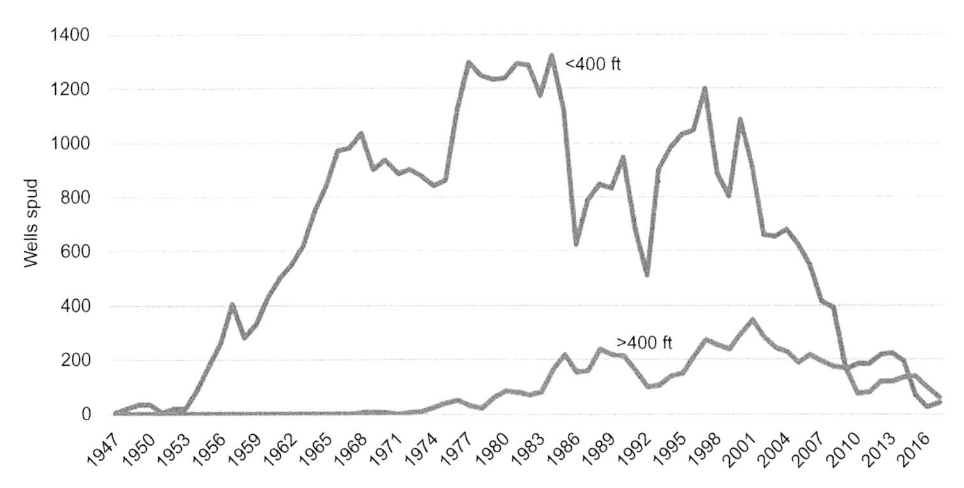

Fig. 6.1 Shallow-water and deepwater wells drilled in the Gulf of Mexico. *(Source: BOEM 2018.)*

significantly, hovering around 200 wells per year from 2009 to 2014 before declining to new lows in recent years. In 2014, 194 wells were spud, 70 in 2015, 26 in 2016, and 39 in 2017.

Deepwater drilling started in the late 1960s and by the mid-1980s reached 200 wells per year (Fig. 6.1). Deepwater drilling activity is much smaller than in shallow water and somewhat less volatile, generally ranging between 100 and 300 wells per year. In 2014, 135 deepwater wells were spud, 139 in 2015, 101 in 2016, and 60 in 2017. In the last 3 years, more deepwater wells have been spud than in shallow water.

6.3 EXPLORATION WELLS

From 1947 to 2017, there were 14,978 exploration wells drilled in shallow water and 3792 exploration wells drilled in deepwater (Table 6.1). Excluding sidetracks, shallow-water and deepwater exploration well counts totaled 12,895 and 2549 wells, respectively, meaning about one in five shallow-water exploratory wells compared with

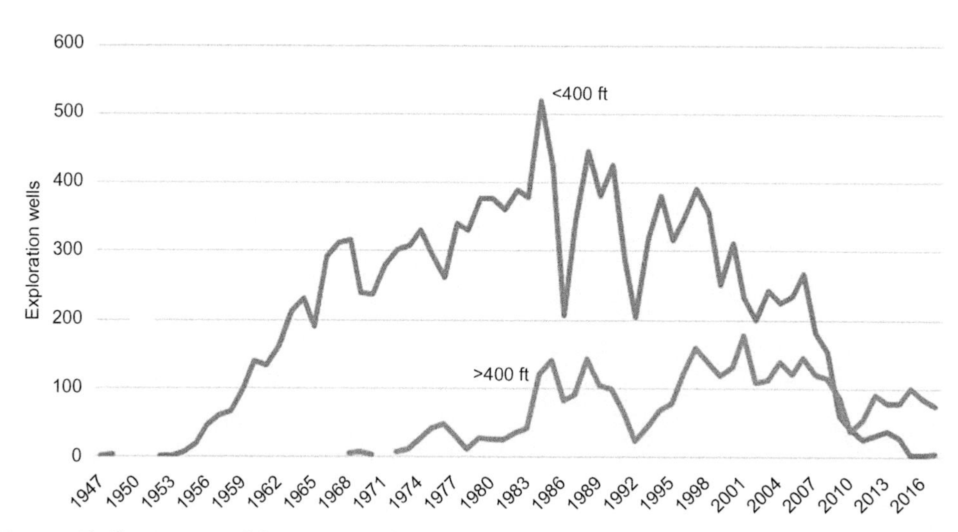

Fig. 6.2 Shallow-water and deepwater exploration wells drilled in the Gulf of Mexico. *(Source: BOEM 2018.)*

nearly half of all deepwater exploratory wells are sidetracked. In deepwater, more complex geology and expensive construction require operators to extract the maximum amount of information from the drilling campaign. Exploration wells are often sidetracked to test nearby features and delineate the reservoir if a discovery is made or to better understand the geology if drilling was unsuccessful.

Exploration activity rapidly expanded in the early years of GoM production, passing 100 wells per year in the late 1950s, 300 wells per year in the mid-1960s, and peaking around 500 wells per year in 1983 (Fig. 6.2). From the early 1960s to 2006, at least 200 shallow-water exploration wells were drilled annually, and in many years, activity levels were twice as high. High levels of volatility characterize exploration activity during this period. Since 2006, exploration well counts have been in steep decline and over the last few years have hit rock bottom. In 2008, 153 exploration wells were drilled followed by 60 wells in 2009 and less than 50 wells per year thereafter. A total of nine exploration wells were drilled from 2015 to 2017.

Deepwater exploratory drilling has never exceeded 200 wells per year and over the past two decades has ranged annually between 50 and 150 wells, contributing about half of drilling activity in the region. The number of deepwater exploration wells has fluctuated within a more narrow range than shallow water, hitting a low in 2010, the year of the Macondo oil spill and the deepwater drilling moratorium before bouncing back. From 2015 to 2017, 260 deepwater exploration wells were drilled, indicating strong interest and prospectivity in the region.

Example. Mississippi Canyon 582 (Medusa) wells and drilling curve

The Medusa field is located in Mississippi Canyon in 2200ft water depth over blocks MC 538 and 582 (Fig. 6.3). The T4b reservoir is one of three major productive intervals and occurs east of a salt weld that bisects the field (Lach et al., 2005). Faults often restrict flow and compartmentalize sands, and thus, reservoirs that are faulted and with poor communication will require more wellbores than better connected reservoir sands. Six dry tree wells A1 through A6 were completed in the initial development phase of Medusa targeting the three main reservoirs (Fig. 6.4) (Chhajlani et al., 2002).

The Medusa discovery well MC 582-1 reached total measured depth of 16,950ft (15,621ft true vertical depth) in September 1999, and four sidetracks were drilled (Fig. 6.5). Measured depth is measured along the wellbore, while TVD is measured straight down. MD is always greater than TVD except in a perfectly straight vertical well. Datums are usually selected at the waterline or the rig floor.

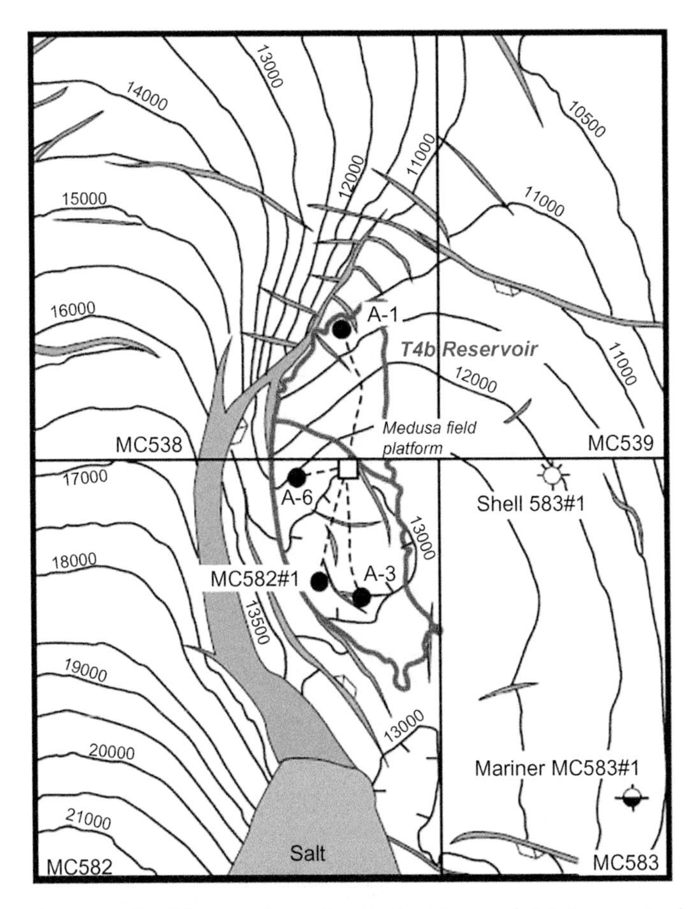

Fig. 6.3 Structural map of the T4b sand formation in the Medusa field. *(Source: Lach, J., McMillen, K., Archer, R., Holland, J., DePauw, R., Ludvigsen, B.E., 2005. Integration of geologic and dynamic models for history matching, Medusa field. In: SPE 95930. SPE Technical Conferences and Exhibition. Dallas, TX, Oct 9–12.)*

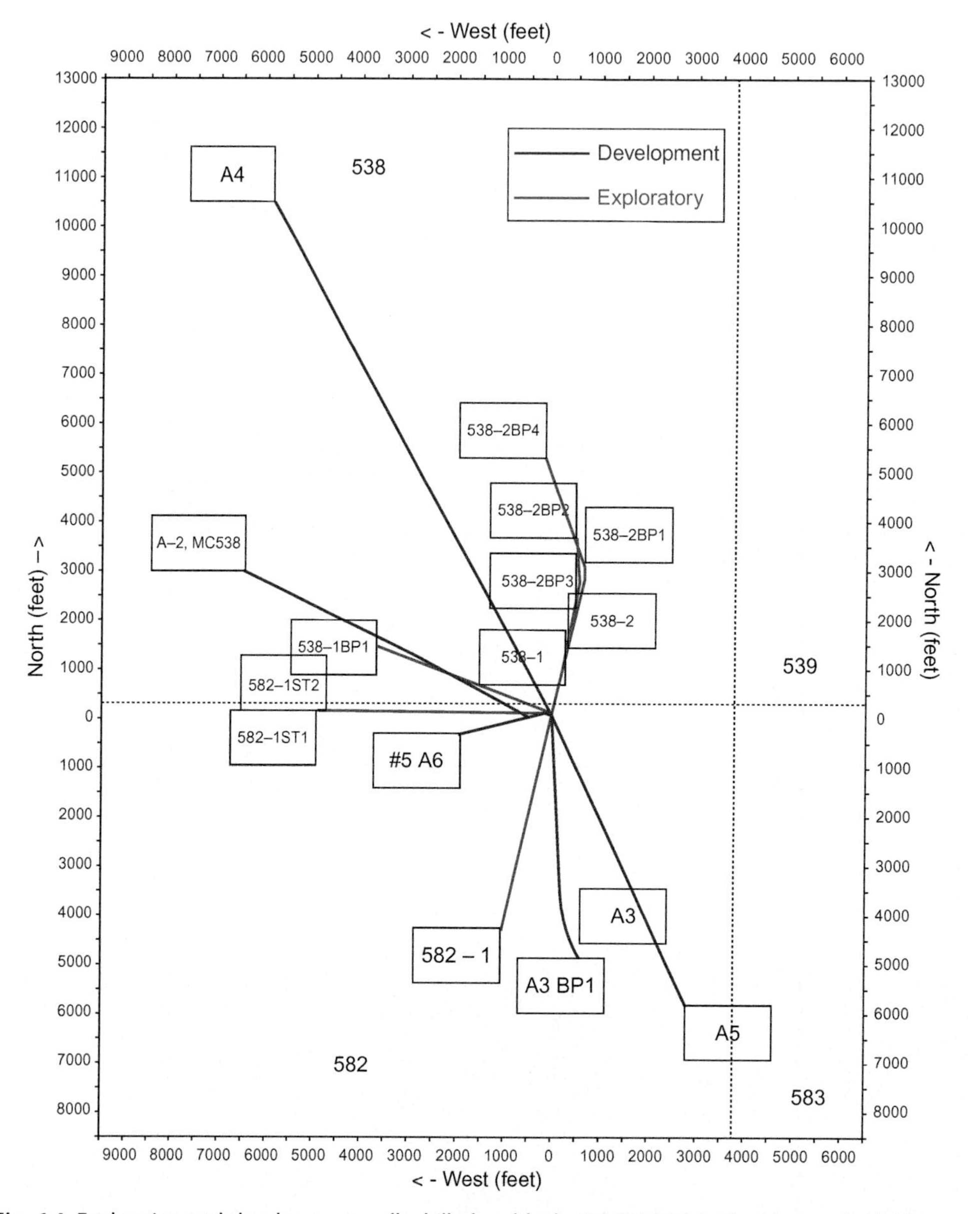

Fig. 6.4 Exploration and development wells drilled on blocks GC 538/582 in the Medusa field. *(Source: Chhajlani, R., Zheng, Z., Mayfield, D., MacArthur, B., 2002. Utilization of geomechanics for Medusa field development, deepwater Gulf of Mexico. In: SPE 77779. SPE Technical Conference and Exhibition. San Antonio, TX Sep 29–Oct 2.)*

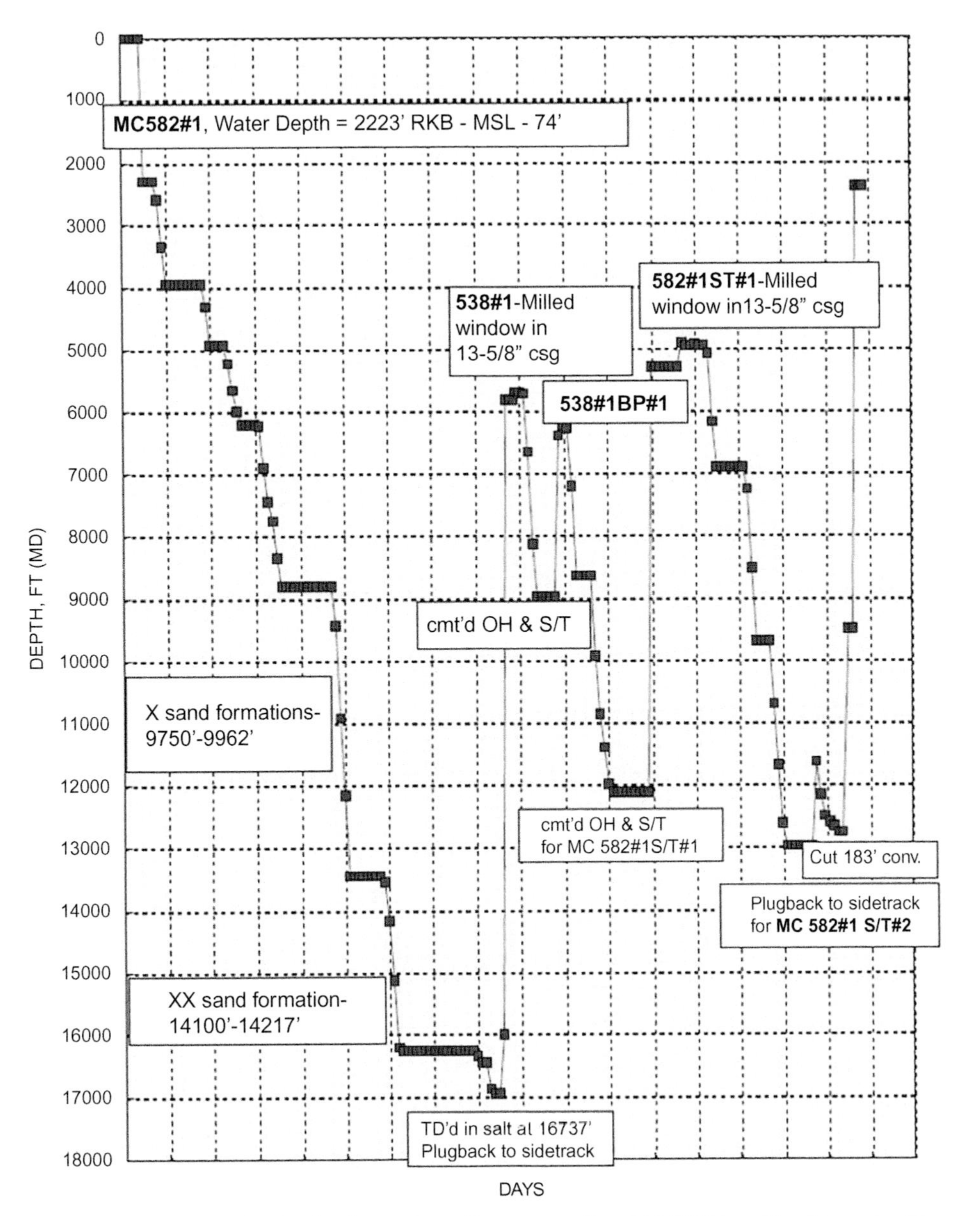

Fig. 6.5 Days versus depth plot for the Medusa discovery well on MC 582#1 and its sidetracks. *(Source: Chhajlani, R., Zheng, Z., Mayfield, D., MacArthur, B., 2002. Utilization of geomechanics for Medusa field development, deepwater Gulf of Mexico. In: SPE 77779. SPE Technical Conference and Exhibition. San Antonio, TX Sep 29–Oct 2.)*

Exploration activity in shallow water has dropped due to a combination of factors, including the lack of new discoveries, sustained low oil and gas prices, and the growth and success of onshore shale plays, especially unconventional shale gas. The lack of new discoveries is due in large part to the maturity of the region. Shallow waters in the western and central GoM have been extensively explored, and the probability of major new (large) discoveries in the region is low. Oil prospects continue to attract attention in and around existing oil fields and deeper in the Miocene section, but success rates and commercial finds over the last several decades have not been significant.

There has been no significant new plays or prospects announced by operators in shallow water in several years, and previous excitement for deep gas plays >15,000 ft TVD (e.g., Davy Jones) has not materialized because of high-cost complicated wells and disappointing production. Deepwater exploration continues to attract attention and capital spending because of continued discoveries and a perception of high regional prospectivity.

Example. Treasure Island play

McMoRan picked up the Blackbeard prospect in the Treasure Island play on the GoM shelf from Exxon along with Davy Jones and other acreage for an undisclosed amount. Exxon abandoned the well in 2006 after spending $200 million and reaching 30,067 ft. McMoRan deepened Blackbeard to 32,997 ft and found Miocene hydrocarbon bearing sands and then drilled six deep wells on lease blocks in South Marsh Island, Eugene Island blocks 26 and 223, South Timbalier 144, and Brazos to delineate the play.

Several problems were encountered when completing the wells including stuck pipe, gummed-up drilling fluids, and inability to flow test wells. Bottom-hole pressures of the wells approached 28,000 psi, and mechanical difficulties eventually prompted the company to plug the holes in 2015 and suspend work. A total of $1.2 billion was reportedly spent before abandoning the endeavor.

6.4 DEVELOPMENT WELLS

The number of wells required in field development depends upon the size and complexity and depth of the reservoir sands, number and location of fault blocks, desired production rates, well type, ownership positions, unitization agreements, development strategies, and other factors. The majority of development wells are usually drilled during the early stages of field development in the traditional spend-produce business model, but development drilling will also occur throughout a field's life as producing zones are plugged back and sidetracks drilled or additional phases of development occur. Phased developments are often the preferred strategy for complex reservoirs or where the operator wants to limit initial development costs.

Circa 2017, a total of 31,148 shallow-water and 3046 deepwater development wells with sidetracks were spud. Excluding sidetracks, well counts total 23,208 shallow-water and 1654 deepwater development wells. About one-third of shallow-water development wells and over 80% of deepwater development wells are sidetracked, about twice the percentages observed in exploration drilling. The increase in sidetracking in development is not surprising, since after a well is drilled, the opportunity to seek additional pockets of reserves at reduced cost is often a favorable business decision.

The number of development wells in shallow water is about twice the exploration well counts, whereas in deepwater, development wells still fall below exploration totals. There are several reasons for these differences. First, in deepwater exploration, successful wells are completed as producers more frequently than in shallow water. Second, wet wells are commonly employed in deepwater development and are usually not sidetracked to the extent of dry tree wells. Third, deepwater reservoirs usually have higher pressures and flow-rate potential than shallow-water reservoirs, and less wells will be used in development.

Development drilling follows successful exploration activity and in shallow water peaked in the mid-1970s and again at lower levels in the late 1990s (Fig. 6.6). From the mid-1960s to 2004, shallow-water development well counts ranged between 400 and 1000 wells per year with smaller levels of activity from 1986 to 1992 due in part to depressed oil prices during the period. In 1978, deepwater production began from the Bourbon and Cognac fields about a decade after the first exploration wells were drilled in the region.

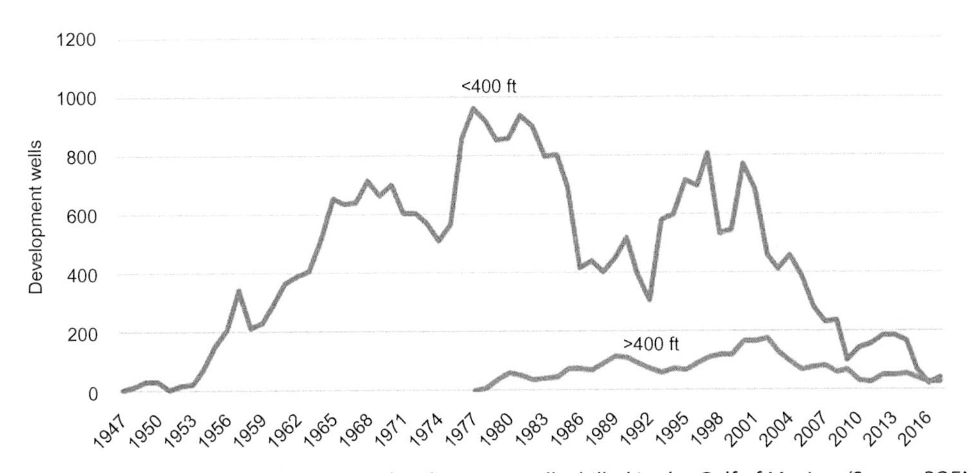

Fig. 6.6 Shallow-water and deepwater development wells drilled in the Gulf of Mexico. *(Source: BOEM 2018.)*

Trends in shallow-water development drilling reflect the general production decline in the region, and since 2008, less than 200 development wells per year have been drilled. The number of deepwater development wells has also declined during this period, but deepwater oil production, after production upsets arising from the 2008 hurricane season and Macondo oil spill in 2010, has actually increased over this time.

In 2000, there were 772 shallow-water and 168 deepwater development wells drilled. In 2017, 43 shallow-water and 26 deepwater development wells were drilled. From 2000 to 2017, 4995 shallow-water and 1414 deepwater development wells were drilled.

Example. Matterhorn

Matterhorn, that majestic and beautiful mountain in Europe created out of the collision of the European and African continental plates, also gives its namesake to a deepwater tension leg platform located in Mississippi Canyon block 243 in approximately 2850 ft water depth. Five principal pay sands were identified during exploration, and the initial development plan included drilling seven dry tree wells and a subsea water injection well approximately a mile from the platform (Hurel and van der Linden, 2004).

The wells were batched drilled to total depth using a semisubmersible drilling rig and temporarily abandoned after running and cementing production casing/liner (Fig. 6.7). All the wells were kicked off in the upper hole to reach the shallow upper intervals of the pay horizon. After releasing the drilling rig, a TLP was installed, and the wells were completed individually after running production risers (Forestier et al., 2004).

All of the wells were deviated with some requiring significant horizontal departure. Aggressive build rates were required for several wells. Batch drilling provided time savings by eliminating blowout preventer and riser trips, and the total time to complete the project was 303 days from the first well spudded until the last well was temporarily abandoned (Forestier et al., 2004).

6.5 ABANDONED WELLS

Through 2017, 52,964 wells have been drilled in the GoM and 27,405 wells have been permanently abandoned, leaving remaining well inventories circa 2017 at 25,559. Permanently abandoned (PA) wells represent the final state of a wellbore, while temporarily abandoned (TA) wells represent a transitory state for wells on their way to production or permanent abandonment.

There are many different pathways that wells may follow during their lifetime depending on why they were drilled, where they were drilled, how they were drilled, and whether the well was a keeper (successful) or not (Fig. 6.8). After wells are drilled, they are temporarily abandoned before structures are installed to protect the wellbore or subsea equipment is ready to handle production. Producing wells that cease production for a long period of time may also be temporarily abandoned if the operator expects to utilize them in the future (e.g., for a sidetrack) or if they are waiting to be permanently abandoned. Permanent abandonment is the final state of all wells. Whenever a sidetrack is drilled, it will create a new well and forward path.

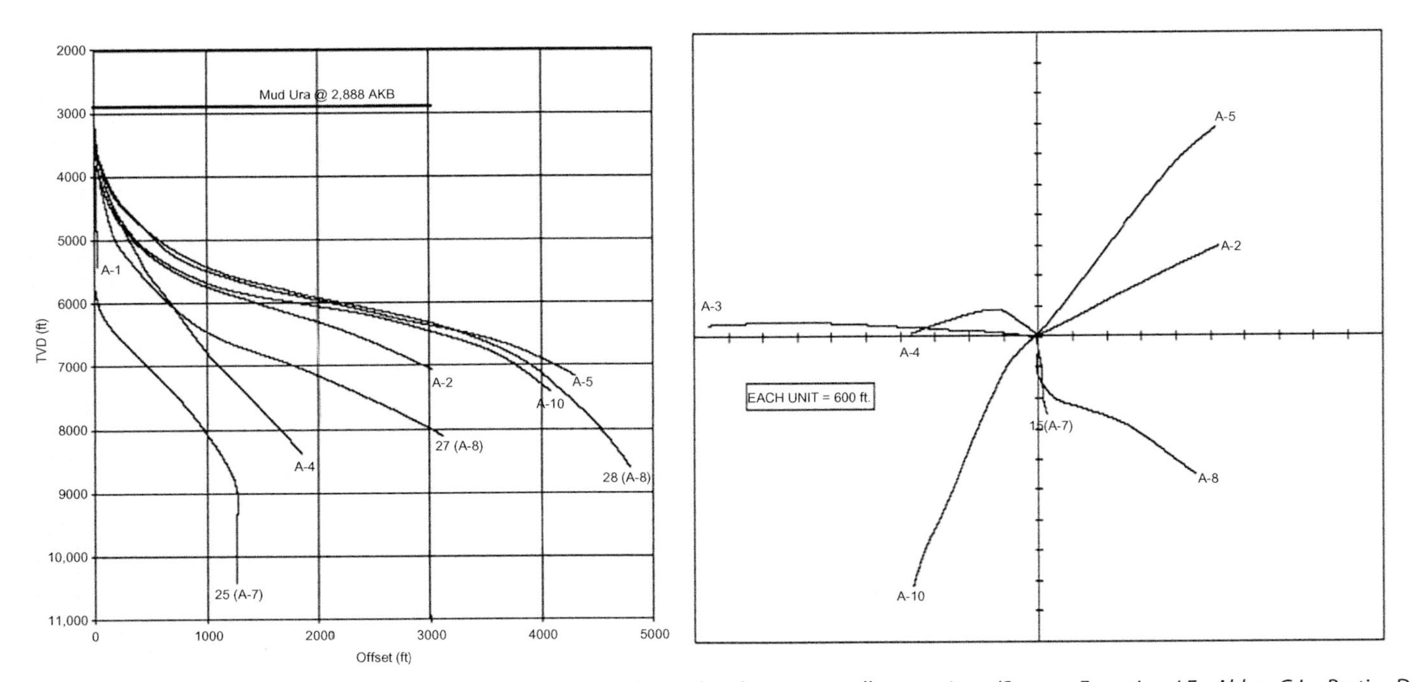

Fig. 6.7 Section view (left) and plan view (right) of the Matterhorn development well campaign. *(Source: Forestier, J.F., Ables, G.L., Bertin, D., Ardignac, D., Hall, C.R., Lucas, M.A., 2004. Matterhorn development drilling-innovation and planning equals deepwater directional success. In: OTC 16607. Offshore Technology Conference. Houston, TX, May 3–6.)*

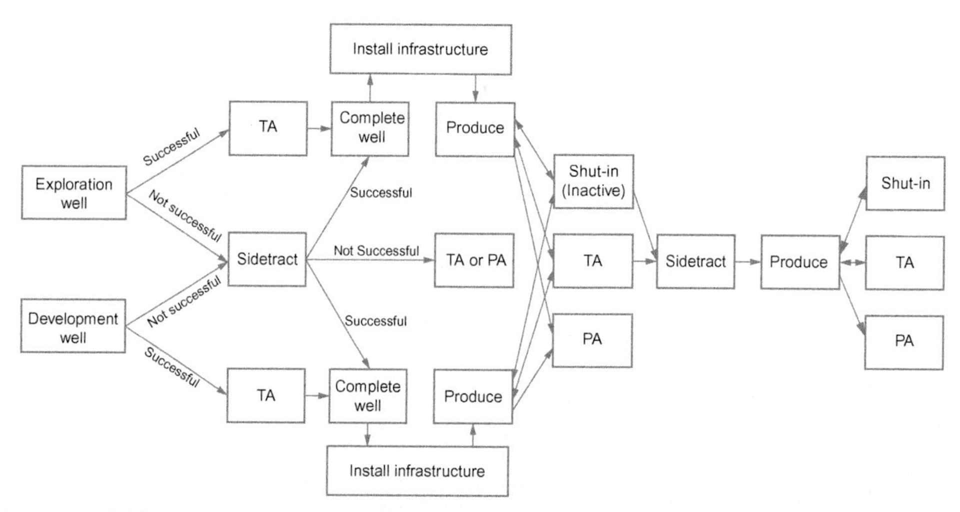

Fig. 6.8 Well life-cycle pathways from spud to abandonment.

Table 6.2 Gulf of Mexico Well Inventory Circa 2017

	< 400 ft	> 400 ft	Total
Drilled	46,243	6733	52,964
Permanently abandoned	25,254	2151	27,405
Remaining	20,989	4582	25,559
Producing	2644	819	3463

Source: BOEM 2018.

Wells that haven't produced for many years are required to be placed in TA status, essentially requiring a cement barrier or mechanical bridge plug or both. TA wells may remain inactive for many years, but all TA wells will eventually be classified as permanently abandoned after final operations are performed and conductors/risers are cut and removed. Under exceptional circumstances (e.g., hurricane destruction), TA wells may be considered the final status of a wellbore.

Permanent abandonment activity and PA wells are unambiguous in the sense that their status denotes the final end state of the well and once entered will not lead to another state unless found to be leaking, in which case the well will be reclassified as TA during remediation. TA activity and TA status wells are more elusive because of their different objectives and transitory nature. Temporarily abandoned wells are neither producing nor permanently abandoned but are on their way to one of these states. Whereas PA inventory can only increase with time, TA inventories can increase or decrease as status codes change and new stock added.

In water depth < 400 ft, there were 25,254 PA wells out of 46,243 wells drilled in the region, leaving 20,989 wells remaining to be permanently abandoned circa 2017 (Table 6.2). In water depth > 400 ft, there were 2151 PA wells out of 6733 wells

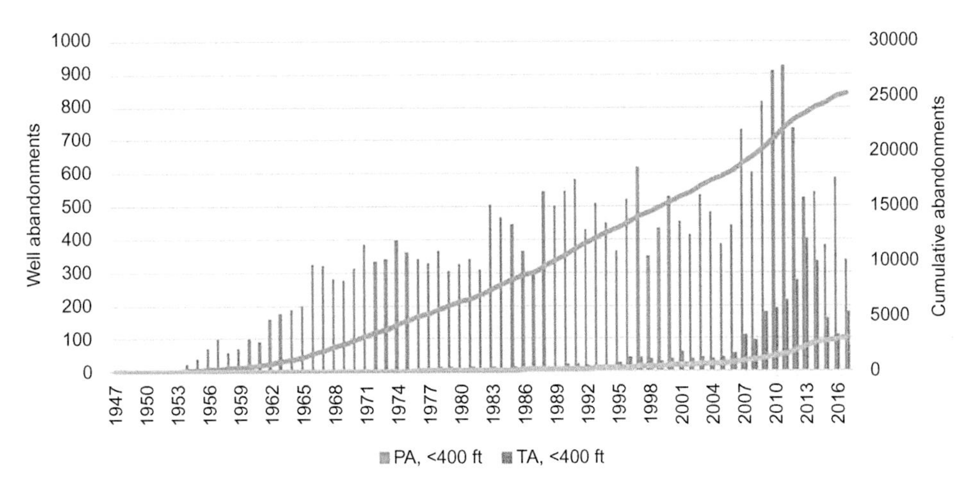

Fig. 6.9 Shallow-water permanent and temporary well abandonments in the Gulf of Mexico. *(Source: BOEM 2018.)*

spud, leaving a well inventory of 4582 deepwater wells remaining to be plugged circa 2017.

In shallow water, plug and activity began soon after the first wells were drilled in the region, either because they were not successful or because of production cessation. By the mid-1960s, activity reached 300 permanent well abandonments per year and has stayed above this level ever since (Fig. 6.9). In later years, larger numbers of well abandonments coincided with higher structure decommissioning activity, reaching 500 permanent abandonments in 1983 and 900 permanently abandoned wells in 2009 and 2010, historic highs that will never be seen again. Permanently abandoned wells and decommissioned structures are correlated (Box 6.2).

BOX 6.2 Abandoned Wells vs. Decommissioned Structures

Wells are abandoned throughout a property's life cycle for a variety of reasons, and it is not immediately obvious that well abandonments and structure decommissioning should be correlated. A closer examination of abandonment activity reveals that operators often perform major well abandonment work near the end of production to maximize development opportunities and delay expenditures, and these are the trends that are picked up with correlative analysis.

Near the end of a structure's life most wells are no longer producing and well abandonment commences. When the structure is decommissioned, TA wells are brought into PA status by cutting and removing the conductors, and all other wellbores are permanently abandoned. In some cases, wellbores may be plugged and the structure sits idle or serves some other purpose for several years before being decommissioned.

Continued

BOX 6.2 Abandoned Wells vs. Decommissioned Structures—cont'd

Before structures are decommissioned, all wells associated with the structure must be permanently abandoned, and all TA wells must be brought into PA status. Shut-in and producing wells will be permanently abandoned, and temporarily abandoned wells will complete isolation and testing requirements and have their conductors cut and removed. Historically, for every structure decommissioned, there have been about three wells permanently abandoned during the same year (Fig. 6.10).

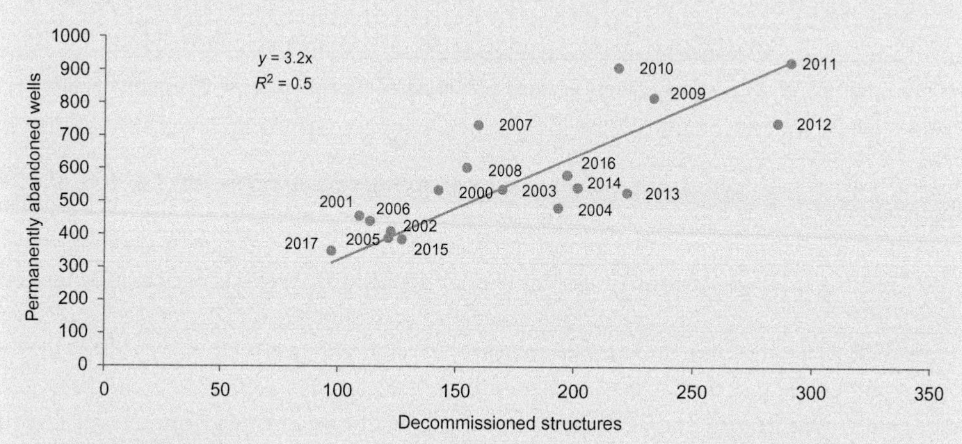

Fig. 6.10 Wells abandoned and structures decommissioned in the Gulf of Mexico, 2000–17. *(Source: BOEM 2018.)*

In deepwater, PA activity is on a much smaller scale because well inventories are smaller, younger, and more valuable. TA activity in deepwater is also higher on a relative basis because of development requirements and optionality associated with boreholes (Fig. 6.11). Circa 2017, about 25% of total deepwater abandonments are TA wells (566/2151), compared with about 12% (3063/25,254) in shallow water.

6.6 PRODUCING AND IDLE WELLS

Well production starts to decline almost immediately after they first produce, and when wells stop producing, they collect in inventory, sometimes returning to production via recompletion or a sidetrack (although strictly speaking a sidetrack is a new wellbore that is producing) and sometimes not returning to production.

Wells fail for any number of reasons and are shut in for remedial work one or more times during their lifetime. Wells will produce more and for longer periods with regular

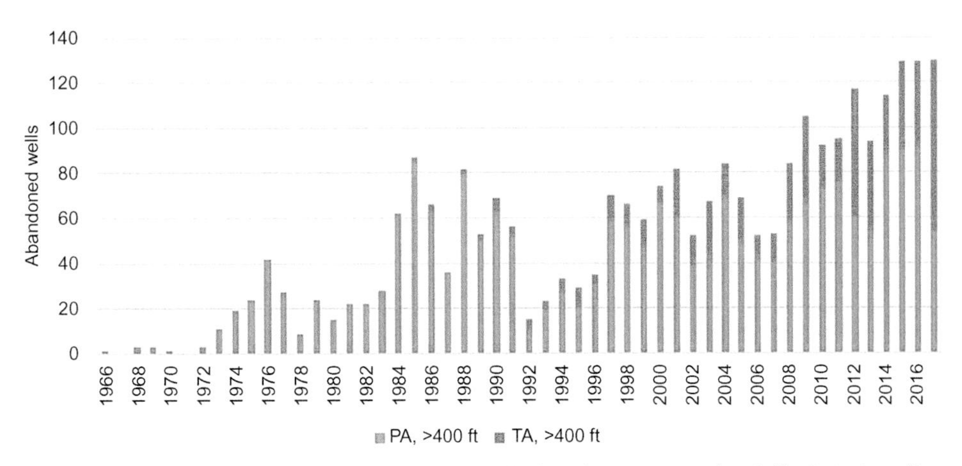

Fig. 6.11 Deepwater permanent and temporary well abandonments in the Gulf of Mexico. *(Source: BOEM 2018.)*

maintenance, which is a primary reason that subsea wells are abandoned at higher economic limits and tend to leave more reserves in the ground than dry tree wells. Some wells require more intervention than others, and the level of intervention is one factor contributing to production volatility. When wells remain inactive for a long period of time (i.e., several years), they have likely depleted their reservoirs or have mechanical problems that are not economically justified to remediate.

In 2017, there were 3463 wells in the GoM that produced hydrocarbons during the last 12 months, 2644 wells in water depth < 400 ft and 819 wells in water depth > 400 ft (Table 6.2). The number of producing wells varies with the size and age of the well inventory, the time of assessment, the number of wells completed during the year, the development status of projects, and, as with all other well categories, a variety of other mostly unobservable factors. The number of producing wells is small relative to the total active well inventory, a mere 13% (2644/20,989) in shallow water and 18% (819/4582) in deepwater, but this is not at all uncommon in offshore development.

The number of producing wells in shallow water has declined markedly over the last three decades (Fig. 6.12). In 1985, there were 7681 shallow-water producing wells, while circa 2017, there were 2644 producing wells. Decline spikes in the years 2004–05, and 2008 is attributed to hurricane activity and response. Every 4 years or so, 1000 wells have dropped out of production. In 1997, over 7000 wells were producing; in 2003, there were ~6000 producing wells; in 2007, there were ~5000 producing wells; in 2011, there were ~4000 producing wells; in 2015, there were ~3000 producing wells. If these dropout trends continue, which seems likely considering the age of producing wells and the lack of replacements, by 2019, one might expect ~2000 producing wells and by 2023 or so ~1000 producing wells in shallow water.

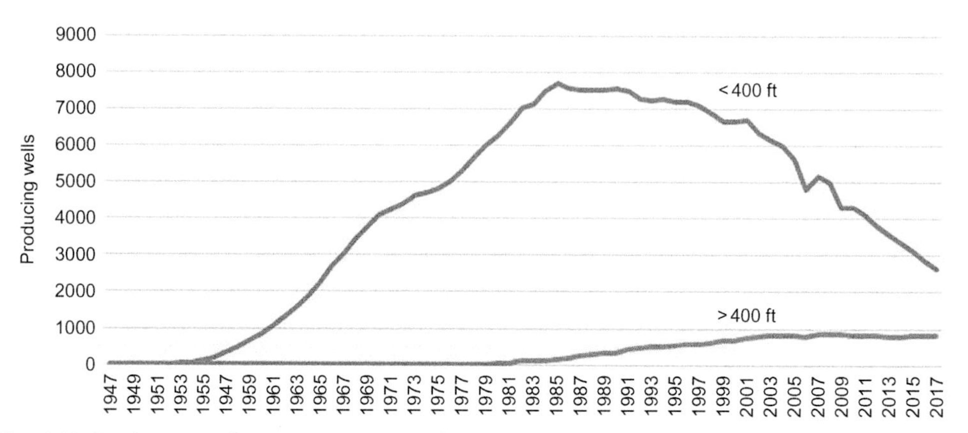

Fig. 6.12 Producing well inventories in the shallow-water and deepwater Gulf of Mexico. *(Source: BOEM 2018.)*

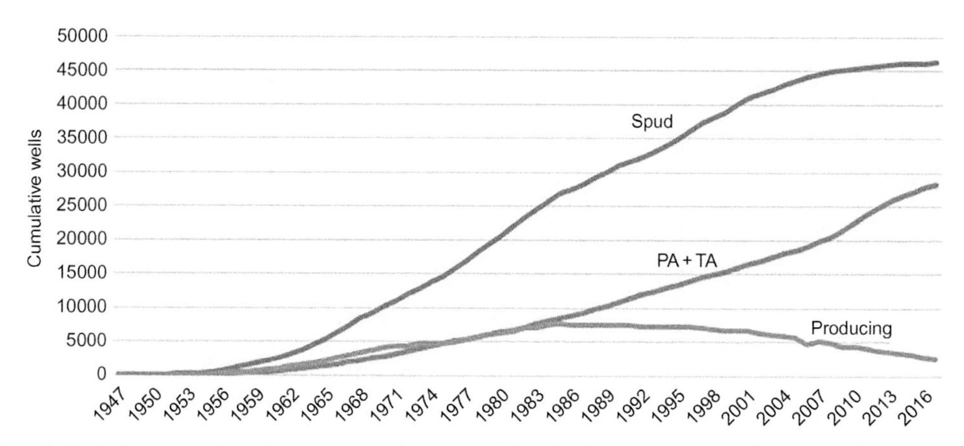

Fig. 6.13 Drilled, abandoned, and producing wells in the shallow-water Gulf of Mexico. *(Source: BOEM 2018.)*

Deepwater producing wells are on a completely different trend than shallow water with new wells replacing those that are shut in (Fig. 6.12). From 2002, there have been 800 or more producing deepwater wells most years, hitting a high in 2007 at 867 and numbering 819 in 2017.

In Figs. 6.13 and 6.14, running totals of wells drilled, abandoned, and producing are depicted for the shallow water and deepwater, respectively. In Fig. 6.15, the difference between cumulative drilled and abandoned wells yields active well inventories in the shallow-water and deepwater regions. Active shallow-water inventories peaked over 25,000 in 2004, while deepwater inventories started to level off at this time. Producing wells are a small subset of the active inventories.

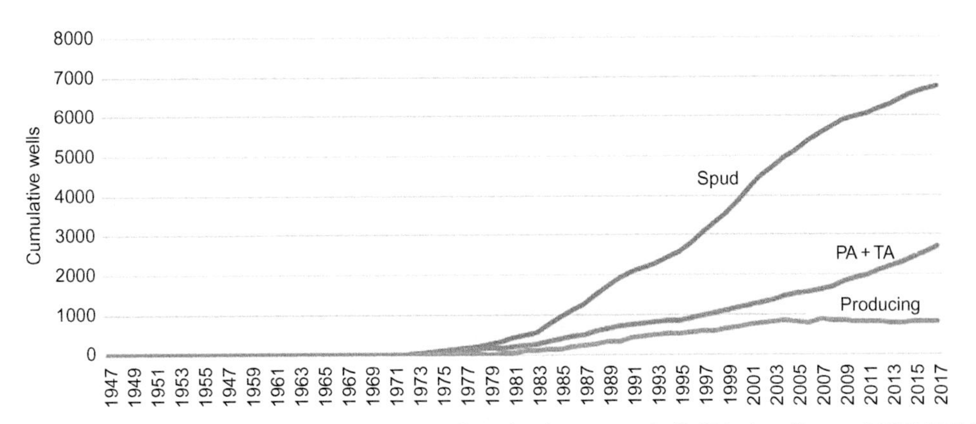

Fig. 6.14 Drilled, abandoned, and producing wells in the deepwater Gulf of Mexico. *(Source: BOEM 2018.)*

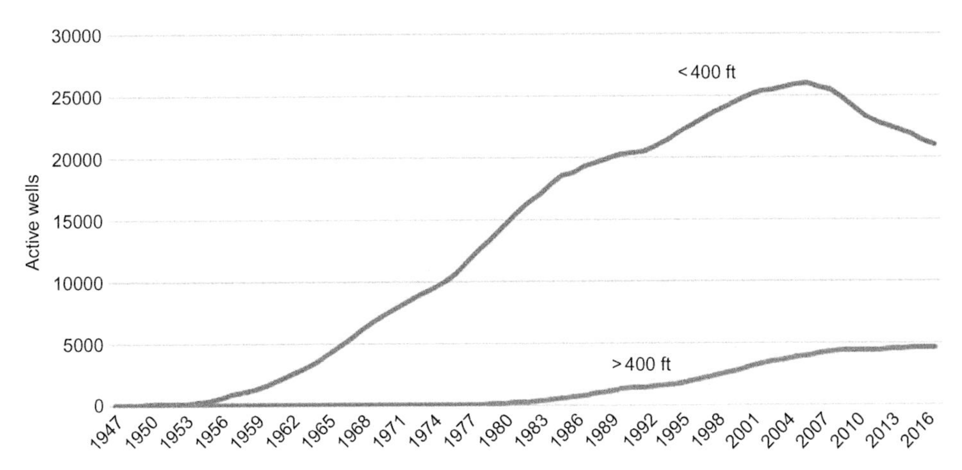

Fig. 6.15 Active well inventories in the shallow-water and deepwater Gulf of Mexico. *(Source: BOEM 2018.)*

6.7 SUBSEA COMPLETIONS

In subsea completions, the tree resides on the seafloor requiring mobilization of an intervention vessel or MODU whenever well work is required. In direct vertical access (DVA) wells, a subclass of wet well, the tree resides on the seafloor but is accessible from a rig on the platform. Subsea wells are identified separate from dry tree wells because they differ in several fundamental ways in terms of their capital expenditures and risks, operational requirements and expenses, recovery rates, and decommissioning liability.

Subsea wells are more expensive to equip, work over, operate, and decommission than dry tree and DVA wells, perhaps an order of magnitude or more, and are abandoned at a higher production rate, for all things equal, because of the back pressure that arises delivering the fluid to the host and less frequent interventions to maintain the well's productivity (see Chapter 10 for additional details). Subsea wells also have greater difficulty flowing with high water cuts because of hydrate formation, and if gas lift or subsea compression is used to flow to a lower abandonment pressure, flow assurance issues such as asphaltene deposition may result. The cost to drill and complete a subsea well, for all things equal, may be comparable with a dry tree well depending on how the wells are drilled and completed (e.g., with or without a platform rig, number of stages, number of casing strings, well complexity, market rates, etc.), but the equipment and construction costs and operational costs and risks of subsea systems are quite different.

The first subsea well in the GoM was drilled in 1958, and operators experimented with subsea production systems in shallow water through the mid-1970s to prove up the technology in anticipation of moving into deeper water (Burkhardt, 1973; Childers and Loth, 1976). The first deepwater subsea well in the GoM was drilled in 1988.

Subsea wells are generally not needed in shallow water because well protectors or fixed platforms can be employed for isolated small reservoirs and there is no advantage using wet wells. In deepwater, subsea wells are an important component in field development, and large numbers began to be drilled in the mid-1990s (Fig. 6.16).

To date, 1443 subsea wells have been drilled, 112 in shallow water and 1331 in deepwater (Table 6.3). Most shallow-water wet wells are old, and about two-thirds (73/112) of shallow-water wells have been permanently abandoned compared with about 15% (178/1331) for deepwater wells. Since 2008, only three shallow-water subsea wells have been drilled (Fig. 6.17) compared with almost 400 in deepwater (Fig. 6.18). Circa 2017, there were 383 producing subsea wells, eight in shallow water and 375 in deepwater.

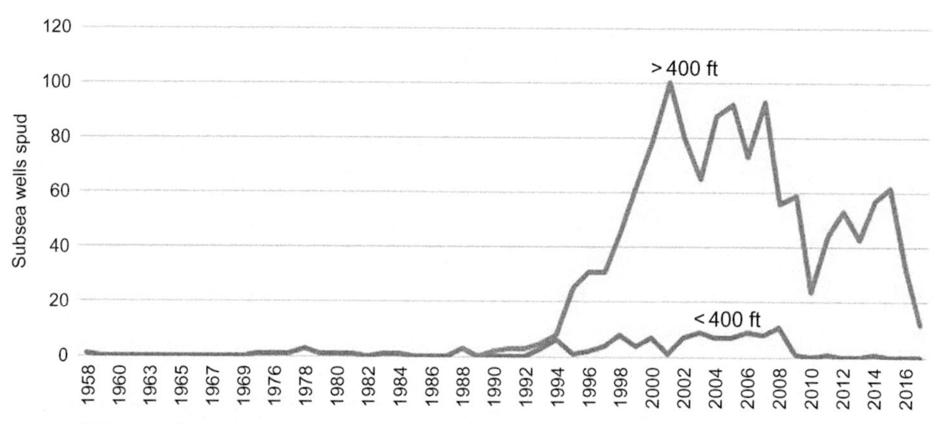

Fig. 6.16 Subsea wells spud in the shallow-water and deepwater Gulf of Mexico. *(Source: BOEM 2018.)*

Table 6.3 Subsea Well Inventory Circa 2017

	< 400 ft	> 400 ft	Total
Drilled	112	1331	1443
Permanently abandoned	73	178	251
Remaining	39	1153	1192
Producing	8	375	383

Source: BOEM 2018.

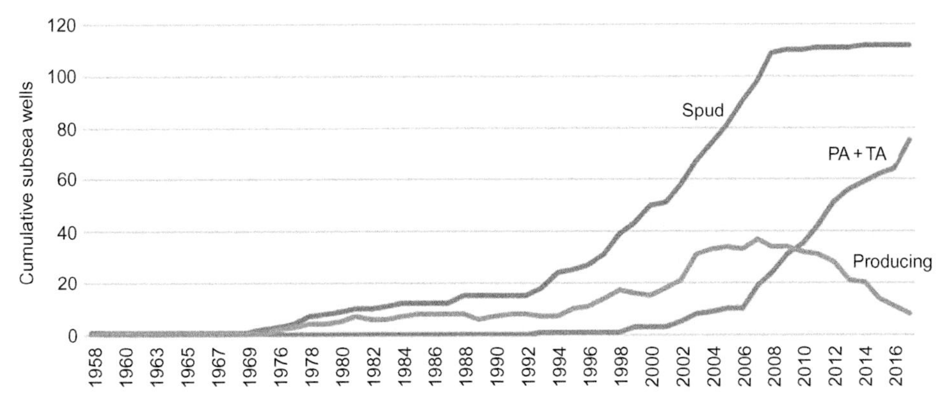

Fig. 6.17 Subsea wells spud, producing and abandoned in the shallow-water Gulf of Mexico. *(Source: BOEM 2018.)*

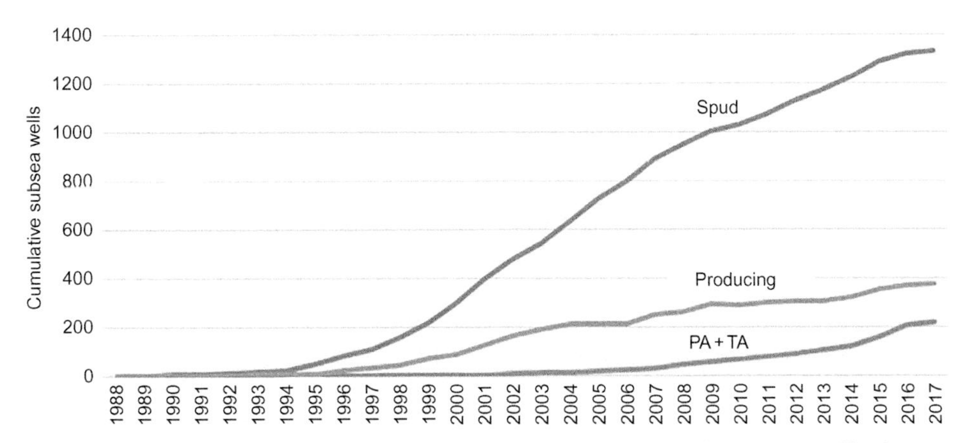

Fig. 6.18 Subsea wells spud, producing and abandoned in the deepwater Gulf of Mexico. *(Source: BOEM 2018.)*

REFERENCES

Attanasi, E.D., Drew, L.J., 1984. Offshore exploration performance and industry changes: the case of the Gulf of Mexico. J. Petrol. Technol. 437–442.
Atwater, G., 1959. Geology and petroleum development of the continental shelf of the Gulf of Mexico. Gulf Coast Assoc. Geol. Soc. Trans. 9, 131–145.
Beu, R.D., 1988. Deepwater exploration patterns in the Gulf of Mexico: an overview and historical perspective of past successes and failures. In: OTC 5694. Offshore Technology Conference. Houston, TX, May 2–5.
Burkhardt, J.A., 1973. Test of submerged production system. In: SPE 4623. Society of Petroleum Engineers of AIME. Las Vegas, NV, Sep 30–Oct 3.
Chhajlani, R., Zheng, Z., Mayfield, D., MacArthur, B., 2002. Utilization of geomechanics for Medusa field development, deepwater Gulf of Mexico. In: SPE 77779. SPE Technical Conference and Exhibition, San Antonio, TX, Sep 29–Oct 2.
Childers, T.W., Loth, W.D., 1976. Test of a submerged production system – progress report. In: SPE 6075. Fall Technical Conference and Exhibition of the Society of Petroleum Engineers of AIME. New Orleans, LA, Oct 3–6.
Forestier, J.F., Ables, G.L., Bertin, D., Ardignac, D., Hall, C.R., Lucas, M.A., 2004. Matterhorn development drilling-innovation and planning equals deepwater directional success. In: OTC 16607. Offshore Technology Conference. Houston, TX, May 3–6.
Francis, D., 2007. 1948–1957: The industry steps out. In: Pike, B. (Ed.), Sixty Years in the Gulf of Mexico. A Supplement to E&P. Hart Energy Publishing, Houston, pp. 13–23.
Hurel, J.P., van der Linden, C., 2004. Matterhorn project. In: OTC 16611. Offshore Technology Conference. Houston, TX, May 3–6.
Lach, J., McMillen, K., Archer, R., Holland, J., DePauw, R., Ludvigsen, B.E., 2005. Integration of geologic and dynamic models for history matching, Medusa field. In: SPE 95930. SPE Technical Conferences and Exhibition. Dallas, TX, Oct 9–12.
National Commission on the BP Deepwater Horizon Oil Spill and Offshore Drilling, 2011. The History of Offshore Oil and Gas in the United States.
Owen, E.W., 1975. Trek of the Oil Finders: A History of Exploration for Petroleum. AAPG, Tulsa, OK.
Priest, T., 2008. Auctioning the ocean: the creation of the federal offshore leasing program, 1954–1962. In: Austin, D., Priest, T., Penney, L., Pratt, J., Pulsipher, A.G., Abel, J., Taylor, J. (Eds.), History of the offshore oil and gas industry in southern Louisiana. vol. 1. U.S. Department of Interior, Minerals Management Service, Gulf of Mexico OCS Region, New Orleans, LA, pp. 93–116. OCS Study MMS 2004–049.

CHAPTER SEVEN

Shallow-Water Structure Inventory

Contents

Abstract

The purpose of this chapter is to review the primary characteristics of shallow-water structures in the Gulf of Mexico. Structures are classified according to their production status, whether they were producing in 2017, did not have any production during the year but previously produced (idle), or were serving in a support role (auxiliary). Production, cumulative production, and revenue are the primary features that describe the producing structure inventory. For idle structures, idle age and idle age at the time of decommissioning are the main features, and for auxiliary structures, structure type and historic installation and removal rates are the relevant characteristics.

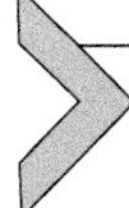

7.1 PRODUCING STRUCTURES

7.1.1 2017 Revenue

In 2017, oil and gas production in the GoM in water depth <400 ft generated approximately \$3.9 billion in gross revenue, \$2.8 billion from oil sales and \$1.1 billion from gas sales (Table 7.1). Per structure, oil producers generated on average about \$5 million and gas producers about \$3.4 million.

Half of the producing inventory generated \$2 million or less during 2017, about one-third of the inventory generated less than \$1 million, and about 20% of structures generated less than \$500,000 (Fig. 7.1). Structures generating half a million dollars a year are still economic since most are unmanned and part of regional operations where labor and logistics costs are shared and allocated across properties, but once revenues fall below a minimum threshold and cash flows are no longer adequate to cover direct cost, operators will find it increasingly difficult to maintain production from marginal producers.

Decommissioning Forecasting and Operating Cost Estimation
https://doi.org/10.1016/B978-0-12-818113-3.00007-2

Table 7.1 Shallow-Water Producing Structures Circa 2017

	Primarily Oil	Primarily Gas
Inventory (#)	563	323
Revenue		
from liquids ($ million)	2522	380
from gas ($ million)	276	708
Total revenue ($ million)	2797	1088

Source: BOEM 2018.

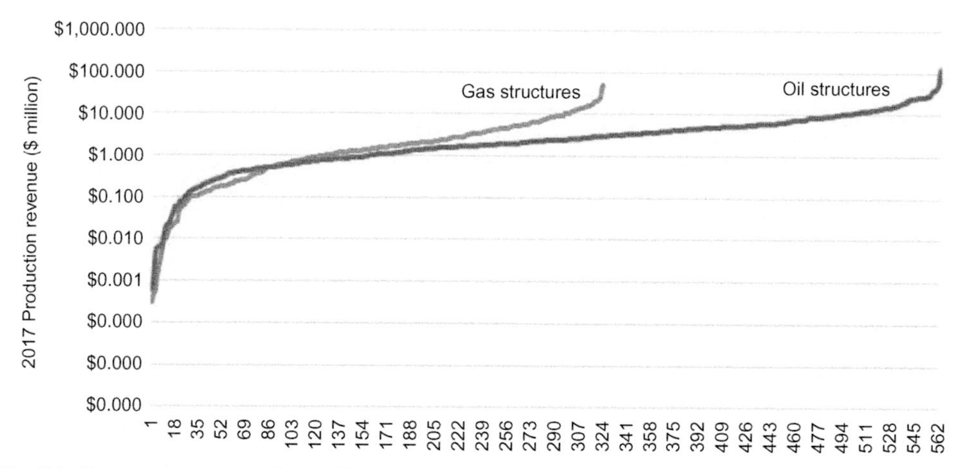

Fig. 7.1 Revenue generated by shallow-water structures in 2017. *(Source: BOEM 2018.)*

Most of the revenue generated from oil and gas production on the shelf is from the top quartile of producers, and platforms hosting multiple wells dominate production (Fig. 7.2). Oil production generates about three-quarters of the revenue for shelf assets, and for the average oil structure associated gas contributes about 10% of sales revenue. For the average gas structure, liquids play a more sizeable role in sales revenue, ranging from about a third if condensate prices are equal to crude oil prices to about a fifth if condensate is sold at a 60% discount to crude oil.

7.1.2 Total Primary Production

Total primary production (i.e., crude oil for oil structures and natural gas for gas structures) is summarized in Table 7.2. On a cumulative basis, producing oil structures extracted 6.2 Bbbl oil through 2017, which for an inventory of 563 structures represents an average cumulative production of about 11 MMbbl per structure. Idle oil structures produced in total about 1.1 Bbbl oil through 2017 or on average 4.5 MMbbl per structure, and all decommissioned oil structures produced 2.8 Bbbl oil or about 2.9 MMbbl per structure.

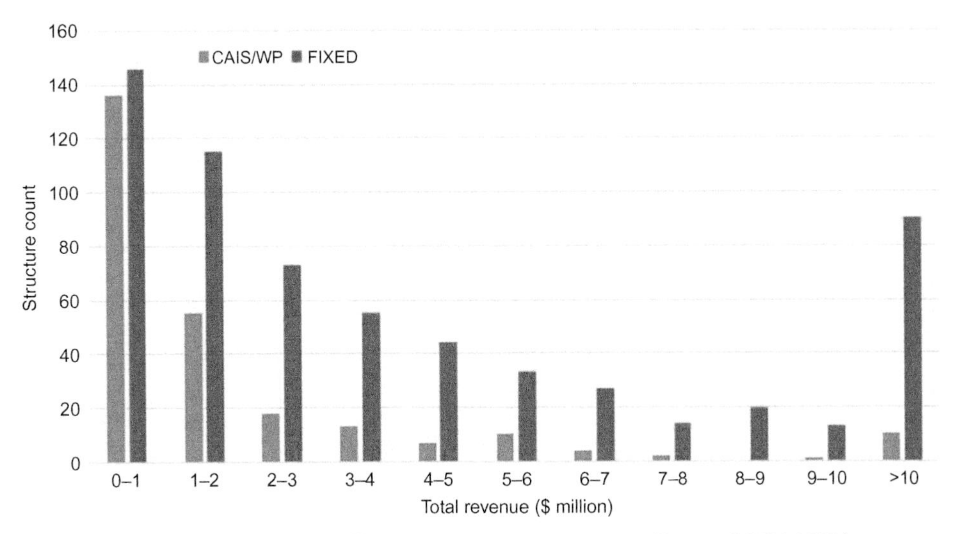

Fig. 7.2 Revenue distribution by shallow-water structures in 2017. *(Source: BOEM 2018.)*

Table 7.2 Production Comparison by Structure Class Circa 2017

	Primarily Oil	**Primarily Gas**
Producing inventory (#)	563	323
Total production circa 2017	6182 MMbbl	23.1 Tcf
Average production	11.0 MMbbl per structure	71.9 Bcf per structure
Idle inventory (#)	237	425
Total production	1074 MMbbl	22.0 Tcf
Average production	4.5 MMbbl per structure	51.7 Bcf per structure
Decommissioned inventory (#)	968	3335
Total production	2834 MMbbl	89.3 Tcf
Average production	2.9 MMbbl per structure	26.8 Bcf per structure

Source: BOEM 2018.

Producing gas structures extracted 23.1 Tcf through 2017 or about 72 Bcf per structure, and idle gas structures produced 22 Tcf gas or 52 Bcf per structure. Decommissioned gas structures have produced the largest volume of gas at 89.3 Tcf, but on a per structure basis is the smallest among the three classes. Average production across both oil and gas structures is lowest for decommissioned structures and highest for producing structures.

In Figs. 7.3 and 7.4, total primary production is depicted for all shallow-water oil and gas producing, idle, and decommissioned structures. Only the primary product is depicted in each graph so that natural volume units (barrels crude and cubic feet gas) are used instead of heat-equivalent units, but the shape of the curves and conclusions

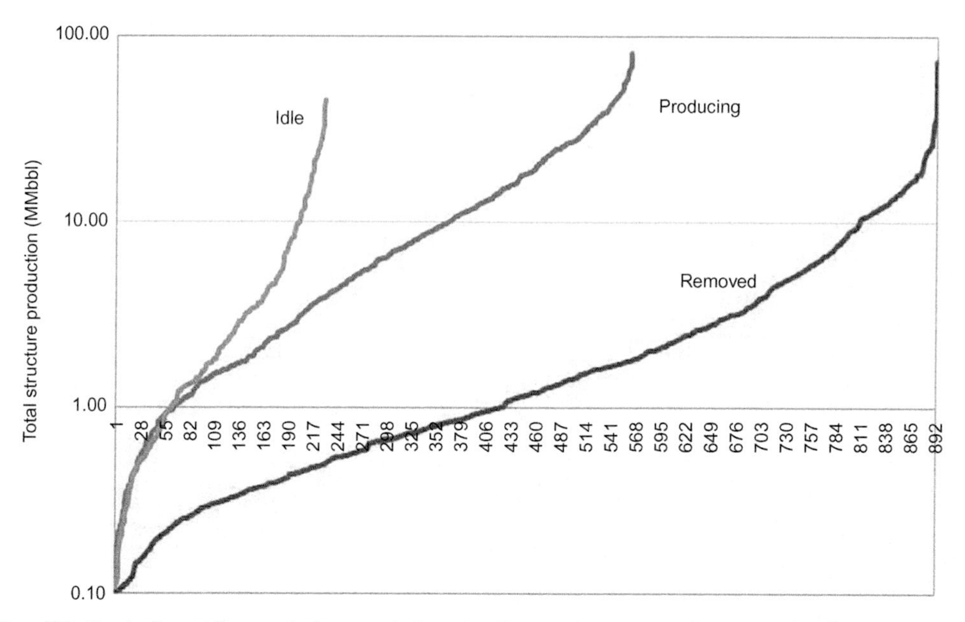

Fig. 7.3 Producing, idle, and decommissioned oil structures <400 ft water depth circa 2017. *(Source: BOEM 2018.)*

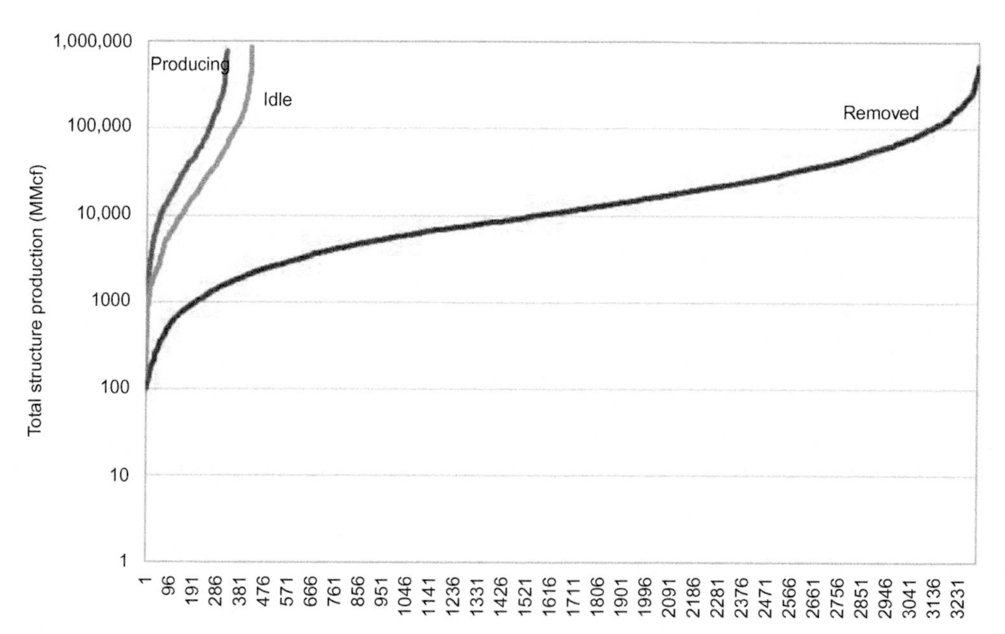

Fig. 7.4 Producing, idle, and decommissioned gas structures <400 ft water depth circa 2017. *(Source: BOEM 2018.)*

are essentially unchanged with heat-equivalent units. Structure counts run along the x-axis. Cumulative production by structure on a log scale is depicted on the ordinate.

Recovery rates in gas reservoirs are significantly higher than oil reservoirs, and so, one might suspect higher average production rates for gas structures relative to oil structures, for all other things equal. Using the heat conversion 6 Mcf/boe yields recoveries of 11.7 MMboe per structure for producing gas structures, 8.2 MMboe for idle gas structures, and 4.5 MMboe for decommissioned gas structures which are larger than the oil structure classes.

7.1.3 Total Cumulative Primary Production

If the production from each structure in Figs. 7.3 and 7.4 is plotted in a cumulative fashion, the graphs in Figs. 7.5 and 7.6 result. The ordinate scales are now much larger, of course, to accommodate cumulative volumes, increasing by two orders of magnitude.

For oil structures, total cumulative primary production is approximately 6.2 Bbbl for producers, 2.8 Bbbl for decommissioned structures, and 1.1 Bbbl for idle structures or about 10 Bbbl total. For gas structures, total cumulative primary production is 89.3 Tcf for decommissioned structures, 23.1 Tcf for producing structures, and 22 Tcf for idle structures, or about 134 Tcf total. Secondary products are excluded from the tally and

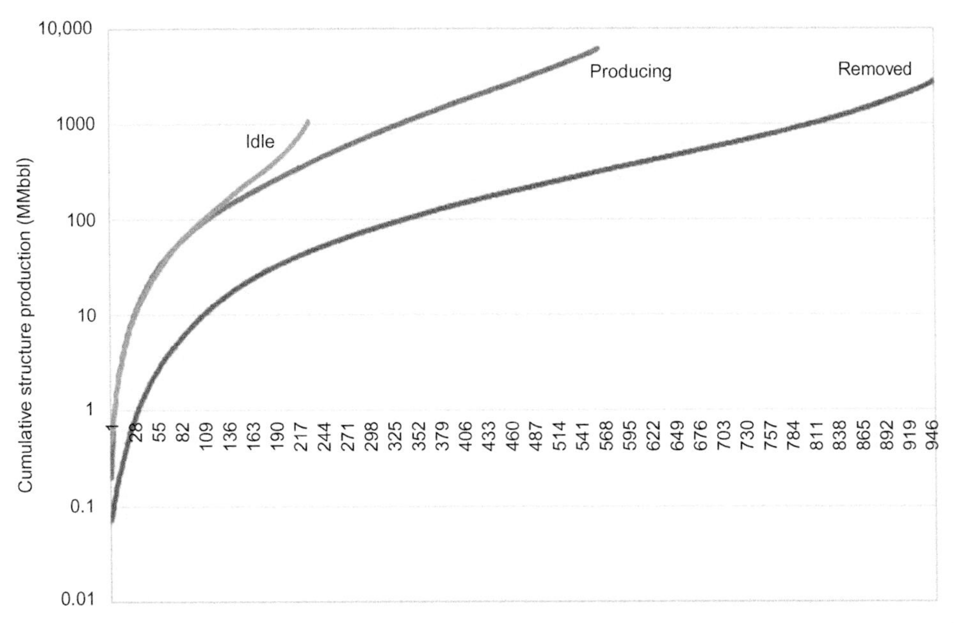

Fig. 7.5 Cumulative production for producing, idle, and decommissioned oil structures <400 ft circa 2017. *(Source: BOEM 2018.)*

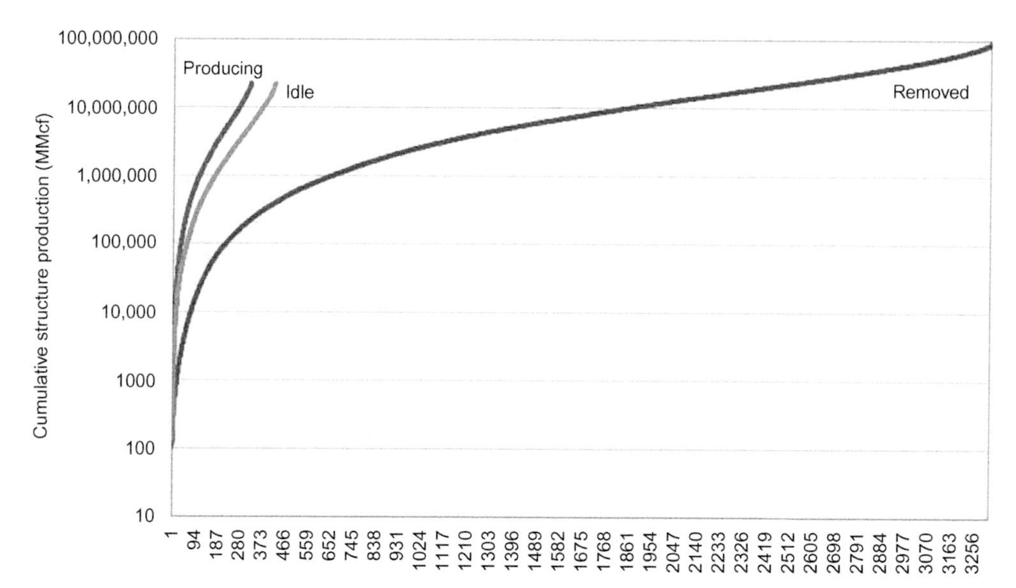

Fig. 7.6 Cumulative production for producing, idle, and decommissioned gas structures <400 ft circa 2017. *(Source: BOEM 2018.)*

account for the incremental difference with total GoM shallow-water production (for oil, 2.1 Bbbl condensate, 12.2 − 10.1 Bbbl; for gas, 52 Tcf associated gas, 186 − 134 Tcf).

7.1.4 Future Dynamics

What will Figs. 7.3 and 7.4 look like in 5 or 10 or 20 years? When a structure stops producing, it will "jump" to either the idle inventory curve or directly to the decommissioned curve depending on the decision of the operator, increasing the structure counts on those curves by one and bringing along its cumulative production that will expand the receiving curve and change its shape while shifting the producing inventory curve to the left and downward since it lost production volume. The volumes are just transferring between the curves.

In recent years, the number of new installations in shallow water has been much smaller than the number of structures that cease production, and assuming this trend continues in the future, the producing structure curve will continue to shrink and shift left over time. Circa 2017, the producing oil structure curve in Fig. 7.3 falls to the right of the idle curve, while the producing gas structure curve in Fig. 7.4 falls on the left of the idle inventory curve having already made this transition.

Idle inventory will expand or shrink depending on the annual net change of transitions, expanding right if additions are greater than decommissioning or moving leftward if additions are less than decommissioning. Idle structures are fed by those structures that stop producing and are reduced by decommissioning activity, but eventually, all idle structures will be absorbed in the decommissioned inventory curve.

Decommissioned structure inventories can only grow in size since decommissioning is an absorbing final state that picks up structures from both producing and idle inventories and does not return any structures. Therefore, the decommissioned structure curves will always shift right and expand upward over time.

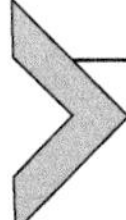

7.2 IDLE STRUCTURES

7.2.1 Idle Inventory

Idle structures are formerly producing structures that have not produced for at least one year. Idle structures may transition back into producing status if inactive wells are reactivated or sidetracking is successful, or if all inactive wells are permanently abandoned, an idle structure may be repurposed to serve another function in field development.

By definition, all idle structures previously produced hydrocarbons, but for some reason—the cause is not publicly reported—the structure was not producing for the previous year. There are many reasons why production may stop, such as well failure, scheduled workovers, third-party problems (i.e., pipeline repair), recompletion, investment review, and hurricane damage.

In 1990, idle structure counts in the shallow-water GoM totaled 505 structures, and up through 2010, every 5 years about 100 idle structures on average were added to the GoM idle inventory (Table 7.3). In 2009, the number of idle structures peaked at 1077 and henceforth has declined every year. In 2017, there were 662 idle structures in the shallow-water GoM.

7.2.2 Idle Age

Idle age describes the number of years since the structure last produced at the time of evaluation (Fig. 7.7). If the last year of production of a structure is denoted by L, then circa 2017, the idle age of a structure is simply 2017–L.

Table 7.3 Idle Inventory in the Shallow-Water Gulf of Mexico

Year	Number
1990	505
1995	578
2000	734
2005	896
2010	995
2015	740

Source: BOEM 2018.

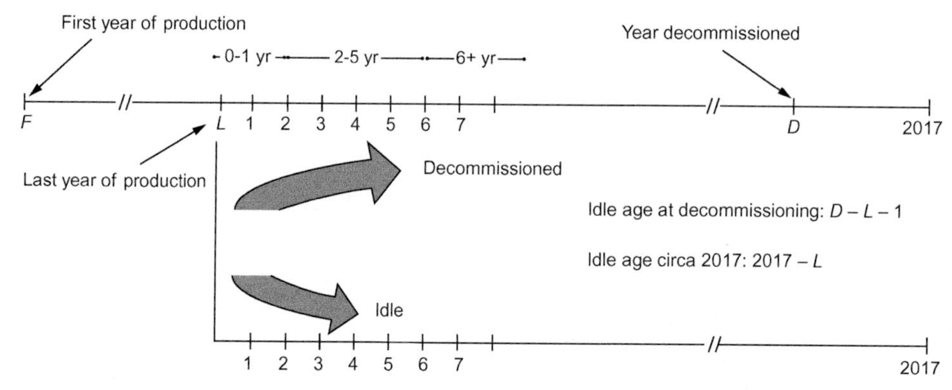

Fig. 7.7 Idle age of a structure at decommissioning and in active inventory circa 2017.

Table 7.4 Idle Age for Shallow-Water Idle Inventory Circa 2017

Idle Age (Year)	Caisson/WP	Fixed Platform	Total
1–2	33	49	82
2–3	29	46	75
3–4	22	37	59
4–5	18	24	42
5–6	21	26	47
6–7	16	14	30
7–8	21	12	33
8–9	10	14	24
9–10	9	9	18
>10	121	131	252
Total	300	362	662

Source: BOEM 2018.

The idle age of GoM shallow-water structures varies widely as one might expect, and in 2017, there were 82 structures <2 years idle and 252 structures >10 years idle (Table 7.4). About 60% of the idle inventory circa 2017 has not produced in >5 years, and about 40% of the inventory hasn't produced in >10 years. Idle structures are roughly evenly distributed between caissons/well protectors and fixed platforms.

The longer a structure sits idle, the less likely it will return to production, but a useful function for the structure may eventually be found if it has not already been repurposed. Because structure function is not publicly reported, a portion of idle structures are expected to be serving a useful purpose in field operations.

Before 2010, structures idle ≤3 years were generally the most populous category, but as idle structures age, they move into older vintage categories if not decommissioned. In recent years, structures idle >10 years are the most common category.

7.2.3 Idle Age at Decommissioning

The idle age of a structure at the time of decommissioning provides information on how long structures are idle before being decommissioned (Fig. 7.7). If the last year of production of a structure is L and the structure was decommissioned in year D, then the age of the structure at decommissioning is simply $D-L-1$. Historic trends in these data provide information on how operators manage their idle structures. Structures idle for <2 years at the time of decommissioning are described as the producing or the 0–1 year class, and other categories include structures idle for 2–5 years, 6–10 years, and 10+ years at the time of decommissioning.

After a structure's application for decommissioning is approved by federal regulators, the operator has 1 year to perform the activity, and because an operator is given a 6-month time window to file its application after the cessation of production, the 0–1 year category is considered a transition direct from production status to decommissioning. The total number of structures and the percentage of structures that fall within each category by decade are shown in Table 7.5.

A total of 4326 producing and formerly producing structures have been decommissioned since 1973. Structures producing and idle for <2 years at the time of decommissioning contributed the majority of the transitions to decommissioning at 1500 total structures, followed by the 2–5 year idle class at 1453 structures, and the ≥6 year idle class at 1373 structures. Hence, on a historical basis, decommissioned structures have an approximately equal probability of arriving from the producing, 2–5 year and ≥6 year categories (Fig. 7.8).

Table 7.5 Structure Count and Class Contribution at Time of Decommissioning by Idle Status

	0–1 year	2–5 years	6–10 years	10+ years	Total
1973–76	17	22	12	3	54
1977–86	75	110	36	32	253
1987–96	374	230	105	146	855
1997–2006	483	386	150	178	1197
2007–16	536	660	320	362	1878
2017	15	45	16	13	89
Total	1500	1453	639	734	4326
1973–76	31%	41%	22%	6%	100%
1977–86	30%	43%	14%	13%	100%
1987–96	44%	27%	12%	17%	100%
1997–2006	41%	32%	12%	15%	100%
2007–16	29%	35%	17%	19%	100%
2017	17%	51%	18%	15%	100%
Average	35%	33%	15%	17%	100%

Source: BOEM 2018.

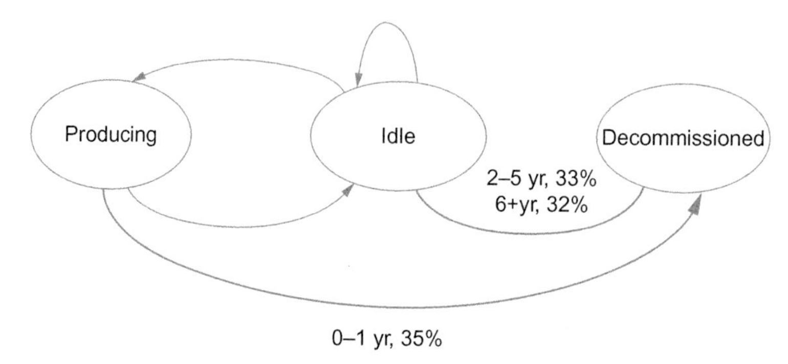

Fig. 7.8 Percentage of structures producing and idle at time of decommissioning, 1973–2017.

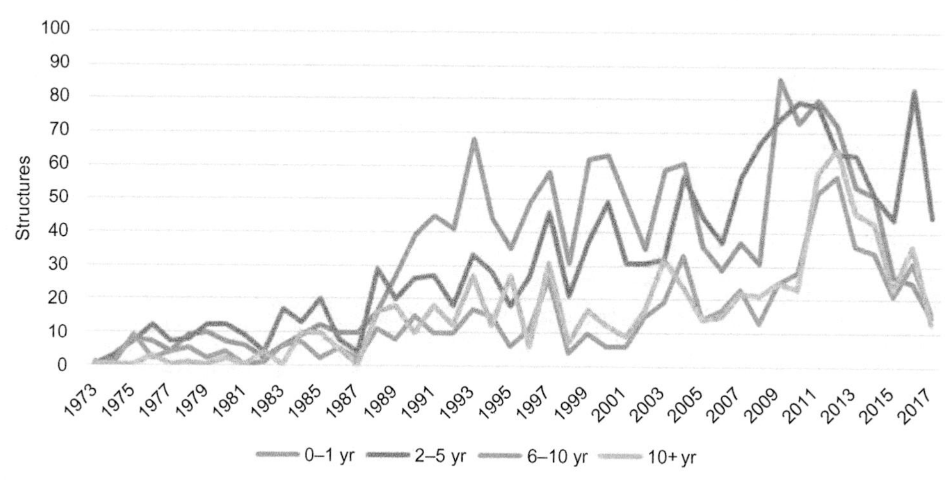

Fig. 7.9 Structures decommissioned by idle age group, 1973–2017. *(Source: BOEM 2018.)*

The idle age of structures at the time of abandonment has changed over the past two decades. From 1997 to 2006, there were 483 structures idle <2 years at the time of decommissioning and 328 structures idle ≥6 years decommissioned, out of a total of 1197 structures. From 2007 to 2016, there were 536 structures idle <2 years decommissioned and 682 structures idle ≥6 years decommissioned, out of a total of 1875 structures. Young idle structure decommissioning as a percent of the total decreased from 41% to 29% over the past two decades, while the proportion of older idle structures increased from 27% to 36%.

Structures decommissioned by idle age group by year are graphed in Fig. 7.9 and in a stacked representation in Fig. 7.10. The 0–1 year group was the most populated in early years followed closely by the 2–5 year group that has dominated in recent years. Producing inventories are shrinking in absolute numbers as the idle classes contribute a larger portion of decommissioning activity, and these trends are expected to continue in the future.

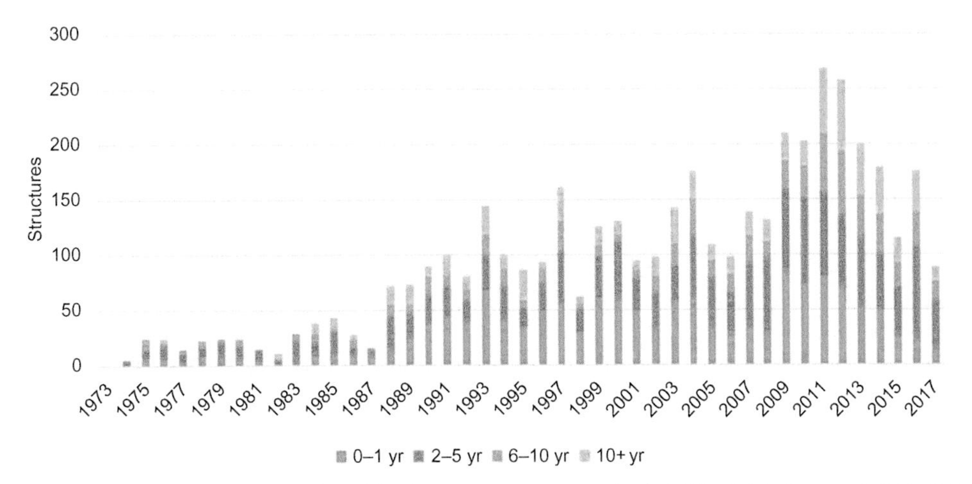

Fig. 7.10 Structures decommissioned by idle age group—stacked, 1973–2017. *(Source: BOEM 2018.)*

7.3 AUXILIARY STRUCTURES

The number of auxiliary structure installations has never exceeded more than 50 structures per year, and decommissioning rates are typically less than half the install rate at between 10 and 25 structures per year. In Fig. 7.11, annual installation and decommissioning activity as well as the active auxiliary inventory are depicted on the left vertical axis, while the cumulative number of installations and removals is denoted on the right vertical axis. The horizontal axis is time in years.

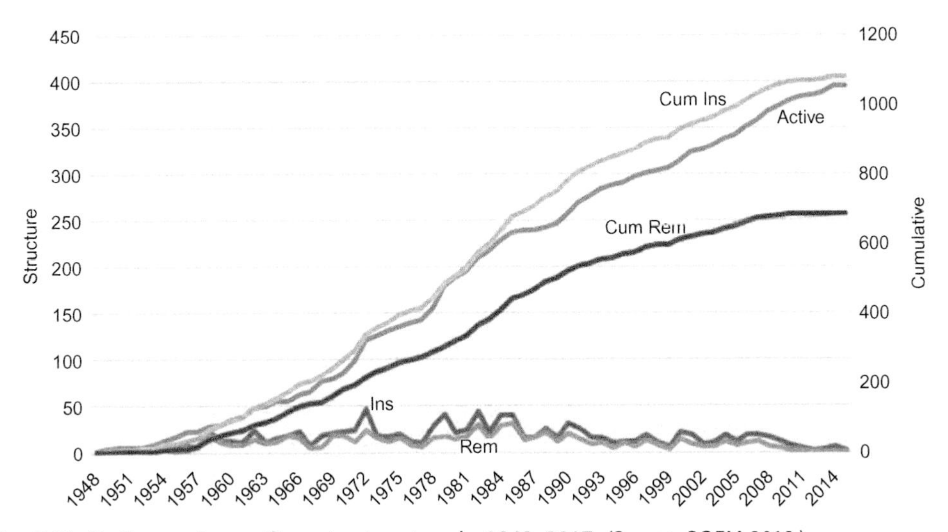

Fig. 7.11 Shallow-water auxiliary structure trends, 1948–2017. *(Source: BOEM 2018.)*

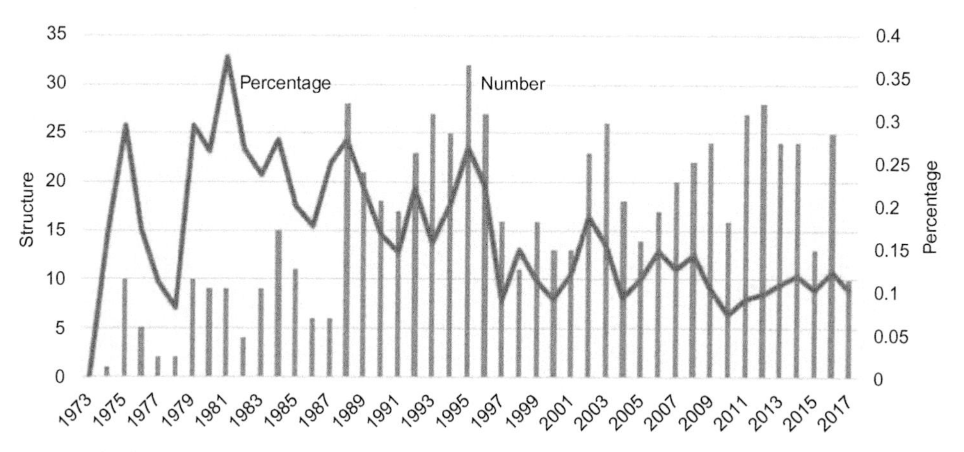

Fig. 7.12 Shallow-water auxiliary structures decommissioned and percentage of total, 1948–2017. *(Source: BOEM 2018.)*

One way for an operator to transform a producing or idle structure into an auxiliary structure is to plug and abandon all the wellbores and run production pipeline from another field/platform across the facility. Another pathway is to recertify a producing or idle structure as a transport and fuel depot hub after abandoning its wells.

Many of the transportation platforms on the shelf were originally anchored by field production and as production was exhausted transformed into auxiliary structures. Two other classes of auxiliary structures on the shelf include those that handle deepwater production and those that support transportation companies. Most but not all of these structures were built for purpose.

Since auxiliary structures are not associated directly with well production, caissons and well protectors are not a common structure type, and the vast majority of auxiliary structures are fixed platforms. Circa 2017, 192 of the 357 auxiliary structures were classified as 24h manned. Over the past two decades, auxiliary structure decommissioning has ranged from 7% to 15% of the annual number of structures decommissioned in the Gulf (Fig. 7.12). The historic average decommissioning rate of auxiliary structures is 14% of total decommissioning activity. Correlating relations exist between idle, auxiliary, and producing structures (Box 7.1).

BOX 7.1 Operator Inventory Correlating Relations

Correlations reveal connections between variables and may lead to new insights. When dealing with data describing offshore activities or the results of those activities, correlations help to confirm and validate relationships between attributes and may also identify problems with data that require further scrutiny.

An operator's inventory of shallow-water producing structures is a good indicator of its idle inventory and auxiliary structure count, as shown in Figs. 7.13 and 7.14 and quantified via the relations:

Continued

BOX 7.1 Operator Inventory Correlating Relations—cont'd

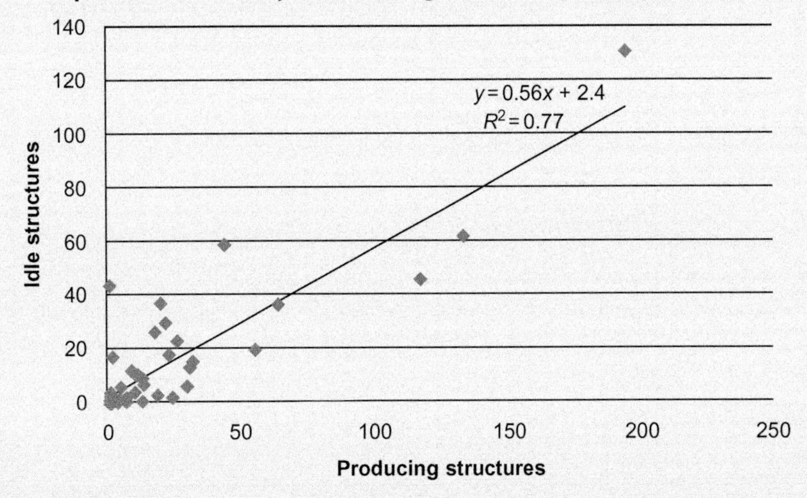

Fig. 7.13 Idle and producing structure shallow-water operator inventories circa 2017. *(Source: BOEM 2018.)*

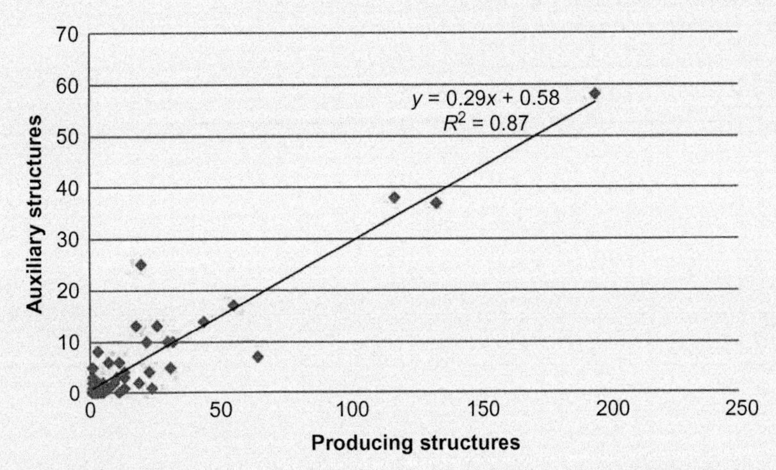

Fig. 7.14 Auxiliary and producing structure shallow-water operator inventories circa 2017. *(Source: BOEM 2018.)*

Number Idle Structures = 0.56 · *Number Producing Structures* + *2.4*, R^2 = 0.78.

Number Auxiliary Structures = 0.29 · *Number Producing Structures* + *0.58*, R^2 = 0.87.

Specifically, for every shallow-water producing structure an operator maintained in 2017, it held 0.56 idle structures and 0.29 auxiliary structures on average in inventory. High model fits indicate that the relations describe industry portfolios reasonably well, reflecting similar requirements of field development, investment strategies, and production among operators.

CHAPTER EIGHT

Shallow-Water Economic Limit Statistics

Contents

Abstract

Net revenue during the last year of a structure's productive life serves as a proxy for the economic limit of production and encapsulates all the relevant cost information publicly available. By reviewing a large group of decommissioned structures, economic limits are quantified and compared using structure attributes to gain insight into shallow-water operating cost thresholds and business practices. Summary statistics are tabulated by primary production, structure type, manned status, and water depth for 3054 decommissioned structures from 1990 to 2017. The P50-adjusted gross revenue the last year of production was $1.2 million for gas structures and $627,000 for oil structures. For gas structures, P20 and P80 economic limits are $282,000 and $3.97 million, for oil structures, $135,000 and $2.01 million, respectively. Factor models distinguish the impact of individual variables and show that majors have economic limits of $820,000 per structure greater than independents holding all other factors constant.

Decommissioning Forecasting and Operating Cost Estimation
https://doi.org/10.1016/B978-0-12-818113-3.00008-4

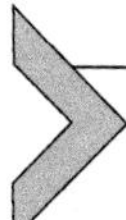

8.1 METHODOLOGY

8.1.1 Revenue Model

For decommissioned structures, the net revenue near the end of its productive life is a proxy of its economic limit and is straightforward to compute since the production output $Q(t)$, royalty rate r, and oil and gas prices $P^o(t)$ and $P^g(t)$ are all known and available historically. If the last month of production is denoted as T_a, then the net revenue the last year of production NR_{LY} determined from monthly oil and gas production volumes and average monthly oil and gas prices the 12 months prior to T_a is determined as

$$NR_{LY} = \sum Q(t)P(t)(1-r).$$

In shallow water, the vast majority of decommissioned structures have royalty rates of 16.67%, and for all practical purposes are constant across the sample. Gross revenues are therefore used as a relative measure of net revenue:

$$GR_{LY} = \sum Q(t)P(t).$$

8.1.2 Categorization

Structures are categorized according to structure type, and oil structures are distinguished from gas structures using a cumulative gas-oil ratio of 10,000 cf/bbl. Manned structures and structures associated with a multistructure complex are identified. Caissons and well protectors are functionally equivalent and are consolidated into one structure category. Three water depth subcategories are applied: <100 ft, 100–200 ft, and 200–400 ft. Major integrated operators (majors) are distinguished from independents.

8.1.3 Sample

The sample consists of 3054 structures that ceased production and were decommissioned between 1990 and 2017. Of the 3054 structures, 512 were decommissioned oil structures (319 C/WP and 193 FP), and 2542 were decommissioned gas structures (1249 C/WP and 1293 FP). The sample was split roughly evenly between major and minor structures with 1568 C/WPs and 1486 FPs, and unmanned facilities were dominant (2947 unmanned and 108 manned). The vast majority of structures resided in water depth <200 ft, and only 281 structures (mostly fixed platforms) were in water depth >200 ft. The sample contained 108 manned structures (27 oil and 81 gas), 114 multistructure platforms, and 289 structures operated by majors. Chevron had the largest volume of decommissioning activity among majors (234 decommissioned structures) followed by BP (23), Shell (18), Texaco (14), and Total (7).

8.1.4 Exclusions

Structures that stopped producing before 1990 and were decommissioned on/after 1990 were excluded from the sample because of the increased uncertainty and potential bias associated with long-term price adjustments. Structures that stopped producing after 1990 and not decommissioned by 2017 were not part of the sample because their ultimate status is not yet determined. All hurricane-destroyed structures identified through official government records were excluded since these represent (at least in part) premature removals that would bias (upward) the economic limit statistics.

8.1.5 Adjusted Gross Revenue

The gross revenue the last year of production is computed using monthly production data and average monthly Henry Hub prices for natural gas and West Texas Intermediate prices for crude oil for one year prior to production cessation. In Figs. 8.1A and 8.1B, production and price data for a typical decommissioned gas structure are illustrated. The product streams and commodity prices are shown across 1-year time windows from the end of production. Crude and condensate prices and gas well gas and casinghead (associated) gas prices are assumed equal. Computed gross revenue is adjusted to 2016 dollars using the CPI.

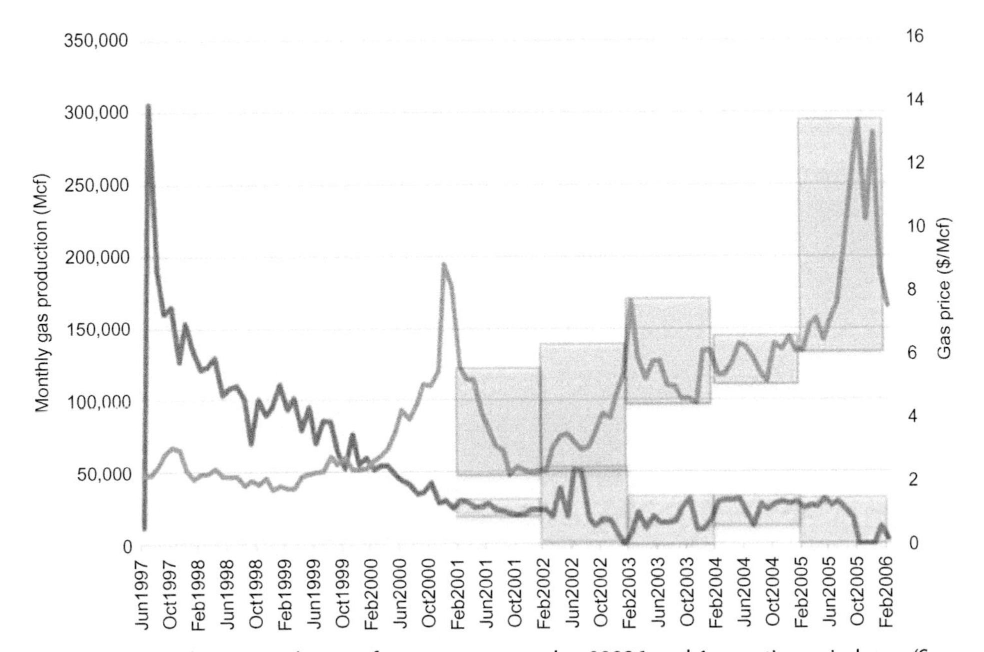

Fig. 8.1A Gas production and prices for structure complex 90026 and 1-year time windows. *(Source: BOEM 2017.)*

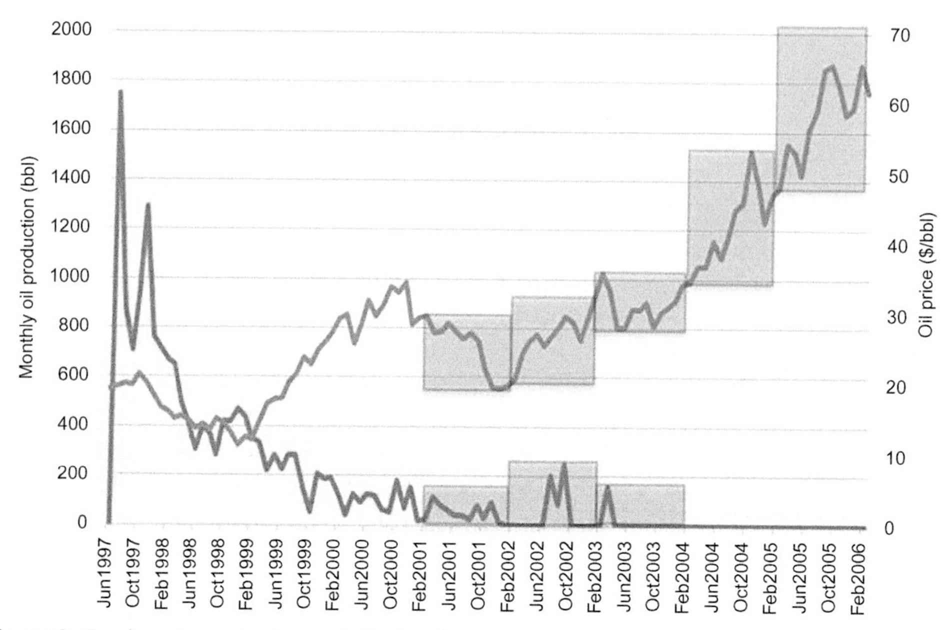

Fig. 8.1B Condensate production and oil prices for structure complex 90026 and 1-year time windows. *(Source: BOEM 2017.)*

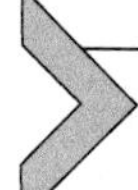

8.2 DISTRIBUTIONS

8.2.1 Oil vs. Gas Structures

The adjusted gross revenue distribution the last year of production for all shallow-water oil and gas structures for all water depths and structure type is summarized in Table 8.1 and depicted in Fig. 8.2 (Kaiser and Narra, 2018). The economic limit for gas structures exceeds the economic limit for oil structures. The median P50 values are $1.23 million

Table 8.1 Economic Limit Statistics by Structure Type, 1990–2017 (Thousand 2016$)

	Gas Structure			Oil Structure		
Structure Type	**P20**	**P50**	**P80**	**P20**	**P50**	**P80**
C/WP	235	1118	3815	81	469	1441
FP	326	1296	4013	297	1017	3348
Manned	651	2045	5379	439	1080	3583
Unmanned	275	1190	3852	122	590	1926
All	282	1220	3970	135	627	2010

Source: Kaiser, M.J., Narra, S., 2018. An empirical evaluation of economic limits in the shallow water U.S. Gulf of Mexico, 1990-2017. J. Petrol. Sci. Eng. 164, 234–244.

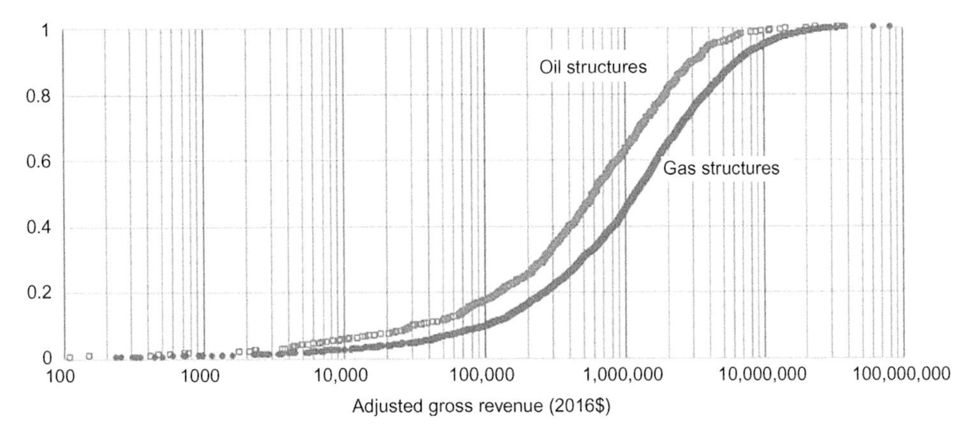

Fig. 8.2 Gross revenues for oil and gas structures the last year of production, 1990–2017. *(Source: Kaiser, M.J., Narra, S., 2018. An empirical evaluation of economic limits in the shallow water U.S. Gulf of Mexico, 1990-2017. J. Petrol. Sci. Eng. 164, 230–244.)*

for gas structures and $627,000 for oil structures. For gas structures, P20 and P80 economic limits range from $282,000 to $3.97 million, respectively. For oil structures, P20 and P80 economic limits range from $135,000 to $2.01 million. The economic limits for some structures can be very low, on the order of a hundred thousand dollars (the approximate labor cost of one offshore worker), and remain profitable (Box 8.1).

The gap between oil and gas structure economic limits is expected to shrink if actual sales prices were applied. Oil and condensate prices and gas well and casinghead prices are assumed equal, but in reality, condensate is usually priced at a discount to crude oil, and casinghead gas is worth more than gas well gas. In other words, the liquid revenue stream from gas wells is priced higher than what will be realized in practice, while for oil wells,

BOX 8.1 Why Are Some Economic Limits So Low?

A hypothetical example is used to illustrate how labor and transportation cost is allocated in regional operations that allow low levels of production at individual structures. As long as operators can recover their direct cost of operations, a structure will continue to produce.

Example. Labor and transportation cost allocation

Field operations involve five platforms, a manned complex with two connected structures P1/P2, and three unmanned remote structures P3, P4, and P5 (Table 8.2). For a permanent six-man production crew working a 1-week tour (1 week on/off), total man-hours are 52,560 h per year, and total annual flight hours for operations are 286 h. Total labor cost at $100,000 per person leads to an annual cost of $1.2 million or a unit cost of $22.8 per man-hour ($1.2 million per 52,560 man-hour).

Continued

BOX 8.1 cont'd

Table 8.2 Labor and Transportation Cost Allocation for Regional Operations

	Annual Man-Hours	Flight Hours	Production (Mboe)	Labor ($1000)	Transport ($1000)	Unit Cost ($/boe)
P1/P2	47,152	104	200	1076	208	6.42
P3	3744	78	97	86	156	2.49
P4	832	52	50	19	104	2.84
P5	832	52	3	19	104	41.00
Total	52,560	286	350	1200	572	5.06

Satellite platform P3 requires a three-man crew visit three times per week, 8h per visit and 15min flight time from the manned complex. This leads to.

$$3 \text{ man} \cdot 8 \text{ h per visit} \cdot 3 \text{ visits per week} \cdot 52 \text{ weeks per year} = 3744 \text{ man-hours per year}$$
$$0.5 \text{ h per trip} \cdot 3 \text{ visits per week} \cdot 52 \text{ weeks per year} = 78 \text{ flight hours per year.}$$

Satellite platforms P4 and P5 require a two-man crew visit once per week, 8h per visit and 30min flight time. Complex C is charged $1.1 million for labor, while platform P3 is allocated $86,000, and platforms P4 and P5 are allocated $19,000 each. Helicopter cost is based on $2000/h flying time. Total transportation cost of $572,000 (286h * $2000/h) is allocated according to flight hours similar to the labor cost allocation.

The total labor and transportation cost ranges from $2.5/bbl at platform P3 to $41/bbl at platform P5. Average operational cost is $5.1/bbl. Allocated labor and transportation cost at platforms P4 and P5 is about $125,000 that would not be able to maintain production outside regional operations. The unit cost for P5 is high but still profitable for the 3000bbl received. If the price of crude oil is less than $41/bbl, platform P5 may be temporarily shut in if cost cannot be reduced. If P5 is decommissioned, the operator saves on transportation cost to the facility, and unit cost will be reduced to $4.8/boe. As high cost structures are shut in, some costs are saved, but remaining expenses will be allocated across fewer properties.

gas production is priced lower than what operators will receive, and these differences will shrink the gap observed. Oil structure revenues will increase slightly from the higher sales price of casinghead gas, while gas structure revenues will decrease from lower condensate sales prices, and the difference in economic limits between oil and gas structures will be reduced.

8.2.2 Structure Type and Manned Status

Structure type and manned status exhibit expected aggregate behavior for each class (Figs. 8.3–8.6). Caissons and well protectors are exclusively unmanned structures with lower fixed operating cost than fixed platforms and are expected to exhibit a lower economic limit in comparison. Manned structures have a higher operating cost

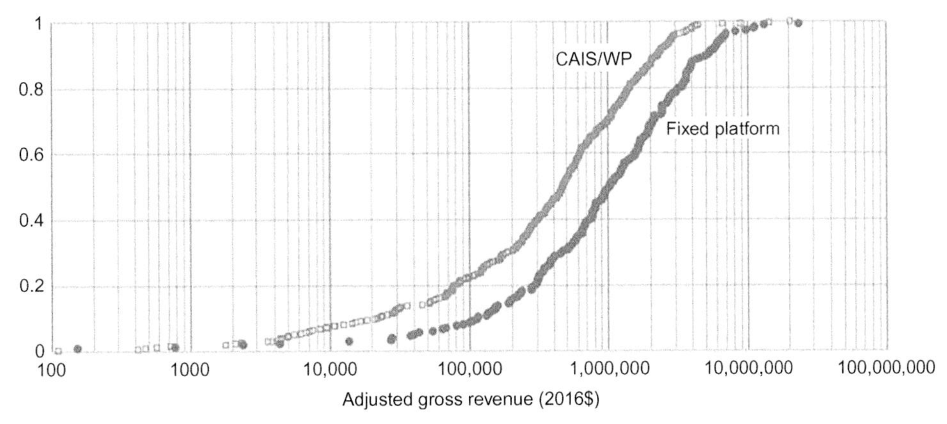

Fig. 8.3 Gross revenues for oil structures last producing year by structure type, 1990–2017. *(Source: Kaiser, M.J., Narra, S., 2018. An empirical evaluation of economic limits in the shallow water U.S. Gulf of Mexico, 1990-2017. J. Petrol. Sci. Eng. 164, 230–244.)*

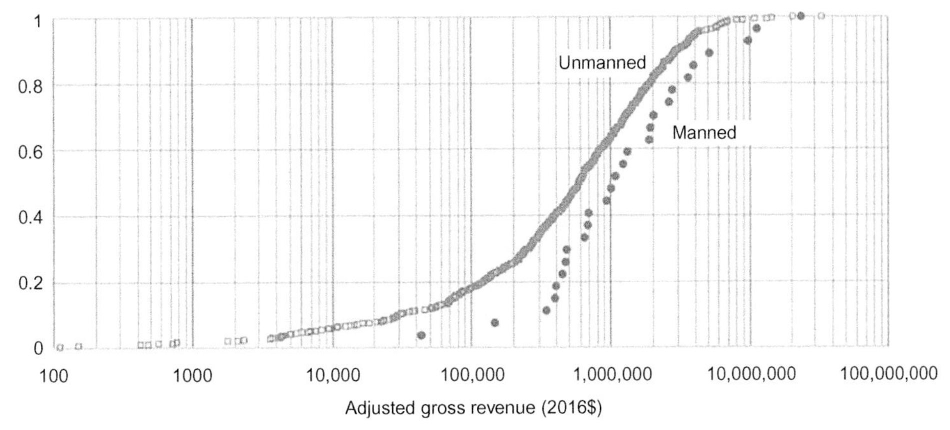

Fig. 8.4 Gross revenues for oil structures last producing year by manning status, 1990–2017. *(Source: Kaiser, M.J., Narra, S., 2018. An empirical evaluation of economic limits in the shallow water U.S. Gulf of Mexico, 1990-2017. J. Petrol. Sci. Eng. 164, 230–244.)*

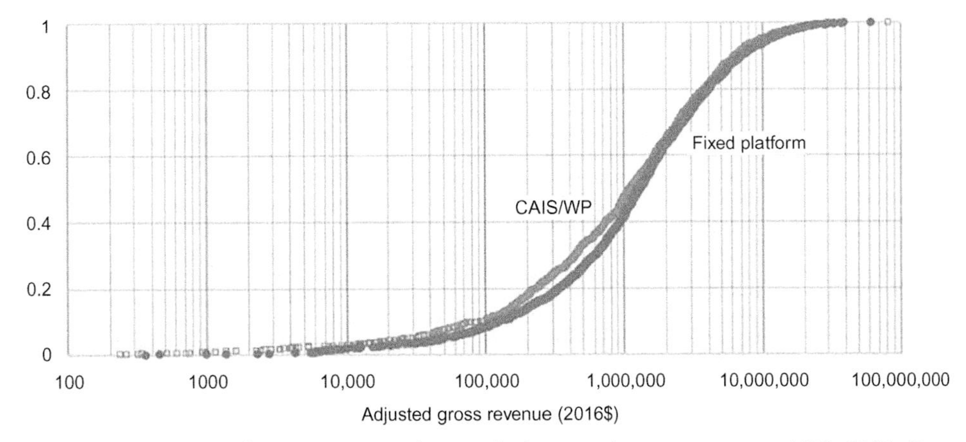

Fig. 8.5 Gross revenues for gas structures last producing year by structure type, 1990–2017. *(Source: Kaiser, M.J., Narra, S., 2018. An empirical evaluation of economic limits in the shallow water U.S. Gulf of Mexico, 1990-2017. J. Petrol. Sci. Eng. 164, 230–244.)*

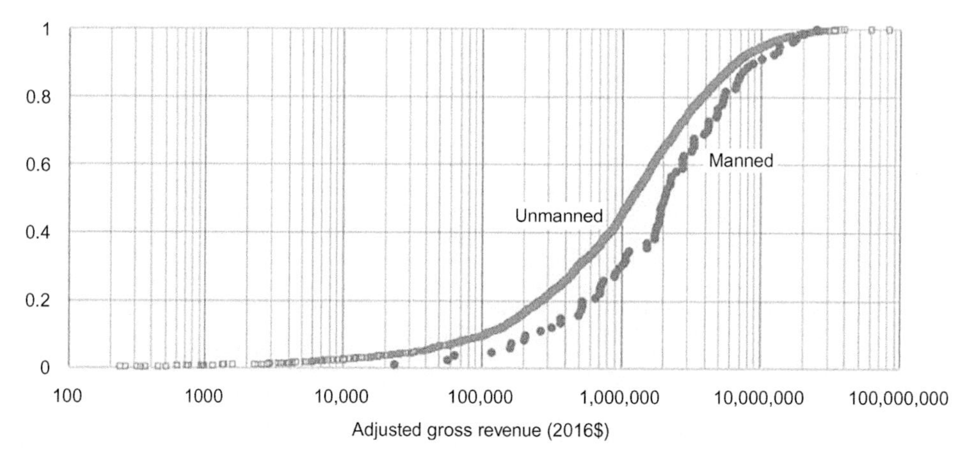

Fig. 8.6 Gross revenues for gas structures last producing year by manning status, 1990–2017. *(Source: Kaiser, M.J., Narra, S., 2018. An empirical evaluation of economic limits in the shallow water U.S. Gulf of Mexico, 1990-2017. J. Petrol. Sci. Eng. 164, 230–244.)*

than unmanned structures, for all other things equal, and are more valuable in regional operations which account for far fewer removals. For gas structures, the difference in economic limits by structure type is less pronounced than for oil structures, indicating similar well counts and cost characteristics at the end of their life. The spread in the distributions between manned and unmanned gas platforms is similar to oil structures.

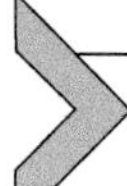

8.3 TIME TRENDS

8.3.1 Structures

The average adjusted annual gross revenue from 1990 to 2017 for shallow-water decommissioned structures is $2.45 million with a standard deviation of $969,000 (Fig. 8.7). From 1990 to 1999, the average gross revenue and standard deviation was $2.55 million ($620,000); from 2000 to 2009, the average gross revenue and standard deviation was $3.0 million ($1.1 million); and from 2010 to 2017, the average gross revenue and standard deviation was $1.59 million ($580,000).

For six years of the 28-year time period, the average adjusted gross revenues exceeded $3 million, and in four of those years, hurricane activity occurred the same or previous year that likely contributed to a knockout affect for damaged but not destroyed structures (Kaiser, 2014). Hurricane activity in 1992 (Andrew), 2005 (Katrina and Rita), and 2008 (Ike and Gustav) appears to have prompted damaged structures to be removed earlier than normal and with higher gross revenues than average.

8.3.2 Oil vs. Gas Structures

For gas structures, the mean adjusted gross revenue and standard deviation from 1990 to 2017 was $2.7 million ($1.1 million), and for oil structures, the mean adjusted gross

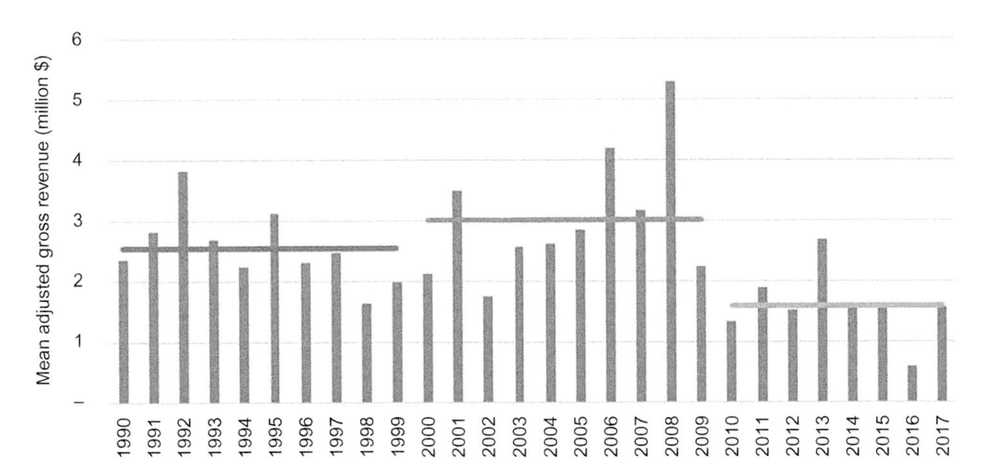

Fig. 8.7 Average gross revenue last year of production, 1990–2017. *(Source: Kaiser, M.J., Narra, S., 2018. An empirical evaluation of economic limits in the shallow water U.S. Gulf of Mexico, 1990-2017. J. Petrol. Sci. Eng. 164, 230–244.)*

revenue and standard deviation was \$1.9 million (\$2.1 million). Before 2007, the average gross revenues of gas structures the last year of production always exceeded the average gross revenues of oil structures, but after 2007, average oil structure gross revenues are usually greater (Fig. 8.8). In 2013, a small number of structures that stopped

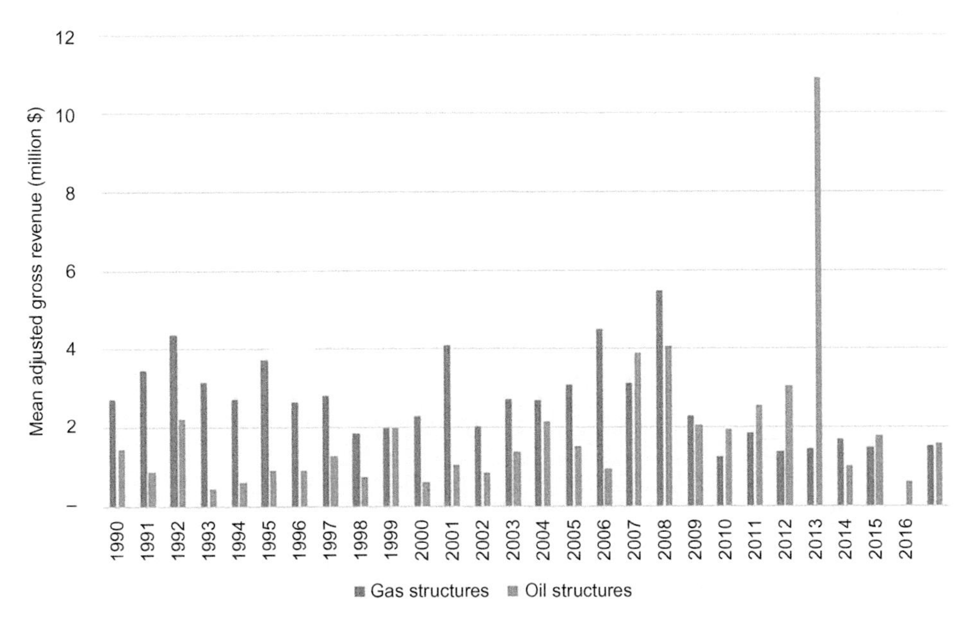

Fig. 8.8 Average gross revenue last year of production for structures. *(Source: Kaiser, M.J., Narra, S., 2018. An empirical evaluation of economic limits in the shallow water U.S. Gulf of Mexico, 1990-2017. J. Petrol. Sci. Eng. 164, 230–244.)*

Table 8.3 Average Gross Revenues for Oil and Gas Structures by Decade, 1990–2017 (Million 2016$)

Structure Type	1990–2017	1990–99	2000–09	2010–17
Gas	2.68 (1.11)	2.94 (0.77)	3.23 (1.12)	1.51 (0.20)
Oil	1.91 (2.07)	1.14 (0.58)	1.85 (1.22)	2.93 (3.31)

Note: Standard deviation in parenthesis.

producing contributed to much higher-than-average gross revenues resulting in a spike. Average gas structure gross revenue was about twice the average oil structure revenue from 1990 to 2009, but in the most recent decade, the trend has reversed (Table 8.3).

8.3.3 Water Depth

The economic limits for oil and gas structures in water depth < 100 ft most often fall at the bottom of the revenue range, while structures in the 200–400 ft water depth category often bound the upper interval (Figs. 8.9 and 8.10). Greater water depth usually corresponds to longer distance to shore and higher logistic cost that at least partially account for the trends shown. Each graph shows a few years of higher-than-average volatility due in part to small sample sizes.

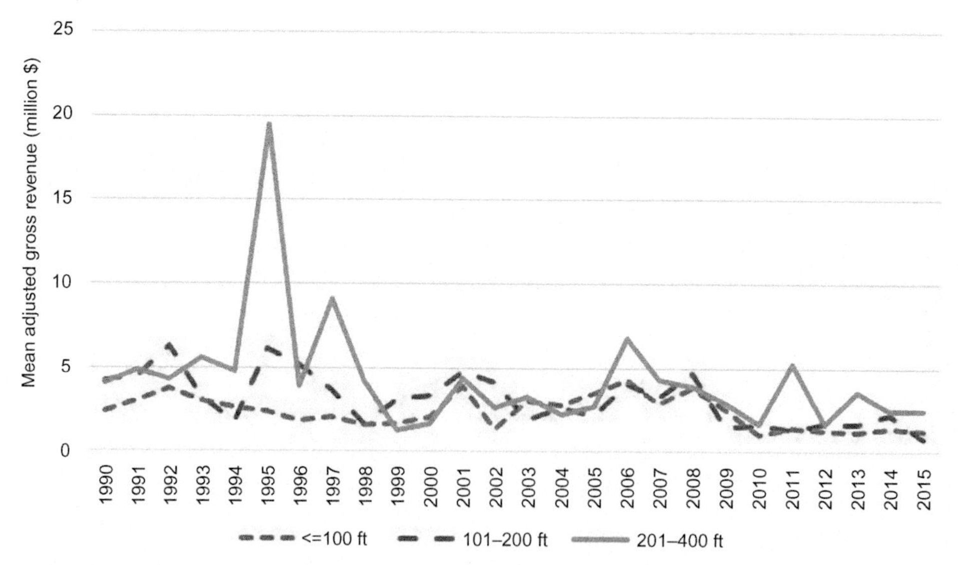

Fig. 8.9 Average gross revenue the last year of production for gas structures, 1990–2017. *(Source: Kaiser, M.J., Narra, S., 2018. An empirical evaluation of economic limits in the shallow water U.S. Gulf of Mexico, 1990-2017. J. Petrol. Sci. Eng. 164, 230–244.)*

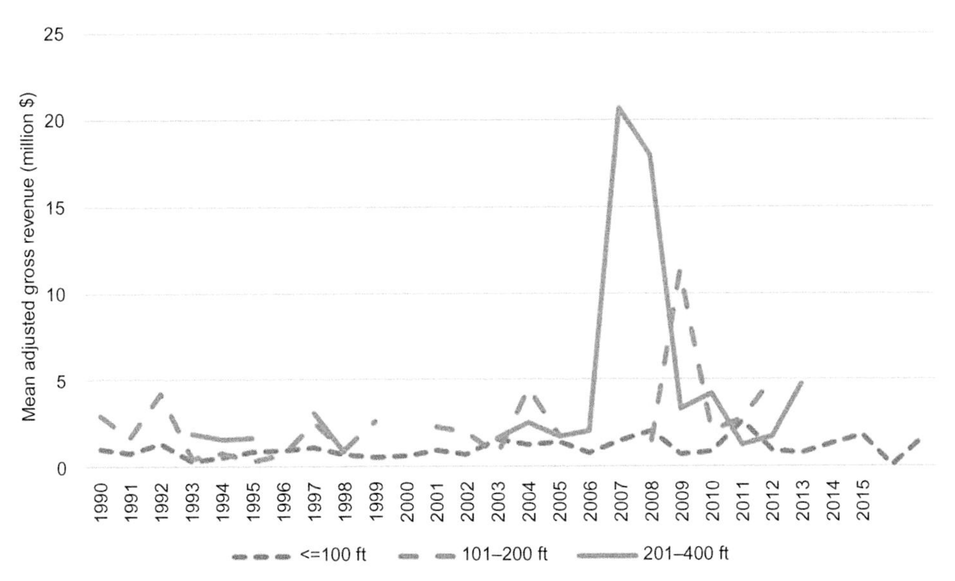

Fig. 8.10 Average gross revenue the last year of production for oil structures, 1990–2017. *(Source: Kaiser, M.J., Narra, S., 2018. An empirical evaluation of economic limits in the shallow water U.S. Gulf of Mexico, 1990-2017. J. Petrol. Sci. Eng. 164, 230–244.)*

8.3.4 Moving Time Windows

The adjusted gross revenue distribution the year before the last year of production and the third and fourth year before cessation is shown in Figs. 8.11 and 8.12, where 0–1 year represents the last year of production, 1–2 years represent the second to last year of

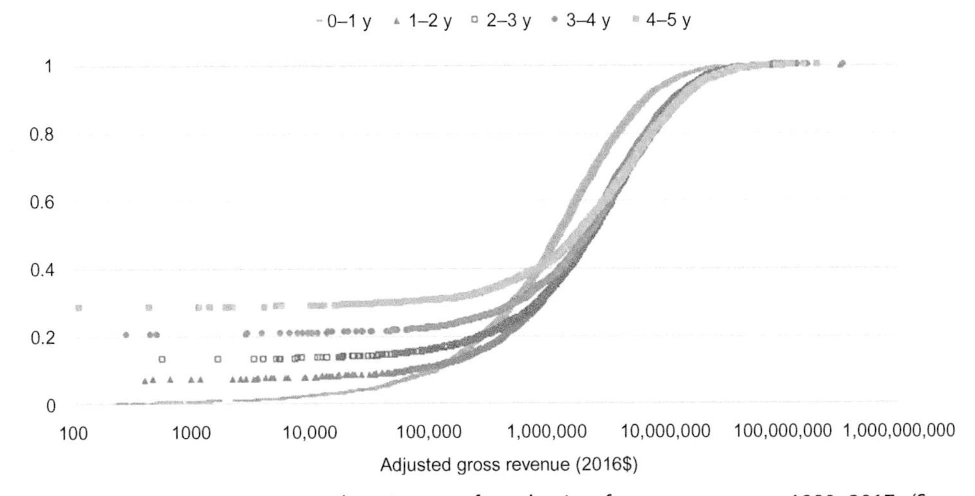

Fig. 8.11 Average gross revenue last 5 years of production for gas structures, 1990–2017. *(Source: Kaiser, M.J., Narra, S., 2018. An empirical evaluation of economic limits in the shallow water U.S. Gulf of Mexico, 1990-2017. J. Petrol. Sci. Eng. 164, 230–244.)*

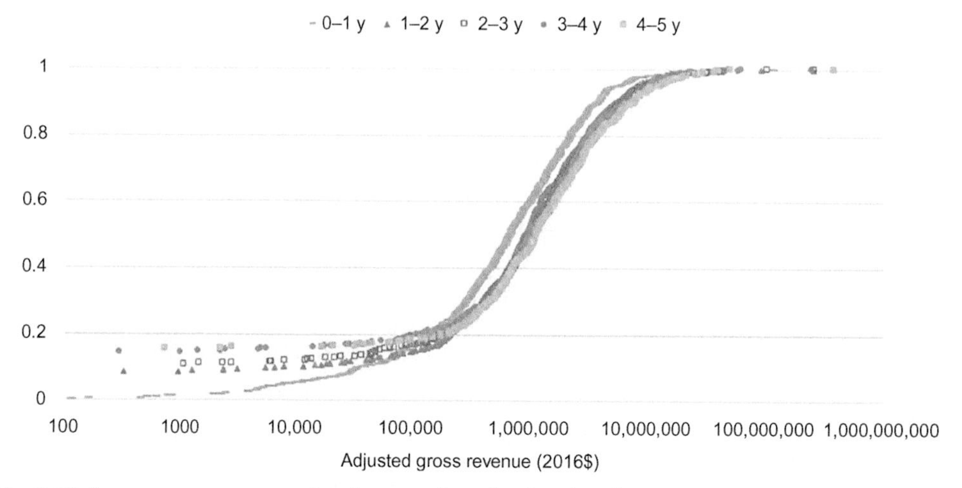

Fig. 8.12 Average gross revenue last 5 years of production for oil structures, 1990–2017. *(Source: Kaiser, M.J., Narra, S., 2018. An empirical evaluation of economic limits in the shallow water U.S. Gulf of Mexico, 1990-2017. J. Petrol. Sci. Eng. 164, 230–244.)*

production, and so on. The revenue distribution differentiates at the low end and is greater at the mid and high end for both oil and gas structures. This is easy to understand since production is usually (although not always) higher before the last year of production, while prices may be higher or lower.

For gas structures, P50 the last year of production was $1.22 million, while for the third and fourth year before production cessation, P50 was $2.34 and $1.97 million, respectively. For oil structures, P50 the last year of production was $627,000, while for the third and fourth year before production cessation, P50 was $1.15 and $1.32 million, respectively. P50 values do not always increase in the years before cessation due to production and price variability but in all cases are greater than the last year of production.

8.4 FACTOR MODEL

8.4.1 Model Specification

A factor model is used to investigate the impact of individual variables on economic limits. A linear model is specified with a fixed term coefficient:

$$EL = \alpha_0 + \sum \alpha_i X_i,$$

where the variables are selected by the user and the factor coefficients are determined via regression. The values of the model coefficients will depend on the period of the evaluation and the selection of the model variables.

Factor variables examined include structure type, production type, water depth, manning status, operator type, and complex type and are defined as indicator variables: *Type* = structure type (0 if C/WP and 1 if FP), *Oil/Gas* = primary production (0 if oil and 1 if gas), *Water Depth* = water depth category (0 if <200 ft and 1 if >200 ft), *Manned* (0 if unmanned and 1 if manned), *Major* = operator type (0 if independent and 1 if major), and *Complex* = complex identification (0 if single-structure complex and 1 if multi-structure complex).

8.4.2 Results and Discussion

The first model constructed is a four-factor model:

$$EL_A = \alpha_0 + \alpha_1\, Type + \alpha_2\, Oil/Gas + \alpha_3\, Water\ Depth + \alpha_4\, Manned$$

All the coefficients are positive and significant (Table 8.4). For an unmanned oil C/WP in <200 ft water depth, for example, the average economic limit is \$1.22 million with a 95% confidence limit between \$821,000 and \$1.62 million. For an unmanned gas fixed platform in <200 ft water depth, *EL* = \$2.65 million. For an unmanned oil fixed platform in <200 ft water depth, *EL* = \$1.33 million.

Among the four descriptor variables, structure type is the least significant, about an order of magnitude smaller than the other variables that are all approximately of the same size. The distinction between oil and gas production is the most significant factor, but only slightly greater than the fixed term coefficient, which can be interpreted as an average fixed cost of operation.

In the second model, a term is added to distinguish operators that are major integrated companies:

$$EL_B = \alpha_0 + \alpha_1\, Type + \alpha_2\, Oil/\text{Gas} + \alpha_3\, Water\ Depth + \alpha_4\, Manned + \alpha_5\, Major$$

For all things equal, one would expect majors to have a greater economic limit than independents because of greater administrative and overhead cost, greater planning

Table 8.4 Economic Limit Regression Models in the Shallow-Water Gulf of Mexico (Million 2016\$)

Variable	Model A	Model B	Model C
Fixed term	1.22 (6.0)	1.07 (5.1)	1.08 (5.2)
Type	0.11 (0.7)	0.17 (1.0)	0.16 (1.0)
Oil/Gas	1.32 (6.2)	1.37 (6.4)	1.37 (6.4)
Water Depth	1.18 (4.0)	1.20 (4.2)	1.20 (4.2)
Manned	1.00 (2.3)	1.00 (2.3)	1.02 (2.3)
Major		0.82 (3.0)	0.82 (3.0)
Complex			−0.31 (−0.8)

Note: t-statistics denoted in parenthesis.
Source: Kaiser, M.J., Narra, S., 2018. An empirical evaluation of economic limits in the shallow water U.S. Gulf of Mexico, 1990-2017. J. Petrol. Sci. Eng. 164, 230–244.

requirements, and the absence of marginal properties in portfolios that contribute to higher average production cost. From Table 8.4, decommissioned structures operated by majors increased the economic limit by \$820,000 and were statistically significant.

Finally, in the third model, a term is added to identify those structures that were part of a multistructure complex at the time of decommissioning:

$$EL_C = \alpha_0 + \alpha_1\, Type + \alpha_2\, Oil/Gas + \alpha_3\, Water\ Depth + \alpha_4\, Manned + \alpha_5 Major + \alpha_6\, Complex$$

One would expect structures that are part of a complex would have a lower economic limit of production compared with isolated structures, and indeed, multicomplex structures on average had a lower economic limit by \$310,000, about one-third as large as the primary terms but a larger contributor than structure type.

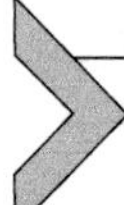

8.5 LIMITATIONS

8.5.1 Generalization

There are few generalized statements that can be made about offshore operations because of the dynamic nature of the industry and broad spectrum of company strategies, sizes, and capabilities within the sector. Conceptual and economic relations allow one to frame issues in a transparent manner and to provide first-order approximations, but because there are usually multiple interacting and overlapping factors that cannot be characterized, it is generally impossible to measure or recognize them all, and special cases and alternative interpretations abound. Conceptual relations and basic regression techniques are usually adequate in analysis but remain constrained in their ability to characterize operations without a deeper understanding of the engineering and technical considerations involved. Using end-of-year revenue as a proxy for the economic limit is a gross approximation.

8.5.2 Gross Revenue Approximation

Gross revenue is an approximation to actual revenues received since sales prices and product quality are unobservable. Market prices are believed to be a reasonable proxy of sales prices, but adjustments for quality (gravity, sulfur content, and heating value), transportation expenses, hedging programs, contract conditions, etc. cannot be performed. Price adjustments for quality and transportation are usually considered second-order effects (i.e., dollars on the barrel) and are frequently neglected but for particular properties may have a significant impact. Royalty rates are essentially constant across the sample, and because they do not differentiate structures, they do not need to be considered. The gross revenue computation is an approximation to net revenue which is an approximation to actual revenue and is believed to be a robust and reliable measure.

8.5.3 Structure Classification

A four-legged jacket structure with two or three wells and minor topside equipment would normally be classified as a well protector, but in some instances, a fixed platform identification may have been adopted, and the distinction between these two structure classes is not always well defined. Caissons and well protectors were consolidated into the same structure category for convenience. Platforms could have been decomposed into minor and major structures, but this was not pursued because of the duplicity of the categorization with structure type. Structure size can be proxied by the number of well slots, or deck size, but was not considered since structure type and water depth provide suitable differentiation and constraints on data availability limit application. Manned multistructure complexes identify all structures as manned and introduce ambiguity on manning status for a subset of the sample but are otherwise considered of negligible impact in the analysis.

8.5.4 Interpretation

Economic limit statistics are straightforward to interpret with minimal levels of uncertainty. A rationale operator will terminate operations at/near its economic limit unless other conditions prevail. Production data are reliable in aggregate although in some cases erroneous well links will impact results. Large samples and application of median statistics help ensure potential errors/bias average out and outlier impacts are small.

8.5.5 For All Other Things Equal

The disclaimer "for all other things equal" was frequently used when discussing or reporting results, one of the favorite monikers of economist. Rarely, however, if ever, and never in fact are all other things equal, especially for offshore developments that are man-made and built at different times and at different locations using different technologies by different operators applying different economic criteria and trade-offs. There are significant commonalities involved in offshore oil and gas development and decision-making, and the interplay between these similarities and differences will impact evaluation and results. Regression models are only able to control for factors in an approximate manner. Although engineering requirements are similar throughout the world and the physics of drilling and fluid flow obey the same laws and properties everywhere, best practices and design choices change over time by region and operator and may or may not impact the unique project-specific nature of development.

8.5.6 Independence

Structures are treated as a statistical ensemble with elements considered independent of one another for the purpose of evaluation, while in practice, significant interrelationships and dependencies are present based on regional operations, pipeline activities, and service (e.g., gas lift) requirements. These dependencies, while interesting and a key feature of

GoM operations, are for most practical purposes difficult to establish and incorporate in models and were not considered. The impact of these conditions may or may not be significant.

8.5.7 Aggregation and Categorization

Structures served as the assessment unit in evaluation, but different units may be applied at higher or lower levels of organization such as a well, lease, or field. Caution should be exercised when establishing correlations between high-level aggregate data if the units do not represent the business/financial groupings of operators. Aggregation consolidates information, stripping away some useful characteristics while smoothing out data variation, and if the data sets are not large enough, moderate levels of correspondence may be "discovered" between unrelated or irrelevant variables. The experience of the analyst should be enough to eliminate irrelevant factors, but this is not a given, nor does it always hold. Consolidation is necessary in many contexts but should not be used without a clear and definitive basis and understanding of operational and logistic requirements. Aggregation may mask or distort relevant information or present trends of a dominant class (i.e., gas structures) at the expense of smaller subcategories.

As a larger number of attributes are applied to delineate structure data, categories become more granular, and the data populating individual categories smaller in size. This creates a trade-off in evaluation since the benefits that accrue when creating more homogenous and refined categories may be lost or reduced with smaller sample sizes that are less representative and introduce greater volatility and outlier data. Sample averages of sparsely populated categories are not representative and usually cannot be considered to support or refute hypothesis.

REFERENCES

Kaiser, M.J., 2014. Hurricane clean-up activity in the Gulf of Mexico, 2004-2013. Mar. Policy 51, 512–526.

Kaiser, M.J., Narra, S., 2018. An empirical evaluation of economic limits in the shallow water U.S. Gulf of Mexico, 1990-2017. J. Petrol. Sci. Eng. 164, 230–244.

CHAPTER NINE

Deepwater Structure Inventory

Contents

Abstract

The deepwater inventory in the GoM consisted of 48 fixed platforms, three compliant towers, and 47 floaters circa 2016. All of the floaters were producing circa 2016, while 12 of the 48 fixed platforms did not produce, and at least five of these structures have been converted to pipeline junctions. In the first part of the chapter, floater equipment capacity and capacity-to-reserves ratios are examined followed by well type, production, reserves, PV-10, and revenue statistics circa 2016. In the second half of the chapter, gross revenue per individual structure provides a snapshot of the sector.

Decommissioning Forecasting and Operating Cost Estimation
https://doi.org/10.1016/B978-0-12-818113-3.00009-6

9.1 FLOATER EQUIPMENT CAPACITY

Topsides include all the equipment to separate, treat, dehydrate, and prepare production for export; to treat the produced water; to compress the gas for treating and export; to provide metering for custody transfer of oil and gas; and utilities, storage, and related system elements. In some cases, water injection and gas injection systems are needed. Equipment capacity is normally designed to match development requirements (right-sized) but in some cases may be built with extra capacity in anticipation of future tiebacks (oversized).

Nameplate equipment capacity describes the maximum oil and gas processing capability of the structure and is described by the gas-to-oil equipment capacity (G/O) ratio measured in cubic feet gas per day (cfpd) per barrel crude per day (bpd). Since the daily rates cancel, the G/O ratio is described in cubic feet of natural gas per barrel crude (cf/bbl) similar to the gas-oil ratio (GOR) describing production. A value $G/O < 5000$ cf/bbl generally indicates an oil structure although thresholds as high as 10,000 cf/bbl may be applied. Values of $G/O < 1000$ cf/bbl imply heavy (black) oil reservoirs.

For GoM floaters, equipment capacity data are publicly available, while fixed platform data are more limited, and so, the focus in this and the next section is on floaters.

Example. Jolliet TLP nameplate capacity

The Jolliet tension leg platform (complex 23583) was the first TLP installed in the GoM in 1990 and was designed for a maximum production rate of 35,000 bpd oil and 50 MMcfd gas. The nameplate G/O capacity ratio is computed as 1429 cf/bbl:

$$G/O = \frac{50\,\text{MMcfpd}}{35{,}000\,\text{bopd}} = 1429\,\text{cf/bbl}$$

Cumulative oil production from Jolliet through 2016 was 36 MMbbl oil and 136 Bcf natural gas or a cumulative gas-oil ratio of CGOR = 3797 cf/bbl. Production of natural gas usually increases as oil reservoir sands deplete.

In Fig. 9.1, the nameplate equipment capacity for all floaters (except Independence Hub) circa 2016 is depicted, and lines of constant slope $G/O = 10$, 5, 2.5, 1, and 0.5 Mcf/bbl are overlaid to characterize reservoir fluids. Structures in construction circa 2017 are shown italicized and in red, and the shaded box represents the most common design capacities. Groupings of equipment capacity that fall along vertical and horizontal lines reflect standardized well and equipment designs.

Most floater G/O capacity falls in the G/O slice between 1 and 2.5 Mcf/bbl, indicating the mostly liquid and oily nature of deepwater reservoirs developed to date. With large production volumes of oil, gas is more than adequate to meet facility needs

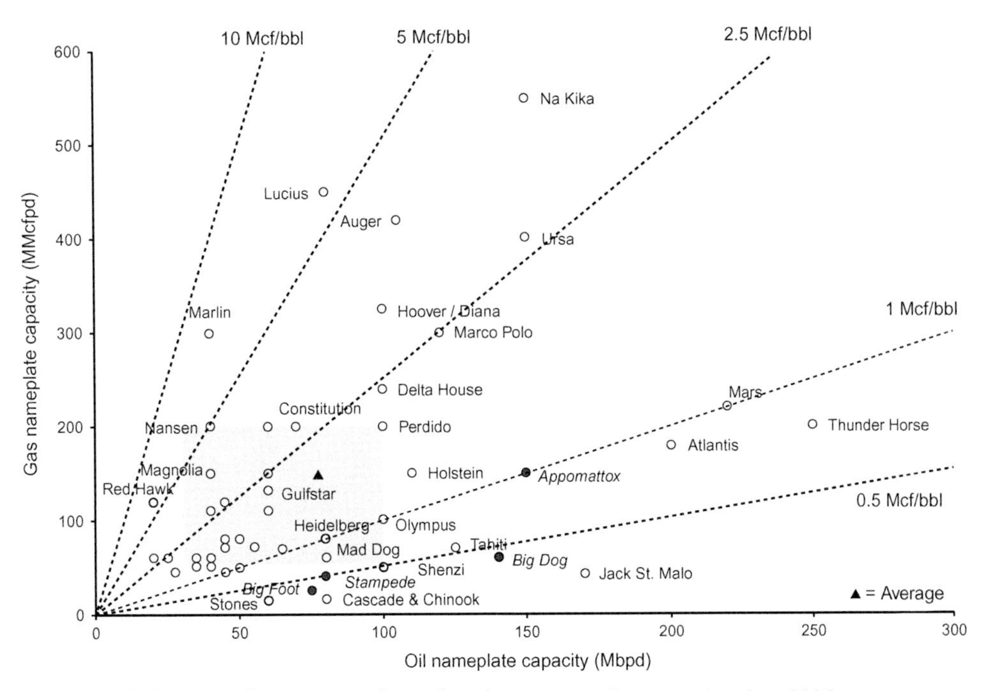

Fig. 9.1 Gulf of Mexico floater nameplate oil and gas processing capacity circa 2016.

and the excess is exported. Many recent developments in the Lower Tertiary trend (e.g., Stones, Jack/St Malo, Cascade and Chinook) are heavy oil developments with equipment capacity G/O<0.5 Mcf/bbl, which present additional engineering challenges and low recovery factors. Independence Hub is the only (dry) gas development in deepwater with a stand-alone structure. Independence Hub's gas processing capacity of 1 Bcfpd and 5000 bpd condensate capacity would plot in the top left slice off the chart in the G/O>10 Mcf/bbl sector.

About half of deepwater floaters have oil processing capacity that ranges between 30 and 100 Mbpd and gas processing capacity between 50 and 200 MMcfpd (shaded box, Fig. 9.1). Average nameplate processing capacity for the floater inventory is 77 Mbpd oil and 150 MMcfpd natural gas (102 Mboepd) with standard deviation of 56 Mboepd. This translates to an annual average processing capacity of 40 MMboe per structure. The most common oil capacity category among floaters is 25–50 Mbopd (Fig. 9.2).

Five structures have oil processing capacity >150 Mbpd (Ursa, Jack/St Malo, Atlantis, Mars, and Thunder Horse), and five structures have gas processing >350 MMcfpd (Devils Tower, Ursa, Auger, Na Kika, and Lucius). Semis have the largest average processing capacity and also the largest variation across structure type reflecting their application in both small and large field developments (recall Fig. 2.8). Projects sanctioned or under construction circa 2017 are summarized in Box 9.1.

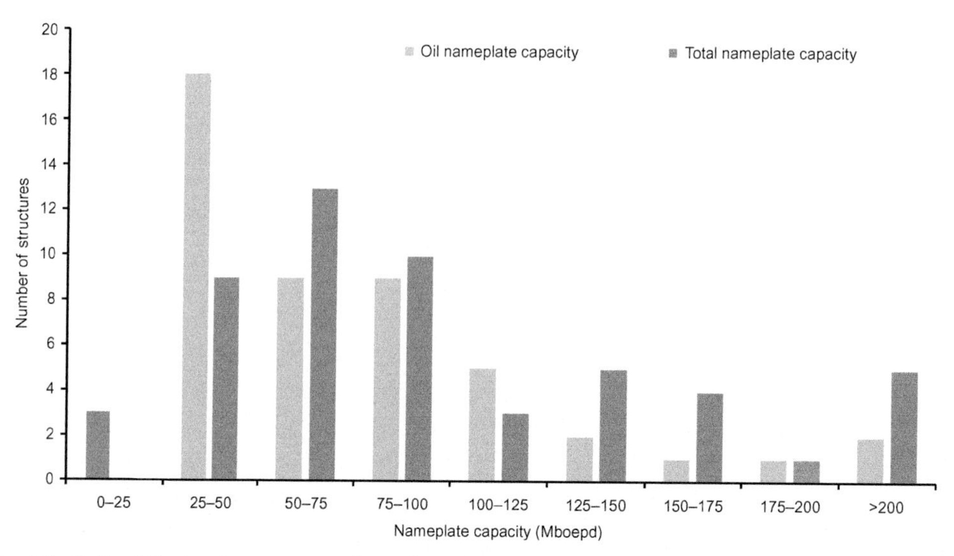

Fig. 9.2 Gulf of Mexico floater nameplate oil and gas processing distribution circa 2016.

BOX 9.1 Projects Sanctioned or Under Construction circa 2017

Deepwater floaters sanctioned or under construction in the GoM circa 2018 include Big Foot (ETLP); Stampede (TLP); Appomattox (semi); and Mad Dog Phase 2 (semi), also known affectionately as Big Dog.

The Big Foot oil field is located in Walker Ridge block 29 in 5200 ft water depth and is estimated to hold more than 200 MMboe reserves. An extended TLP was selected as the development concept, but during installation, nine of its 16 tendons lost buoyancy, and the TLP was severely damaged by a loop current while preparing for hookup in 2015. Production began in November 2018.

Hess installed the Stampede TLP in 2017 in approximately 3500 ft water depth, and first production is expected in early 2018. Stampede is located in Green Canyon blocks 468 and 512 and is estimated to have resources in the range of 300–350 MMboe. Topside processing capacity is approximately 80,000 bpd oil and 40,000 MMcfpd natural gas.

The Appomattox development is located in Mississippi Canyon blocks 348 and 392 in approximately 7400 ft water depth and will be developed with a semisubmersible initially producing from the Appomattox and Vicksburg fields with recovery estimated at 650 MMboe. First production is scheduled for 2020. The project calls for a subsea system featuring six drill centers, 15 producing wells, and five water injection wells. Average peak production capacity is estimated at 200,000 boepd. If the Gettysburg and Rydberg prospects are sanctioned for development, the total estimated recovery would be 800 MMboe.

BP sanctioned the Mad Dog Phase 2 project in December 2016 as a semisubmersible development moored in Green Canyon in 4500 ft water depth about 6 mi southwest of the existing Mad Dog spar. Production capacity is expected at 140 Mbpd crude and 60 MMcfpd natural gas with first production in late 2021.

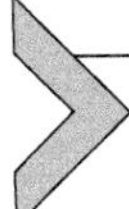

9.2 FLOATER CAPACITY-RESERVES STATISTICS

9.2.1 Capacity-to-Reserves Ratio

At the time of project sanction, the capacity of production equipment and export pipelines are usually known and sized for the well plan and maximum flow rates expected. Capacity-to-reserves (CR) ratio is defined as the oil and gas production handling capacity installed expressed in heat-equivalent barrels on an annual basis to total expected production (i.e., reserves) expressed in barrels of oil equivalent:

$$\text{Capacity-to-reserves ratio} = \frac{\text{Equipment capacity (boe)}}{\text{Total production (boe)}}$$

Equipment capacity at installation does not include water handling, water injection, or gas injection. The denominator needs to be computed (estimated), and it is really total throughput that is measured since tieback fields may bring back additional production to the facility.

Over time, investments may be made to increase processing capacity at the structure, or wells may not produce as expected. In either case, the CR ratio will change, increasing if equipment capacity was increased or expected reserves were not realized, or decreasing if additional production is brought back to the facility. Once equipment is installed, it is rarely downsized, but increasing capacity to handle additional flows from tieback fields is not uncommon.

Equipment capacity is not directly observable unless reported by the operator, and subsequent changes may or may not be reported. Total production changes over time with reserves growth or unanticipated problems or new field tiebacks. Only when the structure is no longer useful and decommissioned is total production known with certainty.

Equipment requirements may change as tieback fields are added, and these are often reported by operators if significant work was required at the facility:

- Mars TLP was originally designed to handle 100 Mbpd oil and 100 MMcfpd gas, but with subsea tiebacks from the King, Europa, and Deimos fields, production capacity was later expanded to 220 Mbopd and 220 MMcfpd.
- Auger TLP processing capacity has been expanded three times since installation, from its original 55/130 Mbpd/MMcfpd capacity to its current 105/420 nameplate.
- Production at the Na Kika hub initially included six fields in 2003, but third-party tiebacks in the Galapagos development—Isabella (2011), Santiago (2014), and Santa Cruz (2015)—required processing capacity to be expanded to 150/550 nameplate.

- The Thunder Hawk semi was designed with 45/70 capacity, but with third-party tiebacks Big Bend and Dantzler in 2015, the facility expanded oil processing capacity to 60 Mbopd and added gas-lift capabilities.

CR ratios have a greater tendency to decrease over time due to reserves growth and tieback fields that add production beyond the design basis. If tiebacks occur when the structure is at or near peak production, capacity will need to be added, while if the tieback occurs late on the decline curve, capacity additions will be limited or not needed.

9.2.2 Capacity-to-Reserves Statistics

The average CR ratio for all floaters circa 2016 (except Heidelberg) is 0.32 with a standard deviation of 0.31 (Fig. 9.3). Heidelberg started producing near the time of evaluation and its reserves could not be reliably computed. About 75% of the floaters have CR ratios <0.4. There are four outliers at the high end:

- Cascade and Chinook produced 29 MMbbl oil and 4.8 Bcf gas through 2016. In 2016, 4.4 MMbbl oil and 0.7 Bcf were produced.
- Prince produced 0.7 MMbbl oil and 1.2 Bcf gas in 2016 from four wells with cumulative production 10.1 MMbbl oil and 11.2 Bcf gas.
- Gulfstar (Tubular Bells) was producing from six wells in 2016 with cumulative production of 17.1 MMbbl oil and 35.6 Bcf gas through 2016.
- Telemark/Mirage/Titan fields produced 13.4 MMbbl and 14.3 Bcf from four producing wells through 2016.

9.3 WELL TYPE

There are two types of offshore wells that are important to distinguish—wet tree wells and dry tree wells. Wet wells have remote subsea wellheads that are connected to their host by steel or flexible catenary riser systems, lazy wave systems, or hybrid risers, and in some cases (e.g., Auger, Mars, Ram Powell, Ursa, Brutus, and Perdido), the wellhead/tree is located on the seabed directly below the structure to reduce topside weight and is referred to as direct vertical access (DVA) wells (Fig. 9.4). DVA wells are subsea wells since the wellhead/tree is on the seabed, but because they allow (direct) rig access from the platform and do not come with all the expensive subsea equipment (e.g., manifolds, flow lines, umbilicals, and jumpers) and flow assurance issues common to wet wells, they are (somewhat) similar to dry tree wells (Reid and Dekker, 2013).

Dry trees reside above the waterline and are connected to the wellbore with a top-tensioned riser. Dry tree and DVA wells allow direct access from the platform, while a wet well requires mobilizing an intervention vessel or MODU to access the well. Wet wells are more expensive to operate, are more expensive to intervene, have a higher economic limit, and have smaller recovery rates relative to dry tree and DVA wells,

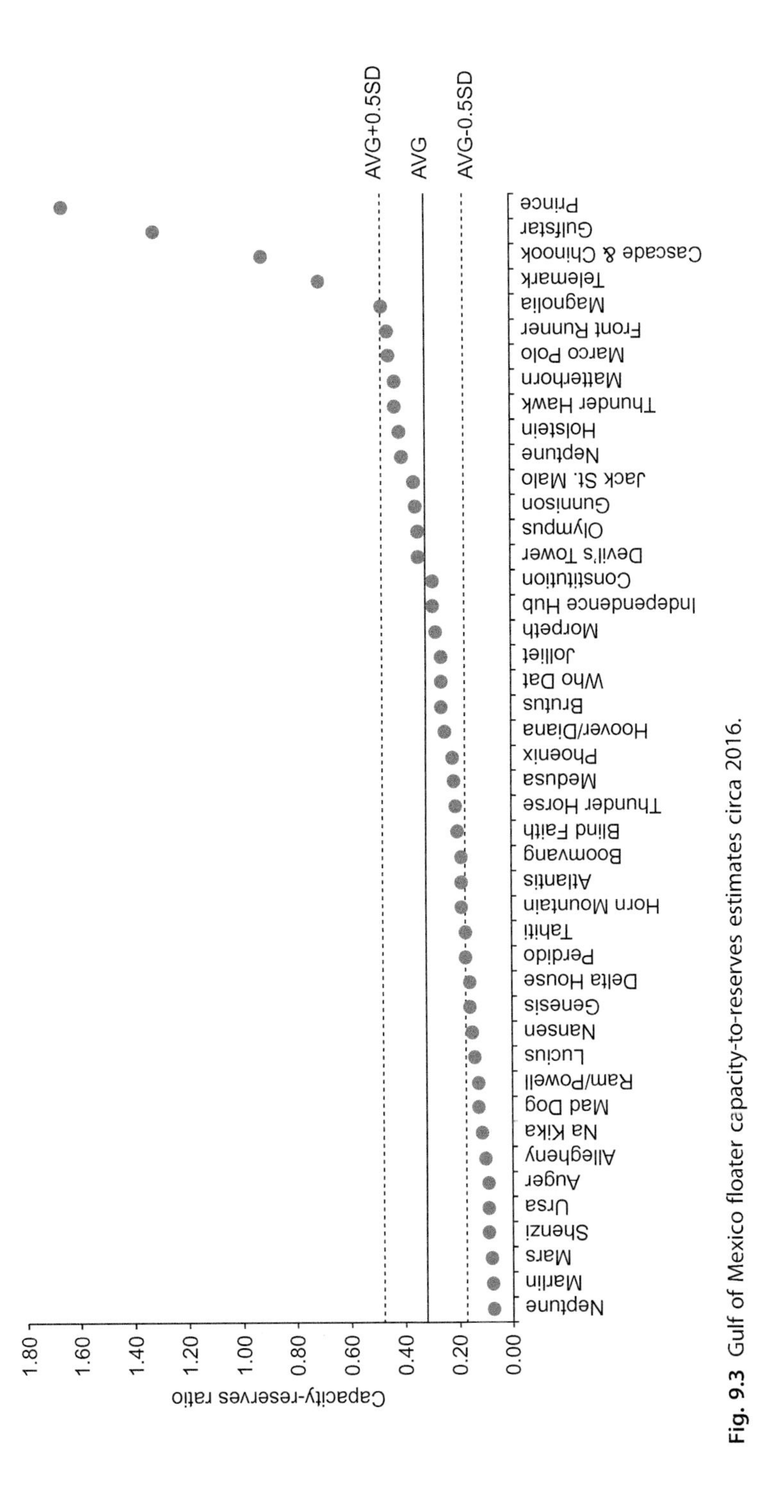

Fig. 9.3 Gulf of Mexico floater capacity-to-reserves estimates circa 2016.

Fig. 9.4 Semisubmersible host with wet tree and direct vertical access capability. *(Source: Reid, D., Dekker, M., 2013. Deepwater development: wet or dry tree? In: OTC 24517. Offshore Technology Conference. Rio de Janeiro, Brazil, Oct 29–31.)*

for all things equal. FPSOs and most semisubmersibles are wet well developments (Lim and Ronalds, 2000), while spars and TLPs permit both dry and wet well tiebacks. Many Shell-operated floaters in the GoM employ DVA wells.

In 2016, there were 52 wet wells tied back to fixed platforms in 400–500 ft water depth compared with 485 dry tree wells, and for fixed platforms and compliant towers in >500 ft water depth, there were 210 wet wells and 906 dry tree wells (Table 9.1). For floaters, wet wells outnumbered dry tree wells 746 to 648. Subsea wells represent a larger percentage of wells in floater developments and a larger percentage of producing wells in deepwater.

9.4 PRODUCTION

In 2016, there were 36 fixed platforms, three compliant towers, and 47 floating structures that were producing. Walter's Coelacanth (2606) and Shell's Stones (2503) facilities achieved first oil in 2016. All the floaters except one (Independence Hub) and the three compliant towers are classified as oil producers (i.e., CGOR >5000 cf/bbl), while the majority of the fixed platforms are also oil producers (30 vs. nine gas producers).

Table 9.1 Gulf of Mexico Deepwater Inventory Statistics Circa 2016

	Fixed Platforms	FPs/CTs	
	400–500 ft	> 500 ft	Floaters
Structures			
Oil (#)	14	20	46
Gas (#)	8	9	1
Cumulative production			
Oil (MMbbl)	360	1838	5732
Gas (Bcf)	3058	6379	11,777
Production			
Oil (MMbbl)	2.5	37	469
Gas (Bcf)	10.5	119	651
Gross revenue ($ million)	129	1781	20,657
Proved reserves			
Oil (MMbbl)	4.3	135	1826
Gas (Bcf)	16.5	447	2312
PV-10 ($ billion)	0.16	4.14	51.3
Wells			
Dry tree (#)	485	906	648
Wet tree (#)	52	210	746
Producing (#)	82	256	464

Note: PV-10 values assume $60/bbl crude oil and $3/Mcf natural gas price, 16.67% royalty rate, and $7/boe operating cost.
Source: BOEM 2017.

Sixteen platforms in 400–500 ft water depth produced 2.5 MMbbl oil and 10.5 Bcf gas in 2016, while the 20 fixed platforms and three compliant towers in >500 ft water depth produced 37 MMbbl oil and 119 Bcf gas (Table 9.1). The 47 structures in the floater class produced 469 MMbbl oil and 651 Bcf gas. The difference in volumes across each class is about an order of magnitude.

Through 2016, deepwater fixed platforms and compliant towers have together produced about 2.2 Bbbl oil and 9.4 Tcf gas over their lifetime, while floating structures have produced 5.7 Bbbl oil and 11.8 Tcf gas (Table 9.1). In Table 9.2, the distribution of production on a heat-equivalent basis is depicted for oil and gas structures with idle structure counts denoted in parenthesis. Floaters populate all of the high-volume oil categories, while fixed platforms are more evenly split between oil and gas.

Table 9.2 Deepwater Structure Cumulative Production in the Gulf of Mexico Circa 2016

Oil Production (MMboe)	Fixed Platforms	Compliant Towers	Floaters
<25	13(3)	0	5
25–50	5	0	4
50–100	5(1)	0	13
100–150	3	1	9
150–200	0	1	4
200–500	4	1	8
>500	1	0	3
Total	31(4)	3	46
Gas Production (Bcfe)	**Fixed Platforms**	**Compliant Towers**	**Floaters**
<50	1(1)	0	0
50–100	3(3)	0	0
100–200	4(1)	0	0
200–300	5(3)	0	0
300–400	1	0	0
400–500	1	0	0
>500	2	0	1
Total	17(8)	0	1

Note that numbers in parenthesis represent structures that did not produce in 2016.
Source: BOEM 2017.

9.5 GROSS REVENUE

Gross revenue is a first-order estimate of revenues received and is computed using the average Light Louisiana Sweet crude ($40.6/bbl) and Henry Hub natural gas price ($2.5/Mcf) in 2016. Condensate is valued at crude oil prices even though it is normally discounted and will generate less revenue per barrel. Associated gas is valued at gas well gas prices even though it will realize additional revenue from natural gas liquid (NGL) sales. No adjustments for product quality are made. Net revenue received by the operator is gross revenue reduced by the royalty payment to the federal government, which for most structures is 16.67% (18.75% royalty applies for leases issued after 2008).

Structures that generate less than a few million dollars annually may be considered marginal or approaching marginal status because the direct operating cost for deepwater structures—production and maintenance crew, helicopter flights, service boats, chemicals, maintenance expenses, etc.—is approximately of this magnitude. However, if structures are part of regional operations or serve third-party production or transportation services, they may be able to operate profitably at lower revenue levels.

Collectively, the 16 producing structures in 400–500 ft water depth generated about $129 million or $8.1 million per structure on average in 2016, compared with

Table 9.3 Deepwater Structure Revenue in the Gulf of Mexico Circa 2016

Gross Revenue (Million $)	Fixed Platforms	Compliant Towers	Floaters
No Production	12		
<5	11		1
5–30	14	1	3
30–100	7		13
100–500	4	2	14
500–1000			8
1000–2000			8
Total	48	3	47

Source: BOEM 2017.

$1.8 billion total or $78 million per structure for structures in >500 ft water depth (Table 9.1). Floaters generated about $21 billion total or about $440 million per structure in 2016.

Twelve fixed platforms were not producing, and 11 fixed platforms and one floater had gross revenue less than $5 million (Table 9.3). Of the 12 nonproducing structures, seven were idle, and five were serving in an auxiliary role.

There were 18 structures that generated between $5 and $30 million, and many of the structures of this class, especially structures at the low end of the revenue category, would be considered as approaching marginal status. There were 20 structures that generated between $30 and $100 million and 20 structures that generated between $100 and $500 million. Eight structures, all floaters, generated between $500 and $1000 million, and another eight floaters each generated more than $1 billion.

The Independence Hub semi (1766) ceased production in 2016 and has subsequently abandoned all its wells. The Mirage/Titan semi (2089) last produced in October 2016 with cumulative production 13.5 MMbbl oil and 14.4 Bcf natural gas. Circa 2017, wells on Mirage/Titan were not abandoned.

9.6 RESERVES

Reserves are estimated using producing well inventories circa 2016 and standard industry models and assumptions on decline curves (e.g., Poston and Poe, 2008). See Chapter 11 for additional discussion and Chapter 12 for two examples. If reserves are computed according to Society of Petroleum Engineers (SPE) guidelines (PRMS, 2007), there is a 90% chance that the estimates will increase from current estimates in the years ahead, and so, the values computed are considered a lower-bound conservative estimate.

Fixed platforms in 400–500 ft water depth are estimated to have proved reserves of 4.3 MMbbl of oil and 16.5 Bcf of gas, compared with 135 MMbbl of oil and 447 Bcf of gas for fixed platforms and compliant towers in >500 ft water depth, and 1.8 Bbbl of oil and 2.3 Tcf of gas for floaters (Table 9.1). Reserves estimates for fixed platforms are more accurate than floaters because of the maturity of the producing wells.

9.7 PV-10

Whereas gross revenue is a snapshot of the value of production at the end of a reporting period, PV-10 provides an indication of the (discounted) value of future production arising from proved reserves. PV-10 denotes the present value of the expected cash flows generated from reserves discounted at 10% on a before-tax basis computed using a standard set of assumptions and procedures on decline rates, oil and gas price, operating cost, and inflation rate. PV-10 requires a cash flow model to compute and is a simple extension of reserves estimation, and although it is not a generally accepted accounting procedure (GAAP) measure, it is still commonly used and reported throughout industry.

The GAAP measure is called the standardized measure and is similar to PV-10 except that it is computed on an after-tax basis (PRMS, 2007). Standardized measures are reported by public companies at the corporate level and are normally consolidated by region or business segment in financial statements.

PV-10 values were computed for each producing structure using a constant oil price of \$60/bbl and gas price of \$3/Mcf, 16.67% royalty rate, and \$7/boe operating cost (Table 9.4). Eighteen fixed platforms had reserves values less than \$10 million, and seven structures had values between \$10 and \$50 million. Twenty structures, mostly floaters, had PV-10 values greater than \$500 million. In total, fixed platforms and compliant towers in water depth > 500 ft had a PV-10 of about \$4.1 billion, compared with the PV-10 of floaters estimated at \$51.3 billion. The PV-10 value of fixed platforms in 400–500 ft water depth was about \$160 million.

Table 9.4 Deepwater Structure PV-10 in the Gulf of Mexico Circa 2016

PV-10 (Million $)	Fixed Platforms	Compliant Towers	Floaters
< 10	18	0	2
10–50	7	1	1
50–100	5	0	5
100–200	2	0	7
200–500	2	2	13
>500	2	0	18
Total	36	3	47

Note: PV-10 values assume oil price of \$60/bbl and gas price of \$3/Mcf, 16.67% royalty rate, and \$7/boe operating cost.
Source: BOEM 2017.

9.8 FIXED PLATFORMS, 400–500 FT

There are 22 fixed platforms that reside in 400–500 ft water depth circa 2016 (Tables 9.5 and 9.6). Sixteen of the 22 fixed platforms were producing circa 2016.

9.8.1 Idle

Six of the 22 fixed platforms were idle circa 2016 with last year of production ranging between 1992 and 2012.

Two of the idle structures are toppled (Taylor Energy's complex 23051 and McMoRan's 23925), and two of the structures serve as pipeline junctions (Manta Ray Gathering 23212 at Ship Shoal 332 and Poseidon Oil Pipeline 23353 at South Marsh Island 205).

Table 9.5 Production and Gross Revenue for Fixed Platforms in 400–500 ft Circa 2016

			2016 Production		Cum. Production		
Operator	**Complex**	**Producing Well**	**Oil (MMbbl)**	**Gas (Bcf)**	**Oil (MMbbl)**	**Gas (Bcf)**	**Gross Revenue (MM $)**
Arena Offshore LP	23848	11		3.4	0.3	201.7	8.6
Bennu Oil & Gas	2027	1		0.1	0.9	2.8	1.0
Chevron U.S.A.	1052	3	0.1	0.1	14.1	120.0	2.3
Energy Res. Techn.	90028	3	0.1	0.2	7.8	87.9	4.7
Energy XXI GOM	22172	11	0.2	0.4	73.3	353.8	7.7
Energy XXI GOM	22685				16.4	29.8	
Energy XXI GOM	23151	6	0.1	0.1	36.9	109.4	3.2
Energy XXI GOM	23893	1	0.1		8.9	649.6	2.3
Fieldwood Energy	1500	7	0.6	1.3	17.0	33.6	27.6
Fieldwood Energy	22224	5	0.1	0.2	32.7	305.6	6.5
Fieldwood Energy	22662	4	0.1	0.2	65.0	241.9	6.3
Fieldwood Energy	23800	16	0.6	1.9	30.0	91.0	27.6
Fieldwood Energy	80015	1			3.7	34.1	
Manta Ray Gathering	70	4	0.4	2.5	8.4	373.6	23.1
Manta Ray Gathering	23212				3.3	75.4	
McMoRan Oil & Gas	23925				5.9	114.7	
Poseidon Oil Pipeline	23353				0.0	88.7	
Renaissance Offshore	1076	2			1.0	1.2	0.6
Tarpon	1165	2	0.1		5.2	3.6	3.3
Taylor Energy	23051				13.8	40.8	
W&T Offshore	1279	5	0.1	0.1	14.7	38.7	4.4
W&T Offshore	10192				0.3	60.6	

Note that structures 23848, 2027, 23893, 1076 and 80015 had oil production <10,000 bbl and/or gas production < 100 MMcf in 2016.

Source: BOEM 2017.

Table 9.6 Fixed Platform Revenue Status in Water Depth 400–500 ft Circa 2016

Operator	No Production (Last Year)	Gross Revenue (Million $)	
		<5	5–30
Arena Offshore			23848
Bennu Oil & Gas		2027	
Chevron U.S.A.		1052	
Energy Res. Tech.		90028	
Energy XXI GOM	22685 (2004)	23151, 23893	22172
Fieldwood Energy		80015	1500, 22224, 22662, 23800
Manta Ray Gathering	23212 (1992)		70
McMoRan Oil & Gas	23925 (2008)		
Poseidon Oil Pipeline	23353 (2007)		
Renaissance Offshore		1076	
Tarpon		1165	
Taylor Energy	23051 (2004)		
W&T Offshore	10192 (2012)	1279	
Total	6	9	7

Source: BOEM 2017.

Taylor Energy's complex 23051 in 475 ft water depth in Mississippi Canyon 20 was destroyed by Hurricane Ivan in 2004 and will likely never be fully decommissioned because of safety and technical issues. The hurricane caused a mudslide in the region, and the platform slid 500 ft downslope, resting on its side partially buried by over 100 ft of mud and sediment. Nine of the 25 wells have been plugged and abandoned circa 2017, the platform deck has been removed, and the oil pipeline has been decommissioned.

Freeport-McMoRan Oil & Gas complex 23925 in Ewing Bank 947 was toppled by Hurricane Ike in 2008. The operator applied for a permit to reef the jacket in place having previously removed the deck, but BSEE did not grant the permit due to uncertainties related to site contamination and instability issues.

Manta Ray Gathering, originally a joint venture between Shell Gas Transmission, Marathon, and Enterprise, acquired complex 23212 in 1992 when production ceased and currently supports the Leviathan Offshore Gathering System and Poseidon Oil and Allegheny Systems.

Energy XXI GOM complex 22685 at South Pass 49 last produced in 2004, and W&T Offshore complex 10192 at High Island 389 stopped producing in 2012.

9.8.2 Gross Revenue <$5 Million

Nine fixed platforms generated less than $5 million in 2016, and producing well counts ranged between one and five wells per structure. Annual production for all nine structures totaled <1 MMbbl crude and 100 MMcf gas, and most structures in this class can be considered marginal or approaching marginal status.

Three structures (Energy XXI GOM 23893, Renaissance Offshore 1076, and Fieldwood Energy 80015) produced <10,000 bbl oil and <100 MMcf gas in 2016 and are probably no longer commercially viable. Gross revenues for these structures are estimated at less than $2 million per structure.

Bennu Oil & Gas complex 2027, Energy XXI GOM complex 23151 and Alabaster (23893), and Fieldwood Energy's complex 80015 all have one producing well circa 2016. Structures with one or two producing wells are in a difficult position since if a well fails or stops producing for any reason, it probably will not be economic to perform a workover.

9.8.3 Gross Revenue $5–30 Million

Seven fixed platforms generated between $5 and $30 million in 2016. Fieldwood Energy complexes 1500 and 23800 generated the largest revenues followed by Manta Ray Gathering complex 70.

9.9 FIXED PLATFORMS AND COMPLIANT TOWERS, >500 FT

There are 26 fixed platforms and three compliant towers in water depth >500 ft circa 2016 (Tables 9.7 and 9.8). These structures contain about twice the number of wells and four times the number of subsea wells and generate more than ten times the revenue of fixed platforms in 400–500 ft.

9.9.1 Idle

There are six idle fixed platforms in water depth >500 ft circa 2016. These structures may be serving an auxiliary role as a pipeline junction such as Triton Gathering's Pimento (23788) and Shell's Boxer (23277) or intend to serve such a role in the future. Idle structures are owned by ATP Oil & Gas (1320, last produced in 2012), Chevron (23760, Tick last produced in 2015), Fieldwood Energy (70016, Spirit last produced in 2013), and Fieldwood SD Development (10242, Tequila last produced in 2009). Production from Chevron's Jack/St Malo field is routed through the Boxer platform.

9.9.2 Gross Revenue <$5 Million

Two structures generated less than $5 million in 2016, Cerveza (10178) and Enchilada (27056). Shell installed Enchilada in 1997 in Garden Banks block 128 to support hub activity in the region and handle inflows from Auger pipeline junctions (Smith and Pilney, 2003). Cerveza and Ligera were installed in 1981 and 1982 for the East Breaks 160 field development.

Table 9.7 Production and Gross Revenue for Fixed Platforms in Water Depth >500ft Circa 2016

Operator	Complex	Producing Well	2016 Production Oil (MMbbl)	2016 Production Gas (Bcf)	Cum. Production Oil (MMbbl)	Cum. Production Gas (Bcf)	Gross Revenue (MM $)
Ankor Energy	1482	15	0.7	1.4	9.1	21.0	30.2
ATP Oil & Gas	1320				0.1	8.6	
Chevron U.S.A.	23760				29.0	263.0	
Chevron U.S.A.	70012	18	4.1	8.0	162.1	261.8	188.1
Eni US	23875	5	2.0	22.2	17.9	419.7	138.3
EnVen Energy	22178	12	0.7	6.1	181.4	808.0	43.1
EnVen Energy	24129	21	2.2	1.9	193.0	172.6	95.9
Exxon Mobil	22840	21	0.4	0.3	63.6	237.4	17.7
Fieldwood Energy	23552	21	3.1	6.7	404.1	686.8	141.5
Fieldwood Energy	70016				3.8	204.8	
Fieldwood SD	10178	5	0.1	0.2	16.4	88.3	4.0
Fieldwood SD	10212	9	0.1	1.2	15.2	186.4	6.7
Fieldwood SD	10242				3.6	190.3	
Fieldwood SD	10297	10	0.4	0.5	38.4	117.2	17.9
Flextrend Dev.	24201	4	0.7	0.5	16.4	125.2	31.6
Flextrend Dev.	27032	7	0.1	6.4	16.5	201.9	21.4
Hess	33039	7	3.1	10.5	124.0	450.5	153.0
MC Offshore	23567_1	3	0.5	2.7	10.9	48.6	28.3
MC Offshore	23567_2	7	0.1	0.1	15.9	16.7	5.9
Shell Offshore	27056	4	0.1	0.5	7.8	135.4	3.5
Shell Offshore	90014	6	8.1	28.4	138.0	553.7	401.2
Shell Oil	23277				9.3	59.6	
Stone Energy	23883	31	1.5	1.7	83.7	95.9	64.5
Stone Energy	24130	25	5.8	11.7	159.8	493.0	264.4
Triton Gathering	23788				0.0	248.5	
W&T Energy VI	113	7	0.1	3.0	4.0	119.0	12.1
W&T Offshore	147	6	1.1	3.5	18.0	43.0	53.4
Walter Oil & Gas	2606	2	0.7	0.5	0.7	0.5	31.4
Whistler Energy II	23503	10	0.6	0.7	95.4	121.2	26.5

Note that complex 2606 was installed in 2016.
Source: BOEM 2017.

9.9.3 Gross Revenue $5–30 Million

Eight structures generated between $5 and $30 million in 2016, including ExxonMobil's compliant tower Lena (22840), Fieldwood's Ligera (10212) and Tequila (10297), MC Offshore Petroleum's Marquette two complex (23567-1 and 23567-2), W&T Energy VI's Virgo (113), and Whistler Energy II's Boxer (23503).

Exxon began talks with the Louisiana Artificial Reef Council in 2015 to reef Lena in place and is undergoing environmental studies and federal review (Truchon et al., 2015).

Table 9.8 Fixed Platform and Compliant Tower Revenue Status in Water Depth >500 ft Circa 2016

Operator	No Production (Last Year)	Gross Revenue (Million $)			
		<5	5–30	30–100	100–500
Ankor Energy				1482	
ATP Oil & Gas	1320 (2012)				
Chevron U.S.A.	23760 (2015)				70012
Eni US					23875
EnVen Energy Ventures				22178, 24129	
Exxon Mobil			22840		
Fieldwood Energy	70016 (2013)				23552
Fieldwood SD Offshore	10242 (2009)	10178	10212, 10297		
Flextrend Development			27032	24201	
Hess					33039
MC Offshore Petroleum			23567-1, 23567-2		
Shell Offshore		27056			90014
Shell Oil	23277 (2013)				
Stone Energy				23883	24130
Triton Gathering	23788 (2003)				
W&T Energy VI			113		
W&T Offshore				147	
Walter Oil & Gas				2606	
Whistler Energy II			23,503		
Total	6	2	8	7	6

Source: BOEM 2017.

The Marquette two-platform complex was installed in 1989 to develop the Jolliet field and reservoirs in a 15-block unit at Green Canyon block 52. The platforms process production from the Jolliet TLP, about 9 mi south, and the GC 52 unit production (Tillinghast, 1990). MC Offshore Petroleum is the operator of Marquette and Jolliet.

9.9.4 Gross Revenue $30–100 Million

Seven structures generated between $30 and $100 million in 2016, including Ankor Energy's Simba (1482), EnVen Energy Ventures Cognac (22178) and Lobster (24129), Flextrend Development's Phar Lap Shallo (24201), Stone Energy's Amberjack (23883), W&T Offshore's complex 147, and Walter's most recently installed Coelacanth (2606).

9.9.5 Gross Revenue $100–500 Million

Shell's Salsa platform (90014) and Stone Energy's Pompano (24130) were the largest producers, generating about $400 million and $260 million, respectively, followed by

Chevron's compliant tower Petronius (70012) and Hess's compliant tower Baldpate (33039), Fieldwood's Bullwinkle (23552), and Eni's Corrla (23875).

9.10 FLOATERS

There are 47 floating structures in the GoM circa 2016, and all but one are primarily oil producers (Tables 9.9 and 9.10). These structures produced almost 90% of the crude oil in the GoM in 2016 and two-thirds of its natural gas production, and these percentages are expected to increase in the future since the vast majority of reserves are located at fields developed by floaters and subsea tiebacks. Eight floaters generated more than $1 billion each during 2016.

Table 9.9 Production and Gross Revenue for Floating Structures Circa 2016

				2016 Production		Cum. Production		
Operator	Complex	Type	Producing Well	Oil (MMbbl)	Gas (Bcf)	Oil (MMbbl)	Gas (Bcf)	Gross Revenue (MM $)
Anadarko	821	SPAR	5	0.8	9.5	74.0	547.9	57.4
Anadarko	822	SPAR	11	1.5	2.0	88.0	255.9	66.2
Anadarko	1288	SPAR	7	0.6	1.7	36.1	214.8	30.0
Anadarko	1323	TLP	15	7.8	5.6	97.5	77.5	331.6
Anadarko	1665	SPAR	15	17.5	17.4	108.5	108.6	752.6
Anadarko	1766	SEMI	3	0.0	0.0	0.3	1263.4	0.0
Anadarko	2576	SPAR	9	27.6	135.9	50.4	239.9	1461.4
Anadarko	2597	SPAR	4	4.7	2.2	4.7	2.2	195.6
Bennu Oil & Gas	2089	SEMI	4	1.1	1.1	13.4	14.3	46.7
BHP Billiton	1799	MTLP	7	2.9	2.2	36.0	29.0	121.2
BHP Billiton	1899	TLP	18	26.3	10.7	251.7	100.5	1092.5
BP E&P	1001	SEMI	18	16.8	27.1	309.2	786.6	749.6
BP E&P	1101	SEMI	16	34.1	30.7	314.3	273.1	1461.5
BP E&P	1215	SPAR	8	20.2	5.6	167.7	50.0	831.7
BP E&P	1223	SEMI	20	37.3	25.0	287.7	187.0	1575.4
Chevron U.S.A.	67	SPAR	11	1.3	1.7	117.6	184.5	56.3
Chevron U.S.A.	1819	SPAR	12	20.1	11.3	218.7	118.4	842.7
Chevron U.S.A.	1930	SEMI	4	1.8	1.5	72.7	56.0	76.2
Chevron U.S.A.	2440	SEMI	11	36.0	8.4	58.0	13.9	1481.8
ConocoPhillips	1218	TLP	5	1.1	2.1	37.1	100.9	48.4
Energy Res. Tech.	2133	MOPU	8	3.3	5.3	61.9	95.7	147.4
Eni US	251	MTLP	6	1.8	2.5	62.4	285.0	77.7
Eni US	1175	SPAR	10	5.4	4.6	84.6	201.6	230.9
Eni US	70020	MTLP	4	0.7	0.5	44.6	39.4	29.6
EnVen Energy	811	TLP	4	0.7	1.2	10.1	11.2	31.2
Exxon Mobil	183	SPAR	9	2.4	2.1	116.1	580.8	101.5

Table 9.9 Production and Gross Revenue for Floating Structures Circa 2016—cont'd

Operator	Complex	Type	Producing Well	2016 Production Oil (MMbbl)	2016 Production Gas (Bcf)	Cum. Production Oil (MMbbl)	Cum. Production Gas (Bcf)	Gross Revenue (MM $)
McMoRan	235	TLP	11	8.8	12.5	181.1	546.5	389.7
McMoRan	876	SPAR	8	3.6	5.1	114.3	100.9	159.5
McMoRan	1035	SPAR	14	4.9	3.5	96.8	93.8	208.9
Hess	2498	SPAR	6	7.4	14.4	17.1	35.6	337.7
LLOG Exp.	2424	SEMI	10	9.3	18.5	44.7	81.1	424.9
LLOG Exp.	2513	SEMI	11	24.8	62.1	37.5	90.0	1161.1
MC Offshore	23583	TLP	11	0.2	0.4	36.0	136.5	9.7
Murphy E&P	1090	SPAR	11	3.0	3.2	67.5	74.9	130.9
Murphy E&P	1290	SPAR	10	2.0	2.1	47.7	59.7	87.7
Murphy E&P	2045	SEMI	7	12.9	10.6	39.4	39.9	550.2
Noble Energy	24235	SPAR	3	1.7	2.1	91.8	245.2	75.0
Petrobras Amer.	2229	FPSO	4	4.4	0.7	28.7	4.8	181.9
Shell Offshore	420	TLP	12	6.3	5.5	143.9	182.4	271.1
Shell Offshore	2008	SPAR	15	23.2	52.8	138.8	245.2	1071.9
Shell Offshore	2385	TLP	6	15.0	18.7	42.9	49.8	654.1
Shell Offshore	2503	FPSO	2	0.7	0.1	0.7	0.1	30.1
Shell Offshore	24080	TLP	17	19.2	48.6	401.8	1253.0	902.5
Shell Offshore	24199	TLP	25	19.8	28.6	801.6	918.7	876.8
Shell Offshore	24229	TLP	9	1.0	6.2	98.4	897.2	56.7
Shell Offshore	70004	TLP	19	25.9	35.3	550.6	831.9	1139.0
W&T Energy	1088	MTLP	9	0.9	2.1	27.5	51.7	40.7

Note that complex 2503 was installed in 2016.
Source: BOEM 2017.

9.10.1 Idle

There were no idle floating structures circa 2016, but in 2017, the Independence Hub semi (1766) and the Mirage/Titan semi (2089) were no longer producing.

9.10.2 Gross Revenue <$5 Million

Anadarko's Independence Hub semisubmersible (1766) was installed in 2007 to develop 10 subsea gas fields and after 10 years had exhausted its reserves. In 2016, Independence Hub generated about $100,000, and the last of its wells ceased production and were permanently abandoned. Gas export from LLOG's Who Dat development is currently being routed through the Independence Trail export line.

9.10.3 Gross Revenue $5–30 Million

Three structures generated between $5 and $30 million in 2016: Anadarko's Gunnison spar (1288), Eni's Morpeth MTLP (70020), and MC Offshore's Jolliet TLP (23583). If

Table 9.10 Floating Structure Revenue Status Circa 2016

	Gross Revenue (Million $)					
Operator	**<5**	**5–30**	**30–100**	**100–500**	**500–1000**	**1000–2000**
Anadarko Petroleum	1766	1288	821, 822	1323, 2597	1665	2576
Bennu Oil & Gas			2089			
BHP Billiton Petroleum				1799		1899
BP E&P					1001, 1215	1101, 1223
Chevron U.S.A.			67, 1930		1819	2440
ConocoPhillips			1218			
Energy Res. Tech.				2133		
Eni US		70020	251	1175		
EnVen Energy Ventures			811			
Exxon Mobil				183		
McMoRan Oil & Gas				235, 876, 1035		
Hess				2498		
LLOG Exp.				2424		2513
MC Offshore		23583				
Murphy E&P			1290	1090	2045	
Noble Energy Inc.			24235			
Petrobras America				2229		
Shell Offshore			24229, 2503	420	2385, 24080, 24199	2008, 7004
W&T Energy VI			1088			
Total	1	3	13	14	8	8

Note that complex 2503 was installed in 2016.
Source: BOEM 2017.

tiebacks or alternative uses are not found for these structures, they will comprise the next batch of floater decommissioning and Rigs-to-Reefs projects.

9.10.4 Gross Revenue $30–100 Million

Thirteen structures generated between $30 and $100 million in 2016, including Anadarko's Nansen (821) and Boomvang (822) spars, Bennu's Mirage/Titan semi (2089), Chevron's Genesis spar (67) and Blind Faith semi (1930), ConocoPhillips Magnolia TLP (1218), Eni's Allegheny MTLP (251), EnVen Energy Ventures Prince MTLP (811), Murphy's Front Runner spar (1290), Noble Energy's Neptune spar (24235), Shell's Ram Powell TLP (24229), and W&T Energy's Matterhorn MTLP (1088).

9.10.5 Gross Revenue $100–500 Million

Fourteen structures generated between $100 and $500 million in 2016, including Anadarko's Marco Polo TLP (1323) and Heidelberg spar (2597), BHP's Neptune MTLP (1799), Energy Resource Technology's Helix MOPU (2133), Eni's Devil Tower spar

(1175), ExxonMobil's Hoover/Diana spar (183), Freeport-McMoRan's Marlin TLP (235), Horn Mountain (876) and Holstein (1035) spars that were sold to Anadarko in 2017, Hess' Gulfstar spar (2498, aka Tubular Bells), LLOG's Who Dat semi (2424), Murphy's Medusa spar (1090), Petrobras' Cascade and Chinook FPSO (2229), and Shell's Brutus TLP (420).

9.10.6 Gross Revenue $500–1000 Million

Eight structures generated between $500 and $1000 million in 2016: Anadarko's Constitution (1665), BP's Na Kika (1001) and Mad Dog (1215), Chevron's Tahiti (1819), Murphy's Thunder Hawk (2045), Shell's Olympus (2385, aka Mars B), Auger (24080), and Mars (24199).

9.10.7 Gross Revenue >$1000 Million

Eight structures generated more than $1 billion each in 2016, about $10 billion in total. Structures included both new installations and older fields: Anadarko's Lucius (2576, installed 2014), BHP's Shenzi (1899, installed 2008), BP's Thunder Horse (1101, installed 2005) and Atlantis (1223, installed 2007), Chevron's Jack/St Malo (2440, installed 2014), LLOG's Delta House (2513, installed 2014), and Shell's Perdido (2008, installed 2009) and Ursa (70004, installed 1998).

Lucius and Delta House are relatively new installations with virgin wells high on their production curves, while Ursa and Thunder Horse are much older fields that have been revitalized with tieback and redevelopment activity. Ursa is part of the giant Mars-Ursa basin and is the largest field in the GoM at 1.85 Bboe reserves. Atlantis is the 13th largest field at 411 MMboe reserves, and North Thunder Horse comes in as the 33rd largest field with 320 MMboe reserves.

REFERENCES

Lim, E.F.H., Ronalds, B.F., 2000. Evolution of the production semisubmersible. In: SPE 63036. SPE Annual Technical Conference and Exhibition. Dallas, TX, Oct 1–4.

Petroleum Resources Management System, 2007. Society of Petroleum Engineers.

Poston, S.W., Poe Jr., B.D., 2008. Analysis of Production Decline Curves. Society of Petroleum Engineers, Richardson, TX.

Reid, D., Dekker, M., 2013. Deepwater development: wet or dry tree? In: OTC 24517. Offshore Technology Conference. Rio de Janeiro, Brazil, Oct 29–31.

Smith, C.D., Pilney, D.A., 2003. Enchilada: from border town to cowtown. In: OTC 15110. Offshore Technology Conference. Houston, TX, May 5–8.

Tillinghast, W.S., 1990. The deepwater pipeline system on the Jolliet project. In: OTC 6403. Offshore Technology Conference. Houston, TX, May 7–10.

Truchon, S., Brzuzy, L., Fonseca, M., 2015. Innovative assessments for selecting offshore platform decommissioning alternatives. In: SPE 173519. SPE E&P Health Safety and Environmental Conference. Denver, CO, Mar 16–18.

CHAPTER TEN

Deepwater Economic Limit Statistics

Contents

Abstract

Deepwater economic limits are expected to be greater than shallow-water economic limits because deepwater structures are larger, more complex, and further from shore, and almost all structures are manned. Unlike shallow-water structures, deepwater platforms typically adopt a combination of dry tree and wet tree wells, the latter being significantly more expensive to operate and maintain than dry tree wells. In this chapter, production and net revenue statistics are evaluated the last year of production for 23 decommissioned structures and 486 permanently abandoned wells from 1977 to 2017 in water depth >400 ft. The average inflation-adjusted net revenue the last year of production of all deepwater structures was $13.4 million with a median value of $3.1 million. The average and median inflation-adjusted gross revenues for deepwater fixed platforms were $9.5 million and $2.2 million, respectively. Deepwater dry tree wells exhibited an average economic limit of $3.8 million compared with $17.8 million for wet tree wells in the last year of production, and gas wells generated $3.2 million at the end of their life compared with $5 million for oil wells.

Decommissioning Forecasting and Operating Cost Estimation
https://doi.org/10.1016/B978-0-12-818113-3.00010-2

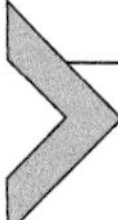

10.1 METHODOLOGY

10.1.1 Revenue Model

For decommissioned wells and structures, the net revenue near the end of its productive life is a proxy of economic limit and is straightforward to compute since the production profile $Q(t)$, royalty rate r, and oil and gas prices $P^o(t)$ and $P^g(t)$ are all known and available historically. If the last month of production is denoted as T_a, then the net revenue the last year of production NR_{LY} is determined from the cumulative sum of monthly oil and gas production volumes and average monthly oil and gas prices one year prior to T_a:

$$NR_{LY} = \sum\nolimits_{t=T_a-12}^{T_a} Q(t)P(t)(1-r)$$

In the deepwater GoM, royalty rates are either 12.5%, 16.67%, or 18.75%.

10.1.2 Primary Product

Wells are differentiated according to their primary product using the cumulative gas-oil ratio (CGOR) threshold of 10,000 cf/bbl. Oil wells are defined by CGOR <10,000 cf/bbl and gas wells by CGOR >10,000 cf/bbl. Structures are categorized as primarily oil or primarily gas using production from all wells boarding the structure and the same threshold.

For oil wells, crude oil and associated gas are combined into a heat-equivalent boe stream, while for gas wells, natural gas and condensate are described as a cubic feet equivalent (cfe) stream using the standard 6000 cf = 1 bbl = 1 boe conversion.

10.1.3 Adjusted Gross Revenue

The gross revenue the last year of production is computed using monthly production data and average monthly Henry Hub prices for natural gas and Brent crude prices for crude oil for one year prior to production cessation. Crude and condensate prices are assumed equal, and gas well gas and casinghead (associated) gas prices are assumed equal. Computed gross revenue is adjusted to 2017 dollars using the CPI.

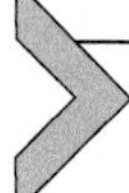

10.2 DECOMMISSIONED STRUCTURES

10.2.1 Sample

Twenty-three deepwater structures have been decommissioned in the GoM through 2017, five floaters and 18 fixed platforms (Table 10.1). Sixteen of the 18 decommissioned fixed platforms were primarily gas producers and four of the five floating structures were oil producers.

The decommissioned fixed platforms produced 42.1 MMbbl oil and 1.8 Tcf natural gas over their lifetime, and the floaters produced 30.5 MMbbl oil and 210 Bcf natural gas, yielding average cumulative production of 19 MMboe per fixed platform and 13 MMboe

Table 10.1 Gross Revenue Statistics for Deepwater Decommissioned Structures the Last two Years of Production

Complex	Structure Type	Cum. Oil (Mbbl)	Cum. Gas (MMcf)	CGOR (Mcf/bbl)	Last Year			Second to the Last Year		
					Oil (Mbbl)	Gas (MMcf)	Revenue (MM$)	Oil (Mbbl)	Gas (MMcf)	Revenue (MM$)
85	FP	3779	6436	1.7	33.8	96.8	3.9	122	331	14.1
735	MTLP	2244	3016	1.3	64.5	113	3.7	1003	1476	50.8
1055	FP	4675	22,745	4.9	322	2058	63.9	400	4246	68.7
1320	FP	76	8605	114		11.9	0.04	20.4	148	2.9
1384	SPAR	137	125,094	915	3.0	8293	87.8	39.6	37,654	314
1771	SEMI	19,877	64,313	3.2	305	584	32.2	1248	4557	139
10242	FP	3605	190,272	52.8	16.6	461	3.0	62.4	4662	59.6
22274	FP	1001	158,194	158		1.4	0.01		13.7	0.1
22372	FP	290	304,770	1053		511	4.1	2.6	2186	9.2
22583	FP	1267	316,731	250	1.5	356	3.6		338	2.5
22705	FP	209	77,913	3712	14.3	5813	67.6	42.4	10,422	88.9
22846	FP	1737	33,297	19.2		39.5	0.3	0.6	335	1.3
23004	FP	3457	61,043	17.7	41.3	63.5	2.2	45.8	88.5	2.1
23308	FP	3251	80,186	24.7	28.0	351	7.7	40.9	670	9.0
23543	SEMI	532	4979	9.4	14.2	344	2.5	452	4259	42.6
23581	FP	11,538	196,131	17.0	3.4	157	1.4	50.3	2611	22.1
23859	FP	1548	136,441	88.1	5.1	681	8.1	12.7	1744	15.4
23891	FP	1133	72,371	63.9		0.1			1500	7.9
24021	FP	2967	41,941	14.1	4.0	12.3	0.3	15.9	728	5.8
24079	SEMI	7711	12,582	1.6	427	565	11.1	1823	2888	49.2
24087	FP	0.05	27,207	555,240		109	0.9		985	4.2
27014	FP	1102	16,377	14.9	3.5	18.3	0.2	22.3	308	2.2
28033	FP	454	76,087	168	6.1	203	3.1	12.3	342	3.8

Note: Gross revenues adjusted to 2017 dollars.

per floater. All the fixed platforms had a royalty rate of 16.67% except for complex 22705 that had a sliding scale and complex 23859 that had a 12.5% royalty. All the floaters were subject to 12.5% royalty.

The five floaters that have been decommissioned through 2017 include three semisubmersibles, one spar and one mini tension leg platform (MTLP). Chevron's Typhoon MTLP suffered catastrophic damage from Hurricane Rita in September 2005 and was decommissioned in June 2006. After ATP Oil & Gas declared bankruptcy in 2013, and no buyers for its Gomez semisubmersible could be found, it was decommissioned in March 2014.

The Red Hawk spar was installed in 2004, and four years later, the wells had watered out after producing 125 Bcf natural gas, about half of the original reserves estimates (Lamey et al., 2005). Anadarko spent several years shopping the facility around, but with more competitive options available and no viable tieback opportunities, the decision to decommission was made (Furlow, 2015). In 2011, the wells were permanently abandoned, and after the topsides were removed in 2014, the 564 ft-long, 64 ft-diameter hull was wet towed vertically 70 mi to the 430 ft-deep Eugene Island 384 reef site and laid on its side (Box 17.1).

10.2.2 Aggregate Economic Limits

The average inflation-adjusted revenue the last year and second to last year of production was \$13.4 million and \$40 million, respectively, with median values \$3.1 million and \$9.2 million. The large difference between the average and median values is due to the occurrence of three structures producing in excess of \$60 million in their last year of production (complex 1055, \$64 million; complex 1384, \$88 million; and complex 22705, \$68 million). The gross revenue distribution the last year and second to last year of production shows the majority of structures reaching their economic limit below \$5 million (Fig. 10.1).

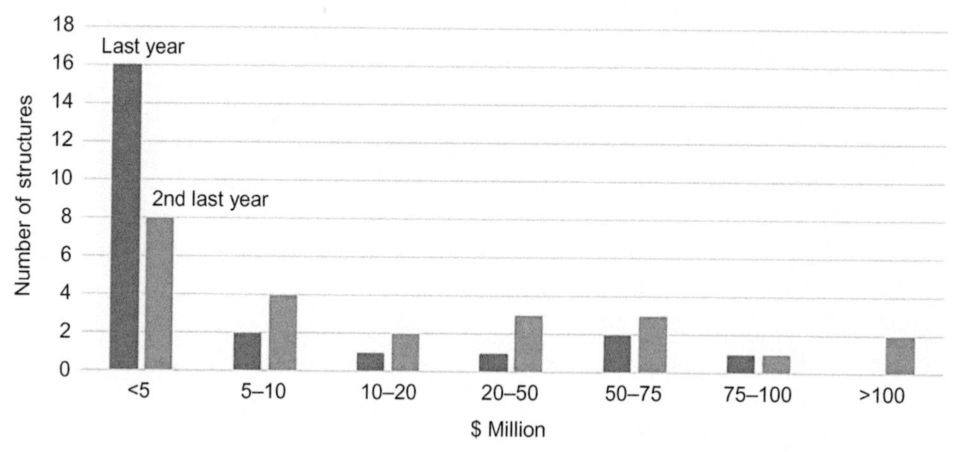

Fig. 10.1 Deepwater decommissioned structure gross revenue near the end of production.

10.2.3 Economic Limits by Structure Type

The average and median inflation-adjusted economic limits for fixed platforms were $9.5 million and $2.2 million, respectively. In total, there were 13 fixed platforms with economic limit less than $4 million and 16 platforms with economic limit less than $10 million. For oil platforms, the average last year gross revenue was $34 million and for gas platforms $6.5 million. With only two oil platforms in the sample, the statistical measures for oil platforms are not representative.

The five floaters had an average inflation-adjusted revenue the last year of production of $28 million and a median value of $11.2 million, about three times larger than fixed platforms. If the Typhoon MTLP is removed from the sample because of its exceptional nature, the average economic limit of the group increases to $33 million. For the four floating oil structures, the average last year gross revenue was $12.4 million and, for the sole floater gas structure, $88 million.

10.3 BOTTOM HOLE FLOWING PRESSURE

Wells reach their economic limit at a particular production rate or bottom-hole flowing pressure (BHP). For subsea developments, where tubing head pressure may have to be several thousand pounds per square inch (psi) to drive produced fluids from the subsea wellhead to the receiving facility several miles away, economic limit BHPs will be higher than for dry tree wells where only the water column has to be overcome, for all other things equal. Operators may install gas lift in the production tubing or at the wellhead to reduce BHP and thereby increase ultimate recovery or may install subsea separation, compression, or pumping equipment, although the later technology is not yet widely adopted. Flowing a well at low abandonment pressure increases the potential for asphaltene deposition, and below the bubble point, well productivity will be reduced (see Chapter 4 for flow assurance discussion).

Wells located downslope from the host in deeper water require more energy to reach the platform because of the hydrostatic head to arrive at station relative to wells located uphill that flow downward. Interestingly, operators prefer downslope well hookups for operational reasons and will locate the host accordingly, and most subsea wells flow upslope. Upslope wells involve more design challenges and present greater operational risk (Dykhno et al., 2003).

Example. Well abandonment pressure at Gemini

The Gemini field is located in Mississippi Canyon 292 in 3400 ft water depth and was developed as a subsea development tied back 27.5 mi to the Viosca Knoll 900 platform. Engineers determined that a minimum gas flow rate of 10–15 MMcfd (~3.7 Bcf per year) is required to lift the fluids in order to prevent the liquids from falling back into the well and killing the well (Kashou et al., 2001). Below 10 Mcfd, the well is expected to die and will be unable to produce. A well abandonment pressure was calculated to be about 1500–1600 psia and translates into a flowing bottom-hole pressure of 1350–1450 psia.

10.4 SUBSEA WELL INTERVENTION

Subsea wells are high-risk high-cost assets compared with dry tree and direct vertical access (DVA) wells due to the complexities and difficulties of access and the costs associated with intervention (Zabaras and Mehta, 2004). A well intervention is defined as any physical connection made to a completed well to alter production or address safety issues (Box 10.1). Unlike dry tree wells, subsea wells are not accessible via a platform rig, and this lack of access has implications for performance and hydrocarbon recovery. Since all wells require intervention to achieve their maximum recovery and since under normal circumstances dry tree and DVA wells receive regular and planned interventions, one would expect subsea wells to have lower performance.

Recovery factors for subsea wells are expected to be lower than dry tree and DVA wells, and economic limits are expected to be higher. Although the former hypothesis is not easily tested, the later claim can be evaluated. The low frequency of interventions and back pressure that arises due to the distance to the host facility are the primary reasons for the lower performance. Operators often forego production-improvement workovers on subsea wells because the combination of high-cost and uncertain improvement fails to justify capital investment. Reservoirs with delayed pressure maintenance or nonproactive intervention programs will have recoveries less than equivalent wells with regular maintenance.

BOX 10.1 Subsea Well Intervention

Subsea interventions are classified according to the methods used to connect to the well, the activities performed in the well, and the type of vessel used in the operation (Nelson and McLeroy, 2014).

Well Operations

Typical subsea well operations include pumping, wellhead/tree maintenance, slickline and wireline, coiled tubing, snubbing, and tubing retrieval:

- Pumping. Pumping chemicals downhole to improve the production rate represent one of the simplest well operations.
- Wellhead/tree maintenance. Operations vary depending on the condition of the equipment and manufacturers' recommended maintenance procedures.
- Slickline. A single-strand wire used to run tools in the wellbore for placement or removal.
- Wireline. A braided line used to lower or retrieve heavier equipment and for logging and perforation.
- Coiled tubing. Coiled tubing is metal pipe used to pump chemicals for circulation, logging, drilling, and production operations.

Continued

BOX 10.1 Subsea Well Intervention—cont'd

- Snubbing (hydraulic workover). Snubbing operations involve running a bottom-hole assembly on a drill string against well pressure to perform the desired tasks.
- Tubing retrieval. Pulling and replacing the tubing hanger and production tubing due to performance deterioration or a new completion.

Classification and Vessel Requirements

Light well interventions (also called Type I or Class A) are serviceable using a variety of equipment that can be deployed from numerous types of vessels. Examples include borehole survey logging, fluid displacement, gas-lift valve repair, perforating, sand washing, pulling tubing plugs, stimulation, and zonal isolation.

Medium well interventions (Type II or Class B) are more demanding than light interventions and typically require more complex equipment and vessels. Examples include casing leak repairs, fishing, well abandonment, remedial cementing, sand control gravel packing, surface controlled subsea safety valve failure, water shutoffs, paraffins, asphaltenes, and hydrates.

Heavy well interventions (Type III or Class C) are typically associated with the deployment of drilling rigs with a full-size BOP stack. Examples include tubing packer failure, electric submersible pump replacement, horizontal well sand control, completion change out, milling, fishing, reentry sidetracks, and subsea tree change out.

Light well interventions are usually the cheapest and least risky operations to perform and often apply riserless wireline technologies or derivatives of this technology to enter the wellbore. The market is mature, and the vessels deployed are usually small with a temporarily installed package deployed over the side using a subsea lubricator or other means to access the well. The technology may be used in combination with more advanced technology such as coiled tubing (CT) operations on larger dedicated intervention vessels.

Medium well interventions typically employ CT operations where fluids are pumped through coiled tubing into the wellbore or used to drive certain components (Zijderveld et al., 2012). CT units require more space than wireline packages, and the two techniques are often used together. If well returns come back to the surface and need to be cleaned or treated on the vessel, a riser-based system will be required to create a flow path. For heavy well interventions, large vessels are needed to pull out production tubing or the tubing hanger.

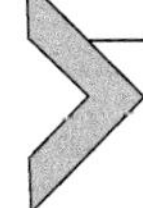

10.5 PERMANENTLY ABANDONED WELLS

10.5.1 Sample

All permanently abandoned wells in the GoM in water depth > 400 ft from 1977 to 2016 were evaluated. A total of 486 deepwater wells were permanently abandoned from 1977 to 2017, 149 oil wells and 337 gas wells. There were 91 dry tree and 58 wet tree oil wells, and 174 dry tree and 163 wet tree gas wells. For wet wells, distance to host (as a bird flies) and the elevation difference between well and host were also examined for their potential impact on economic limits.

Table 10.2 Permanently Abandoned Deepwater Well Statistics, 1977–2016

Statistic	Revenue ($ Million)	Production (boe)
P10	0.11	3190
P20	0.34	10,600
P50	2.76	66,900
P80	12.1	332,000
P90	24.8	649,000
Average	10.2	282,000

Source: BOEM 2017.

The average net revenue and heat-equivalent volumes across the entire sample the last year of production was $10.2 million and 282,000 boe, respectively (Table 10.2). The median (P50) values are $2.8 million and 67,000 boe, respectively, and the large difference with the average due to a handful of high-producing and high-revenue wells at the end of their life.

10.5.2 Exclusions

Well that have ceased production at the time of evaluation are not part of the sample because being shut-in or temporarily abandoned the ultimate status of the well is not yet determined. All hurricane-destroyed wells were identified through MMS/BOEM official records and were not considered since these represent (at least in part) premature removals that would bias (upward) the economic limit statistics.

Three extremely prolific wells were excluded in the assessment to improve the graphic presentations. These three wells generated between $275 and $510 million the last year of their life, which is far beyond the range for the rest of the sample. No other data points were excluded.

In the graphic representations, the entire time period was employed, while the most recent 20-year time period was employed in most of the tables because these data are considered more representative by reducing the bias associated with early developments and the uncertainty associated with adjusting revenue over a long time horizon. With the reduced time horizon, the sample loses about a quarter of its original size and drops from 486 to 392 wells.

10.5.3 Dry Tree vs. Wet Tree

There is a significant difference between dry tree and wet tree economic limits. Dry tree wells are expected to produce at smaller last year production and revenue compared with wet tree wells because they are easier and cheaper to maintain with direct rig access. The expectation is that dry tree wells can produce longer and at lower production and revenue levels compared with wet wells, and these expectations are confirmed.

Dry tree wells produce lower production levels compared with wet tree wells implying lower production cost and greater recovery efficiencies. Average last year production for all dry tree wells in the sample was 109,000 boe compared with 486,000 boe for wet tree wells, and in terms of revenue, dry tree wells produced down to \$3.8 million compared with \$17.8 million for wet tree wells (Table 10.3). Median values were 33,400 boe and \$1.2 million for dry tree wells compared with 184,000 boe and \$6.5 million for wet tree wells, a difference of about four to five for both production and revenue (Figs. 10.2 and 10.3). Time trends illustrate the higher average revenues of wet wells in individual years (Fig. 10.4).

Breaking out wells in terms of their primary product, average last year production for dry tree oil wells was 120,000 boe compared with 370,000 boe for wet tree oil wells and for gas wells 0.61 Bcfe for dry tree wells versus 3.3 Bcfe for wet tree wells (Table 10.4). Similar differences arise in terms of revenue generated. Wet oil wells stopped producing at about four times the revenue of dry tree oil wells (\$21 million vs. \$5 million), while for wet gas wells, the difference was about six times greater, \$18 million versus \$3 million (Table 10.4).

The differences in economic limits that arise between dry and wet tree wells can be translated to hypothetical (potential) reserves by subtracting the two production limit values and dividing by an assumed 10% decline rate for the average well. The cumulative production formula N_p for an exponentially declining well with annual decline rate D is given by

Table 10.3 Dry Tree vs. Wet Tree End-of-Life Production Statistics, 1977–2016

	Dry Tree			Wet Tree		
Statistic	Revenue (\$ Million)	Production (boe)	Water (bbl)	Revenue (\$ Million)	Production (boe)	Water (bbl)
P10	0.05	1890		0.47	13,500	
P20	0.19	5540	454	1.33	40,100	454
P50	1.22	33,400	1290	6.51	184,000	12,900
P80	5.22	156,000	102,000	23.6	622,000	111,000
P90	9.59	250,000	232,000	43.6	1,240,000	268,000
Average	3.81	109,000	77,900	17.8	486,000	86,200

Source: BOEM 2017.

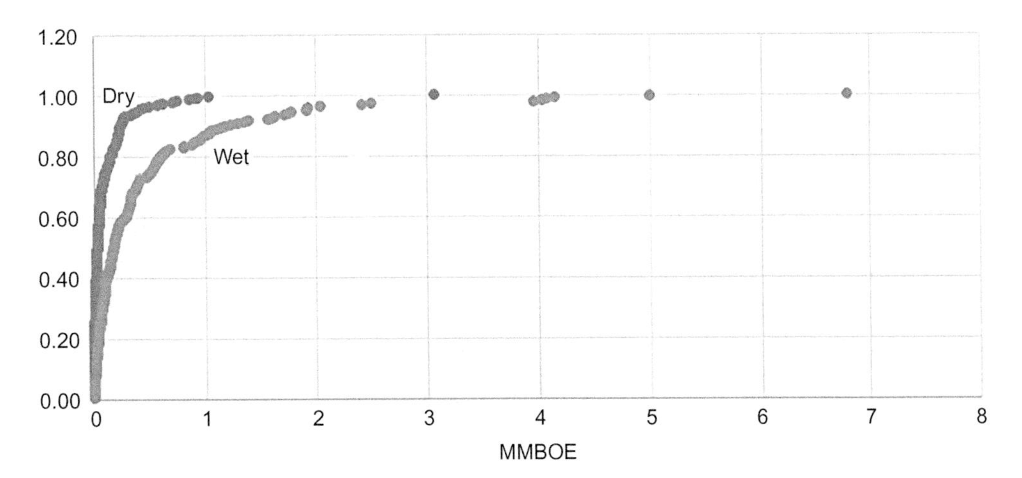

Fig. 10.2 Deepwater dry tree and wet tree wells last year production, 1977–2016. *(Source: BOEM 2017.)*

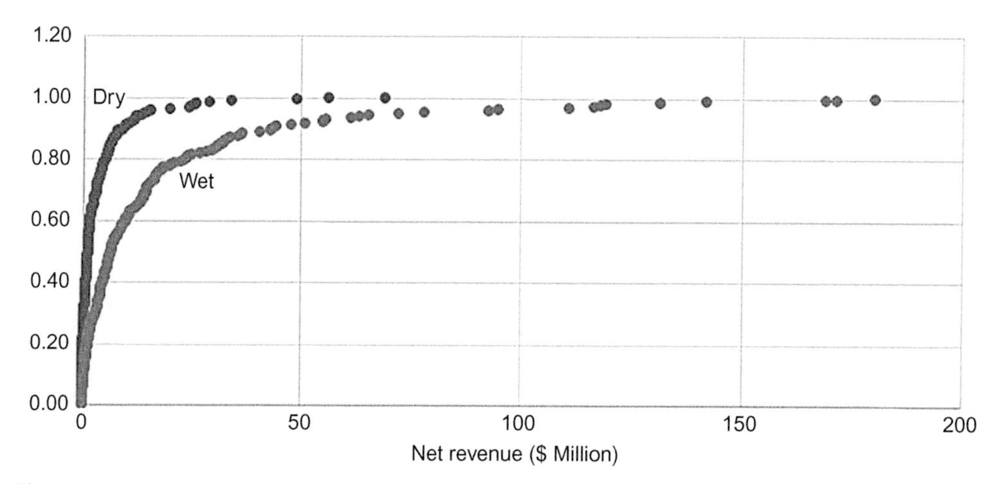

Fig. 10.3 Deepwater dry tree and wet tree wells last year net revenue, 1977–2016. *(Source: BOEM 2017.)*

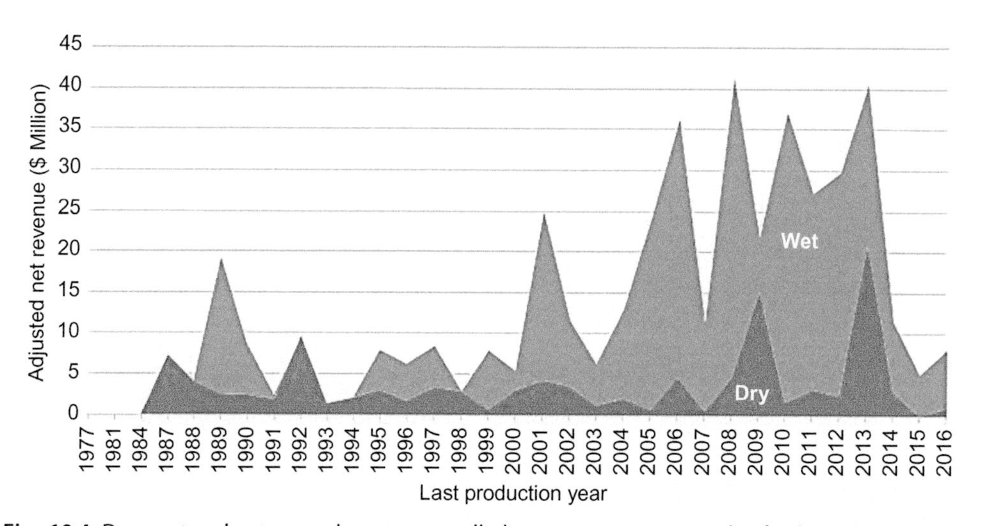

Fig. 10.4 Deepwater dry tree and wet tree wells last year net revenue by final production year, 1977–2016. *(Source: BOEM 2017.)*

$$N_p = \frac{q_i - q_{el}}{D}$$

where q_i and q_{el} are the economic limit and reduced economic limit rates, respectively. This calculation yields 2.3 MMboe for the average dry tree oil well and 26.3 Bcfe for dry tree gas wells that would have been realized by operating to a lower economic limit.

Table 10.4 Dry Tree and Wet Tree End-of-Life Production and Revenue by Primary Product, 1998–2016

Oil Wells	Unit	Dry Tree	Wet Tree
Production	MMboe	0.12	0.37
Revenue	$ million	5.0	21.1
Sample	#	70	50
Gas Wells	**Unit**	**Dry Tree**	**Wet Tree**
Production	Bcfe	0.61	3.31
Revenue	$ million	3.2	17.8
Sample	#	115	157
All Wells	**Unit**	**Dry Tree**	**Wet Tree**
Water Prod.	MMbbl	0.10	0.08
Sample	#	185	207

Source: BOEM 2017.

10.5.4 Water Cuts

Dry tree wells are expected to produce at higher water cuts than wet tree wells, and this is readily verified. On average, the difference is about 20% and may not look especially significant—100,000 bbl annual water production for dry tree wells compared with 80,000 bbl water for wet wells (Table 10.4)—but when combined with the lower production levels achieved at dry tree wells, the average oil well water cuts are 45% for dry tree wells versus 18% for wet wells. On a heat-equivalent basis, gas wells had similar water cuts, 50% for dry tree wells compared with 13% for wet wells. It is widely recognized that wet tree wells and subsea architecture have more difficulty handling water production than dry tree wells because flow assurance conditions are more difficult and complex and require higher operating cost to manage.

10.5.5 Oil vs. Gas Wells

Gas wells are expected to produce at lower production levels and revenues than oil wells because gas wells are subject to less significant flow assurance issues relative to oil wells and are generally cheaper to operate. However, because crude oil is valued higher than natural gas on a heat-equivalent basis, historically, about twice the value on a boe basis, revenue comparisons need to be made carefully. Heat-equivalent comparisons are not perfect and only provide an indication of relative differences. Revenues are believed to be a more useful and direct measure than production when comparing oil and gas wells since they better reflect cost/benefit considerations.

Wet tree gas wells can be produced at greater tieback distance compared with wet tree oil wells due to less restrictive flow assurance considerations. On a heat-equivalent basis, oil and gas wells both produce to about 100,000 boe for dry tree wells, but on a revenue basis, gas wells produce to \$3.2 million on average compared with \$5 million for oil wells (Table 10.4). For wet wells, oil wells produce to 0.37 MMboe compared with 0.55 MMboe for wet gas wells, and on a revenue basis, gas wells produce to \$17.8 million compared with \$21.1 million for oil wells (Table 10.4). Dry tree oil and gas wells produce to about four to five times lower revenue and production than wet tree oil and gas wells.

In Figs. 10.5 and 10.6, end-of-life hydrocarbon production and net revenue are depicted for oil and gas wells. The similarity between oil and gas well boe production and net revenue is striking (Figs. 10.5 and 10.6), but water production levels are different, as expected (Fig. 10.7). Gas wells handle less water production on average than oil wells due to the nature of the reservoir drive. Black oils in particular are commonly supported

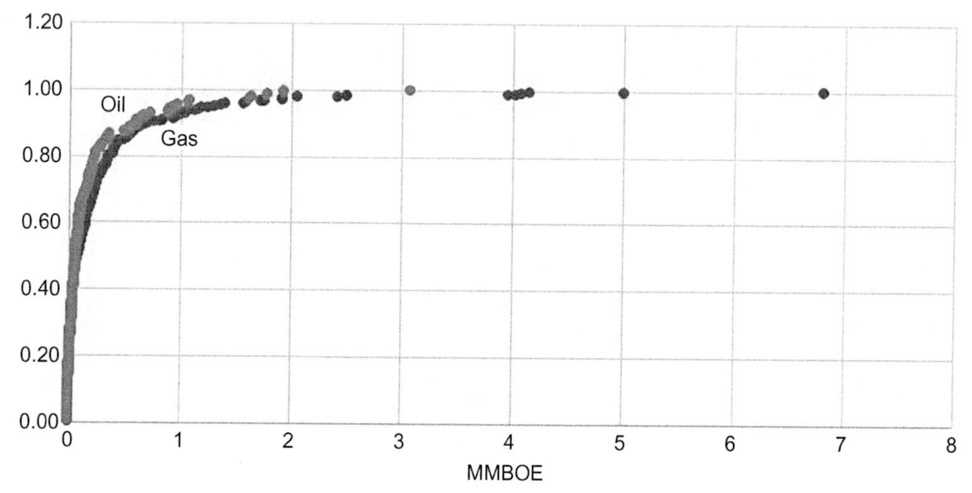

Fig. 10.5 Deepwater oil and gas wells last year production, 1977–2016. *(Source: BOEM 2017.)*

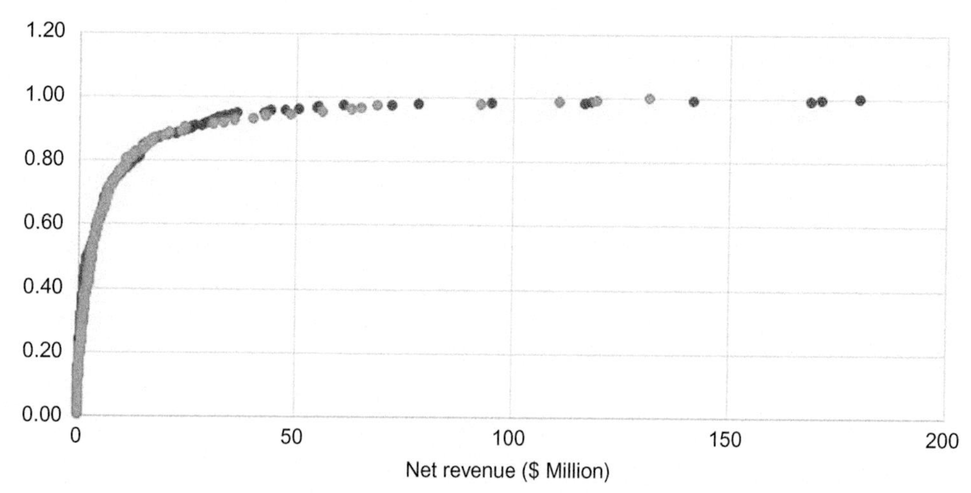

Fig. 10.6 Deepwater oil and gas wells last year net revenue, 1977–2016. *(Source: BOEM 2017.)*

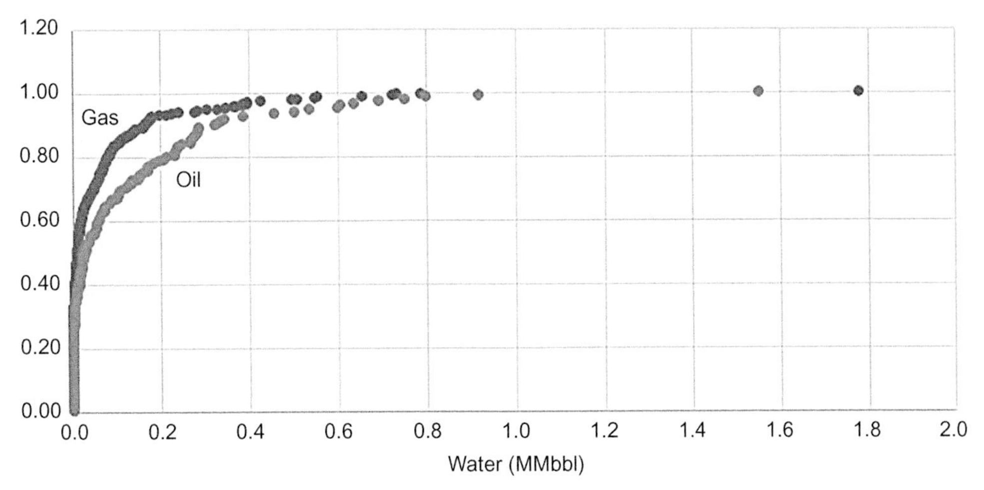

Fig. 10.7 Deepwater oil and gas wells last year water production, 1977–2016. *(Source: BOEM 2017.)*

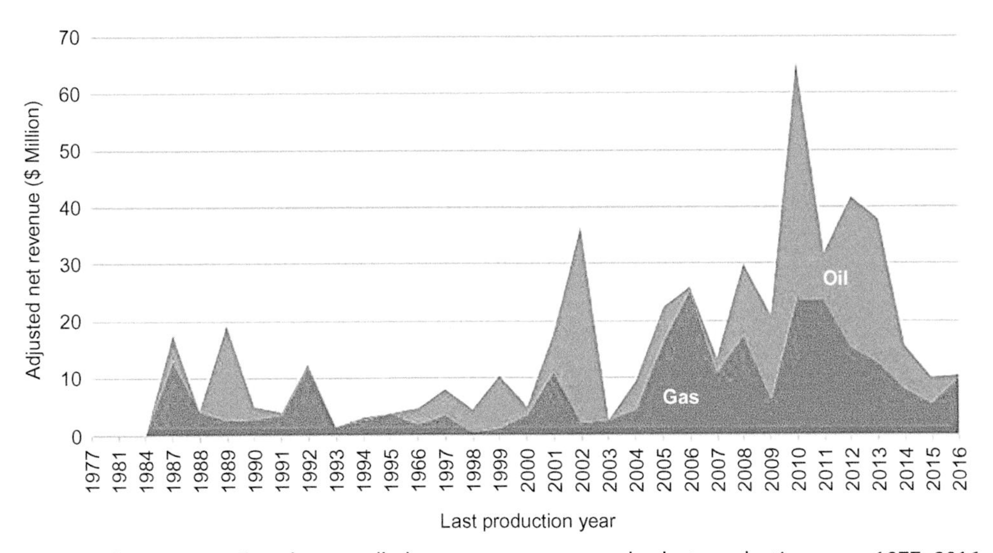

Fig. 10.8 Deepwater oil and gas wells last year net revenue by last production year, 1977–2016. *(Source: BOEM 2017.)*

by aquifers, which normally increase water cuts late in life. Time trends for oil and gas wells show oil wells terminating at consistently higher revenue than gas wells reflecting higher operating expenditures (Fig. 10.8).

10.5.6 Wet Tree Wells

Distance to Host

As the distance between wet wells and hosts increases, economic limits are expected to increase because, for all other things equal, increasing distance creates greater back

Table 10.5 Average Last Year Production and Revenue Statistics for Oil and Gas Wet Wells as a Function of Distance to Host, 1998–2016

Oil Wells	**Unit**	**<5 mi**	**5–10 mi**	**>10 mi**	**All**
Production	MMboe	0.43	0.33	0.27	0.37
Revenue	$ million	25.5	16.0	16.0	21.1
Sample	#	27	10	13	50
Gas Wells	**Unit**	**<5 mi**	**5–10 mi**	**>10 mi**	**All**
Production	Bcfe	2.4	1.2	4.8	3.3
Revenue	$ million	14.2	6.9	25.1	17.8
Sample	#	35	42	80	157

Source: BOEM 2017.

pressure and more difficult flow assurance issues that increase operating costs and create problems in maintaining commercial production.

For oil wells, only the <5 mi category is sufficiently populated to ensure statistical significance; the 5-10 mi and >10 mi categories are too small to be reliable (Table 10.5). For gas wells, data do not increase consistently; however, if the <5 mi and 5–10 mi categories are combined, the resultant statistics show that wells >10 mi from their host exhibit economic limits two to three times the production and revenue levels of wells <10 mi from their host (Table 10.5).

Last year revenues for gas wells <5 mi from their host were less than oil wells by about half. If the one-half ratio represents the true ratio (i.e., one that would be observed for representative samples), then since gas revenues for >10 mi were $25 million, one might expect oil revenues in the >10 mi category would be about twice this value ($50 million instead of the $16 million from the current sample). This is speculative, of course, and will only be determined when more deepwater oil wells are decommissioned.

No significant differences in production or net revenue between wet oil and wet gas wells were found (Figs. 10.9 and 10.10), but differences arise in water production as one would expect from reservoir considerations (Fig. 10.11).

Elevation Difference

Large elevation differences between wet wells and their hosts will create additional back pressure compared with wells with smaller differences. The greater the elevation difference, the higher one would expect the economic limit for all other things equal.

Economic limits for wells with elevation differences <500 ft versus elevation differences >500 ft were distinguishable by well types, $18.6 million versus $20.3 million for oil wells and $15.4 million versus $22.7 million for gas wells (Table 10.6). For elevation differences less than and >250 ft, the difference in the economic limits was essentially indistinguishable, and as the elevation difference threshold increased to 1000 ft, the differences between the economic limits were greater for the larger elevation category.

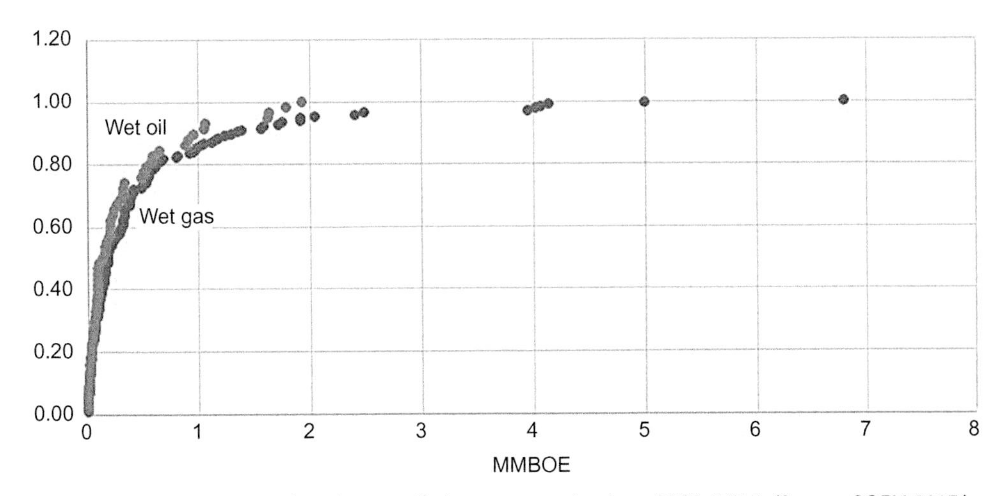

Fig. 10.9 Deepwater wet oil and gas wells last year production, 1977–2016. *(Source: BOEM 2017.)*

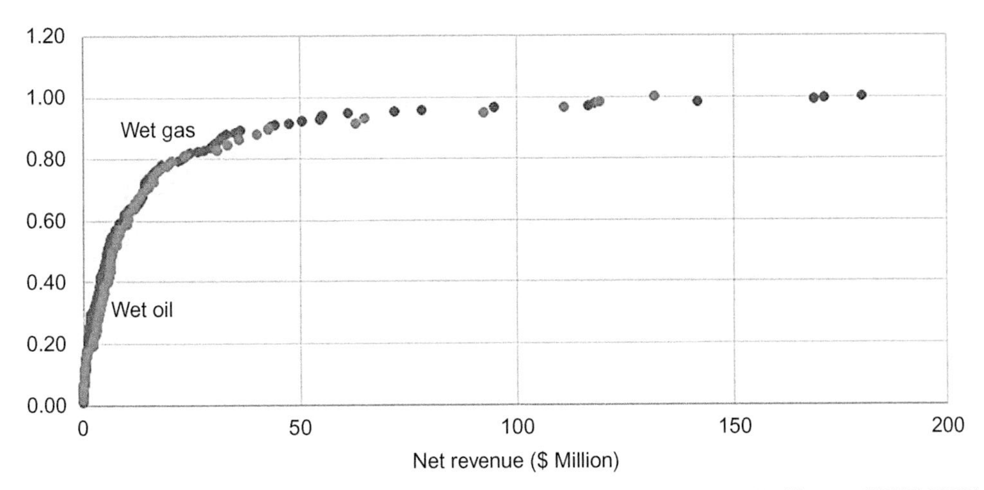

Fig. 10.10 Deepwater wet oil and gas wells last year net revenue, 1977–2016. *(Source: BOEM 2017.)*

10.6 LIMITATIONS

The limitations of economic limit analysis were described in Chapter 8 and are not repeated here. In all data categories, standard deviations were on the order of the mean and illustrates the wide variability in statistical measures. For small samples (<10 wells per category), caution in interpretation is advised because of sample size error. In small samples, expected trends may or may not hold, but neither should be construed as evidence to support or refute hypothesis.

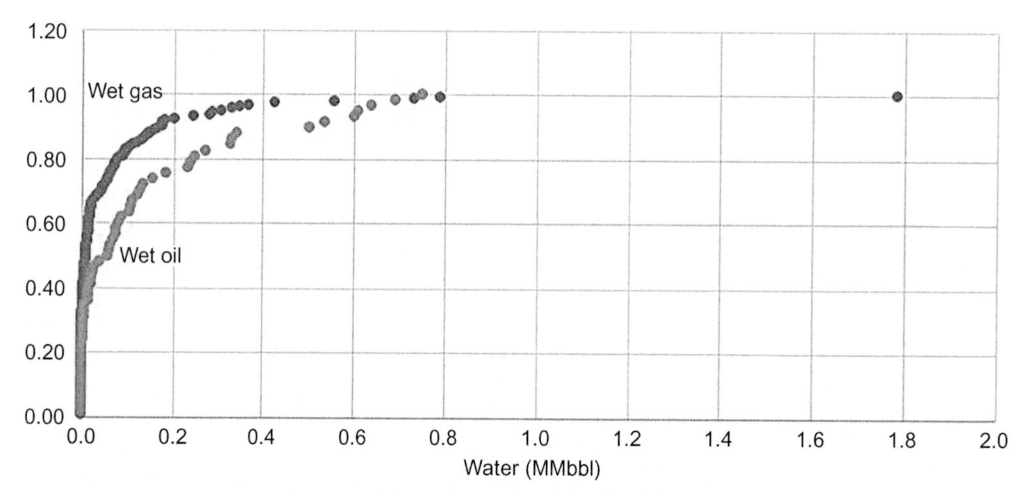

Fig. 10.11 Deepwater wet oil and gas wells last year water production, 1977–2016. *(Source: BOEM 2017.)*

Table 10.6 Average Last Year Revenue of Wet Wells by Elevation Difference Category, 1998–2016

Elevation Difference (ft)	Oil Wells ($ Million)	Gas Wells ($ Million)	Water Production (MMbbl)
<250	21.0 (25)	17.5 (62)	0.11 (87)
>250	21.3 (25)	17.8 (93)	0.07 (118)
All	21.1 (50)	17.8 (157)	0.08 (207)
<500	18.6 (39)	15.4 (84)	0.11 (123)
>500	20.3 (19)	22.7 (79)	0.06 (99)
All	19.1 (58)	19.0 (163)	0.08 (221)
<1000	21.0 (44)	15.9 (107)	0.10 (151)
>1000	22.2 (6)	21.9 (50)	0.04 (56)
All	21.1 (50)	17.8 (157)	0.08 (207)

Note: Sample sizes denoted in parenthesis.
Source: BOEM 2017.

It is important to realize that under some circumstances an operator may operate an asset when net revenues fall below operating cost for strategic reasons or choose to abandon a property or well before its economic limit is reached, for example, if destroyed or damaged by a hurricane. Economic limits therefore cannot be considered an unambiguous statistic of offshore operations, merely a proxy of average end-of-life conditions (Kaiser and Narra, 2019).

A greater level of granularity could be used to delineate fluid types and reservoir conditions (Box 4.1), but this is unlikely to lead to useful results because of the gross nature of the data. See Dindoruk and Christman (2002) for detailed information on GoM oil correlations and properties.

The elevation difference between wet well and host was applied in this evaluation because it is an easily computed measure. A better indicator might be the average depth along the route or the maximum depth, which in most cases will be significantly different than the evaluation difference and may lead to different results.

REFERENCES

Dindoruk, B., Christman, P.G., 2002. PVT properties and viscosity correlations for Gulf of Mexico oils. In: SPE 71633. SPE Annual Technical Conference and Exhibition. New Orleans, LA, Sep 30-Oct 3.

Dykhno, L.A., Jayawardena, S.S., Schoppa, W., 2003. Blowdown feasibility for downhill flowlines. In: OTC 15256. Offshore Technology Conference. Houston, TX, May 5–8.

Furlow, W., 2015. Lessons learned as world's first cell spar laid to rest. SPE Oil Gas Facilit. 2, 16–19.

Kaiser, M.J., Siddhartha, N., 2019. An empirical evaluation of deepwater economic limits in the Gulf of Mexico. J. Natural Gas Sci. Eng. 63, 1–14.

Kashou, S., Matthews, P., Song, S., Peterson, B., Descant, F., 2001. Gas condensate subsea systems, transient modeling: a necessity for flow assurance and operability. In: OTC 13078. Offshore Technology Conference. Houston, TX, Apr 30–May 3.

Lamey, M., Hawlay, P., Maher, J., 2005. Red Hawk project: overview and project management. In: OTC 17213. Offshore Technology Conference. Houston, TX, May 2–5.

Nelson, M.E., McLeroy, P.G., 2014. Subsea well intervention: recent developments and recommendations to increase overall project returns. In: SPE 170980. SPE Technical Conference and Exhibition. Amsterdam, The Netherlands, Oct 27–29.

Zabaras, G.J., Mehta, A.P., 2004. Effectiveness of bullheading operations for hydrate management in DVA and subsea wells. In: OTC 16689. Offshore Technology Conference. Houston, TX, May 3–6.

Zijderveld, G.H.T., Tiebout, J.J., Hendriks, S.M., Poldervaart, L., 2012. Subsea well intervention vessel and systems. In: OTC 23161. Offshore Technology Conference. Houston, TX, Apr 30–May 3.

PART THREE

Decommissioning Forecast

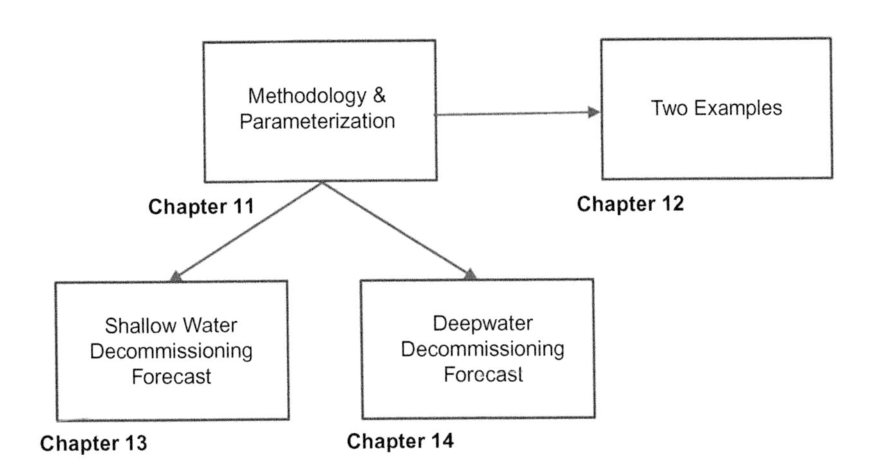

Methodology and Parameterization

Contents

Abstract

Decommissioning forecasts requires the use of different models to describe the behavior of different structure classes. The problem is challenging because offshore structures are diverse, numerous, and used in several different ways. For producing structures, decommissioning forecasts are developed using economic models based on decline curves and cash flow analysis, but for structures that are not currently producing or have never produced, production data obviously cannot be employed. A user-defined methodology is adopted to schedule idle structure decommissioning, and statistical methods are used to characterize auxiliary structure removal rates. Differences between shallow-water and deepwater structure inventories lead to variation in model implementation and are explained. The chapter is organized around the three structure classifications and concludes with the parameterization of structure installation rates.

Decommissioning Forecasting and Operating Cost Estimation
https://doi.org/10.1016/B978-0-12-818113-3.00011-4

11.1 INTRODUCTION

11.1.1 Overview

As with other large-scale development activities on public lands, offshore exploration and development in the US Outer Continental Shelf require environmental impact statements to be performed on a regular basis. The environmental impacts of leasing, development, and production are evaluated as part of BOEM's compliance with the National Environmental Policy Act, which requires federal agencies to study the environmental impacts of their decision-making (Luther, 2005). One significant source of environmental impacts is the installation, operation, and decommissioning of offshore structures, and to perform these studies, it is necessary to quantify how structure inventories are expected to change with changes in business conditions and leasing programs.

Forecast models for offshore decommissioning have not been an especially popular avenue for research, but progress has been made over the years in improving the methods and results, beginning with heuristic techniques described in the National Research Council's report (National Research Council, 1996) to econometric approaches by Pulsipher et al. (2001), structure decline methods (Kaiser, 2008), simulation techniques (Kemp and Stephen, 2011), and current models (Kaiser and Liu, 2018; Kaiser and Narra, 2018). Commercial reports in the sector are published sporadically, but methods are rarely described.

11.1.2 Challenges

Building decommissioning models in the shallow-water GoM is challenging for several reasons. First, the structure inventory is diverse and needs to be characterized prior to evaluation, and because the size of the inventory is relatively large, automated approaches are needed. The deepwater GoM hosts significantly fewer structures and installation and decommissioning activity that can be counted on one hand most years. Second, public databases only describe primary attributes, so classification systems must be defined and applied before initiating work. A significant amount of preprocessing is required, and organizing the system elements is a major effort that draws upon several databases. Third, all modeling is subject to uncertainty and must broadly reflect the reality of limited information and probabilistic outcomes. The probability of producing shallow-water structures transitioning to auxiliary status, for example, is small but nonzero; in deepwater, transitions to auxiliary status are expected to be higher because of the greater inherent value of structures. Operators seek every opportunity to maximize the value of their infrastructure and delay decommissioning.

11.1.3 Shallow Water vs. Deepwater

At the end of 2016, there were 2009 standing structures in water depth <400 ft in the Gulf of Mexico—960 producing structures, 675 idle structures, and 374 auxiliary structures (Table 11.1). Each class is an important contribution to the total with about half of

Table 11.1 Gulf of Mexico Structure Inventories by Production Class Circa 2016

Class	Shallow Water (%)	Deepwater (%)
Producing	960 (48%)	86 (88%)
Idle	675 (34%)	7 (7%)
Auxiliary	374 (19%)	5 (5%)

Source: BOEM 2017.

the shallow-water inventory producing, one-third idle, and one-in-five structures auxiliary. This means that each of the structure classes in shallow water needs to be considered in evaluation, and the uncertainty of the model results will be determined by the relative contribution of the uncertainty of the three models. Both oil and gas structures and major and minor structures are common in shallow water, and almost all wells are dry tree wells.

In 1978, Shell installed fixed platforms in water depth 423 ft (Bourbon) and 1023 ft (Cognac), and is generally recognized as the beginning of deepwater production in the region. In the deepwater GoM circa 2016, there were 48 fixed platforms, three compliant towers, and 47 floaters in water depth >400 ft. Of the 98 active structures, 86 were producing, about 90% of the total; seven were idle; and at least five served in auxiliary roles. In deepwater, producing structures are the most important structure class and production forecasting is the dominant feature in decommissioning models. Most deepwater structures and all floaters are primarily oil producers, which further narrows the parameter space, but the high utilization of wet tree wells complicate modeling efforts. Deepwater wells and structures are much more expensive to construct than shallow-water wells and structures, and both maintain a higher residual value late in life.

11.2 MODEL FRAMEWORK

Three models are used in shallow water to characterize decommissioning activity and represent the links between the producing, idle, and auxiliary structure subgroups and the final state (Figs. 11.1 and 11.2). In Fig. 11.1, dashed and dotted lines represent less likely transitions between states (smaller transition probabilities) than solid lines. In deepwater, idle, auxiliary, and marginal structures (structures that produce below their economic limit) are scheduled together because they are fewer in number (Fig. 11.3). In both shallow-water and deepwater procedures, the submodules are combined using a scenario-based procedure that incorporates exogenous factors such as oil and gas prices, royalty rates, and model-specific parameters. When forecasting active structure inventories, the user must assume future installation rates in each water depth category.

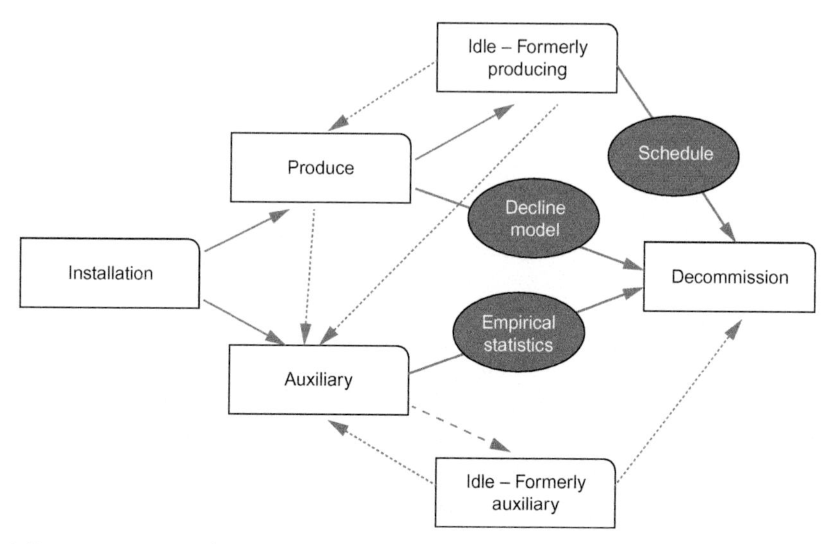

Fig. 11.1 Decommissioning forecast and schedule models in shallow water.

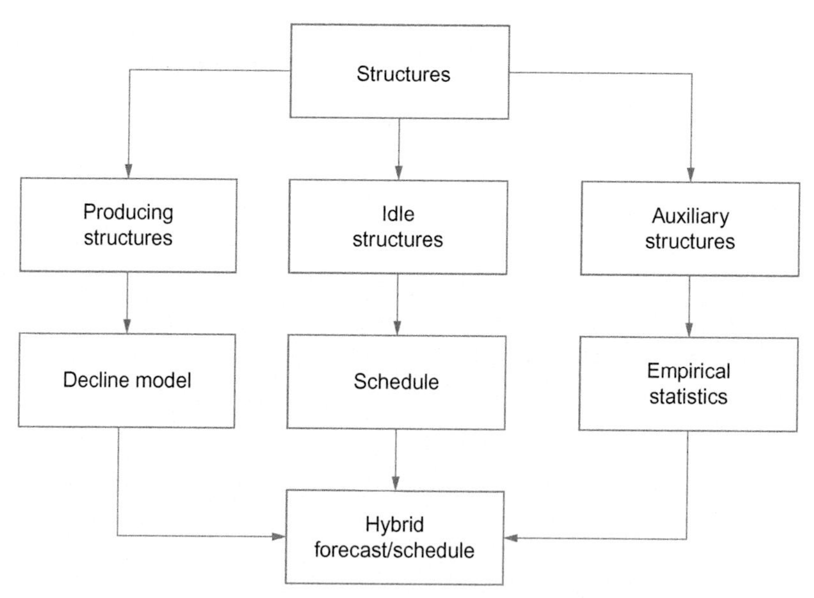

Fig. 11.2 Shallow-water structure decommissioning forecast flowchart.

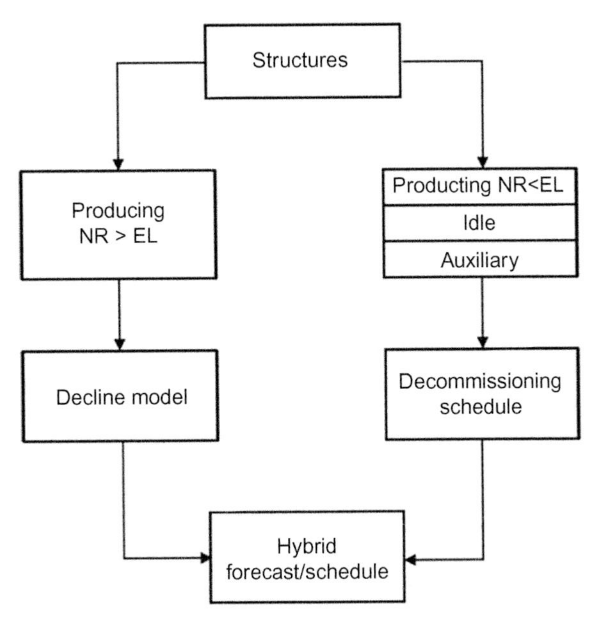

Fig. 11.3 Deepwater structure decommissioning forecast flowchart.

11.2.1 Producing Structures

Well production and reserves are forecast using established industry and regulatory guidelines available from the Society of Petroleum Engineers (SPE) and the US Security and Exchange Commission (SEC) (US SEC, 2008; PRMS, 2007). For producing structures, well forecasts are aggregated to the structure level and used to predict when production will reach its economic limit or abandonment time. Producing structures subsequently transition to idle status, auxiliary status, or decommissioning with probabilities that sum to one and assumed constant over time.

Decline curve models and cash flow analysis are used with an assumed economic limit to determine the time when a structure is no longer economic, that is, generating net revenue greater than its direct operating cost. There are choices in how the procedure is implemented, and differences will arise from the choices that are made, but these differences are not expected to result in wide variation because the procedures are constrained by best practice. Hence, the procedures are repeatable with broadly similar results. The major constraint limiting the reliability of the method is the impact of the assumption of "constant reservoir and investment conditions" and how the structure will transition during the next stage in its life cycle.

11.2.2 Idle Structures

For idle structures, decline curves and cash flow models cannot be applied because the structure has not produced for at least one year. In practice, a small percentage of these structures will be repurposed and their useful lives extended, and a small number may also return to production.

Two modeling approaches can be adopted. The characteristics of idle structures that have been decommissioned can be used to infer the conditions prevalent at the time of decommissioning to infer future activity, or a user-defined model can be used to reflect expected characteristics of the decommissioning schedule. Available information and how much processing one wants to perform dictate which approach to apply, but for practical purposes, the level of uncertainty of the two approaches is believed to be similar.

Using physical insight and knowledge of idle structure inventories, simple decommissioning schedules with well-defined assumptions are performed without sweating the details. Allocation models are introduced that are deliberately simple but rich enough to capture activity characteristics. Allocation methods are based entirely on user-defined assumptions, however, and have no economic basis, and since we cannot be certain user-defined model parameters capture all the important factors (of course, it cannot), development of alternative models and sensitivity analysis is used to gain confidence in the methods and results.

For idle structures, decommissioning is assumed to occur uniformly over a specified future time period determined by the user, for example, 10 years. We then postulate that the number of removals (*NR*) within a specific idle structure subgroup *G*, *NR(G)*, is uniformly distributed over the future time horizon *T*, $NR(G) \sim U(T)$. The idle structure subgroups are defined by idle age, but other choices such as well status can also be adopted. There is no complicated math, only careful physical reasoning based on inventory data.

11.2.3 Auxiliary Structures

Auxiliary structures are distinct from idle structures since they were installed (or have transitioned) to support a nondrilling function and are frequently physically connected to one or more producing platforms. Recent installations of auxiliary structures in shallow water have been used to provide pipeline pumping and compression pressure for deepwater operations.

To schedule auxiliary structure decommissioning, an approach similar to idle scheduling can be used where a user-defined schedule allocates decommissioning activity, or historic activity statistics can be applied. Both methods are believed to have comparable levels of uncertainty. Empirical statistics are used in shallow water to illustrate a technique different than the one employed for idle structures. In deepwater, operators rarely employ auxiliary structures in development because of the extra cost of a nonproducing structure, but after a field has been exhausted, every effort will be made to find an alternative use which may entail an auxiliary role.

11.3 PRODUCING STRUCTURE DECOMMISSIONING MODEL

For all producing structures, the main model parameters include the decline curves constructed for each producing well associated with the structure, oil and gas prices, economic limits, and regulatory requirements (Table 11.2). Each model parameter has a number of additional assumptions that are highlighted in this section, and implementation differences in the shallow water and deepwater are described.

11.3.1 Oil Wells vs. Gas Wells

Oil wells and gas wells are distinguished according to the relative quantities of hydrocarbon liquid and vapor produced using the cumulative gas oil ratio (CGOR). Oil wells, after initial separation, produce crude oil and associated (also called casinghead) gas. Gas wells, after initial separation, produce gas well (nonassociated) gas and condensate. The gas from both types of wells is normally processed onshore to produce residue gas (which is mostly methane) and natural gas liquids (NGLs), which consist of ethane, propane, butanes, and natural gasoline.

11.3.2 Commodity Prices

Crude oil and condensate are liquids at normal conditions. Condensate is a very light crude oil (>50° API) with a narrow distillation spectrum, and in the US Gulf Coast has historically been priced at a 40% to 60% discount to West Texas Intermediate and Louisiana Light Sweet crude oil. Crude oils generally have a gravity in the range of 20–45° API. Heavier crudes (low API gravity) have higher sulfur content and viscosity and greater levels of impurities such as metals and are priced at a discount to lighter sweeter crudes, generally a few dollars per °API and percent sulfur, depending on local market conditions, regional refinery configurations, and other factors.

Gas well gas and casinghead gas are used offshore for a structure's energy requirements and are used throughout the GoM for gas lift in oil wells. Gas production usually only needs to be dehydrated for export. Gas export lines can handle a wide variety of gas composition as long as high levels of carbon dioxide or hydrogen sulfide are not present. Both gas streams are processed onshore at gas plants and fractionators to produce residue gas and NGLs. There are many different types of contracts involved in gas plant processing

Table 11.2 Producing Structure Decommissioning Forecast Model Parameters

Parameter	Notation	Range, Type, or Value
Decline model	$q(b, D)$	Exponential (D), hyperbolic (b, D)
Commodity price	P^{oil}, P^{cond}; P^{gas}, P^{ass}	\$40, \$60, \$80/bbl; \$2, \$3, \$4/Mcf
Economic limit	EL	\$200, \$300, \$400, \$500, \$1000 M
Regulatory	τ; roy	1 year; 12.5%, 16.67%, 18.75%

and fractionation, and producers are often involved in the ownership and operation of gathering systems and gas processing plants, which means they receive a portion of the revenue associated with gas plant processing directly or through their subsidiary or affiliate. These additional revenue streams are not considered in the model.

Associated gas is worth more than gas well gas because in the reservoir it will pick up heavier hydrocarbons from the crude oil that are stripped out and sold as NGL streams. Associated gas has a higher heat content (expressed as Btu/Mcf) than dry gas, which translates to greater market value. On a volumetric basis, there may be a 10% to 50% premium for casinghead gas relative to gas well gas depending on market conditions.

11.3.3 Well Forecasting

Production forecasts employ wells as the basic unit of analysis, and the best forecast will consider the oil and gas streams separately but will not perform independent forecasts of each because the behavior of the two streams is related to the reservoir sands and to each other and correlates differently depending on the reservoir drive and fluid type and stage of its production cycle. For black oil reservoirs, for example, after the pressure in the reservoir falls below the bubble-point pressure of the fluid, more natural gas will come out of solution, and the relative volumes of oil and gas will change, which should be reflected in the production forecast. The GOR that was previously constant will begin to increase and should be reflected in the forecast for the individual product streams.

For oil producers, crude oil is the primary product and the main forecast variable, and the derived secondary product is associated gas. For gas producers, gas is the forecast variable, and condensate is the secondary derived product. Best practice is to forecast the dominant (primary) well stream using decline curve analysis and to use the forecast production curve of the primary stream with an extrapolated CGOR or cumulative condensate gas ratio (CCGR) trend for the well to forecast the secondary stream (Yu, 2014). In this manner, the secondary stream is derived from the primary stream rather than being forecast independently, which leads to better and more accurate forecasts (Fig. 11.4).

Decline curve models are fit to historic production data, and the model with the highest fit parameter is normally selected as the best predictor for future production (Poston and Poe Jr, 2008). For wells in different stages of their life cycle producing from different reservoirs and formation conditions, the ability of decline curves to reflect future production trends may be challenging because it fails to account for standard business practices that attempt to maintain production via investment and intervention for as long as possible.

In shallow water, both exponential and hyperbolic decline curves are applied to all producing wells beginning from their peak production or their last local peak if the model fit from first peak is not adequate. Exponential decline models describe a fast pace reduction in production, whereas hyperbolic decline curves provide for a much slower decline rate. For the 960 producing shallow-water structures, reality is expected to be bound between these

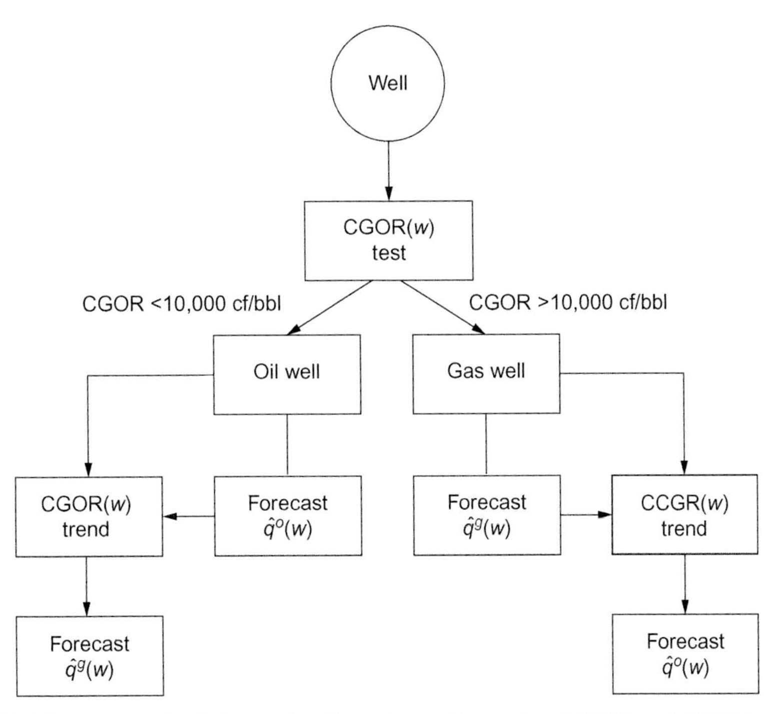

Fig. 11.4 Well forecast methodology using the primary forecast and CGOR and CCGR trends.

two model extremes. Since we cannot predict for an individual well which approach will result in a more accurate forecast a priori and there are such a large numbers of wells, we don't try and apply both models to delineate ranges. In deepwater, the best-fit decline curve model is applied to the primary stream with the secondary stream derived from the primary forecast.

11.3.4 Constant Reservoir and Investment Conditions

All production forecast require boundary conditions and model assumptions. "Constant" (aka status quo) conditions at the time of evaluation are the standard industry and regulatory approach. Constant conditions refer to a broad set of assumptions that imply no production problems in the future, no new sidetracks drilled, no change in operating conditions due to mechanical problems, no processing constraints, no significant spending outside normal operating and maintenance requirements, and no operational changes due to commodity price variation. In other words, all factors beyond the physical conditions of the reservoir that may materially impact well production are fixed at the time of evaluation.

Obviously, constant reservoir and investment conditions do not reflect reality since the business strategy of companies is to maintain profitable production for as long as possible via judicious use of capital spending. In reality, operators are always intervening in one way or another to maintain optimal conditions, but because an operator's future capital spending is unknown and generally unobservable and because outcomes are also unknown, they are not permitted by regulatory agencies or standard industry practice in the estimation of reserves because of their speculative nature. Under this assumption, new wells are not allowed, and investment to recover additional production is not permitted in reserves reporting. The implication is that reserves under status quo conditions will always be underestimated and structure/well abandonment time will always arise earlier-than-expected and realized in practice.

In shallow water, the status quo assumption is balanced by applying both slow (hyperbolic) and fast (exponential) decline curve models to bound the expected decommissioning times. In deepwater, the status quo assumption is balanced in a different way by adding a fixed time to the economic limit year results (Fig. 11.5). For deepwater structures, the time delay is estimated using historic statistics between the last year of production and the actual time of decommissioning, which has ranged between 2 and 10 years. Both approaches act to balance the earlier-than-expected decommissioning forecast predicated on the status quo assumption.

11.3.5 Gross Revenue

For oil wells, future gross revenue is computed using assumed oil and associated gas sale prices and forecast volumes. Gross revenue in year *t* is computed as the sum of the product of monthly production and price for each hydrocarbon stream:

$$GR_t^o(w) = \sum_{i=1}^{12} q_i^{oil}(w)P^{oil} + \sum_{i=1}^{12} q_i^{ass}(w)P^{ass}$$

where oil prices are denoted as P^{oil} and associated (casinghead) gas prices are denoted as P^{ass}. Oil and associated gas volumes are denoted by $q_i^{oil}(w)$ and $q_i^{ass}(w)$. For gas wells, gas and condensate sale prices P^{gas} and P^{cond} and volumes $q_i^{gas}(w)$ and $q_i^{cond}(w)$ are applied:

$$GR_t^g(w) = \sum_{i=1}^{12} q_i^{gas}(w)P^{gas} + \sum_{i=1}^{12} q_i^{cond}(w)P^{cond}$$

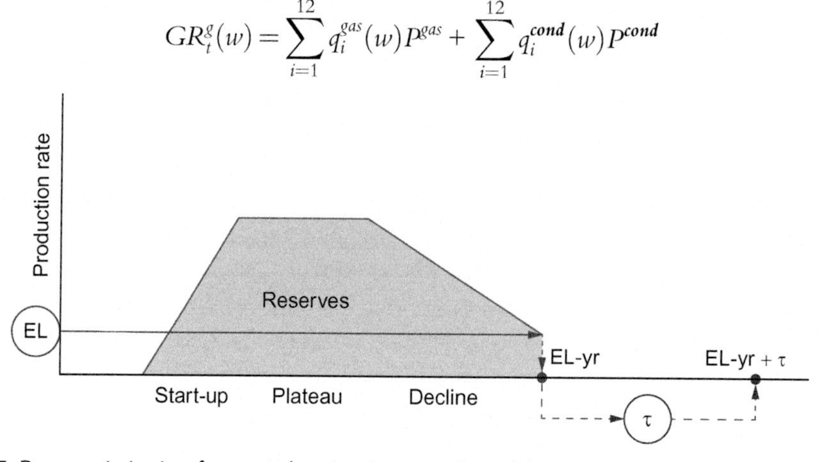

Fig. 11.5 Decommissioning forecast showing input and model parameters.

In production modeling, oil and condensate liquids are commonly combined, and prices are assumed equal ($P^{oil} = P^{cond} = P^{o}$), and similarly, gas well gas and casinghead gas volumes are added together, and their prices are assumed equal ($P^{gas} = P^{ass} = P^{g}$). It is easy to differentiate volumes and prices should the need arise. The end result is the gross revenue estimate for a well:

$$GR_t(w) = q_t^o(w)P^o + q_t^g(w)P^g$$

Gross revenue is an approximation to actual revenues received since sales prices are much more difficult to incorporate. Market-based prices are believed to be a reasonable proxy of sale prices, but adjustments for quality (gravity, sulfur content, and heating value), transportation expenses, the use of hedging programs, contract conditions, etc. cannot be performed using public data.

11.3.6 Structure Production

Every well is associated with a structure that processes production from one or more wells before export to shore, so the oil and gas production and gross revenue associated with a structure are simply the sum of all its producing wells:

$$q_t^o(s) = \sum_w q_t^o(w), q_t^g(s) = \sum_w q_t^g(w), GR_t(s) = \sum_w GR_t(w.)$$

Small structures tend to be associated with small reserves, and the longer a structure is in place, the greater the expected production (Box 11.1).

Platform (dry tree) wells are usually owned in the same proportion as the structure's working interest owners, but subsea wells frequently involve different ownership. When well and structure ownership positions are different, companies arrange a Production Handling Agreement (PHA) to process at the host platform and pay for these services separately (see Chapter 22). PHAs are not modeled nor are the ownership positions in wells and structures considered, which may impact decommissioning timing decisions. In shallow water, subsea wells are not common, and these additional complexities do not arise; in deepwater, subsea wells contribute a significant share of production, and these issues play a larger role.

11.3.7 Net Revenue

Net revenue is the revenue realized after subtracting the royalty payment to the mineral rights owner. In the GoM, royalty payments are defined through the royalty clause as a percentage or share of production proceeds that the lessee pays to the lessor computed from the gross revenue of production after deduction for transportation and processing fees, if applicable:

$$ROY(s, t) = GR(s, t)roy(s, t)$$

where $ROY(s,t)$ denotes the royalty payment, $GR(s,t)$ is the gross revenue, and $roy(s,t)$ is the royalty rate of structure s in year t. For off-lease wells, royalty rates for each well need

BOX 11.1 Main Sequence Curves

The total production of all decommissioned shallow-water oil and gas structures versus their age at the cessation of production is shown in Figs. 11.6 and 11.7. A handful of structures produced for 50 to 60 years but lifetimes between 20 and 30 years are the most common.

Production spreads across the sample are large, on the order of two or three magnitudes. Small structures tend to be associated with small reserves, and caissons/well protectors should be more prevalent at the low end and underside of the band. The longer a structure is in place, the greater the expected production, and the trend should slope upward.

The main sequence curve describes the average production volume at cessation and serves as a general descriptor of the time the average structure produces its reserves. The triangles depicted in Figs. 11.6 and 11.7 represent the average production per age group. For example, a gas structure with 10 Bcfe gas reserves is expected to require 8 years on average to exhaust its production. Changes in technology and the impacts of different geologic deposits and other factors are not controlled.

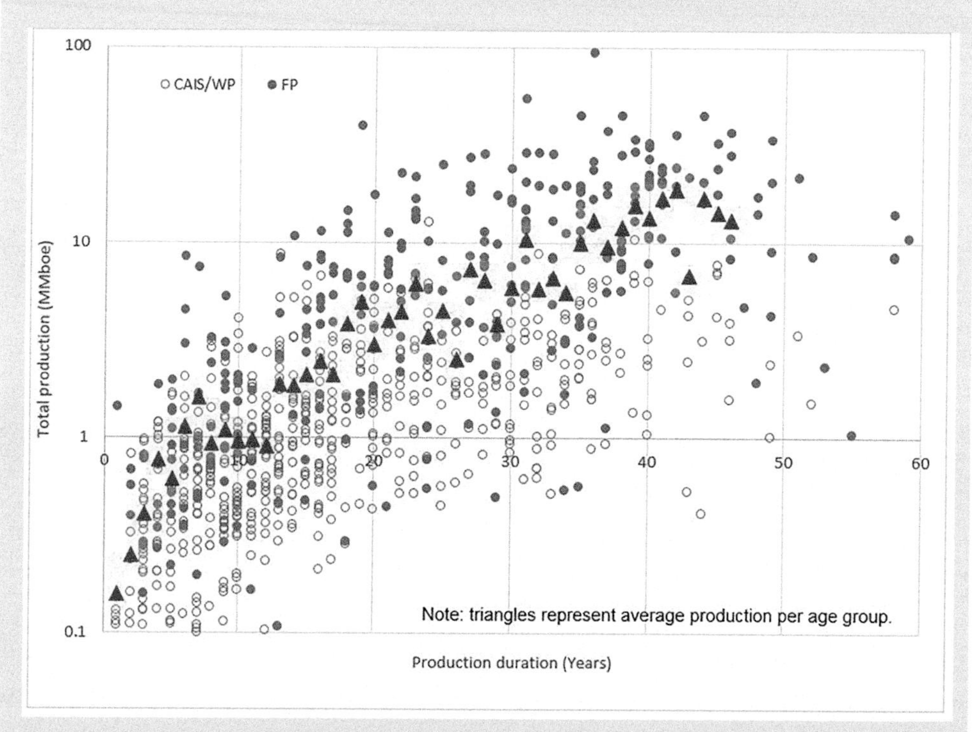

Fig. 11.6 Decommissioned oil structure cumulative production at time of cessation, shallow-water GoM circa 2016. *(Source: BOEM 2017.)*

Continued

BOX 11.1 Main Sequence Curves—cont'd

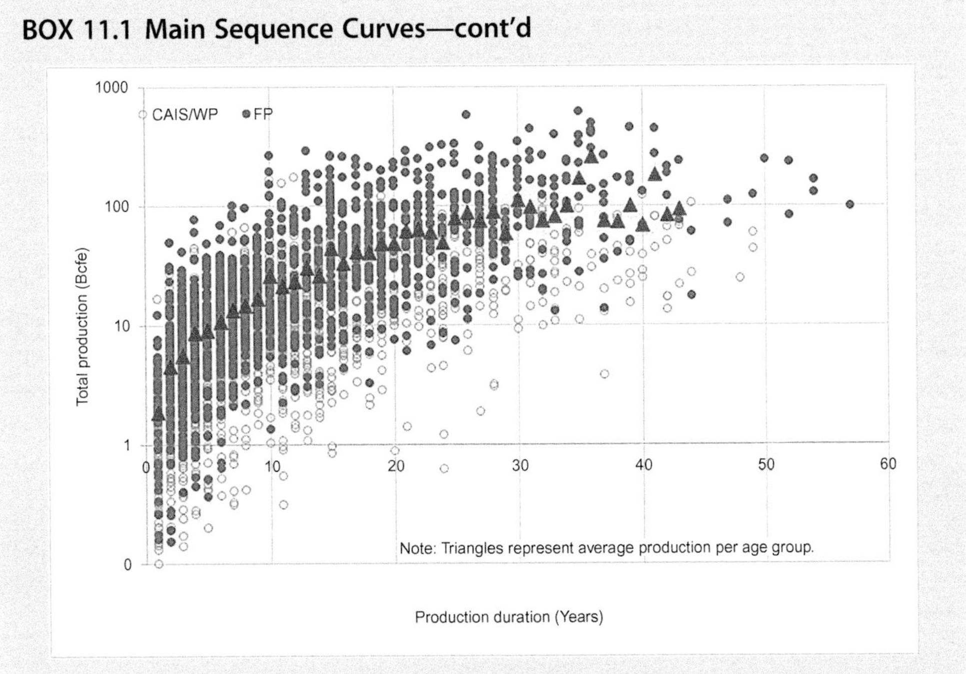

Fig. 11.7 Decommissioned gas structure cumulative production at the time of cessation, shallow-water GoM circa 2016. *(Source: BOEM 2017.)*

to be considered. In the United States, royalty rates are fixed throughout production and for offshore leases in federal waters are either 12.5%, 16.67%, or 18.75% depending on the time of sale and water depth of the lease.

From 1954 to 1982, areas for leasing in the GoM were nominated by oil and gas companies, lease terms were for 5 years, and the royalty rate was 1/6 (16.67%). In 1983, area-wide leasing was introduced that made available all unleased and available blocks in a program area and modified some of the lease conditions. The first lease sales held from 1983 to 1986 had a \$3/ac rent and a 16.67% royalty rate in water depth <200 m. From 2008 to 2016, the royalty rates increased to 18.75% with deep gas relief provided for wells drilled >15,000 ft subsea. In 2017, royalty was reduced to 12.5% in water depth <200 m.

11.3.8 Economic Limit

The economic limit is the production rate beyond which the net operating cash flows (net revenue minus direct operating cost) are negative. Direct operating costs normally include all direct cost to maintain operations, property-specific fixed overhead charges, and production and property taxes but exclude depreciation, abandonment costs, and income tax. Since direct operating cost is not publicly available or reported for offshore structures, historic data were used to infer economic limits.

In Chapter 8, the median-adjusted net revenue the last year of production for shallow-water structures decommissioned from 1990 to 2017 was computed to be $1.23 million for gas structures and $627,000 for oil structures. For oil structures, the adjusted net revenue the last year of production was smaller for minor structures ($352,000 for caissons and well protectors) and larger for larger structures ($873,000 for fixed platforms), while for gas structures, there was essentially no difference between structure types.

In Chapter 10, the average and median inflation-adjusted economic limits for deep-water fixed platforms were shown to be $9.5 million and $3.1 million, respectively. Five decommissioned floaters had an average inflation-adjusted revenue the last year of production of $28 million and a median value of $11.2 million.

11.3.9 Abandonment and Decommissioning Time

A structure is assumed to cease production when net revenue falls below its direct operating expense or economic limit $EL(s)$. When $NR(s) < EL(s)$, production is no longer commercial, and the economic limit year T_{EL} or EL-yr is achieved:

$$T_{EL} = \min \{t \mid NR(s) < EL(s)\}$$

In shallow water, structures are assumed to be decommissioned one year after they reach their economic limit (Fig. 11.8). In shallow water, the decommissioning time T_a is therefore determined as

$$T_a = T_{EL} + 1$$

In deepwater, a more general decommissioning timing equation is applied:

$$T_a = T_{EL} + \tau$$

where the EL-yr is added to a user-defined time period τ parameterized by historic data. For fixed platforms, the time period τ is selected as 3, 5, or 10 years, which bounds historic activity statistics. For floaters, the time period τ is selected as 2 years.

11.4 IDLE STRUCTURE DECOMMISSIONING SCHEDULE MODEL

11.4.1 Parameter Models

The simplest decommissioning schedule for idle structures is to assume the entire idle inventory is removed at a uniform rate over T years. For example, since the size of the shallow-water GoM idle inventory I circa 2016 numbered 675 structures, if the user assumes a 10-year future horizon for decommissioning, then $I/T = 675/10 = 67.5$ idle structures per year will be scheduled for removal. If the user postulates a larger value

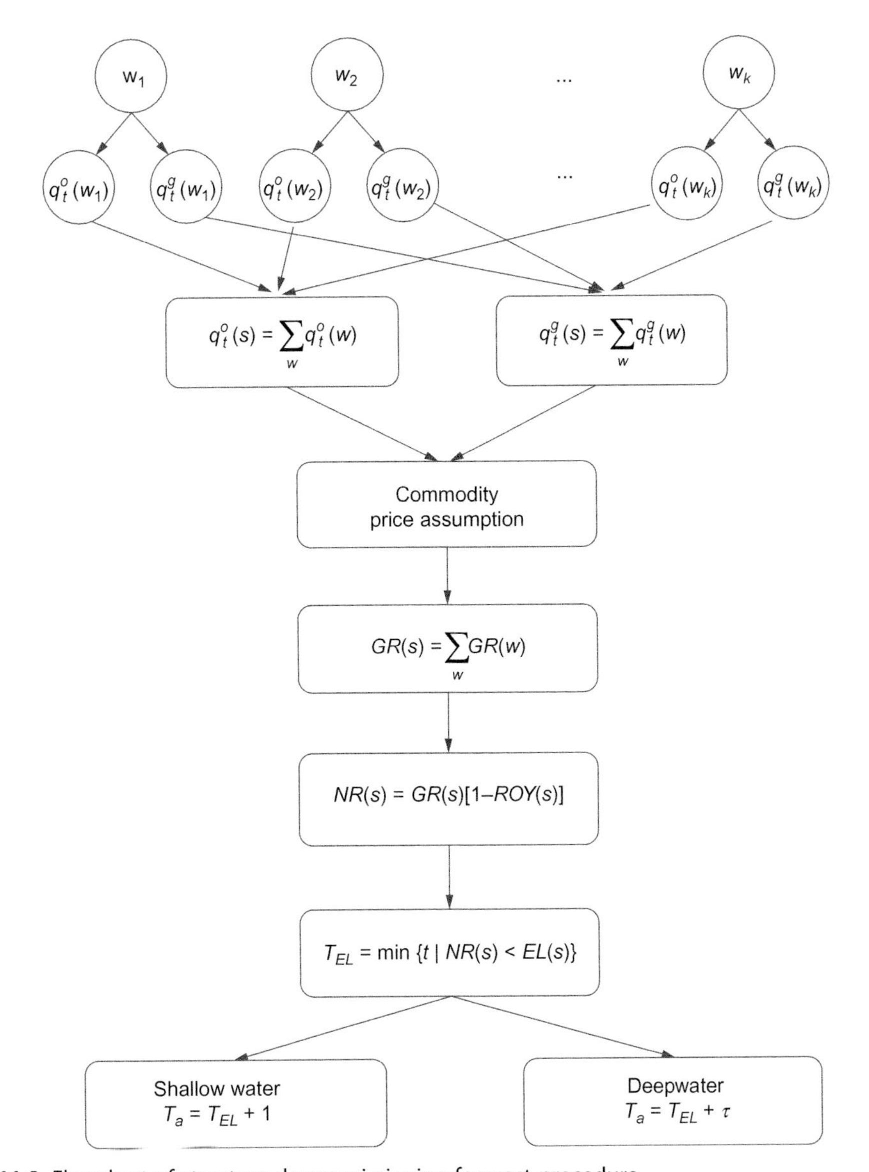

Fig. 11.8 Flowchart of structure decommissioning forecast procedure.

of T (e.g., 20 years), annual activity will be smaller and occur over a longer period. This model has appeal due to its simplicity and may be appropriate in particular circumstances, but it fails to capture age-specific decommissioning characteristics.

A three-parameter model generalizes the one-parameter approach and uses information on idle age and user preference to perform the decommissioning schedule. Two mutually exclusive age groups are described by the number of years since the structures

Table 11.3 Idle Inventory in the Shallow-Water Gulf of Mexico by Idle Age Circa 2016

Idle Age (yr)	Idle Structures	p	$I(\leq p)$	$I(> p)$
1–2	111	2	111	564
2–3	84	3	195	480
3–4	49	4	244	431
4–5	51	5	295	380
5–6	33	6	328	347
6–7	35	7	363	312
7–8	25	8	388	287
8–9	21	9	409	266
9–10	22	10	431	244
> 10	244		675	

Source: BOEM 2017.

last produced. After selecting the idle age group parameter p, the structure set I is decomposed into idle structures that are less than or equal to p years idle at the time of evaluation, and those idle structures that are greater than p years idle:

$$I = \{I\,(\text{Idle age} \leq p\,\text{years})\} \cup \{I\,(\text{Idle age} > p\,\text{years})\}$$

The number of structures that fall within each subgroup is identified by the symbols $I(\leq p)$ and $I(> p)$. The value of I is simply the inventory size at the time of evaluation, while the value of p is user-defined and is a positive integer denoting years. For example, for $p = 7$ years, $I(\leq 7) = 363$, and $I(> 7) = 312$ as shown in Table 11.3. In recent years, a greater number of older idle structures have been removed and are associated with a higher probability of decommissioning relative to younger idle structures. Dividing the idle inventory into two subgroups allows different removal rates per group to be modeled.

After the user posits the horizon for decommissioning, all the structures within the subgroup are assumed to be decommissioned within the period using a uniform mechanism. For example, for $T_1 = 4$ years, the structures of $I(< p)$ would be removed at an annual rate of $I(< p)/T_1$ over each of the next 4 years.

11.4.2 Scenarios

Subgroup $I(\leq p)$ inventory is allocated uniformly over a time period T_1, and subgroup $I(>p)$ inventory is allocated over time period T_2. The values of T_1 and T_2 are user-defined and with the selection of p determine a three-parameter schedule denoted $S = (p, T_1, T_2)$. Normally, T_1 and T_2 would be selected so that $T_1 \geq T_2$ to reflect the expected faster removal rate of older structures, but this is not a requirement of the formulation. Selection of the model parameter T in the one-parameter model and (p, T_1, T_2) in the three-parameter model is referred to as a scenario and after selection completely determines the decommissioning schedule for the idle structure inventory.

11.4.3 Model Equations

For the idle inventory subgroup $I(\leq p)$, the time horizon T_1 sets the annual number of decommissioned structures and the time to clear the inventory. Assuming uniform activity per year, activity in year i is determined as c_i and is constant for each year through T_1, $i = 1, 2, \ldots, T_1$:

$$c_i = \frac{I(\leq p)}{T_1}, i = 1, 2, \ldots, T_1$$

Using a vector notation, the removal schedule is written $R_s(I > p)$:

$$R_S(I \leq p) = (c_1, c_2, \ldots, c_{T_1})$$

Similarly, for the structures in the idle inventory subgroup $I(>p)$, T_2 determines the annual removal activity in year i as d_i that is constant for each year through T_2, $i = 1, 2, \ldots, T_2$:

$$d_i = \frac{I(> p)}{T_2}, i = 1, 2, \ldots, T_2.$$

In vector notation

$$R_S(I > p) = (d_1, d_2, \ldots, d_{T_1})$$

The decommissioning schedule for the entire idle inventory is determined by summing the vectors $R_s(I \leq p)$ and $R_s(I > p)$:

$$R_s(I) = R_s(I \leq p) + R_s(I > p)$$

If $T_1 > T_2$, the composite vector for the removal scenario $R_S(I)$ will appear as

$$R_S(I) = (c_1 + d_1, c_2 + d_2, \ldots, c_{T_2} + d_{T_2}, c_{T_2} + 1, \ldots, c_{T_1})$$

The vector terms sum to the size of the inventory, and the last element of the vector is determined by the structures in the $I(\leq p)$ subgroup since $T_1 > T_2$. If $T_1 = T_2$, c_i and d_i will terminate simultaneously.

The value of the model parameters (p, T_1, T_2) can be bound between upper and lower values to incorporate uncertainty in the assessment, and a phase space approach can be used to generalize the procedure. To compute the average of two or more scenarios, a normalization procedure is required.

Example 1. $S = \{p = 10, T_1 = 10, T_2 = 5\}$

Using Table 11.3 with $p=10$ yields subgroup counts of $I(\leq 10)=431$ and $I(>10)=244$. The decommissioning schedule vectors for this scenario are written as

$$R_S(I \leq 10) = (43.1, 43.1, 43.1, 43.1, 43.1, 43.1, 43.1, 43.1, 43.1, 43.1)$$
$$R_S(I > 10) = (48.8, 48.8, 48.8, 48.8, 48.8)$$

The decommissioning schedule is determined by adding the $R_S(I \leq 10)$ and $R_S(I > 10)$ vectors together:

$$R_S(I) = (91.9, 91.9, 91.9, 91.9, 91.9, 43.1, 43.1, 43.1, 43.1, 43.1)$$

Note that the $R_S(I \leq 10)$ vector occurs over 10 years, while $R_S(I > 10)$ occurs over 5 years, and the sum of all the elements in $R_S(I)$ is 675, the size of the idle inventory. The period of the schedule is 10 years.

Example 2. $S = \{p = 10, T_1 = 20, T_2 = 10\}$

Using the same value $p=10$ as Example 1, the subgroup counts of $I(\leq 10)=431$ and $I(>10)=244$ are the same, but now, each of the time periods is doubled so that $T_1=20$ years and $T_2=10$ years. In this case, the vector elements are halved; $c_i=431/20=21.6$ for $i=1, 2, \ldots, 20$, and $d_i=244/10=24.4$ for $i=1, 2, \ldots, 10$. The individual vectors and aggregate schedule are written as

$$R_S\,(I \leq 10) = (21.6, 21.6, 21.6, 21.6, 21.6, 21.6, 21.6, 21.6, 21.6, 21.6, 21.6, 21.6, 21.6, 21.6, 21.6, 21.6, 21.6, 21.6, 21.6, 21.6)$$
$$R_S\,(I > 10) = (24.4, 24.4, 24.4, 24.4, 24.4, 24.4, 24.4, 24.4, 24.4, 24.4)$$
$$R_S\,(I) = (46, 46, 46, 46, 46, 46, 46, 46, 46, 46, 22, 22, 22, 22, 22, 22, 22, 22, 22, 22)$$

The multiplication works using noninteger values as long as the same multiplier is applied to both T_1 and T_2 and values are rounded to the nearest integer.

Example 3. $S = \{p = 10, T_1 = 10, T_2 = 5\}, q = 1.4$

Using the multiplier $q=1.4$, $T_1=1.4(10)=14$ years, and $T_2=1.4(5) \approx 8$ years. The effect will be to expand each time horizon by 40%. In this case, $c_i=431/14=30.8$ for $i=1, 2, \ldots, 14$, and $d_i=244/18=30.5$ for $i=1, 2, \ldots, 8$. In terms of the individual vectors,

$$R_S\,(I \leq 10) = (30.8, 30.8, 30.8, 30.8, 30.8, 30.8, 30.8, 30.8, 30.8, 30.8, 30.8, 30.8, 30.8, 30.8)$$
$$R_S\,(I > 10) = (30.5, 30.5, 30.5, 30.5, 30.5, 30.5, 30.5, 30.5)$$
$$R_S\,(I) = (61.3, 61.3, 61.3, 61.3, 61.3, 61.3, 61.3, 61.3, 30.8, 30.8, 30.8, 30.8, 30.8, 30.8)$$

Note that $61.3(8)+30.8(6)=675$. The period of the schedule is 14 years.

Example 4. $S = \{p = 7, T_1 = 5, T_2 = 3\}$

Using Table 11.3 with $p=7$, $I(\leq 7)=363$, and $I(>7)=312$. The values of $T_1=5$ and $T_2=3$ determine removal schedules over 5 years and 3 years, respectively, as follows:

$$R_S(I \leq 7) = (72.6, 72.6, 72.6, 72.6, 72.6)$$
$$R_S(I > 7) = (104, 104, 104)$$
$$R_S(I) = (176.6, 176.6, 176.6, 72.6, 72.6)$$

11.4.4 Normalization

Normalization ensures that decommissioning schedules are combined in a manner consistent with the model output. In normalization, all the decommissioning schedules are averaged to yield a composite schedule. For example, if the scenario $(p, T_1, T_2) = (10, 10, 5)$ yields the decommissioning schedule vector $R_{(10,10,5)}$ over 10 years and scenario (7, 5, 3) yields a decommissioning schedule vector $R_{(7,5,3)}$ over 5 years, the average of these two scenarios is computed by dividing the total annual activity by total activity and then using the percentage vector as a multiplying factor (Box 11.2).

Example 5. Vector normalization

If the decommissioning schedule vectors computed in Examples 1 and 4 are normalized on a percentage basis by dividing each entry by the row sum total (i.e., 675), then

$$r_{(10,10,5)} = (0.136, 0.136, 0.136, 0.136, 0.136, 0.064, 0.064, 0.064, 0.064, 0.064)$$
$$r_{(7,5,3)} = (0.262, 0.262, 0.262, 0.108, 0.108).$$

The average aggregate schedule is the average of these two vectors computed termwise:

$$\bar{r} = \frac{1}{2}\left(r_{(10,10,5)} + r_{(7,5,3)}\right) = (0.20, 0.20, 0.20, 0.12, 0.12, 0.03, 0.03, 0.03, 0.03, 0.03)$$

The allocation percentage for the decommissioning schedule is denoted by $\bar{r}$ and when multiplied by the idle inventory yields the average idle structure schedule $\bar{R}$:

$$\bar{R} = 675 \cdot \bar{r} = (135, 135, 135, 81, 81, 20, 20, 20, 20, 20)$$

Example 6. Tableau normalization

Tableau normalization is equivalent to vector normalization except that a table is used to organize the calculations (Table 11.4). The decommissioning schedules to be normalized are placed in separate rows with columns representing years. In each scenario, 675 structures are removed as indicated by the first two row sums, and column sums show the total number of removals each year for the combined

Table 11.4 Normalization of Two Scenarios for Idle Schedule

Scenario	1	2	3	4	5	6	7	8	9	10	Row Sum
(10, 10, 5)	91.9	91.9	91.9	91.9	91.9	43.1	43.1	43.1	43.1	43.1	675
(7, 5, 3)	176.6	176.6	176.6	72.6	72.6						675
Column sum	268.5	268.5	268.5	164.5	164.5	43.1	43.1	43.1	43.1	43.1	1350
% Activity	0.20	0.20	0.20	0.12	0.12	0.03	0.03	0.03	0.03	0.03	1.00

scenarios. The total number of structures represented in the two scenarios (1350=2*675) is the normalizing factor and, when divided into the column sums, yields the normalized decommissioning activity percentage for each year. The elements of the percentage vector sum to one and when multiplied by the idle structure count yield the average decommissioning schedule.

BOX 11.2 A Phase Space Approach for Uncertainty Bounds

The concept of a model phase space is an abstract space in which each state of a system is represented as a unique point bound between specific values. The phase space is discretized into cells, and each cell represents a particular decommissioning schedule. There are three independent variables (p, T_1, T_2) per cell with each variable required to be a positive integer. Constraints may exist between the variables (i.e., $T_1 > T_2$) to reflect user preferences and/or system constraints. Each possible state is represented by a unique point.

Two different averages can be computed within the phase space, the "bounded" average and the "ensemble" average. The bounded average $\mathcal{R}_b$ represents the average of the boundary points of each variable. Since there are three variables and two boundary conditions per variable, eight decommissioning schedules are required in the computation, and the normalized average from these eight schedules yields the bounded average. We demonstrate the bounded average computations for a specific example.

Example 7. $\{\mathcal{R}_b \mid p = 3, 10; T_1 = 5, 10; T_2 = 3, 7\}$

From Table 11.3, the GoM idle structure inventory circa 2016 yields $I(\leq 3)=195$ and $I(>3)=480$. For $p=10$, $I(\leq 10)=431$, and $I(>10)=244$. Tableaus are computed describing the decommissioning schedules for each set of triplets. The idle inventories yield four decommissioning schedules denoted R_{ij}^p; $i, j=1, 2$ for each pair of (T_1, T_2) values (Table 11.5).

Table 11.5 Bounded Average Schedule Organization

$p=3$	$I(\leq 3)$, $T_1 = 5$	$I(>3)$, $T_1 = 10$
$T_2=3$	R_{11}^p	R_{12}^p
$T_2=7$	R_{21}^p	R_{22}^p

For $p=3$, each cell of the tableau requires a separate computation and is composed of a single decommissioning schedule vector (Table 11.6). For $p=10$, the four decommissioning schedules are shown in Table 11.7.

Continued

BOX 11.2 A Phase Space Approach for Uncertainty Bounds—cont'd

Table 11.6 Bounded Average Decommissioning Schedule Vectors for $p=3$, $T_1=\{5, 10\}$, and $T_2=\{3, 7\}$

$T_1=5, T_2=3$	$R_{11}^3=(199, 199, 199, 39, 39)$
$T_1=5, T_2=7$	$R_{21}^3=(108, 108, 108, 108, 108, 68, 68)$
$T_1=10, T_2=3$	$R_{12}^3=(180, 180, 180, 20, 20, 20, 20, 20, 20, 20)$
$T_1=10, T_2=7$	$R_{22}^3=(88, 88, 88, 88, 88, 88, 88, 20, 20, 20)$

Table 11.7 Bounded Average Decommissioning Schedule Vectors for $p=10$, $T_1=\{5, 10\}$, and $T_2=\{3, 7\}$

$T_1=5, T_2=3$	$R_{11}^{10}=(168, 168, 168, 86, 86)$
$T_1=5, T_2=7$	$R_{21}^{10}=(147, 147, 147, 147, 147, 61, 61)$
$T_1=10, T_2=3$	$R_{12}^{10}=(124, 124, 124, 43, 43, 43, 43, 43, 43, 43)$
$T_1=10, T_2=7$	$R_{22}^{10}=(78, 78, 78, 78, 78, 78, 78, 43, 43, 43)$

When all eight vectors are summed and normalized, the average bounded allocation vector in absolute and percentage terms is as follows:

(137, 137, 137, 76, 76, 45, 45, 16, 16, 16)

(0.20, 0.20, 0.20, 0.11, 0.11, 0.07, 0.07, 0.02, 0.02, 0.02)

In shorthand notation, the bounded average approach can be described as

$$\bar{r}=\frac{1}{8T}\sum_{p}\sum_{i,j=1}^{2}R_{ij}^{p}$$

$$\mathcal{R}_b=\bar{r}T.$$

The ensemble average $\mathcal{R}$ is more computationally challenging since it computes every possible (integer) schedule in the phase space discretization. Phase spaces are useful because they encapsulate all possible combinations of events/scenarios within the ranges defined by the user and thus represent the average or expected outcome determined by user bounds.

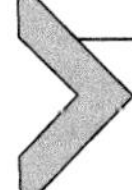

11.5 AUXILIARY STRUCTURE DECOMMISSIONING SCHEDULE MODEL

Decommissioning forecasts for auxiliary structures entail the same difficulty in forecasting as idle structures because of the lack of a production stream. In principle, for those cases where a link with one or more producing structures can be made, it would be a simple matter to apply the correspondence for the auxiliary structure, but this would not by itself ensure a more reliable forecast. In deepwater, nonproducing structures are more likely to be repurposed in an auxiliary role because of their greater capital costs and potential use in support activities, and operator reluctance to decommission valuable assets.

In shallow water, a simple aggregate model is adopted that captures the global characteristics of the auxiliary structure class without recourse to additional linkages by using historic average activity. Another way to estimate future decommissioning is to base activity on the historic percentage/range relative to total producing and idle structures. Historically, an average of 20 auxiliary structures in shallow water have been decommissioned per year over the past two decades, representing between 7% and 15% of total structure decommissioning activity over this time.

In deepwater, only a few auxiliary structures were in inventory circa 2016, and these structures can be either ignored due to their small number or decommissioned according to an assumed schedule. Most of the deepwater auxiliary structures are pipeline junctions that are expected to remain useful for a significant period of time.

11.6 INSTALLED STRUCTURES

The size of the active inventory at any point in time is based on the difference between the cumulative number of installed and decommissioned structures at the time of evaluation:

$$Active_t = CumInstalled_t - CumDecom_t$$

Since new structures are continually being installed in both the shallow-water and deepwater GoM and are expected to continue to be installed in the future, to forecast active inventories it is necessary to assume an installation rate for each water depth region and add this to the decommissioning forecast results. Assuming a future installation rate is a speculative exercise, of course, since no one knows—or should pretend to know for that matter—what the future will bring, but under status quo conditions, a reasonable case can be made using historic activity as a guide.

There has been a dramatic decline in the number of shallow-water structures installed in recent years. Caissons, well protectors, and fixed platforms all show similar installation trends across time, peaking in the 1977–86 period at an average rate of 170 structures per year and declining to 123 structures per year from 1997 to 2006 (recall Table 3.4). Over the past decade, only 28 shallow-water structures on average have been installed per year, and over the past 5 years, 11 shallow-water structures per year have been installed.

For fixed platforms in water depth >400 ft, declining installations have been dramatic, while floater installations have held reasonably steady. From 1987 to 1996, 2.6 fixed platforms were installed per year on average, which declined to 1.8 structures per year from 1997 to 2006, to 0.2 structures per year from 2007 to 2016 (recall Table 3.6). Over the past decade, they were on average about two floaters installed per year compared with three floaters per year over the preceding decade. Over the 30-year period from 1987 to 2016, 54 floaters were installed or about 1.8 structures per year on average.

REFERENCES

Kaiser, M.J., 2008. Modeling regulatory policies associated with offshore structure removal requirements in the Gulf of Mexico. Energy 33 (7), 1038–1054.

Kaiser, M.J., Liu, M., 2018. A scenario-based deepwater decommissioning forecast in the U.S. Gulf of Mexico. J. Petrol. Sci. Eng. 165, 913–945.

Kaiser, M.J., Narra, S., 2018. A hybrid scenario-based decommissioning forecast for the shallow water U.S. Gulf of Mexico, 2018-2038. Energy 163, 1150–1177.

Kemp, A.G., Stephen, L., 2011. Prospective Decommissioning Activity and Infrastructure Availability in the UKCS: 1–80. Aberdeen, Scotland.

Luther, L., 2005. The National Environmental Policy Act: Background and Implementation. Congressional Research Service, Washington, DC.

National Research Council, 1996. An Assessment of Techniques for Removing Offshore Structures. The National Academies Press, Washington, DC.

Petroleum Resources Management System, 2007. Society of Petroleum Engineers.

Poston, S.W., Poe Jr., B.D., 2008. Analysis of Production Decline Curves. Society of Petroleum Engineers, Richardson, TX.

Pulsipher, A.G., Iledare, O.O., Mesyanzhinov, D.V., Dupont, A., Zhu, Q.L., 2001. Forecasting the number of offshore platforms on the Gulf of Mexico OCS to the year 2023. U.S. Dept. of Interior, Minerals Management Service, New Orleans, LA, OCS Study MMS 2001-013.

U.S. Securities and Exchange Commission, 2008. Modernization of the oil and gas reporting requirements. Confirming version (proposed rule), 17 CFR Parts 210, 211, 229, and 249, Release Nos. 33-8995; 34-59192; FR-78; File No. S7-15-08, RIN 3235-AK00.

Yu, S., 2014. A new methodology to predict condensate production in tight/shale retrograde gas reservoirs. In: SPE 168964. SPE Saudi Unconventional Resources Conference. The Woodlands, TX, Apr 1–3.

CHAPTER TWELVE

Two Examples

Contents

Abstract

Two examples are used to illustrate the computational steps involved in decommissioning forecast models for producing structures. Chevron's deepwater Tick platform is evaluated circa 2012, and Anadarko's Horn Mountain spar is evaluated circa 2016. The examples are presented using the same organizational framework but different evaluation times. The sensitivity of decommissioning timing and reserves to oil and gas price variation is examined.

Decommissioning Forecasting and Operating Cost Estimation
https://doi.org/10.1016/B978-0-12-818113-3.00012-6

12.1 TICK AND LADYBUG

The Tick and Ladybug projects were evaluated in March 2013 using production data through 2012. A postscript updates the status of the structure circa 2018.

12.1.1 Development (Fig. 12.1)

In 1991, Texaco, Inc. installed the Tick platform in Garden Banks block 189 in 720 ft water depth as a one-piece jacket over three predrilled wells (Curtis and Gilmore, 1992). Three subsea wells owned and operated by ATP Oil & Gas Corp. were drilled at the nearby Ladybug field and tied back to Tick in 1997 and 2006.

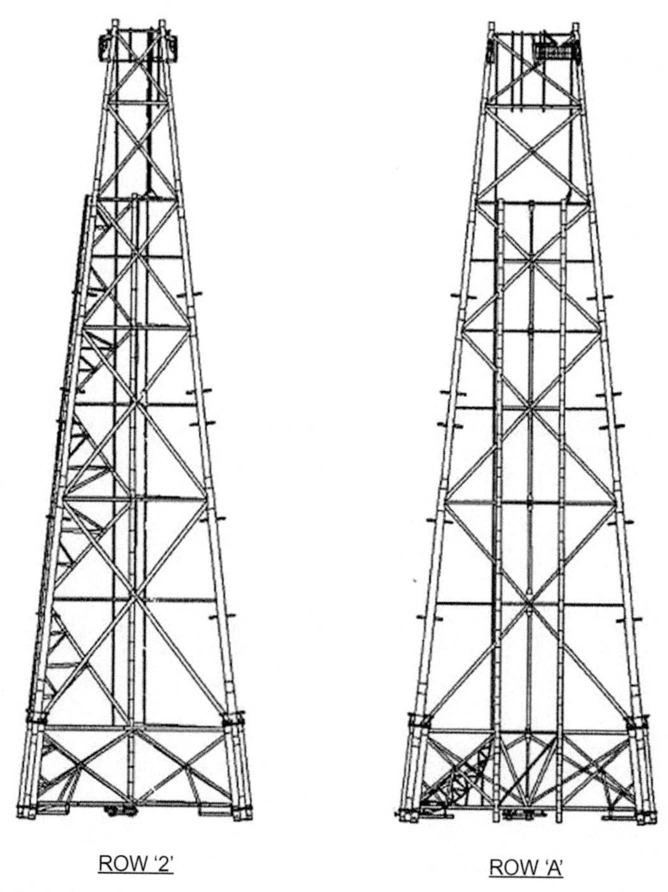

Fig. 12.1 The Tick platform was installed in 1991 in 720 ft water depth. *(Source: Chevron.)*

12.1.2 Structure Production (Figs. 12.2 and 12.3)

In 2012, Tick produced 97 Mbbl oil and 1.67 Bcf natural gas from five producing wells and generated gross revenue of approximately $15 million during the year. Cumulative production through 2012 was 28.6 MMbbl oil and 259 Bcf natural gas. Ladybug contributed 11.2 MMbbl oil and 11.9 Bcf gas, about half of Tick's total production circa 2012. A cumulative gas-oil ratio CGOR of 9056 cf/bbl classifies the Tick platform as primarily oil.

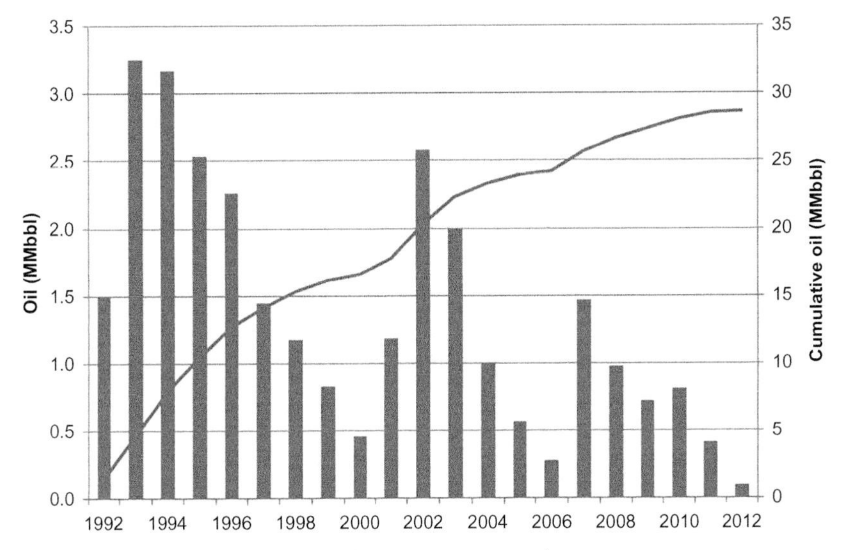

Fig. 12.2 Tick oil production, 1992–2012. *(Source: BOEM 2013.)*

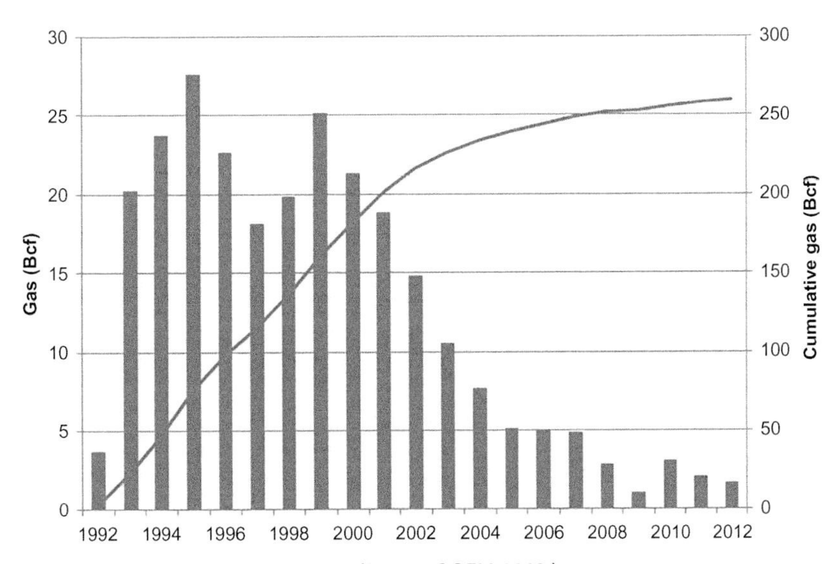

Fig. 12.3 Tick gas production, 1992–2012. *(Source: BOEM 2013.)*

12.1.3 Well Inventory (Table 12.1, Fig. 12.4)

A total of 27 wells, eight of which are sidetracks, were drilled at Tick through 2012. Two wells were temporarily abandoned, 19 wells were idle, and three wells were producing circa 2012, one oil well and two gas wells. Three subsea wells have been drilled at the Ladybug field, and two oil wells were producing circa 2012.

Table 12.1 Tick Well Inventory Circa January 2013

API Number	First Production	Status/Wet	Cumulative Production		
			Oil (Mbbl)	Gas (MMcf)	GOR (Mcf/bbl)
608074005100	1992	Idle	683	1905	3
608074005500		Dry hole			
608074005501	1998	Idle	29	14,009	486
608074005900		Dry hole			
608074005901	1992	Idle	165	2091	13
608074005902	1997	Idle	139	2714	20
608074009200	1992	Idle	250	2189	9
608074009201	1994	Producing	266	38,076	143
608074009500		TA			
608074009501	1992	TA	493	6255	13
608074009600		Dry hole			
608074009601	1992	Idle		10,824	69,829
608074010700	1992	Idle	2317	19,036	8
608074010800	1992	Producing	4887	16,498	3
608074010900	1992	Idle	2430	10,607	4
608074011000	1992	Idle	660	12,002	18
608074011100	1992	Idle	540	18,057	33
608074011200	1993	Idle	75	5144	69
608074011400	1993	Idle	303	24,040	79
608074011500	1993	Producing	187	16,975	91
608074011600	1993	Idle	100	23,632	236
608074012000		Dry hole			
608074012001	1993	Idle	1658	6700	4
608074012400	1993	Idle	188	933	5
608074012600	1994	Idle	1187	1801	2
608074012700		Dry hole			
608074012701	1994	Idle	911	14,043	15
608074016300	2001	Producing/wet	6912	7303	1
608074063500	1997	Idle/wet	1632	1453	1
608074063501	2007	Producing/wet	2634	3126	1
Total			28,648	259,414	9

Source: BOEM 2013.

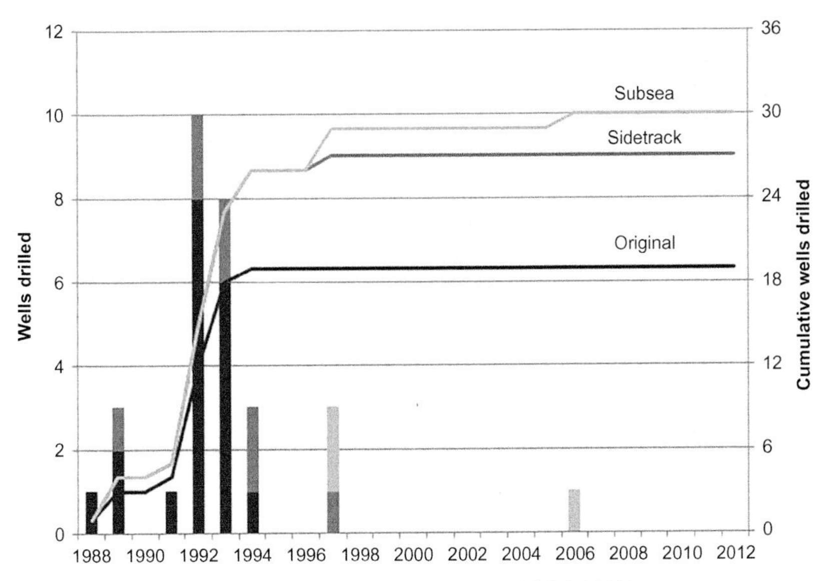

Fig. 12.4 Tick drilling and subsea tieback schedule. *(Source: BOEM 2013.)*

12.1.4 Sidetrack Production (Fig. 12.5)

Sidetrack wells are drilled in search of additional production after the original wellbore no longer produces. Sidetrack wells cost less than spudding a new well but also usually deliver less than the original wellbore since they often target smaller pockets or less promising sands. The risk in drilling is that the sidetrack may not provide a return on capital deployed. From an investment point of view, it is the total cost and performance for the sidetrack program that is important in creating value.

12.1.5 Subsea Production (Fig. 12.6)

Subsea wells may be drilled and tied back as part of the initial field development or as discoveries are made in the vicinity of the facility. For third-party fields, Production Handling Agreements are negotiated to allow production to be processed at another operator's facility (see Chapter 22). In 1997, ATP drilled two subsea wells that were tied back to Tick. Flowline issues in the oil line delayed oil receipt until 2000.

12.1.6 Decline Curve Specification (Table 12.2)

Each producing well is curve fit based on the historic primary product stream profile at the time of evaluation using a hyperbolic decline curve model. For primarily oil wells, oil

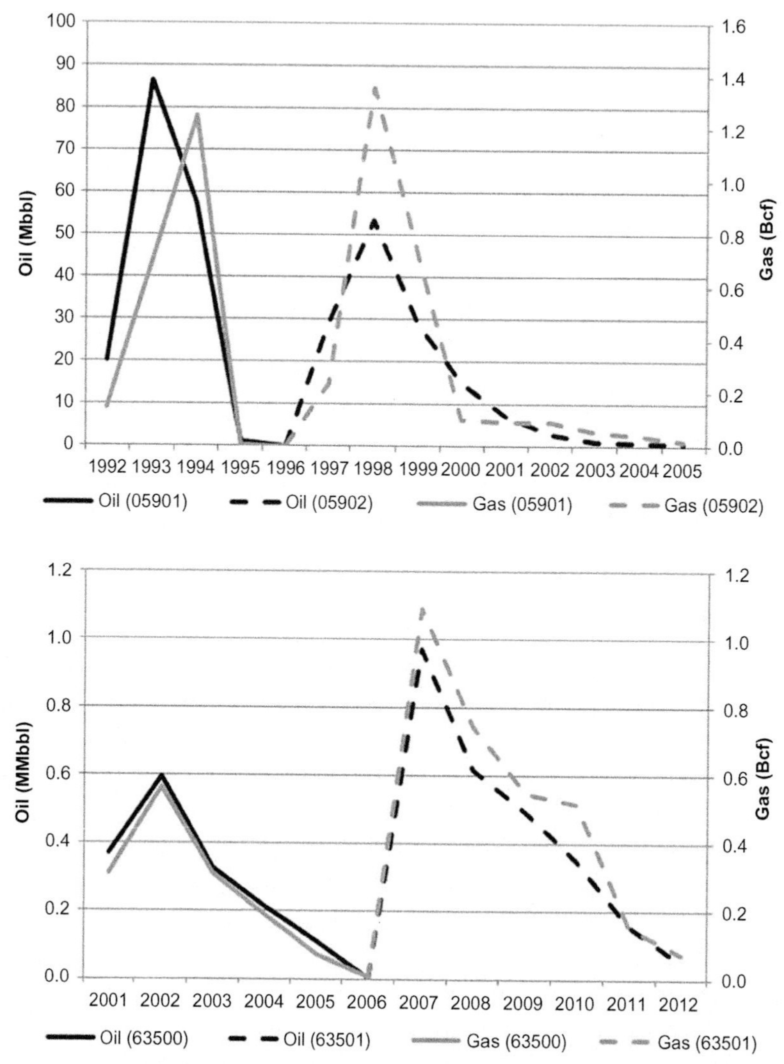

Fig. 12.5 Tick production profiles for wells 05901 and 63500 and sidetracks 05902 and 63501. *(Source: BOEM 2013.)*

production is forecast based on the best-fit model parameters, and the secondary product (associated gas) is forecast by extrapolating the cumulative gas-oil ratio versus cumulative oil trend. For primarily gas wells, gas production is forecast using the best-fit model parameters, and the secondary product (condensate) is forecast using the cumulative condensate-gas ratio versus cumulative gas trend.

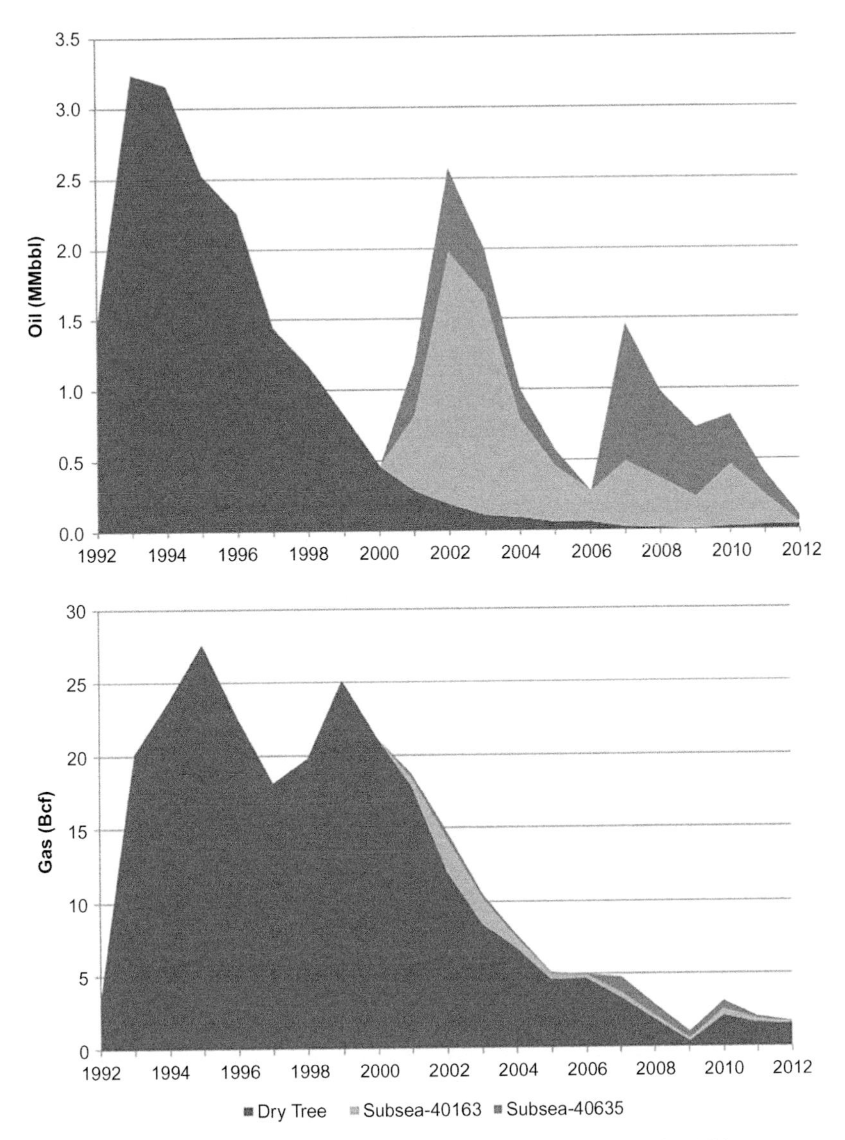

Fig. 12.6 Tick and Ladybug subsea well production profile. *(Source: BOEM 2013.)*

12.1.7 Primary Production Forecast (Table 12.3)

There are two gas and one oil dry tree producing wells at Tick circa 2012, and the two producing Ladybug subsea wells are both classified as primarily oil. Production and price determine gross revenue, and reduction by the royalty rate yields the net revenue.

Table 12.2 Tick's Producing Wells and Estimated Reserves Circa 2012

	Parameters			Remaining Production			Total Production		
API Number	b	D	R^2	Oil (Mbbl)	Gas (MMcf)	BOE (Mboe)	Oil (Mbbl)	Gas (MMcf)	BOE (Mboe)
608074009201	0.13	0.19	0.96	5	1350	230	271	39,426	6842
608074010800	0.00	0.23	0.94	35	121	56	4923	16,619	7692
608074011500	1.00	0.35	0.88	5	453	81	192	17,428	3097
608074016300	0.39	0.45	0.89	15	16	18	6927	7319	8147
608074063501	0.00	0.40	0.97	208	248	249	2842	3374	3405
Total				269	2188	633	15,155	84,167	29,183

Note: Remaining production estimates are based on $100/bbl oil price and $4/Mcf gas price.

Table 12.3 Tick Primary Product Stream Forecast Circa 2012

Well	09201	10800	11500	16300	63501	Structure	
Primary	Gas	Oil	Gas	Oil	Oil	Oil	Gas
Unit	MMcf	Mbbl	MMcf	Mbbl	Mbbl	Mbbl	MMcf
1992		235				235	
1993		530	24			530	24
1994	1264	447	46			447	1310
1995	2179	627	168			627	2347
1996	3377	649	129			649	3506
1997	4187	616	134			616	4321
1998	4571	585	241			585	4812
1999	4182	430	2083			430	6266
2000	3529	198	2641			198	6170
2001	2724	160	2111	526		686	4835
2002	2081	108	1641	1789		1897	3722
2003	1612	82	1290	1551		1633	2902
2004	1578	68	889	700		768	2466
2005	1172	41	722	396		436	1894
2006	1031	39	950	220		259	1981
2007	1006	12	798	463	973	1448	1804
2008	773	2	527	350	616	968	1300
2009	164		121	230	490	721	286
2010	1067	7	974	445	349	801	2041
2011	809	22	765	225	157	404	1574
2012	770	30	720	17	49	96	1490
2013	409	12	308	9	86	108	716
2014	356	10	146	6	58	73	502
2015	311	8			38	46	311
2016	273	6			26	32	273
Cumulative	38,076	4887	16,975	6912	2634	14,433	55,052
Remaining	1350	35	453	15	208	258	1803
Total	39,426	4923	17,428	6927	2842	14,692	56,854

Note: Oil price = $100/bbl; gas price = $4/Mcf.
Source: BOEM 2013.

The economic limit is assumed to be $750,000 per well and $5 million at the structure level, whichever comes first. Production ceases when net revenue falls below the economic limit. As prices change, wellbore revenue will change, which will impact the time production ceases and the ultimate recovery of each well. At $100/bbl oil and $4/Mcf gas, wells 11500 and 16300 were expected to reach their economic limit in 2014 and the remaining wells in 2016. Remaining reserves for the primary product streams are estimated at 258 Mbbl oil and 1.8 Bcf natural gas.

12.1.8 CGOR and CCGR Trends (Fig. 12.7)

The relative contribution of oil and gas production at each well changes over time with changes in reservoir and operating conditions. For black oil wells, CGOR is often flat, while for retrograde and condensate wells, CGOR often increases with time. For dry gas wells, CCGR is often flat and decreasing with time for wet gas wells. Trends in CGOR and CCGR are used to forecast the secondary product stream for each producing well. Logarithmic transformations are applied to smooth out the cumulative profiles. If the secondary product streams follow historical trends, the secondary product forecast will be

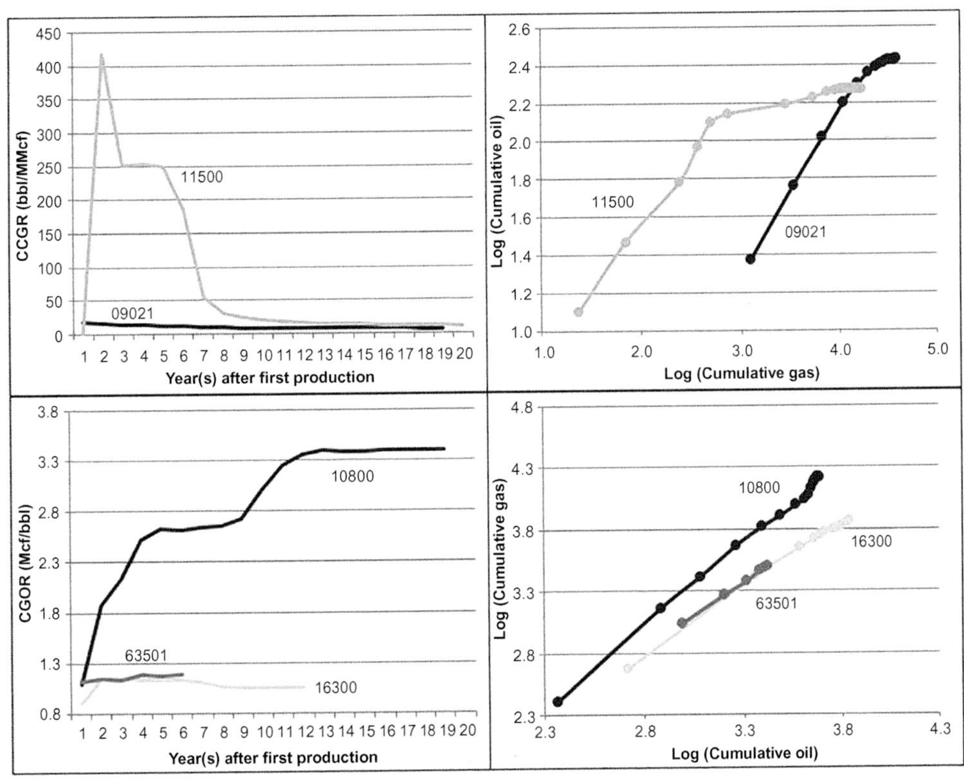

Fig. 12.7 Cumulative gas-oil and condensate-gas ratios for Tick's five producing wells circa 2012. *(Source: BOEM 2013.)*

robust; otherwise, errors will arise in the projections. However, because secondary products are usually significantly smaller in volume and revenue contribution than primary products, the impact of model errors is usually not significant.

12.1.9 Secondary Product Forecast (Table 12.4)

The secondary product stream is forecast for each producing well based on the CGOR and CCGR trend relations and the primary product forecast. The primary product is first forecast using the best-fit decline curve parameters, and then, CGOR or CCGR trends are applied to obtain the secondary product forecast. Remaining reserves for the

Table 12.4 Tick Secondary Product Stream Forecast Circa 2012

Well	09201	10800	11500	16300	63501	Structure	
Secondary	Oil	Gas	Oil	Gas	Gas	Oil	Gas
Unit	Mbbl	MMcf	Mbbl	MMcf	MMcf	Mbbl	MMcf
1992		257					257
1993		1181	12			12	1181
1994	23	1147	17			40	1147
1995	34	2057	31			65	2057
1996	46	1883	33			79	1883
1997	54	1572	32			86	1572
1998	40	1653	14			54	1653
1999	28	1203	17			45	1203
2000	16	801	14			30	801
2001	9	1735	9	476		18	2211
2002	5	1359	4	2194		9	3553
2003	3	793	1	1773		4	2566
2004	3	438		733		3	1171
2005	1	84		466		1	550
2006	1	130		266		1	396
2007	1	53		335	1090	1	1478
2008	1	24	1	192	742	1	958
2009		4		182	546		732
2010	1	12		414	517	1	942
2011		41		242	155	1	438
2012		74		31	75		179
2013	2	42	3	10	103	5	154
2014	1	33	2	6	69	3	108
2015	1	26			46	1	72
2016	1	21			31	1	51
Cumulative	266	16,498	187	7303	3126	453	26,927
Remaining	5	121	5	16	248	10	386
Total	271	16,619	192	7319	3374	463	27,313

Note: Oil price = $100/bbl; gas price = $4/Mcf.
Source: BOEM 2013.

secondary products at Tick and Ladybug wells are estimated to be 10 Mbbl oil and 386 MMcf natural gas, about 10% of primary reserves. Liquids from the secondary streams represent about 3% of the total crude oil (453/14,433) remaining to be produced, while gas from the secondary streams are about half of the expected remaining gas production (27/55).

12.1.10 Structure Production Forecast (Fig. 12.8)

Structure production forecasts are based on the sum of the production forecast of each producing well assuming constant (status quo) conditions at the time of evaluation. When

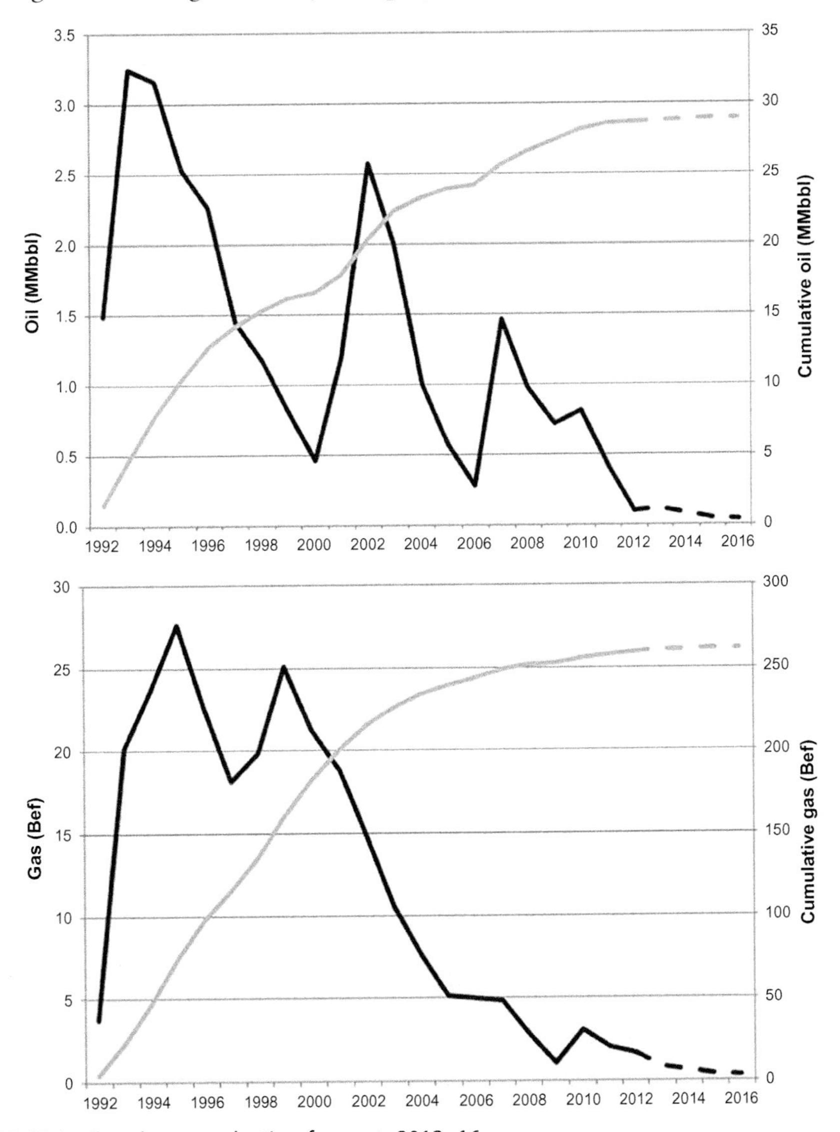

Fig. 12.8 Tick oil and gas production forecast, 2012–16.

production revenue is not adequate to cover direct operating expense, the structure is no longer commercial and will cease operations. Primary oil and gas production dominates the secondary streams (oil, 258 Mbbl vs. 10 Mbbl; gas, 1803 MMcf vs. 386 MMcf), and the total remaining oil and gas production is estimated to be 268 Mbbl oil and 2.2 Bcf gas at $100/bbl oil and $4/Mcf gas prices. Remaining reserves represent a small percentage of the total oil and gas extracted through 2012 (<1% oil and <1% gas) and are a primary indicator that the structure is near its economic limit.

12.1.11 Net Revenue Forecast (Table 12.5)

Future net revenue is determined by multiplying the primary and secondary production forecasts with assumed future oil and gas prices and reducing the sum by the royalty rate.

Table 12.5 Tick Revenue Forecast per Wellbore Circa 2012 (Million $)

Well Unit	09201 Million $	10800 Million $	11500 Million $	16300 Million $	63501 Million $	Total Million $
1992		5.3				5.3
1993		12.3	0.3			12.6
1994	2.8	9.9	0.4			13.1
1995	4.3	15.0	0.8			20.2
1996	10.3	19.6	1.1			31.0
1997	11.7	16.7	1.0			29.4
1998	10.1	11.9	0.7			22.7
1999	10.0	11.0	5.0			26.1
2000	15.4	9.4	11.6			36.4
2001	11.3	11.2	8.8	15.6		46.9
2002	7.1	7.3	5.6	54.1		74.1
2003	9.1	7.0	7.3	58.2		81.6
2004	9.3	5.3	5.2	32.5		52.3
2005	10.3	2.9	6.3	24.6		44.2
2001	7.0	3.4	6.4	15.9		32.7
2007	7.0	1.2	5.6	36.1	78.5	128.5
2008	6.9	0.4	4.7	38.2	70.9	121.2
2009	0.6	0.0	0.5	14.8	32.2	48.1
2010	4.7	0.6	4.3	37.1	29.9	76.7
2011	3.3	2.6	3.1	25.3	17.6	52.0
2012	2.1	3.4	2.0	1.9	5.5	15.0
2013	1.8	1.4	1.6	1.0	9.0	14.8
2014	1.6	1.1	0.7	0.6	6.0	10.0
2015	1.4	0.9			4.0	6.3
2016	1.2	0.7			2.7	4.6
Cumulative	143.7	156.4	80.8	354.3	234.7	969.9
Remaining	5.9	4.0	2.3	1.6	21.8	35.6
Total	149.6	160.4	83.1	355.9	256.5	1005.5

Note: Oil price = $100/bbl; gas price = $4/Mcf.
Source: BOEM 2013.

Wells cease production at their economic limit or when structure revenue falls below $5 million. At $100/bbl oil and $4/Mcf gas, the economic limit is forecast to occur in 2016. Wells 09021, 10800, and 63501 are still expected to be producing in 2016, but because their combined revenue falls below $5 million, the structure is no longer considered commercially viable. Undiscounted future revenue circa 2012 is estimated at $36 million, $23 million of which is expected to arise from the Ladybug wells. Tick operators will receive a processing fee for handling Ladybug production from ATP Oil & Gas.

12.1.12 Economic Limit Year Sensitivity (Table 12.6)

It is useful to evaluate the sensitivity of the economic limit year to changing oil and gas prices since the well inventory and decline trends are fixed per the status quo conditions. Since Tick wells are far along their decline curves and remaining reserves are small, changing oil and gas prices are not expected to have a significant impact on production volumes or the timing (year) of the economic limit. According to the model, at $60/bbl oil and $2/Mcf gas, the economic limit is estimated to occur in 2015, while at $120/bbl oil and $8/Mcf gas, the economic limit is estimated to occur two years later in 2017.

12.1.13 Proved Reserves Sensitivity (Table 12.7)

Reserves are a function of oil and gas prices, the economic limit threshold, capital spending, and geologic conditions. Capital spending and geology are the most important factors but cannot be evaluated vis-à-vis the status quo requirements, so what's remaining is the impact of prices on reserves and their valuation. As commodity prices increase, revenues increase, and economic limits are delayed which lead to additional incremental production beyond the base case assumptions. At $60/bbl and $2/Mcf gas, Tick is estimated to produce 528 Mboe from its 2012 inventory of wells. At $120/bbl oil

Table 12.6 Economic Limit Year Sensitivity Analysis Circa 2012

	Gas Price ($/Mcf)			
Oil Price ($/bbl)	**2**	**4**	**6**	**8**
60	2015	2015	2016	2016
80	2015	2016	2016	2017
100	2016	2016	2017	2017
120	2016	2017	2017	2017

Table 12.7 Proved Reserves Sensitivity Analysis Circa 2012 (Mboe)

	Gas Price ($/Mcf)			
Oil Price ($/bbl)	**2**	**4**	**6**	**8**
60	528	528	630	630
80	546	624	647	708
100	587	633	708	715
120	633	715	715	715

Table 12.8 Reserves Valuation Sensitivity Analysis Circa 2012 (Million $)

	Gas Price ($/Mcf)			
Oil Price ($/bbl)	2	4	6	8
60	7	10	14	18
80	11	15	19	24
100	16	20	25	29
120	21	26	30	34

Note: Assumes 10% discount rate, 16.7% royalty rate, and $15/boe operating cost.

and $8/Mcf gas, reserves increase to 715 Mboe. As oil prices double and gas prices quadruple, reserves increase by about a third.

12.1.14 Reserves Valuation Sensitivity (Table 12.8)

Tick's reserves circa 2012 are estimated to be worth between $7 and $34 million for oil prices between 60 and 120 $/bbl and gas prices between 2 and 8 $/Mcf, assuming a 10% discount rate, 16.67% royalty rate, and $15/boe operating cost. At $2/Mcf gas, reserves triple in value as oil prices double, while for gas at $8/Mcf, the value of reserves doubles as oil prices double.

12.1.15 Postscript Circa 2018

In 2012, Tick had already exhausted most of its reserves and reached the point when it was unlikely to be redeveloped or sold unless Chevron paid another company to account for its decommissioning liability. In 2013, ATP declared bankruptcy, and no bidders were found for Ladybug, which sealed Tick's fate. In 2015, Tick ceased production, and alternative uses for Tick are under review. Deepwater production can be directed across Tick after all its wells are plugged if a suitable development is available and it improves operational flexibility and reliability.

12.2 HORN MOUNTAIN

The Horn Mountain development was evaluated in June 2017 using production data through 2016.

12.2.1 Development (Fig. 12.9)

The Horn Mountain field was discovered in 1999 in Mississippi Canyon 126 in about 5400 ft of water and has been in production since 2002 (Dijkhuizen et al., 2003). Two Middle Miocene reservoirs were developed with eight producing and two water injection dry tree wells. All 10 wells were predrilled using a semisubmersible and batch completed from the spar. In 2015, two subsea wells were tied back to the platform. In 2012, BP sold its interest to Plains Exploration that was subsequently acquired by Freeport McMoRan who later sold their interest to Anadarko in 2016.

Fig. 12.9 Horn Mountain was installed in 2002 in Mississippi Canyon 126 in 5400 ft water depth. *(Source: BP.)*

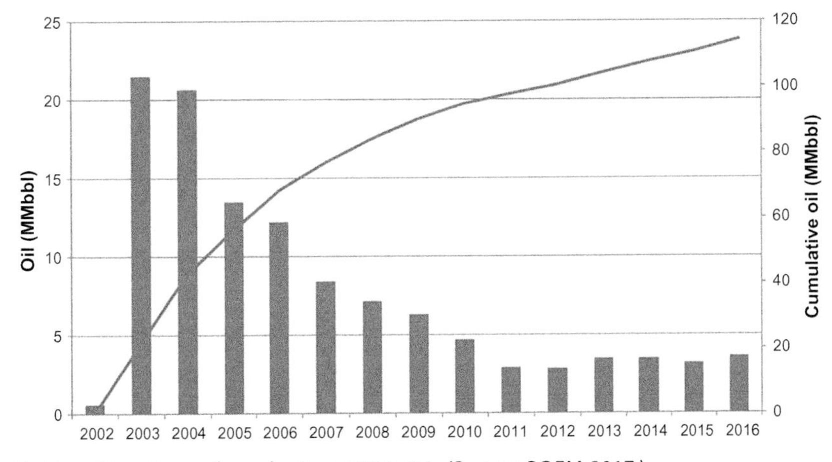

Fig. 12.10 Horn Mountain oil production, 2002–16. *(Source: BOEM 2017.)*

12.2.2 Structure Production (Figs. 12.10 and 12.11)

Horn Mountain production peaked a year after first oil in 2003 at 59 Mbopd oil and 51 MMcfpd natural gas. In 2016, Horn Mountain produced 3.6 MMbbl oil and 5.1 Bcf natural gas from eight producing wells, generating gross revenue of about $159 million

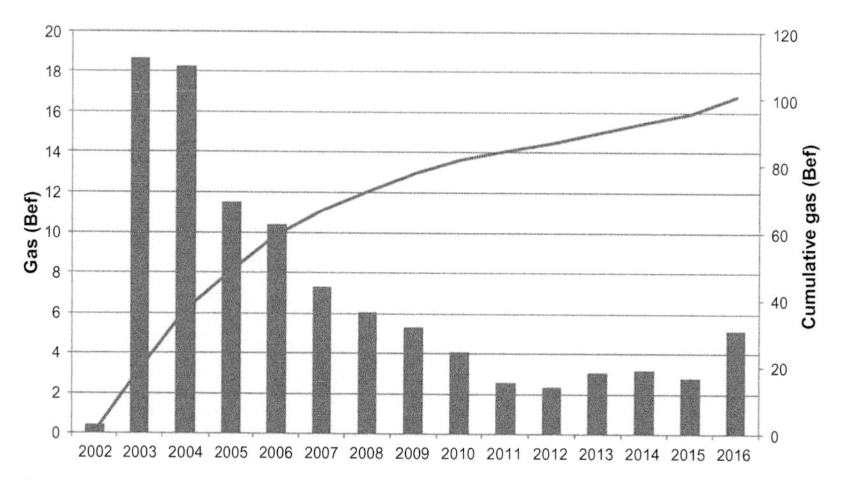

Fig. 12.11 Horn Mountain gas production, 2002–16. *(Source: BOEM 2017.)*

Table 12.9 Horn Mountain Well Inventory Circa 2016

API Number	First Production	Status/Wet	Well Type	Cumulative Production		GOR (Mcf/bbl)
				Oil (MMbbl)	Gas (Bcf)	
608174092700	2003	Producing	Oil	17.6	16.0	0.91
608174093000	2002	Idle	Oil	7.6	6.3	0.83
608174093100		Dry hole				
608174093101	2003	Producing	Oil	8.2	7.9	0.97
608174093200	2003	Producing	Oil	15.1	12.0	0.79
608174093300	2003	Producing	Oil	9.7	11.1	1.14
608174093400	2003	Idle	Oil	8.7	8.1	0.92
608174093500		Dry hole				
608174093600	2002	Producing	Oil	22.6	17.4	0.77
608174093700	2003	Producing	Oil	24.0	19.4	0.81
608174095200		Dry hole				
608174130600	2016	Producing/wet	Oil	0.2	1.1	4.64
608174131100	2016	Producing/wet	Oil	0.6	1.6	2.76
608174131900		Dry hole/wet				

Source: BOEM 2017.

during the year. Cumulative production through 2016 was 114 MMbbl oil and 101 Bcf natural gas valued at $7.3 billion undiscounted.

12.2.3 Well Inventory (Table 12.9, Fig. 12.12)

Eleven dry tree wells and three subsea wells have been drilled at Horn Mountain. In 2016, there were eight oil-producing wells, six dry tree wells, and two wet wells. There have been four dry holes drilled, one of which was sidetracked and is producing, and circa

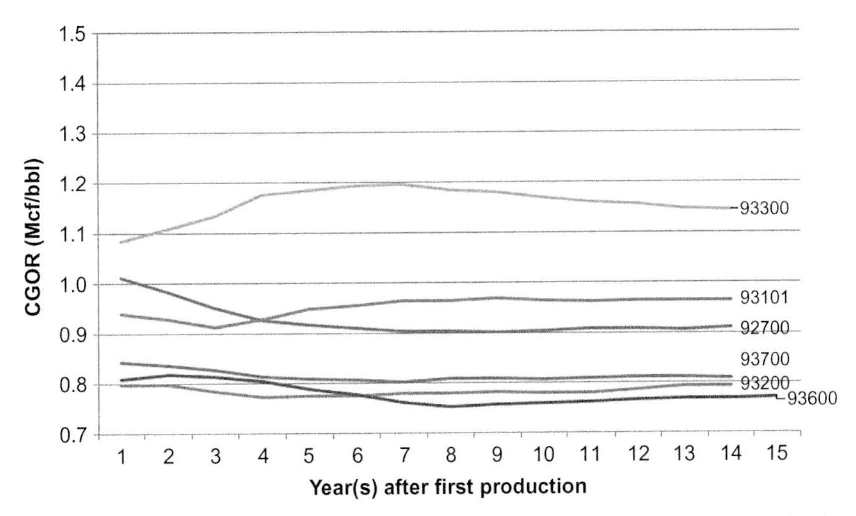

Fig. 12.12 Cumulative gas-oil ratio trends for the producing wells on Horn Mountain circa 2016. *(Source: BOEM 2017.)*

Table 12.10 Summary of Best-Fit Parameters, Remaining and Total Production for Producing Wells Circa 2016

	Parameters			Remaining Production			Total Production		
API Number	b	D	R^2	Oil (MMbbl)	Gas (Bcf)	BOE (MMboe)	Oil (Mbbl)	Gas (MMcf)	BOE (Mboe)
608174092700	0.78	0.40	0.96	0.67	0.56	0.76	18.23	16.56	20.99
608174093101	0.48	0.57	0.98	0.91	0.86	1.05	9.07	8.73	10.52
608174093200	0.00	0.32	0.96	0.18	0.13	0.20	15.26	12.11	17.27
608174093300	1.00	0.66	0.97	2.34	2.79	2.81	12.05	13.89	14.36
608174093600	0.00	0.12	0.86	6.73	4.97	7.56	29.35	22.42	33.09
608174093700	0.24	0.12	0.95	6.70	5.25	7.57	30.70	24.65	34.81
608174130600	0.35	0.47		0.59	2.73	1.04	0.83	3.85	1.47
608174131100	0.35	0.47		1.46	4.02	2.13	2.02	5.58	2.95
Total				19.58	21.31	23.13	117.51	107.79	135.47

Note: Best-fit parameters for subsea wells 608174130600 and 608174131100 apply average b and D parameters from dry tree wells.

2016, two wells are idle. No new wells or sidetracks have been drilled from the spar since 2003. Subsea well production began in 2016.

12.2.4 Decline Curve Specification (Table 12.10)

Each producing well circa 2016 is curve fit based on its primary product stream using a hyperbolic decline curve model. For the two subsea wells that started producing in 2016, the average decline curve parameters from the producing dry tree wells and average cumulative gas-oil ratio trends are used to forecast production.

12.2.5 Primary Production Forecast (Table 12.11)

Economic limits are assumed to be $750,000 per well and $5 million per structure, whichever is reached first. At $60/bbl oil and $3/Mcf gas, well 93200 reaches its economic limit in 2022; well 92700 in 2028; well 30600 in 2029; well 31100 in 2035; well 93101 in 2044; and wells 93300, 93600, and 93700 in 2047. Subsea well forecasts are

Table 12.11 Horn Mountain Primary Product Stream Forecast Circa 2016

Well Primary Unit	92700 Oil Mbbl	93101 Oil Mbbl	93200 Oil Mbbl	93300 Oil Mbbl	93600 Oil Mbbl	93700 Oil Mbbl	30600 Oil Mbbl	31100 Oil Mbbl	Structure Oil Mbbl
2002					67				67
2003	2229	1335	2553	1479	3088	3092			13,777
2004	3286	2068	3522	2071	2499	3046			16,492
2005	2348	1421	2822	984	2109	2486			12,170
2006	1974	812	2662	972	2292	2387			11,100
2007	1313	389	1397	763	2101	2047			8010
2008	1095	368	692	591	2250	1701			6697
2009	1054	353	554	529	1940	1542			5973
2010	771	365	171	418	1305	1449			4479
2011	464	212	260	265	625	942			2768
2012	298	199	178	311	693	1021			2701
2013	717	129	26	363	1056	1137			3427
2014	713	175	109	317	999	1181			3493
2015	698	174	100	363	822	1004			3162
2016	602	156	32	276	777	963	242	566	3615
2017	150	87	57	223	695	700	156	366	2433
2018	118	77	41	187	626	630	107	250	2036
2019	93	68	30	161	564	567	77	179	1739
2020	74	61	22	141	508	510	57	133	1505
2021	58	55	16	126	458	458	43	101	1315
2022	46	50	12	113	412	412	34	79	1158
2023	36	46		103	371	371	27	63	1017
2024	28	42		95	334	333	22	51	906
2025	22	38		88	301	300	18	42	809
2026	18	35		81	271	270	15	35	725
2027	14	32		76	245	242	13	29	651
2028	11	30		71	220	218	11	25	586
2029		28		67	198	196	9	21	520
2030		26		63	179	176		19	463
2031		24		60	161	159		16	420

Continued

Table 12.11 Horn Mountain Primary Product Stream Forecast Circa 2016—cont'd

Well	92700	93101	93200	93300	93600	93700	30600	31100	Structure
Primary	Oil	Oil	Oil	Oil	Oil	Oil	Oil	Oil	Oil
Unit	Mbbl	Mbbl	Mbbl	Mbbl	Mbbl	Mbbl	Mbbl	Mbbl	Mbbl
2032		23		57	145	143		14	382
2033		21		54	131	128		12	347
2034		20		52	118	115		11	316
2035		19		50	106	104		10	288
2036		18		48	96	93			254
2037		17		46	86	84			233
2038		16		44	78	75			213
2039		15		42	70	68			195
2040		14		41	63	61			179
2041		13		40	57	55			165
2042		13		38	51	49			151
2043		12		37	46	44			140
2044		12		36	42	40			129
2045				35	37	36			108
2046				34	34	32			100
2047				33	30	29			92
Cumulative	17,562	8157	15,078	9704	22,622	23,998	242	566	97,930
Remaining	669	911	177	2342	6733	6700	588	1457	19,576
Total	18,231	9068	15,255	12,046	29,355	30,698	830	2023	117,506

Note: Oil price = \$60/bbl; gas price = \$3/Mcf.
Source: BOEM 2017.

highly uncertain at the time of the evaluation because of the lack of production history. Remaining reserves are estimated at 19.6 MMbbl oil representing about 15% of the total oil expected to be produced at the facility.

12.2.6 CGOR Trends (Figs. 12.12 and 12.13)

The relative contribution of oil and gas production changes with changes in the reservoir conditions and drive mechanisms. For black oil wells, CGOR trends are usually flat, which simplifies the secondary stream forecasts. CGOR trends for each well are used to forecast the secondary product stream based on the primary product forecast. Logarithmic transformations were applied, but linear relations are also suitable for these wells.

12.2.7 Secondary Production Forecast (Table 12.12)

The secondary product stream is forecast for each producing well based on individual CGOR trend relations and the primary product forecast. CGOR extrapolated trends are multiplied by the primary product forecast to yield the secondary product forecast. Remaining reserves for all secondary streams are estimated to be 21.3 Bcf natural gas.

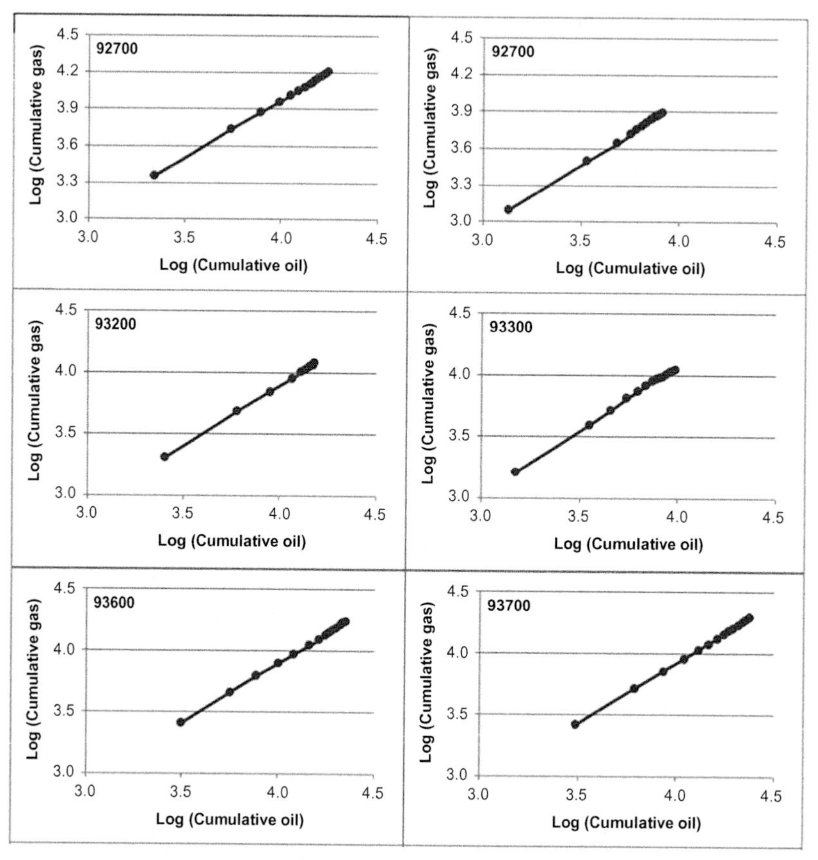

Fig. 12.13 Gas-oil ratio trends for producing wells at Horn Mountain circa 2016. *(Source: BOEM 2017.)*

Table 12.12 Horn Mountain Secondary Product Stream Forecast Circa 2016

Well Secondary Unit	92700 Gas MMcf	93101 Gas MMcf	93200 Gas MMcf	93300 Gas MMcf	93600 Gas MMcf	93700 Gas MMcf	30600 Gas MMcf	31100 Gas MMcf	Structure Gas MMcf
2002					54				54
2003	2253	1254	2037	1605	2525	2607			12,282
2004	3165	1905	2805	2331	2017	2530			14,753
2005	2051	1243	2139	1200	1647	1994			10,273
2006	1640	833	1943	1340	1683	1826			9266
2007	1114	480	1117	948	1517	1622			6798
2008	912	390	555	773	1542	1325			5497
2009	890	398	490	638	1328	1188			4931
2010	676	362	126	428	1067	1251			3909
2011	409	228	246	263	502	774			2422

Continued

Table 12.12 Horn Mountain Secondary Product Stream Forecast Circa 2016—cont'd

Well Secondary Unit	92700 Gas MMcf	93101 Gas MMcf	93200 Gas MMcf	93300 Gas MMcf	93600 Gas MMcf	93700 Gas MMcf	30600 Gas MMcf	31100 Gas MMcf	Structure Gas MMcf
2012	291	170	101	282	550	805			2199
2013	706	93	220	348	895	952			2995
2014	658	194	200	327	820	987			3187
2015	617	172	181	333	663	828			2794
2016	611	149	35	289	634	711	1125	1561	5115
2017	127	82	43	265	515	550	726	1008	3315
2018	100	73	31	222	464	495	497	690	2570
2019	79	65	23	191	417	445	356	494	2068
2020	62	58	17	168	376	400	264	366	1709
2021	49	52	12	149	338	359	201	279	1440
2022	39	47	9	135	305	323	157	218	1232
2023	30	43		123	274	290	125	174	1060
2024	24	39		113	247	261	101	141	926
2025	19	36		104	222	235	83	116	815
2026	15	33		97	200	211	69	96	722
2027	12	31		90	180	190	58	81	642
2028	9	28		85	162	171	50	69	574
2029		26		80	146	153	43	59	508
2030		25		76	132	138		51	421
2031		23		72	119	124		44	382
2032		21		68	107	112		39	347
2033		20		65	96	100		34	316
2034		19		62	87	90		30	288
2035		18		59	78	81		27	263
2036		17		57	70	73			217
2037		16		55	63	66			199
2038		15		52	57	59			183
2039		14		51	51	53			169
2040		13		49	46	48			156
2041		13		47	42	43			144
2042		12		46	38	39			134
2043		11		44	34	35			124
2044		11		43	31	31			115
2045				41	28	28			97
2046				40	25	25			90
2047				39	22	23			84
Cumulative	15,994	7871	11,975	11,104	17,444	19,400	1125	1561	86,474
Remaining	563	861	134	2785	4973	5249	2730	4016	21,313
Total	16,557	8733	12,110	13,889	22,418	24,649	3854	5577	107,787

Note: Oil price = \$60/bbl; gas price = \$3/Mcf.
Source: BOEM 2017.

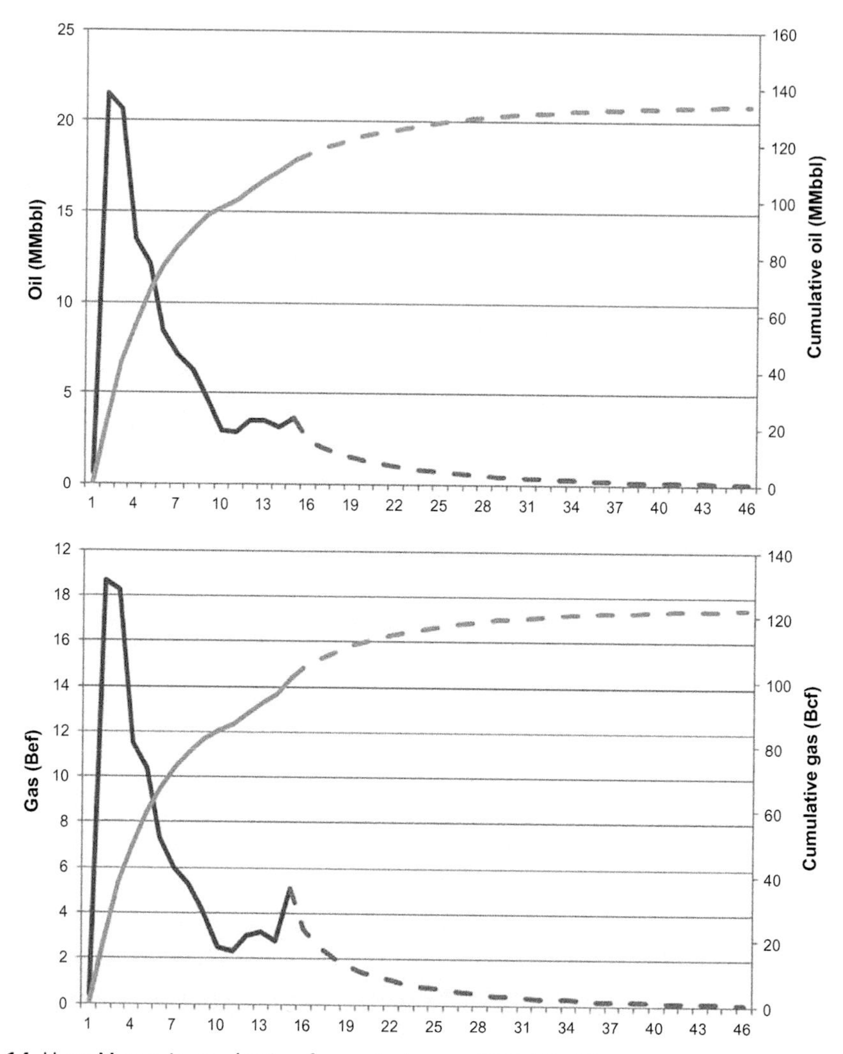

Fig. 12.14 Horn Mountain production forecast. *(Source: BOEM 2017.)*

12.2.8 Structure Production Forecast (Fig. 12.14)

Structure production is forecast based on the sum of all the producing well forecasts assuming status quo conditions at the time of evaluation and $60/bbl oil price and $3/Mcf gas price. Total cumulative production (proved reserves) circa 2016 is estimated to be 117 MMbbl oil and 108 Bcf natural gas.

12.2.9 Revenue Forecast (Table 12.13)

Gross revenue is determined in the customary way using assumed oil and gas prices and production streams. Wells cease production at their economic limit or when the structure

Table 12.13 Horn Mountain Revenue Forecast per Wellbore (Million $)

Well Primary Unit	92700 Oil Mbbl	93101 Oil Mbbl	93200 Oil Mbbl	93300 Oil Mbbl	93600 Oil Mbbl	93700 Oil Mbbl	30600 Oil Mbbl	31100 Oil Mbbl	Structure Oil Mbbl
2002					1.9				1.9
2003	81.9	48.5	90.8	55.0	110.1	110.7			497.1
2004	150.7	94.3	158.2	97.0	112.4	137.4			749.9
2005	140.0	84.8	165.4	61.7	124.0	146.7			722.5
2006	137.5	57.6	183.6	71.3	158.2	165.2			773.5
2007	103.5	31.7	109.6	62.2	163.8	160.6			631.4
2008	122.5	41.9	77.2	68.6	248.8	189.5			748.5
2009	68.1	23.2	35.8	34.9	124.0	99.1			385.1
2010	64.1	30.5	14.1	35.0	108.1	120.4			372.2
2011	52.0	24.0	29.1	29.8	69.8	105.2			310.0
2012	32.8	21.9	19.5	34.3	76.0	112.0			296.4
2013	78.9	14.0	2.8	39.9	115.6	124.4			375.5
2014	70.2	17.3	11.2	31.3	97.9	115.8			343.6
2015	35.1	8.8	5.3	18.3	41.1	50.3			158.8
2016	26.0	6.7	1.4	11.9	33.1	40.8	12.7	26.9	159.5
2017	9.4	5.5	3.5	14.2	43.2	43.7	11.6	25.0	156.0
2018	7.4	4.8	2.6	11.9	38.9	39.3	7.9	17.1	129.9
2019	5.8	4.3	1.9	10.2	35.1	35.3	5.7	12.2	110.5
2020	4.6	3.9	1.4	9.0	31.6	31.8	4.2	9.1	95.4
2021	3.6	3.5	1.0	8.0	28.5	28.6	3.2	6.9	83.2
2022	2.9	3.1	0.7	7.2	25.6	25.7	2.5	5.4	73.2
2023	2.3	2.9		6.6	23.1	23.1	2.0	4.3	64.2
2024	1.8	2.6		6.0	20.8	20.8	1.6	3.5	57.1
2025	1.4	2.4		5.6	18.7	18.7	1.3	2.9	51.0
2026	1.1	2.2		5.2	16.9	16.8	1.1	2.4	45.7
2027	0.9	2.0		4.8	15.2	15.1	0.9	2.0	41.0
2028	0.7	1.9		4.5	13.7	13.6	0.8	1.7	36.9
2029		1.8		4.3	12.3	12.2	0.7	1.5	32.7
2030		1.6		4.0	11.1	11.0		1.3	29.0
2031		1.5		3.8	10.0	9.9		1.1	26.4
2032		1.4		3.6	9.0	8.9		1.0	23.9
2033		1.3		3.5	8.1	8.0		0.9	21.8
2034		1.3		3.3	7.3	7.2		0.8	19.8
2035		1.2		3.2	6.6	6.5		0.7	18.1
2036		1.1		3.0	5.9	5.8			15.9
2037		1.0		2.9	5.4	5.2			14.5
2038		1.0		2.8	4.8	4.7			13.3
2039		0.9		2.7	4.3	4.2			12.2
2040		0.9		2.6	3.9	3.8			11.2
2041		0.8		2.5	3.5	3.4			10.3
2042		0.8		2.4	3.2	3.1			9.5

Continued

Table 12.13 Horn Mountain Revenue Forecast per Wellbore (Million $)—cont'd

Well Primary Unit	92700 Oil Mbbl	93101 Oil Mbbl	93200 Oil Mbbl	93300 Oil Mbbl	93600 Oil Mbbl	93700 Oil Mbbl	30600 Oil Mbbl	31100 Oil Mbbl	Structure Oil Mbbl
2043		0.8		2.4	2.9	2.8			8.7
2044		0.7		2.3	2.6	2.5			8.1
2045				2.2	2.3	2.2			6.8
2046				2.1	2.1	2.0			6.3
2047				2.1	1.9	1.8			5.8
Cumulative	1163	505	904	651	1585	1678	13	27	6526
Remaining	42	57	11	149	419	418	43	99	1239
Total	1205	563	915	800	2004	2096	56	126	7764

Note: Oil price = $60/bbl; gas price = $3/Mcf.

Table 12.14 Economic Limit Year Sensitivity Analysis

	Gas Price ($/Mcf)			
Oil Price ($/bbl)	**2**	**3**	**4**	**5**
40	2042	2042	2043	2043
60	2047	2047	2047	2047
80	2051	2051	2051	2051
100	2056	2056	2056	2056

revenue falls below a specified commercial threshold. In 2048, net revenue falls below $5 million and is no longer economic. At $60/bbl oil and $3/Mcf gas, undiscounted revenue from the future production streams is calculated at $1.2 billion.

12.2.10 Economic Limit Sensitivity (Table 12.14)

At $40/bbl oil and $2/Mcf gas, the economic limit for Horn Mountain occurs in 2042. At $100/bbl oil and $5/Mcf gas, the economic limit is forecast to occur in 2056. Since Horn Mountain is an oil structure, oil price is the primary determinant in decommissioning timing, and changing gas prices are expected to have minor/negligible impact. If high oil prices prompt new drilling and additional reserves are recovered or a tieback opportunity arises, the lifetime of the structure will likely be extended, but such considerations are specifically excluded by the status quo assumptions.

12.2.11 Reserves Sensitivity (Table 12.15)

Reserves are a function of reservoir and well characteristics, oil and gas prices, investment decisions, economic limits, and other factors. As commodity prices increase, economic limits are delayed, which lead to additional incremental production. At $40/bbl and $2/ Mcf gas, Horn Mountain reserves are estimated at 22.2 MMboe; at $100/bbl oil and $5/ Mcf gas, reserves increase about 2 MMboe to 24.0 MMboe. The change in reserves is small (~10%) relative to total production and arises solely from 2016 well inventories.

12.2.12 Reserves Valuation Sensitivity (Table 12.16)

Horn Mountain's remaining reserves circa 2016 are estimated to be worth between $199 and $816 million for oil and gas prices between $40 and $100/bbl and $2 and $5/Mcf, a 10% discount rate, a 16.67% royalty rate, and $15/boe operating cost.

Table 12.15 Proved Reserves Sensitivity Analysis (Mboe)

	Gas Price ($/Mcf)			
Oil Price ($/bbl)	**2**	**3**	**4**	**5**
40	22,205	22,283	22,429	22,466
60	23,098	23,128	23,128	23,142
80	23,609	23,609	23,621	23,621
100	24,038	24,038	24,038	24,046

Table 12.16 Reserves Valuation Sensitivity Analysis (Million $)

	Gas Price ($/Mcf)			
Oil Price ($/bbl)	**2**	**3**	**4**	**5**
40	199	211	222	233
60	393	405	416	427
80	588	599	610	621
100	782	793	805	816

Note: Assumes 10% discount rate, 16.67% royalty rate, and $15/boe operating cost.

BOX 12.1 Eugene Island 330 Field Case Study

The Eugene Island 330 field is the second largest field in the GoM covering seven lease blocks in water depth ranging from 210 to 266 ft. The field was discovered in 1970 by Pennzoil and Shell on blocks EI 330 and 331 using the new technology of "bright spot" identification (Holland et al., 1990). By the end of 1971, two platforms had been set, and during 1972 four more platforms were installed and first production began. Circa 1990, nine drilling and production platforms and four satellite platforms were installed, with 192 development wells and 26 exploratory wells drilled across the field (Fig. 12.15).

Continued

BOX 12.1 Eugene Island 330 Field Case Study—cont'd

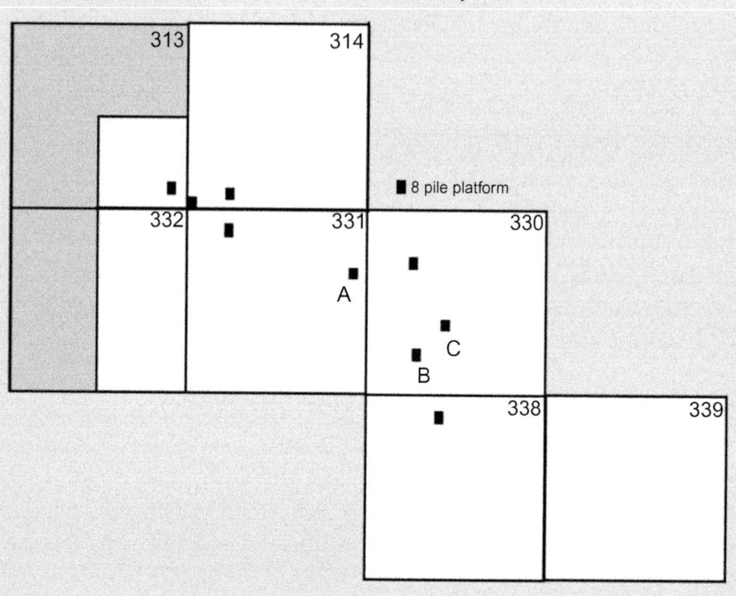

Fig. 12.15 Eugene Island 330 field lease position circa 1984.

Hydrocarbon accumulations occur in more than 25 Pliocene-Pleistocene sandstones, and trapping mechanisms are combinations of structural and stratigraphic varieties, including four-way dip closure, fault closure, and facies change (Holland et al., 1990). The field is a rollover anticline formed on the downthrown side of a large, northwest-trending, salt-diaper-related growth fault. The reservoir energy results from a combination of water-drive and gas-expansion system. There are 10 major reservoir sand series that range from 55 to 400 ft in gross thickness with net pays from 27 to 49 ft. Oil gravity ranges from 23 to 36° API. Eight of 10 reservoir sandstone units are predominately oil producing, with productive areas ranging from 6500 acres (2633 ha) to less than 200 acres (81 ha). Most of the sandstones are cut by faults that break them into several distinct reservoirs and require separate wells.

Eighteen- and 24-slot eight-pile platforms were selected due to the large number of directional wells required. On block 330, three drilling platforms and a central four-pile platform were installed for production equipment. On blocks 331 and 314, two drilling platforms were connected to production platforms, and on blocks 313 and 338, one drilling platform was set on each block.

In December 1978, average daily production peaked at 61,000 bbl of oil, 9000 bbl of condensate, and 400 MMcf of gas (Lewis and Dupur, 1983), and by 1983, cumulative production reached 224 MMbbl oil and 1.03 Tcf gas, and water injection began on blocks 331 and 314 to slow decline rates. Through September 2017, cumulative production was 455 MMbbl of liquid

Continued

BOX 12.1 Eugene Island 330 Field Case Study—cont'd

hydrocarbons and 1.9 Tcf of gas. In 2015, field production was 5 MMbbl of oil and 9 Bcf of gas with declines in 2006 and 2009 due to hurricane impacts (Fig. 12.16). Remaining reserves as of December 2015 were estimated to be 11.7 MMbbl oil and 22.6 Bcf gas.

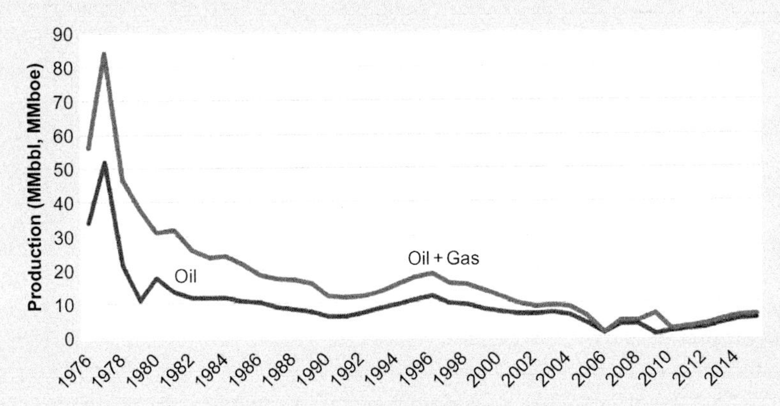

Fig. 12.16 Eugene Island 330 field oil and gas production plot, 1976–2015. *(Source: BOEM 2017.)*

The pressure histories of three reservoir sands GA, LF, and OI provide insight into some of operational issues that arose early in development (Fig. 12.17):

- The GA sandstone is normally pressured with water drive. The reservoir pressure decreased less than 8% after the production of 71% of the recoverable reserves. Ultimate recovery was expected to exceed 37% of the original oil-in-place.

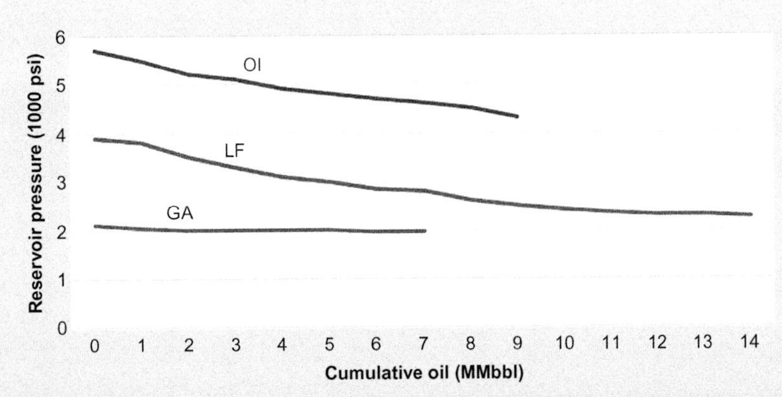

Fig. 12.17 Pressure history of three reservoirs in the Eugene Island 330 field. *(Source: Modified from Holland, D.S., Leedy, J.B., Lammlein, D.R., 1990. Eugene Island block 330 field, offshore Louisiana. In: Beaumont, E.A., Foster, N.H., (Eds.), Structural Traps III, Tectonic Fold and Fault Traps: AAPG Treatise of Petroleum Geology, Atlas of Oil and Gas Fields. pp. 103–143.)*

Continued

BOX 12.1 Eugene Island 330 Field Case Study—cont'd

- The LF sandstone reservoirs have weak water-drive systems, and during early production, the pressure dropped 34% after only 8 MMbbl of oil had been produced. The operator revised its production plan to allow water influx to approximately match production withdrawal, which resulted in the production of the next 7 MMbbl with only an additional 7% pressure drop.
- The OI sandstone reservoir in a fault block exhibited a classic gravity-segregation drive mechanism. A gas cap developed after a 16% decrease in pressure occurred prompting an immediate response for a gas injection program that began in 1979.

In 2005, Hurricane Rita passed through the area causing significant damage to EI 330B and EI 330C platforms and claimed connecting platform EI 330S. In order to prevent damage from future storms, Devon and partners in May 2006 sanctioned raising the decks on EI 330B and EI 330C platforms 14 ft. Operations were successfully completed using the Versabar (Van Kirk and Day, 2007).

In 2008, Hurricane Ike passed through the field and caused significant damage to Shell's EI 331A platform that was towed to an existing Special Artificial Reef Site in EI 313 and reefed (see Section 16.1.5). Eleven pipelines crossing the EI 331A structure had to be rerouted or abandoned.

REFERENCES

Curtis, W.B., Gilmore, L.A., 1992. Garden Banks 189 "A" deepwater jacket project. In: OTC 6952. Offshore Technology Conference. Houston, TX, 4–7 May.

Dijkhuizen, C., Coppens, T., van der Graaf, P., 2003. Installation of the Horn Mountain spar using the enhanced DCV Balder. In: OTC 15367. Offshore Technology Conference. Houston, TX, May 5–9.

Holland, D.S., Leedy, J.B., Lammlein, D.R., 1990. Eugene Island block 330 field, offshore Louisiana. In: Beaumont, E.A., Foster, N.H. (Eds.), Structural Traps III, Tectonic Fold and Fault Traps: AAPG Treatise of Petroleum Geology, Atlas of Oil and Gas Fields. pp. 103–143.

Lewis, R.L., Dupur, H.J., 1983. Eugene Island block 330 field – development and production history. J. Petrol. Technol. 35, November.

Van Kirk, P., Day, M., 2007. Existing platforms raised to increase storm clearance. Offshore Mag. 67(2).

CHAPTER THIRTEEN

Shallow Water Decommissioning Forecast

Contents

Abstract

At the end of 2016, there were 2009 structures in water depth less than 400 ft in the Gulf of Mexico—960 producing structures, 675 idle structures, and 374 auxiliary structures. Shallow-water decommissioning activity is expected to continue at a relatively high rate in the short term with average annual activity between 75 and 150 decommissioned structures per year. Decommissioning forecasts require different models to describe the unique characteristics of producing, idle, and auxiliary structures. The future cash flows of producing structures are estimated using decline curves to predict expected decommissioning timing. Schedule approaches are adopted for idle and auxiliary structures. Hybrid scenarios combine the relevant features of the three classes, and sensitivity analysis is performed. Transition probabilities are used to model structure transitions between states in a quasi-equilibrium manner. From 2017 to 2022, between 474 and 828 structures are expected to be decommissioned in the shallow-water Gulf of Mexico, and by 2027, between 704 and 1199 structures are expected to be decommissioned. A discussion of the limitations of the analysis concludes the chapter.

Decommissioning Forecasting and Operating Cost Estimation
https://doi.org/10.1016/B978-0-12-818113-3.00013-8

13.1 MODEL RECAP

The approach taken is to apply standard economic models where applicable and a scheduled approach where economic models cannot be applied (Kaiser and Narra, 2018). Structures are classified into one of three categories—producing, idle, and auxiliary—and are decommissioned according to models specific to each structure class. Circa 2017, there were 960 producing structures, 675 idle structures, and 374 auxiliary structures in the shallow-water GoM. Producing structures are the largest class, and standard industry models can be applied in decommissioning forecasting. Idle and auxiliary structures together comprise more than half the inventory and require new approaches.

For producing structures, exponential and hyperbolic decline curve models and assumed economic limits and commodity prices determine the time of abandonment under the assumption of no capital spending outside of normal maintenance activities. Two decline models are used to bound activity levels, and decommissioning is assumed to occur one year after the economic limit is reached. Economic limits, oil and gas prices, and royalty rate are model parameters (Table 13.1).

Structures without production are scheduled for decommissioning using user-defined assumptions and historical data on decommissioning activity. For idle structures, a phase-space ensemble average constructs the average decom schedules for scenarios based on decomposing the idle inventory into two idle age categories. The model requires the user to specify upper and lower bounds on model parameters and recognizes the uncertainty inherent in the parameter selection. A simple empirical model based on historical decommissioning activity is used to schedule auxiliary structure decommissioning.

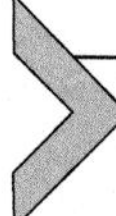

13.2 PRODUCING STRUCTURE DECOMMISSIONING FORECAST

13.2.1 Reference Case

Using hyperbolic decline models and assuming a crude and condensate price of $60/bbl, casinghead and gas well gas of $3/Mcf, an economic limit of $300,000 per structure for all structures, and a one-year delay in which decommissioning is performed after the economic limit is reached, the forecast results are shown in Fig. 13.1.

In 2017, 46 producing structures are expected to fall below their economic limit and be decommissioned, and in subsequent years, activity levels decline to 10–20 structures per year,

Table 13.1 Shallow-Water Decommissioning Model Parameter Summary

Class	Model Parameters	2016 Inventory
Producing	$q(b, D)$, P^o, P^g, EL	960
Idle	(p, T_1, T_2)	675
Auxiliary	AVG, k	374

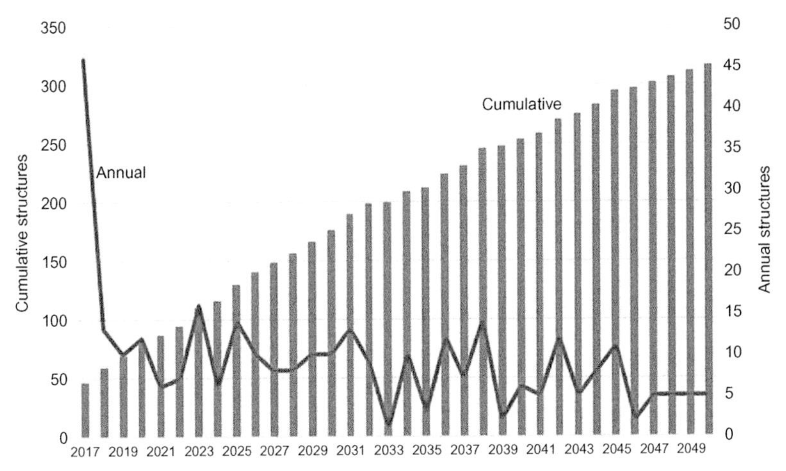

Fig. 13.1 Reference case, hyperbolic decline model decommissioning forecast. *(Source: Kaiser, M.J., Narra, S., 2018. A hybrid scenario-based decommissioning forecast for the shallow water U.S. Gulf of Mexico, 2018-2038. Energy 163, 1150–1177.)*

achieving 148 cumulative decommissioned structures in 2027, 231 cumulative decommissioned structures in 2037, and 301 cumulative decommissioned structures in 2047.

Activity per decade is computed by subtraction, so that between 2027 and 2036, there are 83 (= 231 − 148) decommissioned structures and 70 (= 301 − 231) decommissioned structures from 2037 to 2046. Annual average activity is computed by dividing cumulative totals by the time period, so that ~15 (= 148/10) producing structures on average are decommissioned per year through 2027.

13.2.2 Sensitivity Analysis

Sensitivity analysis is performed to understand how the model results vary with changes in the input parameters. The impact of $40, $60, and $80/bbl oil prices; gas prices of $2, $3, and $4/Mcf; and economic limits of $200, $300, $400, $500, and $1000 thousand dollars will lead to $3 \times 3 \times 5 = 45$ combinations for each decline model assuming $P^{oil} = P^{cond}$ and $P^{gas} = P^{ass}$. Relaxing the commodity price equivalence and because there are two decline models (exponential and hyperbolic) increase the number of combinations another eightfold ($2 \times 2 \times 2$) with the total number of scenarios falling into the hundreds. Obviously, some judicious selections are needed to manage all these combinations.

13.2.3 Hyperbolic vs. Exponential Decline Curve

Intuitively, hyperbolic decline models will predict low/slow levels of decommissioning activity compared with exponential models because of the nature of the decline curve employed. Assuming $3/Mcf gas price and a $300,000 economic limit for all structures, the cumulative number of structures decommissioned for hyperbolic and exponential models is compared in Fig. 13.2 for $60/bbl oil and is depicted at 10-year intervals

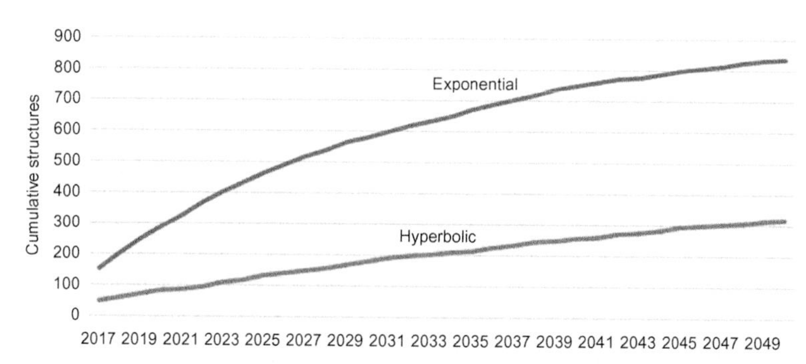

Fig. 13.2 Hyperbolic versus exponential decline cumulative structure decommissioning forecast. *(Source: Kaiser, M.J., Narra, S., 2018. A hybrid scenario-based decommissioning forecast for the shallow water U.S. Gulf of Mexico, 2018-2038. Energy 163, 1150–1177.)*

Table 13.2 Hyperbolic vs. Exponential Cumulative Decommissioned Structures—Oil Price Variation Circa 2016

Year	\$40/bbl	\$60/bbl	\$80/bbl
2027	177, 555	148, 517	121, 478
2037	282, 738	231, 700	202, 670
2047	360, 843	301, 812	263, 789

Note: EL = \$300,000, \$3/Mcf gas.
In table entries, the first element of the pair is the hyperbolic model results followed by the exponential model results.

for three oil prices in Table 13.2. The choice of decline model has a significant impact on the forecast and provides the greatest range of variation among the model variables.

The first entry in the pair of numbers in Table 13.2 is the cumulative number of structures decommissioned according to the hyperbolic model, and the second entry is the cumulative count for the exponential model. For example, using exponential decline curves at \$60/bbl oil, the model predicts that there will be 517 structures decommissioned from 2017 to 2027, 700 by 2037, and 812 by 2047.

Since there are 960 producing structures in inventory circa 2016, the exponential model predicts clearance of most of the inventory by 2047 under the three price assumptions, from 82% (789/960) at high oil prices to about 88% (843/960) at low oil prices. By comparison, the hyperbolic model indicates a much more leisurely removal rate, from 27% (263/960) at high oil prices to 38% (360/960) at low oil prices. The decline curve models bound the decommissioning forecasts for producing structures holding all other factors constant.

13.2.4 Price Variation

Oil and gas prices enter gross revenue linearly, but because structures aggregate production from multiple wells, there will be differences in the relative contribution of oil revenue and gas revenue. Doubling one price while holding the other price fixed will not

Table 13.3 Hyperbolic vs. Exponential Cumulative Decommissioned Structures—Gas Price Variation Circa 2016

Year	\$2/Mcf	\$3/Mcf	\$4/Mcf
2027	172, 533	148, 517	136, 501
2037	257, 719	231, 700	212, 685
2047	327, 823	301, 812	284, 807

Note: $EL = \$300,000$, \$60/bbl oil.
In table entries, the first element of the pair is the hyperbolic model results followed by the exponential model results.

exhibit proportionate behavior in decommissioning counts. Oil price changes will impact oil structures more than gas price changes because the primary product stream (usually but not always) plays the dominant role in structure revenue. A similar expectation holds for gas structures and gas prices.

When comparing the model results for two different oil prices holding all other factors fixed, the number of decommissioned structures at lower oil price should be greater than at higher prices. In 2027, for example, 29 ($= 177 - 148$) additional structures are decommissioned with the hyperbolic model at \$40/bbl oil compared with \$60/bbl, while 38 ($= 555 - 517$) additional structures are decommissioned using exponential decline (Table 13.2). Conversely, at higher oil prices, again holding all other model parameters constant and for all else equal, greater revenue is generated that will delay the economic limit and reduce decommissioning activity compared with lower prices. At \$80/bbl crude, the hyperbolic model predicts 121 structures decommissioned in 2027 compared with 148 structures at \$60/bbl (Table 13.2).

Since about two-thirds of shallow-water structures are oil producers, oil production and oil prices are expected to have the greatest impact on total decommissioning count and is the primary variable of interest. Fixing oil price and varying gas prices will have a smaller impact because there are fewer gas structures and gas contributes less to total revenue (Table 13.3). The variation of oil and gas price on cumulative structure counts is depicted in Figs. 13.3 and 13.4. Doubling oil prices leads to a larger variation in the cumulative count and a wider gap relative to a doubling in gas price because of the greater number of oil structures in inventory.

13.2.5 Economic Limit Variation

In Fig. 13.5, the impact of changing economic limits is shown for hyperbolic decline models and fixed \$60/bbl oil and \$3/Mcf gas prices. As economic limits increase, cumulative counts increase at every time period because a greater number of structures reach their commercial thresholds sooner. The spread increases over time as structure counts accumulate. Similar behavior holds for exponential models and other price combinations.

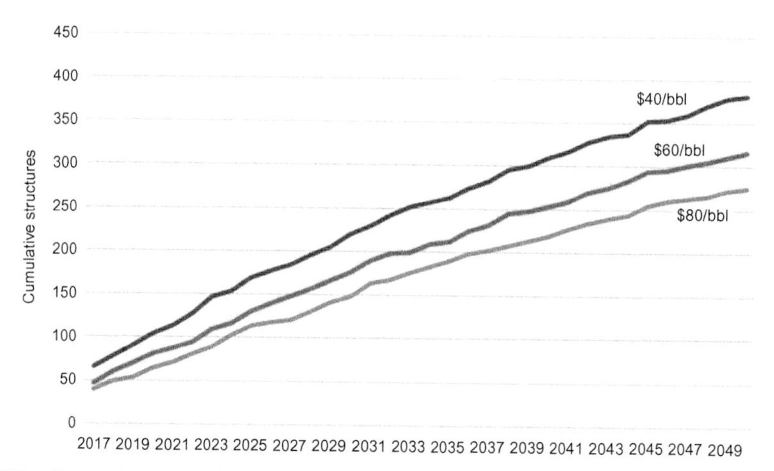

Fig. 13.3 Oil price variation on hyperbolic decline decommissioning forecast. *(Source: Kaiser, M.J., Narra, S., 2018. A hybrid scenario-based decommissioning forecast for the shallow water U.S. Gulf of Mexico, 2018-2038. Energy 163, 1150–1177.)*

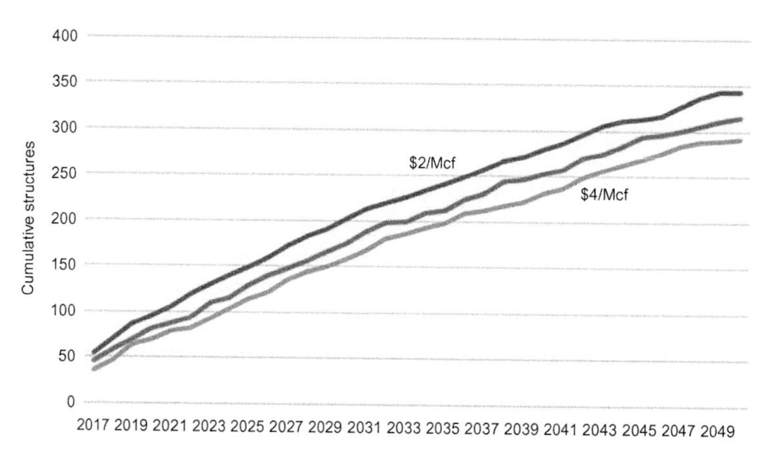

Fig. 13.4 Gas price variation on hyperbolic decline decommissioning forecast. *(Source: Kaiser, M.J., Narra, S., 2018. A hybrid scenario-based decommissioning forecast for the shallow water U.S. Gulf of Mexico, 2018-2038. Energy 163, 1150–1177.)*

13.2.6 Oil vs. Gas Structures

Using the exponential model, the impact of oil and gas prices on the number of decommissioned oil structures and the number of decommissioned gas structures is examined. Oil structures should exhibit a greater sensitivity to oil price than gas prices. The results are depicted in Tables 13.4 and 13.5 and confirm that the model construction behaves according to expectation at a granular level.

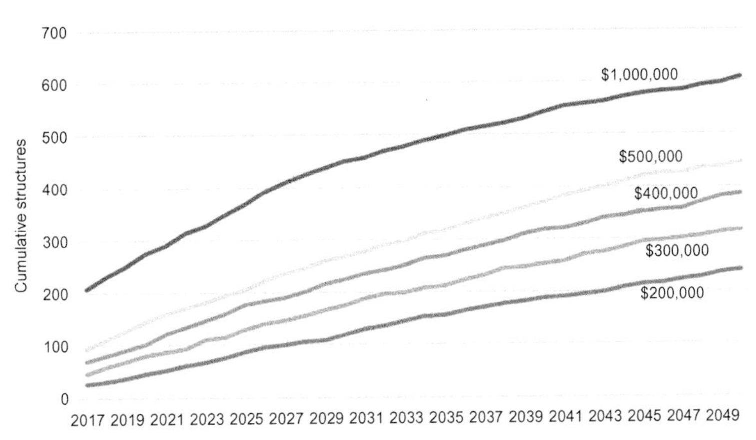

Fig. 13.5 Economic limit variation on hyperbolic decline decommissioning forecast. *(Source: Kaiser, M. J., Narra, S., 2018. A hybrid scenario-based decommissioning forecast for the shallow water U.S. Gulf of Mexico, 2018-2038. Energy 163, 1150–1177.)*

Table 13.4 Oil and Gas Structure Cumulative Decommissioning Count—Oil Price Variation Circa 2016

Year	$40/bbl, $3/Mcf	$60/bbl, $3/Mcf	$80/bbl, $3/Mcf
2027	228, 327	227, 290	219, 259
2037	288, 450	282, 418	277, 393
2047	318, 525	314, 498	312, 477

Note: Exponential decline, $3/Mcf gas, and $EL=\$300,000$.
In table entries, the first element of the pair is the gas structure count followed by the oil structure count.

Table 13.5 Oil and Gas Structure Cumulative Decommissioning Count—Exponential Decline, Gas Price Variation Circa 2016

Year	$60/bbl, $2/Mcf	$60/bbl, $3/Mcf	$60/bbl, $4/Mcf
2027	238, 295	227, 290	217, 284
2037	291, 428	282, 418	272, 413
2047	323, 500	314, 498	310, 497

Note: Exponential decline, $60/bbl oil, and $EL=\$300,000$.
In table entries, the first element of the pair is the gas structure count followed by the oil structure count.

In Table 13.4, doubling oil price from $40 to $80/bbl at $3/Mcf gas has a negligible impact on gas structure counts (first element in each pair entry, read across the row)—about 10 structures at any point in time—and a more significant impact on oil structure counts (the second term of the pair) that vary by 68 in year 2027, 57 at year 2037, and 48 at year 2047.

In Table 13.5, doubling gas price from $2 to $4/Mcf at $60/bbl oil has a small input on oil structure counts and a modest impact on gas structure counts (again, read across the row), ranging from 13 to 21 structures for gas structures, 21 at year 2027, 19 at year 2037, and 13 at year 2047.

13.2.7 Commodity Price Adjustment

Previously, oil prices were assumed equal to condensate prices, and associated and non-associated natural gas prices were considered equivalent in well revenue. Using a more realistic condensate price assumption, gas structures will receive less revenue for their liquid hydrocarbon production that will subsequently accelerate the time of their economic limit, while for oil structures, there will be no impact from the model parameter change.

Condensate prices are assumed equal to 60% of the crude price input. After 10 years, the cumulative difference for $60/bbl crude for oil structures and $36/bbl condensate for gas structures, leads to an average difference of 31 decommissioned structures relative to the base case (Table 13.6). After 20 years, the average difference declines to 19 structures, and at 30 years, the average difference is less than a handful. Over long periods of time, the significance of the modified pricing appears negligible. There is not a uniform increase in the number of decommissioned structures as economic limits increase because of the nonlinear impact of the commodity price change.

13.2.8 Royalty Relief

BOEM offers end-of-life royalty relief for properties that demonstrate economic hardship due to the royalty payment (US DOI, 2010). Removing or reducing the royalty rates will reduce the number of structures that reach their economic limit since net revenues will increase incrementally. Overall impacts are expected to be small, however, because of the relatively small role royalty rates play in cash flows at the end of life of production. Model results support this intuition.

Under the $60/bbl oil and $3/Mcf gas price scenario, the reduction in the structure counts with royalty relief averaged 25 structures over 10 years or 2.5 structures per year. At 20 years, the cumulative change declines to 17 structures and, at 30 years, 11 structures. The conclusion is that removing royalty rate on end-of-life producers represents

Table 13.6 Increase in Decommissioning Count for a Reduced Condensate Price

Economic Limit ($)	2027	2037	2047
200,000	36	15	4
300,000	28	22	5
400,000	24	19	4
500,000	26	23	2
1,000,000	42	16	0
Average	31	19	3

Note: Assumes crude price of $60/bbl for oil structures and a condensate price of $36/bbl for gas structures Gas price is $3/Mcf for both oil and gas structures.

secondary or smaller impacts relative to the impact of other model parameters. End-of-life royal relief may be important for some properties and operators, but the program is not popular, and the model does not capture operator-specific circumstances, so conclusions are limited.

13.3 HYBRID MODEL SCENARIOS

13.3.1 Notation

Since three models are adopted in the decommissioning forecast and each model is described by an independent set of parameters, it is useful to define a shorthand notation to designate the model parameterization. For the producing (*P*), idle (*I*), and auxiliary (*A*) structure subclasses, three vector triplets describe model scenarios:

$$\{(P, I, A)\} = \{(Decline, EL, P^o, P^g), (p, T_1, T_2), (AVG, k)\}$$

The notation helps to organize the parameter space and specifies the variables that define the model inputs.

For producing structures, *Decline* signifies either hyperbolic or exponential model forms, the economic limit EL is defined in \$1000, and the oil and gas prices are specified in standard volumetric units. These four parameters uniquely determine the abandonment time for a producing structure.

For idle structures, the model parameter p subdivides the idle inventory into two classes by idle age, and for each subgroup, user-defined parameters T_1 and T_2 are specified that allocate the inventory classes uniformly over the two periods. All the parameters are specified in years. Scheduling is performed across class members and not on an individual basis.

For auxiliary structures, AVG denotes the average level of decommissioning activity performed over the (historic) time period k in years and is used to specify future decommissioning rates. Similar to the idle class, auxiliary structures are not distinguished individually in forecasting but are considered on a class basis.

13.3.2 Scenario Parameterization

The following scenario parameterization is applied in the model construction:

$$(P, I, A) = \{(Hyp, 300, 60, 3), (7, 20, 10), (20, 20)\}$$

Hyperbolic decline curves, an economic limit of \$300,000, an oil price of \$60/bbl, and a gas price of \$3/Mcf are the model inputs for producing structures. The idle structure decommissioning schedule is based on the model parameters (7, 20, 10) and extended to a 20-year horizon. The auxiliary structure submodule assumes 20 structures per year are removed over the forecast period based on a 20-year average rate.

13.3.3 Decommissioning Scenarios

The results of combining the three submodules for producing, idle, and auxiliary structure decommissioning are shown in Figs. 13.6–13.9. In Fig. 13.6, the results of the hyperbolic decline model are depicted for producing structures under reference case conditions and schedules for idle and auxiliary structures. In Fig. 13.7, the decommissioning model results are broken out by structure class. The constant 20 auxiliary structure removals per year are clearly visible. In Fig. 13.8, the impact of using an exponential model instead of hyperbolic decline curves yields an upper bound on activity levels, which is depicted on a cumulative basis in Fig. 13.9.

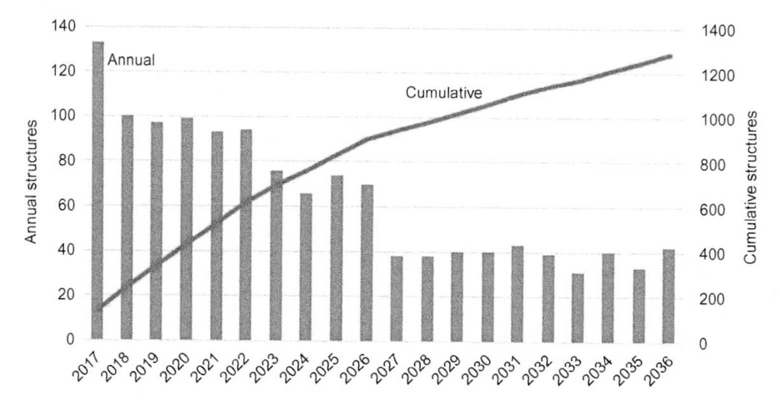

Fig. 13.6 Shallow-water decommissioning forecast, 2017–36. *(Source: Kaiser, M.J., Narra, S., 2018. A hybrid scenario-based decommissioning forecast for the shallow water U.S. Gulf of Mexico, 2018-2038. Energy 163, 1150–1177.)*

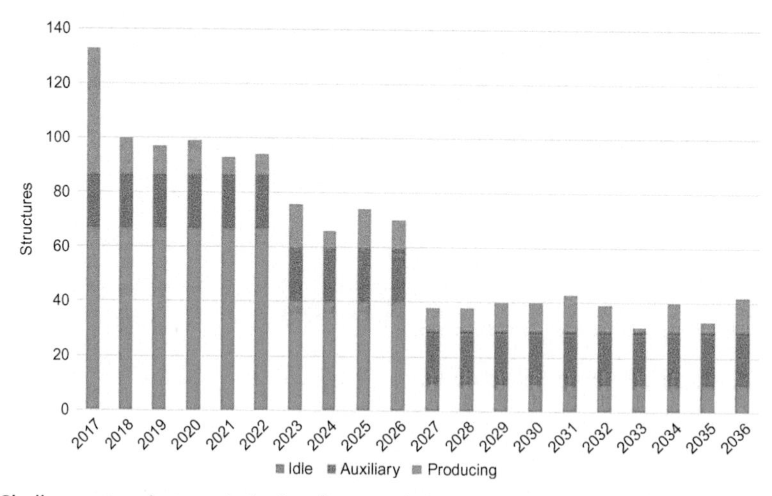

Fig. 13.7 Shallow-water decommissioning forecast by structure class, 2017–36. *(Source: Kaiser, M.J., Narra, S., 2018. A hybrid scenario-based decommissioning forecast for the shallow water U.S. Gulf of Mexico, 2018-2038. Energy 163, 1150–1177.)*

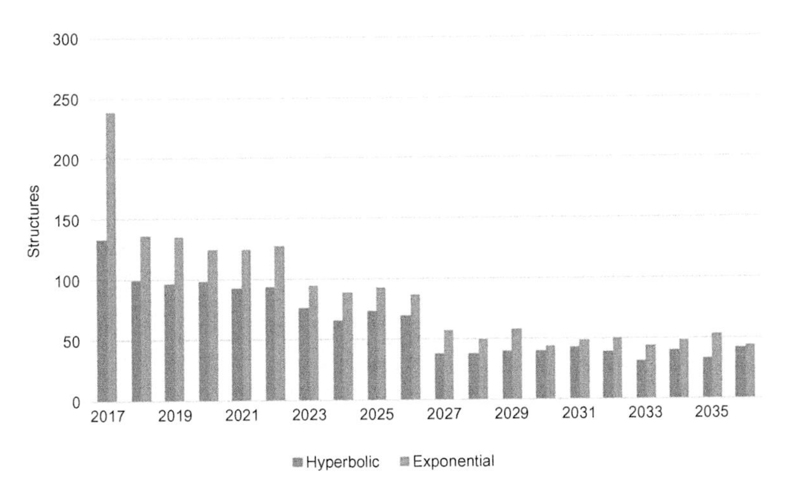

Fig. 13.8 Hyperbolic versus exponential decline decommissioning forecast, 2017–36. *(Source: Kaiser, M.J., Narra, S., 2018. A hybrid scenario-based decommissioning forecast for the shallow water U.S. Gulf of Mexico, 2018-2038. Energy 163, 1150–1177.)*

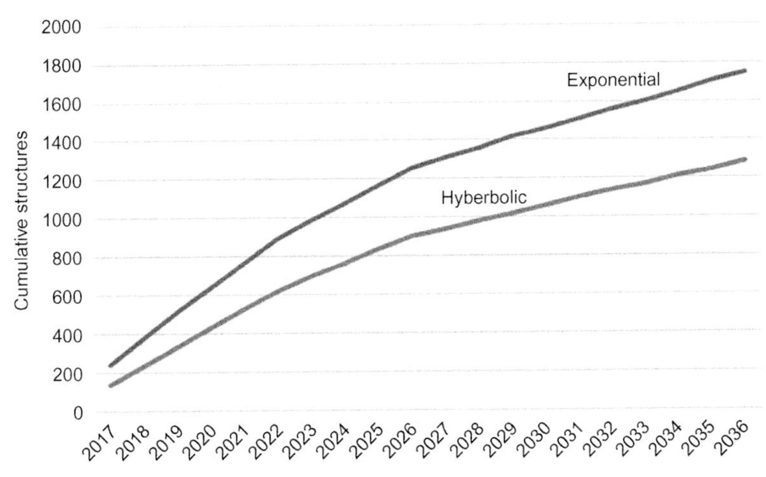

Fig. 13.9 Hyperbolic versus exponential decline cumulative decommissioning forecast, 2017–36. *(Source: Kaiser, M.J., Narra, S., 2018. A hybrid scenario-based decommissioning forecast for the shallow water U.S. Gulf of Mexico, 2018-2038. Energy 163, 1150–1177.)*

13.3.4 Class Transitions

Structures that are allowed to transition between classes prior to decommissioning will yield a more accurate model reflective of operational characteristics. Transition probabilities are used in a quasi-equilibrium manner to capture this behavior. Producing platforms will become idle when their reservoirs are depleted or transition directly into decommissioned status. Both transitions are believed to be equally likely and more

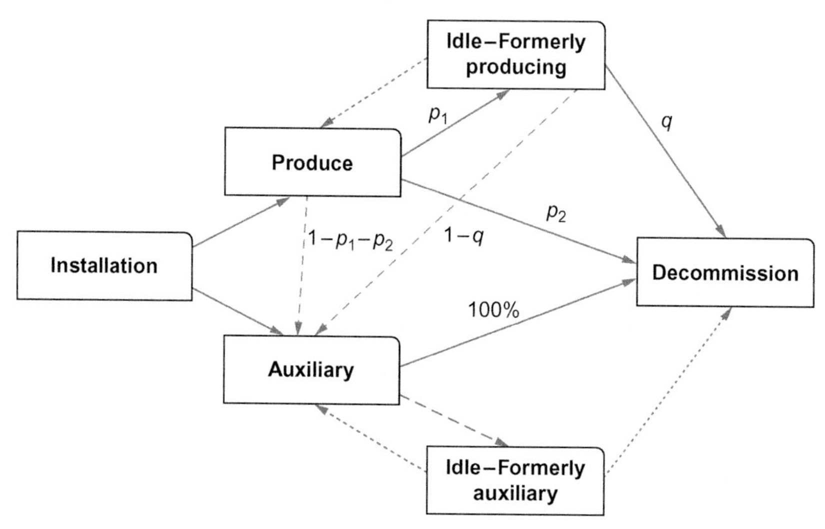

Fig. 13.10 Transition probabilities defined for the structure classes.

common than a transition to auxiliary status. Idle structures may be repurposed to serve an auxiliary role if they are well positioned and gain regulatory approval, but the number of these transitions is relatively few, perhaps between 5 and 15% of structures entering the class. The exact values of the transition probabilities are not known.

Producing structures are assumed to transition to the idle, decommissioned, and auxiliary classes with probability p_1, p_2, and $1-p_1-p_2$, respectively, assumed constant over time (Fig. 13.10). Idle structures are assumed to transition to the decommissioned and auxiliary class with probability q and $1-q$, respectively. Auxiliary structures are assumed to transition direct to decommissioning. Idle structures are not allowed to return to production status because the only way an idle structure could return to production is via capital investment in wells, and this particular type of spending to add/increase production is not permitted in the model framework. Capital investment to change a structure's function is acceptable.[1]

Using the transition probabilities $(p_1, p_2, 1-p_1-p_2)=(0.4, 0.4, 0.2)$ for structures exiting the producing class and $(q, 1-q)=(0.9, 0.1)$ for structures exiting the idle class, the number of structures entering decommissioning is reduced in a probabilistic quasi-equilibrium fashion as shown in Fig. 13.11 for the exponential and hyperbolic decline model results.

[1] This may seem like a quirk of the model, but it is a central assumption in all reserves estimates and production forecasts (see Section 11.2.5).

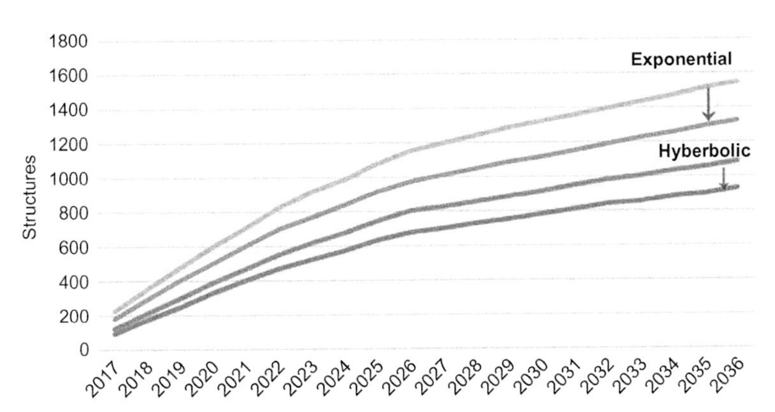

Fig. 13.11 Cumulative decommissioning forecast with and without transition probabilities, 2017–36. *(Source: Kaiser, M.J., Narra, S., 2018. A hybrid scenario-based decommissioning forecast for the shallow water U.S. Gulf of Mexico, 2018-2038. Energy 163, 1150–1177.)*

According to the hybrid model bounds, from 2017 to 2022, between 556 and 828 structures will be decommissioned, and by 2027, between 830 and 1199 structures are expected to be decommissioned. With the probability transitions incorporated in the evaluation, future activity levels will be reduced and range between 474 and 702 decommissioned structures from 2017 to 2022 and between 704 and 1014 structures from 2017 to 2027. In both cases, decommissioning activity is reduced up to about 80% total activity over the forecast period. Note that decommissioning activity that is delayed is simply pushed forward in time and beyond the horizon of the forecast.

13.4 ACTIVE INVENTORY SCENARIO

There has been a dramatic decline in the number of shallow-water structures installed in recent years. Caissons, well protectors, and fixed platforms all show similar installation trends, peaking in the 1977–86 period at an average rate of 170 structures per year and declining to 123 structures per year from 1997 to 2006. Over the past decade, however, only 28 shallow-water structures on average have been installed per year, and over the past 5 years, about 11 shallow-water structures per year have been installed.

In Fig. 13.12, the active shallow-water GoM inventory forecast is depicted for assumed installation rates of 5, 20, and 40 structures per year. Actual activity levels are most likely at the low end of the range between 5 and 20 structures per year. Under the exponential model, reduction in active inventories is more dramatic (Fig. 13.13).

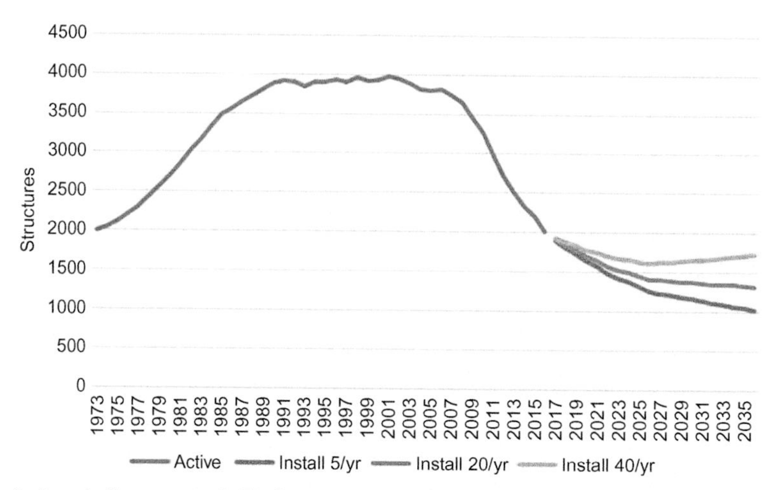

Fig. 13.12 Active shallow-water Gulf of Mexico inventory forecast using hyperbolic model, 2017–36. *(Source: Kaiser, M.J., Narra, S., 2018. A hybrid scenario-based decommissioning forecast for the shallow water U.S. Gulf of Mexico, 2018-2038. Energy 163, 1150–1177.)*

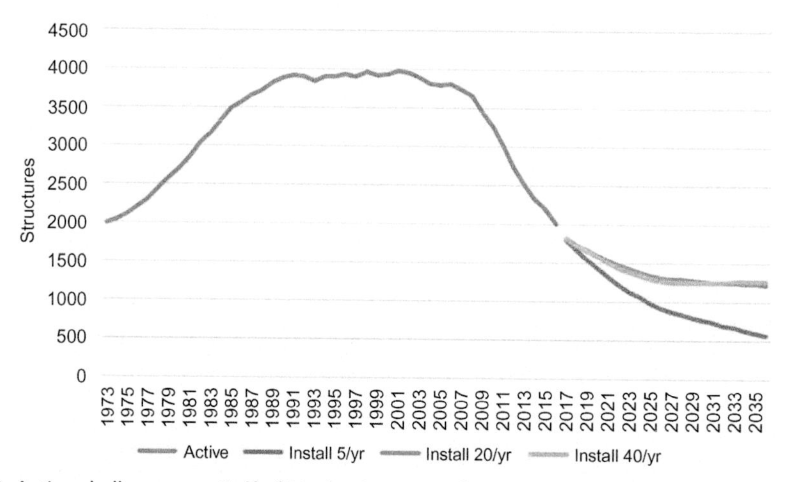

Fig. 13.13 Active shallow-water Gulf of Mexico inventory forecast using exponential model, 2017–36. *(Source: Kaiser, M.J., Narra, S., 2018. A hybrid scenario-based decommissioning forecast for the shallow water U.S. Gulf of Mexico, 2018-2038. Energy 163, 1150–1177.)*

According to the hyperbolic model, assuming installation rates between 5 and 20 structures per year, in 2027, active inventories will range between 1234 and 1399 structures. By 2036, active inventories are expected to range between 1023 and 1323 structures. These are upper bound limits. Using the exponential model for lower bound limits, by 2027, active structures will range between 865 and 1293 structures, and by 2036, active inventories will range between 562 and 1217 structures.

13.5 DISCUSSION

Shallow-water decommissioning activity is expected to continue at a relatively high rate in the short term with average annual activity between 75 and 150 decommissioned structures per year. The rate of decrease is expected to slow in the years ahead but remain significant and large, probably about an order of magnitude greater than installation rates, which are expected to range below 10 to 20 structures per year.

Active inventories depend on installation rates but will continue to decline following the general downward trend observed over the past decade. It is possible that active inventories will level out if new plays are discovered and developed in the region, but it is much more likely that the declining trends will continue at rates similar to historical activity over the next 5–10 years.

Decommissioning forecasting requires the use of different models for different structure classes and is intended to reflect primary characteristics of decommissioning activity. The producing structure model is capable of providing good quantitative agreement with operations under the assumptions specified. If model assumptions depart from the realized future, model results will vary from actual activity. Scheduling models are intended to permit general correspondence and are used mostly as an accounting mechanism to organize the data. All three models are required in evaluation because each structure class represents a significant percentage of the total inventory.

Producing and idle structures comprise about two-thirds of the structure inventory. The accuracy and reliability of the producing structure forecast are presumably better than the idle structure schedule, and so, the uncertainty of the composite results is expected to be determined by the uncertainty of the idle schedule. Auxiliary structure inventories are less significant than idle structures, and the reliability of their decommissioning schedule is expected to have a smaller impact on the overall results.

The model fits are unique based upon the model assumptions, but several model parameters are unknown or speculative. There is a general tendency to focus on uncertainty for explicit model parameters, but this is often misleading because of the large impact of parameters that are not modeled.

13.6 LIMITATIONS

Structures in the shallow-water GoM comprise a system whose elements were modeled independent of one another, while in practice, interrelationships and dependencies are present based on regional operations, multistructure complexes, pipeline activities, and service (e.g., gas-lift) functions. These dependencies are difficult to establish and were not considered in the evaluation.

Decommissioning forecasting for producing structures is based on rigorous and objective economic models and cash flow analysis but is still subject to severe modeling restrictions that will yield higher levels of decommissioning than expected in practice due to the constant investment and reservoir condition assumption employed in the model construction.

Unlike the physical laws of nature, the equations used to predict well production are strongly constrained by assumptions. There is no unique expression for well production that can be applied because capital spending decisions cannot be reliably modeled. So although there are elegant mathematical expressions that can be derived using historical data to extrapolate well production trends into the future, in practice, these expressions are limited in their ability to model production outside of the assumption space. Strategic considerations and regional operations and other unobservable conditions complicate evaluation. All of these complexities, interactions, and uncertainties mean that approximation methods are a central feature in model construction and interpretation.

Scheduling is a subjective procedure because the user schedules the structures for decommissioning according to a priori assumptions, either to reflect statistical characteristics of decommissioning activity or user preferences. The uncertainty levels of these models are difficult to quantify, and the models are less satisfying in the sense that there is no fundamental knowledge involved and insights are limited.

There are many complications involved in attempting to develop a more granular and precise scheduling model. For example, if an auxiliary structure is part of a producing complex, it is likely that when the complex is no longer producing, the structures of the complex will have limited utility, in which case the link between the producing and auxiliary structures of the complex could be used to activate decommissioning for the auxiliary structures. If the structure is not part of a producing complex, it may already serve a useful function for pipeline transmission. Unfortunately, modeling these linkages is both tedious and highly uncertain and thus of questionable utility. An aggregate approach was adopted that captures the global characteristics of the structure class without recourse to structure linkages.

We did not consider the differences that exist between structure types. Caissons and well protectors, for example, have a greater chance of removal after becoming idle and a smaller chance of being used for further operations relative to fixed platforms for all other things equal. No distinction was made between structure type, manning status, or production complexes in the models presented. Operator differences were not considered, and a single economic limit was applied across all structures. Economic limits characterize individual structures, and although it is easy to link economic limits to structure attributes, its application is not warranted by the model uncertainty.

Finally, decommissioning models are region-specific to account for country-specific regulatory, fiscal, and ownership issues, and generally speaking, models do not transfer between countries. For example, in the United Kingdom, North Sea operators are required to provide information regarding decommissioning plans and timing to regulators, and therefore, there is less need for independent forecasts. In other offshore regions such as Southeast Asia, title to structures often transfers to the government, and government's ability and interest to decommission will enter the evaluation. The issues with aging infrastructure and marginal production are broadly similar everywhere, of course, but the models used in their forecasting are site-specific.

BOX 13.1 Grand Isle 9 Reef

The Grand Isle sulfur mine was the world's first offshore sulfur mine that began production in 1960 (Fig. 13.14). Eight large jackets were erected 70 ft above sea level, and numerous smaller tower jackets were used to support bridgework. In total, 29 jacket structures were used in development.

Fig. 13.14 Grand Isle block 18 sulfur mining complex. *(Source: Freeport Sulfur Mining.)*

Continued

BOX 13.1 Grand Isle 9 Reef—cont'd

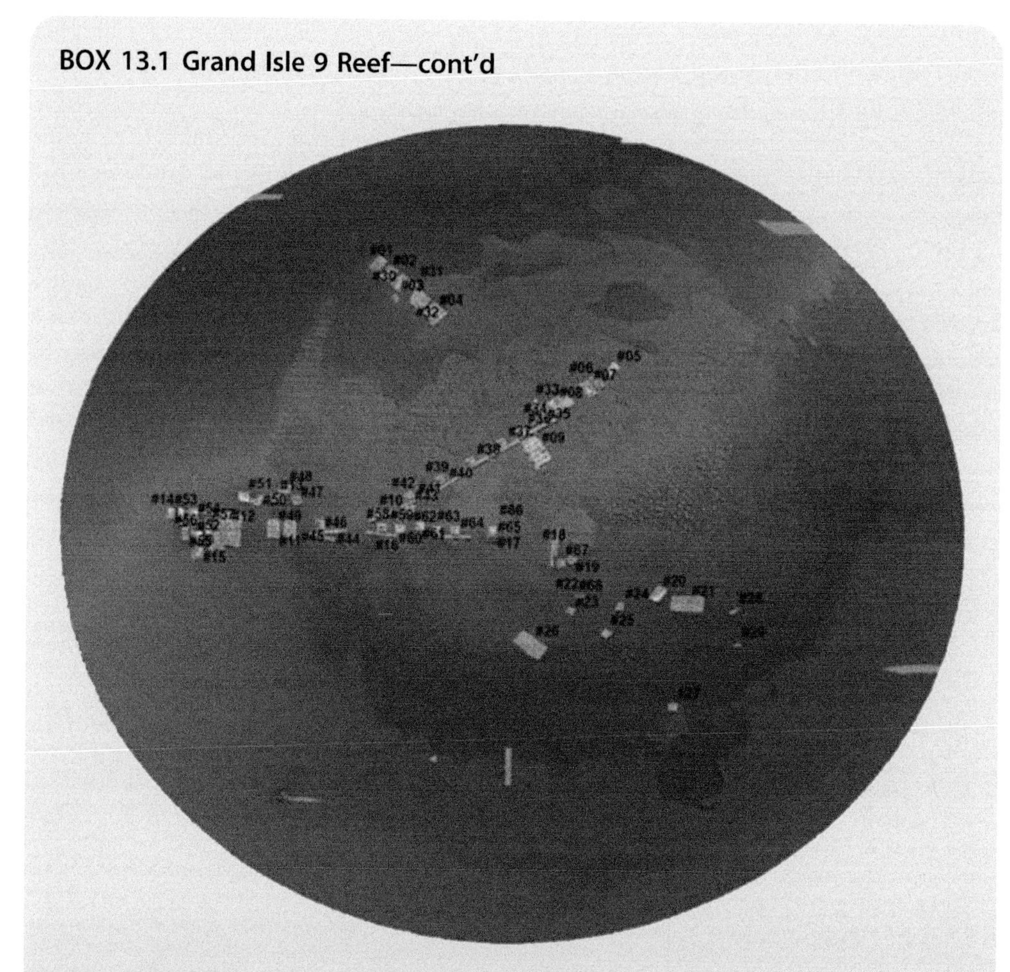

Fig. 13.15 Grand Isle 18 artificial reef site. *(Source: LDWF.)*

By the late 1990s, operations were no longer commercial, and in 1999, the complex was reefed in place. After clearance requirements were satisfied, a total of 81 structures were placed on the seabed (Fig. 13.15). The water depth at the site ranges from 42 to 50 ft, and the reef has a 27 ft clearance. Five lighted buoys are emplaced for safety of navigation.

REFERENCES

Kaiser, M.J., Narra, S., 2018. A hybrid scenario-based decommissioning forecast for the shallow water U.S. Gulf of Mexico, 2018-2038. Energy 163, 1150–1177.

U.S. Department of the Interior, 2010. Minerals Management Service Appendix II to NTL No. 2010-N03. Guidelines for the application, review, approval, and administration of royalty relief for end-of-life leases.

CHAPTER FOURTEEN

Deepwater Decommissioning Forecast

Contents

Abstract

A total of 23 structures in the Gulf of Mexico have been decommissioned in water depth >400 ft through 2017, leaving an active deepwater inventory of 46 fixed platforms, three compliant towers, and 47 floating structures circa 2017. Installation and decommissioning activity in deepwater are both small and sporadic, and these trends are unlikely to change in the future because of the significant capital requirements and planning involved and residual value of structures. All of the floaters and about half of the fixed platforms were held by production, and about a dozen fixed platforms were idle, and at least five of these structures have been converted to auxiliary roles as pipeline junctions. Several fixed platforms and a few floaters are long on their decline curve and are expected to be decommissioned within the next few years unless tieback opportunities materialize or alternative uses are found. Model results predict between 27 and 51 deepwater structures will be decommissioned through 2031, and between 12 and 25 removals are expected from 2017 to 2022.

14.1 MODEL RECAP

The approach taken is to apply economic models where applicable and a scheduled approach where economic criteria cannot be applied, similar to the shallow water model but with differences in implementation. All idle structures and structures with net revenue less than their economic limit $NR < EL$ at the time of evaluation are scheduled for decommissioning using user-defined model assumptions, while all structures with $NR > EL$ employ a production forecast model. Deepwater GoM inventory circa 2016 is applied where there were 13 idle structures, 15 (marginal) producers with $NR < EL$, 26 platforms with $NR > EL$, and 44 floaters with $NR > EL$.

Three scenarios are used to tie the model results together: a base (expected) case, a slow (or low) case, and a fast (or high) case. The selection of the slow/low and fast/high

Decommissioning Forecasting and Operating Cost Estimation
https://doi.org/10.1016/B978-0-12-818113-3.00014-X

model parameters is relative to the base case and refers to the impact on decommissioning activity, not to the model parameters themselves. For example, for producing structures, the slow/low scenario assumes an oil price of \$80/bbl, whereas the fast/high scenario applies an oil price of \$40/bbl. High oil prices will generate greater revenue than low prices, and for a given production decline will delay decommissioning, resulting in a slow/low decommissioning trajectory relative to the base case.

For producing structures, an assumed economic limit determines the *EL-yr* assuming no capital spending outside of normal maintenance activities, which is then adjusted by the user-defined parameter τ to determine decommissioning timing (Fig. 14.1). *EL* and τ are model inputs parameterized using historical data. The parameter τ helps to balance the model results against the early decommissioning forecast predicted through the application of the status quo assumption. Base-case parameters apply average data on economic limits, removal times, and \$60/bbl oil and \$3/Mcf gas price (Table 14.1).

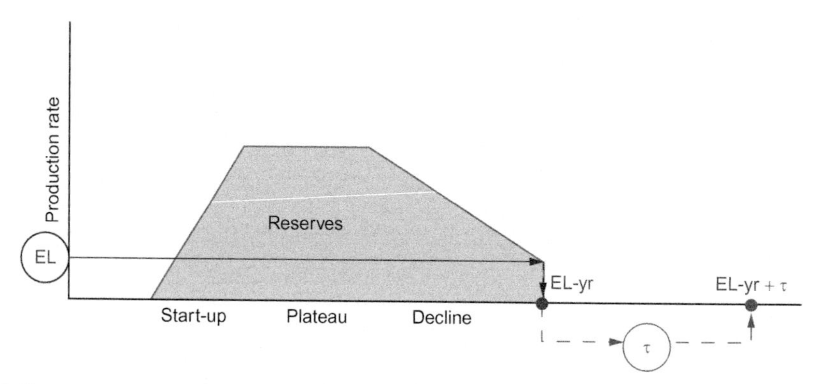

Fig. 14.1 Decommissioning forecast showing input and model parameters.

Table 14.1 Structure Class Scenarios and Model Assumptions

	Base	Slow/Low	Fast/High
Idle structure removal rate	U(5)	U(10)	U(3)
Platforms with $NR < EL$	U(5)	U(10)	U(3)
Platforms with $NR > EL$	*EL-yr* + 5	*EL-yr* + 10	*EL-yr* + 3
Floaters with $NR > EL$	*EL-yr* + 2	*EL-yr* + 2	*EL-yr* + 2
EL (\$ million)	2.7	0.5	5.0
Oil price (\$/bbl)	60	80	40
Gas price (\$/Mcf)	3	4	2
Roy (%)	16.67	16.67	16.67

Note: U(*T*) represents uniform removal of inventory over *T* years.

The slow/low-case model parameters are selected to schedule idle structures and structures producing below their economic limit slower than the base-case schedule (10 years versus 5 years), and for all other structures, decline curve estimates are used with *EL-yr* delays based on the upper range of historic removal times. For platforms, the *EL-yr* delay is 5 years in the base case and 10 years in the slow/low case. For floaters, the *EL-yr* delay is 2 years across all scenarios.

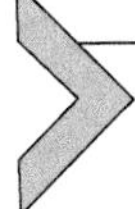

14.2 DECOMMISSIONING FORECAST

14.2.1 Producing Structures

The expected last year of production for producing fixed platforms in 400–500 ft (Table 14.2), fixed platforms and compliant towers in >500 ft (Table 14.3), and floaters (Table 14.4) is presented by commodity price for the primary production stream (Kaiser and Liu, 2018). The secondary product price is held fixed at $3/Mcf for oil producers and $60/bbl for gas producers across all price scenarios. For fixed platforms, oil and gas structures are shown separately since there is a mix of producers, while all floaters (except Independence Hub complex 1766) are oil structures. Independence Hub stopped producing in 2016.

Table 14.2 Expected Last Year of Production for Fixed Platform in Water Depth 400–500 ft

Complex	$40/bbl	$60/bbl	$80/bbl
1052	2016	2016	2016
1076	2016	2016	2016
1165	2016	2018	2024
1279	2016	2016	2016
1500	2022	2026	2039
2027	2016	2016	2016
22172	2017	2018	2026
22224	2017	2018	2028
22662	2017	2019	2025
23151	2016	2016	2016
23800	2021	2026	2035
80015	2016	2016	2016
Complex	**$2/Mcf**	**$3/Mcf**	**$4/Mcf**
70	2021	2023	2036
23848	2017	2020	2026
23893	2016	2016	2016
90028	2016	2016	2016

Note: For oil structures, gas price assumed constant at $3/Mcf for all oil price scenarios. For gas structures, oil price assumed at $60/bbl for all gas price scenarios.

Source: Kaiser, M.J., Liu, M., 2018. A scenario-based deepwater decommissioning forecast in the U.S. Gulf of Mexico J. Petrol. Sci. Eng. 165, 913–945.

Table 14.3 Expected Last Year of Production for Fixed Platforms and Compliant Towers in Water Depth >500 ft

Complex	$40/bbl	$60/bbl	$80/bbl
147	2019	2021	2025
1482	2021	2027	2037
10178	2016	2016	2016
10297	2018	2022	2029
22178	2025	2034	2050
22840	2017	2022	2035
23503	2018	2021	2025
23552	2032	2039	2055
23883	2024	2030	2040
24129	2027	2030	2038
24130	2033	2040	2063
24201	2025	2031	2043
33039	2032	2038	2049
70012	2028	2034	2043
90014	2056	2065	2084
23567–1	2020	2023	2028
23567–2	2016	2016	2016
Complex	**$2/Mcf**	**$3/Mcf**	**$4/Mcf**
113	2019	2024	2036
10212	2017	2018	2034
23875	2028	2031	2037
27032	2021	2025	2057
27056	2016	2016	2016

Note: For oil structures, gas price assumed constant at $3/Mcf for all oil price scenarios. For gas structures, oil price assumed constant at $60/bbl for all gas price scenarios.

Source: Kaiser, M.J., Liu, M., 2018. A scenario-based deepwater decommissioning forecast in the U.S. Gulf of Mexico J. Petrol. Sci. Eng. 165, 913–945.

Table 14.4 Expected Last Year of Production for Floaters

Complex	$40/bbl	$60/bbl	$80/bbl
67	2026	2036	2088
183	2029	2037	2058
235	2052	2061	2067
251	2028	2034	2062
420	2027	2031	2036
811	2021	2028	2039
821	2030	2036	2047
822	2027	2034	2067

Continued

Table 14.4 Expected Last Year of Production for Floaters—cont'd

Complex	$40/bbl	$60/bbl	$80/bbl
876	2038	2048	2090
1001	2046	2055	2070
1035	2027	2034	2046
1088	2020	2022	2034
1090	2033	2041	2082
1101	2069	2086	2090
1175	2031	2038	2076
1215	2053	2067	2090
1218	2023	2028	2045
1223	2053	2069	2090
1288	2022	2024	2028
1290	2023	2027	2031
1323	2039	2046	2059
1665	2028	2030	2041
1766	2016	2016	2016
1799	2031	2036	2045
1819	2040	2046	2056
1899	2062	2076	2090
1930	2029	2033	2041
2008	2041	2046	2053
2045	2030	2048	2090
2089	2021	2025	2035
2133	2036	2045	2063
2229	2023	2024	2029
2385	2038	2043	2049
2424	2043	2051	2090
2440	2049	2057	2076
2498	2021	2026	2044
2513	2090	2090	2090
2576	2069	2078	2089
2597	2047	2057	2071
23583	2016	2018	2021
24080	2031	2035	2042
24199	2052	2060	2073
24229	2025	2029	2035
24235	2029	2036	2060
70004	2058	2077	2090
70020	2023	2029	2041

Note: For oil structures, gas price assumed constant at $3/Mcf for all price scenarios. For gas structure 1766, oil price assumed constant at $60/bbl for all gas price scenarios.

Source: Kaiser, M.J., Liu, M., 2018. A scenario-based deepwater decommissioning forecast in the U.S. Gulf of Mexico J. Petrol. Sci. Eng. 165, 913–945.

14.2.2 Model Scenarios

Using the assumptions described for the base, slow/low, and fast/high scenarios, the expected decommissioning times for each scenario are depicted in Tables 14.5–14.7 and graphed for the slow and fast scenarios in Figs. 14.2 and 14.3.

The base-case scenario falls between the cumulative curves for the slow and fast scenarios as expected, and future decommissioning activity will likely range between the base- and low-case scenarios. According to the model results, over the next 5 years, between 12 and 25 deepwater structures are expected to be decommissioned, and over the next decade, between 25 and 36 structures are expected to be decommissioned. If alternative uses for structures are found, activity levels will be smaller than depicted.

In Tables 14.8 and 14.9, the base-case and slow scenario model results are presented in 5-year time blocks.

Table 14.5 Expected Decommissioning Schedule for Fixed Platforms in Water Depth 400–500 ft

Complex	High	Base	Low
1052	2017–19	2017–21	2017–26
1076	2017–19	2017–21	2017–26
1165	2019	2023	2034
1279	2017–19	2017–21	2017–26
1500	2025	2031	2049
2027	2017–19	2017–21	2017–26
22172	2020	2023	2036
22224	2020	2023	2038
22662	2020	2024	2035
22685	2017–19	2017–21	2017–26
23051	2017–19	2017–21	2017–26
23151	2017–19	2017–21	2017–26
23800	2024	2031	2045
80015	2017–19	2017–21	2017–26
Complex	**High**	**Base**	**Low**
70	2024	2028	2046
10192	2017–19	2017–21	2017–26
23212	2017–19	2017–21	2017–26
23353	2017–19	2017–21	2017–26
23848	2020	2025	2036
23893	2017–19	2017–21	2017–26
23925	2017–19	2017–21	2017–26
90028	2017–19	2017–21	2017–26

Source: Kaiser, M.J., Liu, M., 2018. A scenario-based deepwater decommissioning forecast in the U.S. Gulf of Mexico J. Petrol. Sci. Eng. 165, 913–945.

Table 14.6 Expected Decommissioning Schedule for Fixed Platforms and Compliant Towers in Water Depth >500 ft

Complex	High	Base	Low
147	2022	2026	2035
1482	2024	2032	2047
10178	2017–19	2017–21	2017–26
10297	2021	2027	2039
22178	2028	2039	2060
22840	2020	2027	2045
23277	2017–19	2017–21	2017–26
23503	2021	2026	2035
23552	2035	2044	2065
23760	2017–19	2017–21	2017–26
23883	2027	2035	2050
24129	2030	2035	2048
24130	2036	2045	2073
24201	2028	2036	2053
33039	2035	2043	2059
70012	2031	2039	2053
90014	2059	2070	2094
23567–1	2023	2028	2038
23567–2	2017–19	2017–21	2017–26
Complex	**High**	**Base**	**Low**
113	2022	2029	2046
1320	2017–19	2017–21	2017–26
10212	2020	2023	2044
10242	2017–19	2017–21	2017–26
23788	2017–19	2017–21	2017–26
23875	2031	2036	2047
27032	2024	2030	2067
27056	2017–19	2017–21	2017–26
70016	2017–19	2017–21	2017–26

Source: Kaiser, M.J., Liu, M., 2018. A scenario-based deepwater decommissioning forecast in the U.S. Gulf of Mexico J. Petrol. Sci. Eng. 165, 913–945.

Table 14.7 Expected Decommissioning Schedule for Floaters

Complex	High	Base	Low
67	2028	2038	2090
183	2031	2039	2060
235	2054	2063	2069
251	2030	2036	2064
420	2029	2033	2038
811	2023	2030	2041

Continued

Table 14.7 Expected Decommissioning Schedule for Floaters—cont'd

Complex	High	Base	Low
821	2032	2038	2049
822	2029	2036	2069
876	2040	2050	2092
1001	2048	2057	2072
1035	2029	2036	2048
1088	2022	2024	2036
1090	2035	2043	2084
1101	2071	2088	2092
1175	2033	2040	2078
1215	2055	2069	2092
1218	2025	2030	2047
1223	2055	2071	2092
1288	2024	2026	2030
1290	2025	2029	2033
1323	2041	2048	2061
1665	2030	2032	2043
1766	2017	2021	2026
1799	2033	2038	2047
1819	2042	2048	2058
1899	2064	2078	2092
1930	2031	2035	2043
2008	2043	2048	2055
2045	2032	2050	2092
2089	2023	2027	2037
2133	2038	2047	2065
2229	2025	2026	2031
2385	2040	2045	2051
2424	2045	2053	2092
2440	2051	2059	2078
2498	2023	2028	2046
2513	2092	2092	2092
2576	2071	2080	2091
2597	2049	2059	2073
23583	2018	2020	2023
24080	2033	2037	2044
24199	2054	2062	2075
24229	2027	2031	2037
24235	2031	2038	2062
70004	2060	2079	2092
70020	2025	2031	2043

Source: Kaiser, M.J., Liu, M., 2018. A scenario-based deepwater decommissioning forecast in the U.S. Gulf of Mexico J. Petrol. Sci. Eng. 165, 913–945.

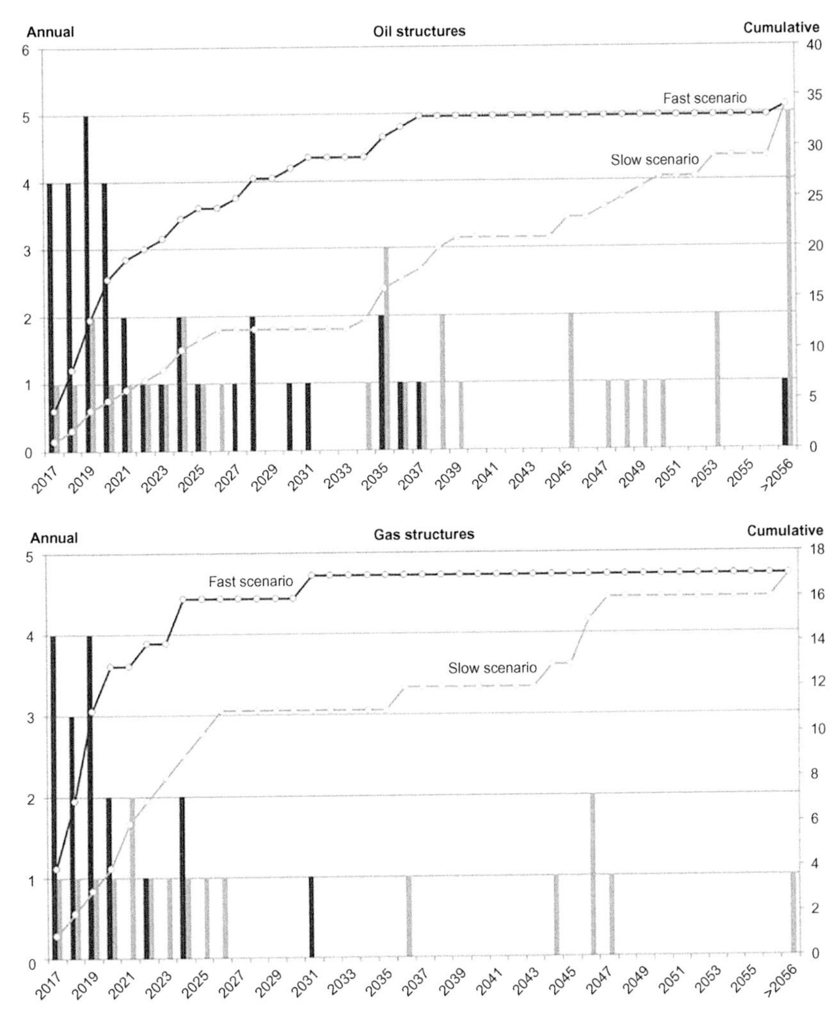

Fig. 14.2 Decommissioning schedules for deepwater fixed platforms—slow and fast scenarios. *(Source: Kaiser, M.J., Liu, M., 2018. A scenario-based deepwater decommissioning forecast in the U.S. Gulf of Mexico J. Petrol. Sci. Eng. 165, 913–945.)*

14.2.3 Sensitivity Analysis

Using the base-case model parameters, royalty rates and economic limits were reduced by half and eliminated and the model results recomputed. In both cases, there was no noticeable difference in the model-predicted decommissioning times except in a handful of structures, and for those structures affected, there was only a difference of a year or two in the expected timing.

Royalty rates and economic limits are generally considered secondary factors in economic modeling and do not have a major impact on value at the end of production

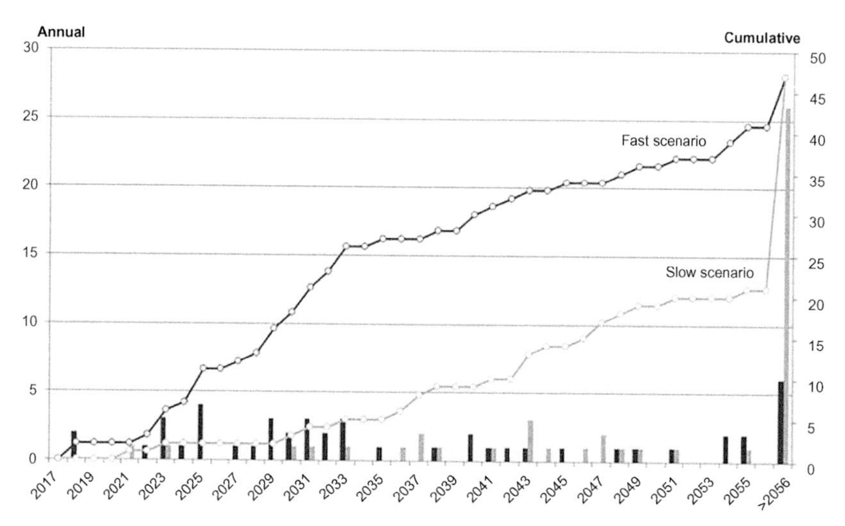

Fig. 14.3 Decommissioning schedule for floating structures—slow and fast scenarios. *(Source: Kaiser, M.J., Liu, M., 2018. A scenario-based deepwater decommissioning forecast in the U.S. Gulf of Mexico J. Petrol. Sci. Eng. 165, 913–945.)*

Table 14.8 Deepwater Decommissioning Forecast in the Gulf of Mexico Under Base Scenario

Water Depth (ft)	2017–21	2022–26	2027–31	2032–36	2037–41	2042–46	2047–51	2052–56	2057–96	Total
400–500	14	5	3	0	0	0	0	0	0	22
500–1500	9	3	6	5	1	3	0	0	1	28
>1500	2	3	6	6	8	3	6	1	13	48
Total	25	11	15	11	9	6	6	1	14	98

Source: Kaiser, M.J., Liu, M., 2018. A scenario-based deepwater decommissioning forecast in the U.S. Gulf of Mexico J. Petrol. Sci. Eng. 165, 913–945.

Table 14.9 Deepwater Decommissioning Forecast in the Gulf of Mexico Under Low Scenario

Water Depth (ft)	2017–22	2022–26	2027–31	2032–36	2037–41	2042–46	2047–51	2052–56	2057–96	Total
400–500	7	7	0	4	1	2	1	0	0	22
500–1500	4	5	0	2	3	4	4	1	5	28
>1500	1	1	2	2	3	5	5	2	27	48
Total	12	13	2	8	7	11	10	3	32	98

Source: Kaiser, M.J., Liu, M., 2018. A scenario-based deepwater decommissioning forecast in the U.S. Gulf of Mexico J. Petrol. Sci. Eng. 165, 913–945.

relative to changes in operating cost and commodity prices. Whereas royalty relief may induce capital spending and incentivize operators to invest in new facilities, the ability of royalty relief to maintain production on marginal properties in a high-cost environment is less likely to be successful because of its incremental impact relative to other more significant factors.

14.3 ACTIVE STRUCTURE FORECAST

Historically, there have only been a handful of deepwater structures installed annually, and there is no reason to believe that future activity will depart from these levels. In the near term, projects sanctioned and under construction provide reliable short-term estimates for activity, but beyond a 5-year outlook speculation is required.

Fixed platform and floater installation activity over the next two decades is assumed to be bound between historic activity rates. Installation rates of fixed platforms are assumed to be bound between 0.2 and 2.6 platforms/year, and new floater installations are assumed to be bound between 2 and 3 floaters/year.

In Fig. 14.4, the active inventory of fixed platforms and compliant towers is depicted, and in Fig. 14.5, the active floater inventory is shown under the slow and base-case scenarios for installation rates specific to each structure class. Fixed platform activity is expected to fall at the low end of the range, while floaters are more likely to fall in the middle of the range.

Under the slow scenario, fewer structures will be decommissioned, which will lead to a larger standing inventory relative to the base-case scenario, and thus, the model results for a given installation rate depict the slow scenario above the base case. Similarly, a lower installation rate will depress active structure counts relative to higher install rates since decommissioned structures will not be replenished.

For fixed platforms, the most likely future scenario shows a decline in the standing structures following the slow scenario with an assumed 0.2 structures/year install rate. For floaters, the most likely scenario yields an increasing inventory that ranges midway

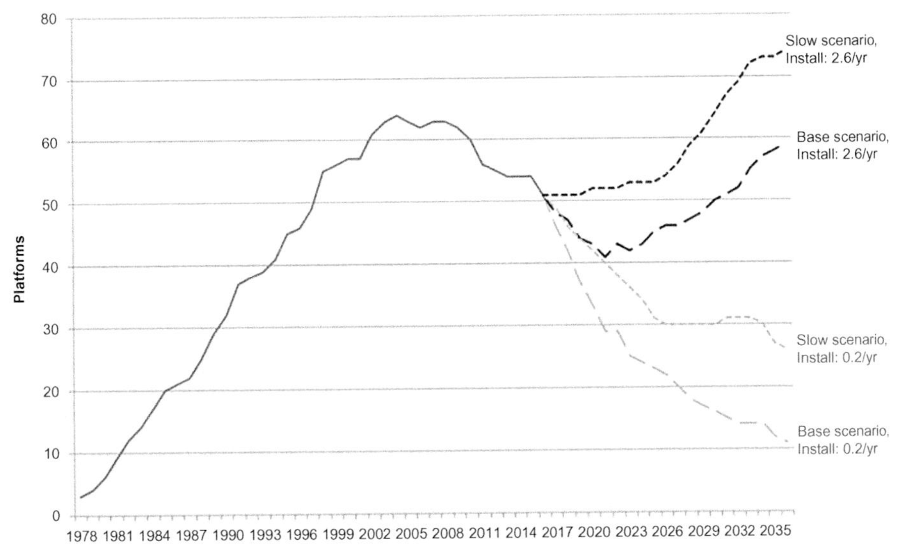

Fig. 14.4 Active inventory of deepwater fixed platforms—slow and base scenarios. *(Source: Kaiser, M.J., Liu, M., 2018. A scenario-based deepwater decommissioning forecast in the U.S. Gulf of Mexico J. Petrol. Sci. Eng. 165, 913–945.)*

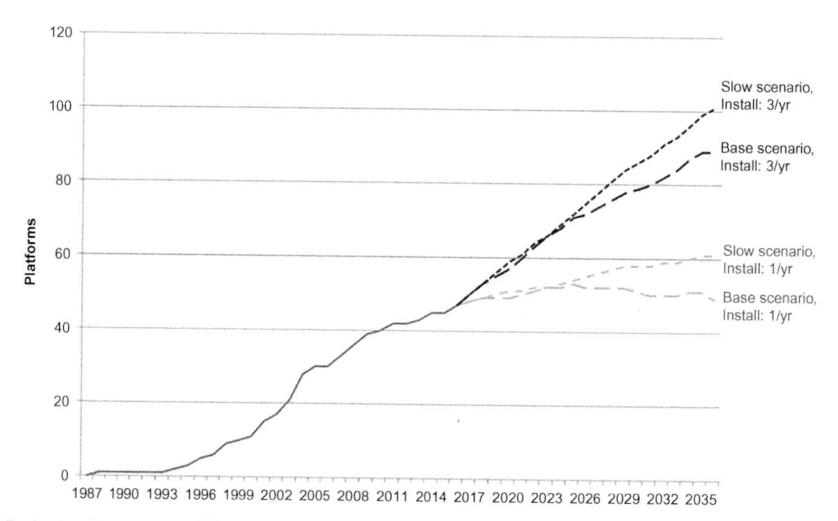

Fig. 14.5 Active inventory of floaters under slow and base scenarios. *(Source: Kaiser, M.J., Liu, M., 2018. A scenario-based deepwater decommissioning forecast in the U.S. Gulf of Mexico J. Petrol. Sci. Eng. 165, 913–945.)*

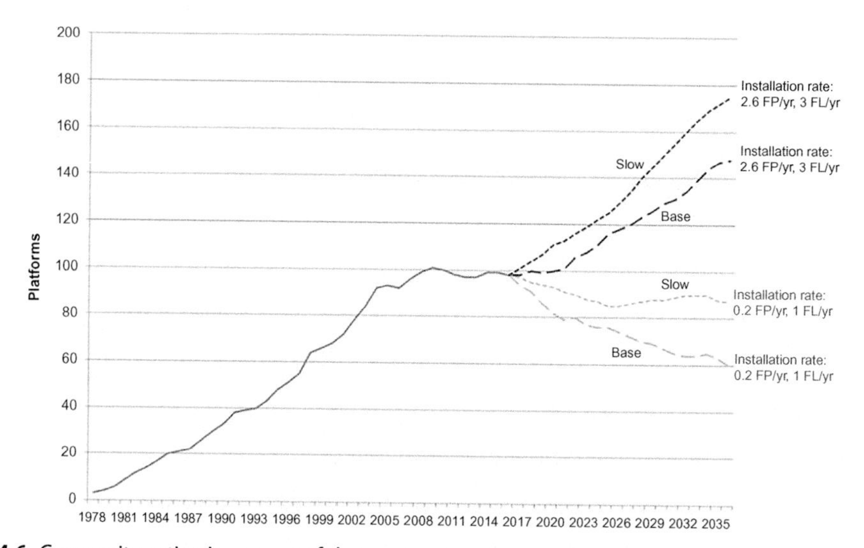

Fig. 14.6 Composite active inventory of deepwater structures under slow and base scenarios. *(Source: Kaiser, M.J., Liu, M., 2018. A scenario-based deepwater decommissioning forecast in the U.S. Gulf of Mexico J. Petrol. Sci. Eng. 165, 913–945.)*

between the slow and the base scenarios under an assumed two floaters per year installation rate.

In Fig. 14.6, the composite active structure forecast is shown with the slow scenario dominating the base-case. Lower installation rates will depress the active inventory since decommissioned structures are not being replaced. The active deepwater structure inventory is expected to maintain its plateau or decline slightly in the years ahead.

14.4 LIMITATIONS

The purpose of decommissioning modeling is to gain insight into system characteristics and to understand the manner in which model assumptions impact predictions. Numbers are important, but the certainty of absolute values are a false premise, and focus should be on activity ranges and general trends. Modeling always involves a number of assumptions, and the desire to make assumptions transparent arises from the need to understand the results in the context of the model construction. Good modeling practice dictates that procedures are well defined and assumptions clear and transparent.

Decline models are expected to yield results consistent with the (conservative) capital spending constraints associated with the status quo assumptions. The decommissioning model for producing structures is reliable relative to the assumption set, but given the large number of factors involved in asset reviews and the limits imposed by the status quo conditions, it is not surprising that numerical models are limited in their ability to predict the exact outcomes of decommissioning. One way to reduce model uncertainty is to aggregate results over time blocks to reflect general trends.

If model parameters are not well known, they can be treated as a free parameter when constructing the forecast, but it is preferable to minimize the number of these assumptions and to use historical data when applicable. For example, the economic limit of production is a model input that may be based on historical data of similar structures in the region (first choice), an estimate of the structure's direct operating cost (second choice), or simply a number that seems reasonable to the user (third choice, free parameter). All are valid, but some choices are better than others. Allowing the parameter to vary over specified ranges may help or hinder the evaluation depending on parameter relevance and how it enters the model.

Wells long on their decline curve are more likely to yield an accurate production forecast relative to younger wells, and consequently, mature reservoirs are more likely to yield an accurate production forecast relative to younger reservoirs. In terms of decommissioning timing, however, because the potential for tiebacks and alternative uses will vary with each structure and is relatively high, perhaps as high as 30% in aggregate, there will always be modeling uncertainty associated with decommissioning forecasts.

Sidetrack drilling can occur anytime and if successful will extend the life of the structure beyond the model forecast. Tiebacks may also occur at any time, on or off peak production. When sidetrack opportunities are no longer available or are too risky relative to the potential rewards, the facility may be repurposed as a compressor or pump station for pipelines or nearby facilities. Operators seek every available opportunity to keep structures in place before decommissioning due to the high residual value of deepwater assets.

For the active structure forecast, installation rates for fixed platforms and floating structures were based on historic statistics over the past two decades. There are several reasons why such bounds are reasonable and are expected to reflect future trends for each structure class.

BOX 14.1 The Continuing Saga of MC-20 Platform Decommissioning

The following account is a summary based on BOEM and USCG press releases and other news reports (Kunzelman, 2018; Schleifstein, 2017).

On 15 September 2004, Hurricane Ivan toppled Taylor Energy's MC-20A platform in 479ft water depth, located approximately 11 mi offshore and host to 28 wells (Fig. 14.7). Wave heights reached 100ft at the storm's eye, repeating every 19s. The fixed leg platform installed in 1984 by Sohio was not designed to withstand such wave heights and periods and subsequently toppled. The hurricane caused a mudslide in the region, and the platform slide about 500ft down slope, resting on its side partially buried by more than 100ft of mud and sediment (Fig. 14.8).

Fig. 14.7 Taylor Energy's Mississippi Canyon 20 eight-pile platform before Hurricane Ivan. *(Source: USCG.)*

The structure was submerged nearly 75% below the mudline, and all the production piping suffered significant structural damage below the mudline. The landslide pulled all the connections to the wells from the platform and bent piping connected to the well by 90°. Well abandonment efforts proved difficult or impossible to perform, and excavation of the structure was considered unsafe, so the jacket is expected to remain in place for the foreseeable future. Through 2016, the platform deck has been removed, the oil pipeline has been decommissioned, and nine of the 25 wells have been plugged and abandoned using intervention wells. Subsea debris in the vicinity of the platform has also been removed.

In 2008, the MC-20 site was identified as the source of discharge and daily sheen, and the US Coast Guard issued Taylor Energy an administrative order to conduct daily overflights of the well site to visually monitor the oil discharges and report the presence and estimated amount of oil. As the responsible party under the Oil Pollution Act of 1990, Taylor Energy is required to pay for oil spill recovery and response costs and has a continuing legal obligation to respond to the

Continued

BOX 14.1 The Continuing Saga of MC-20 Platform Decommissioning—cont'd

Fig. 14.8 Taylor Energy's MC-20 toppled platform was pushed 400 ft downslope from its wells (triangle). *(Source: USCG.)*

ongoing oil discharge. The specific source(s) of discharge at the MC-20 site are not fully known, but because the discharge volume is greater than can reasonably be accounted for by oil released from sediment only, oil is most likely emanating from one or more of the damaged wells.

A subsea containment system was installed in 2009 in an effort to recover the discharge and minimize the amount of oil seepage into the ocean (Fig. 14.9), but by 2013, the last containment dome was removed due to damaged components and the lack of efficiency and recovery of oil. In 2012, the US Coast Guard issued an administrative order that required the design, construction, and installation of a new more effective containment dome, but as of Dec. 2018, a new dome has not been installed.

Since September 2014, oil sheens as large as 1.5 mi wide and 14 mi long have been observed and reported by Taylor Energy. Estimates of daily oil volume are based on observations of percent cover, sheen color, and dimensions and are highly variable and uncertain. Over this period, the daily volume of oil discharging from the MC-20 well site is estimated to have fluctuated between less than one barrel and 55 bbl of oil. The average reported oil volume on the sea surface from September 2014 to April 2015 was about 2 bbl/day.

Taylor had originally been ordered by the Minerals Management Service in October 2007 to permanently plug and abandon all the wells by June 2008, but numerous complexities and safety hazards have prevented completing operations. In November 2009, MMS issued Taylor an order to provide supplemental bonding to guarantee performance of its decommissioning

Continued

BOX 14.1 The Continuing Saga of MC-20 Platform Decommissioning—cont'd

obligations, and the Department of the Interior and Taylor Energy entered into a Trust Agreement wherein Taylor committed $666 million to fulfill obligations under the Outer Continental Shelf Lands Act.

Taylor Energy has reported spending more than $480 million through 2015 on its efforts to decommission the platform and stop the leak. In January 2016, the company sued the federal government to recover $432 million of the $666 million that the company was required to post as a security bond in 2009 for leak response work. Federal authorities rebuffed the company's settlement overtures and ordered it to perform more work at the site.

On 23 Oct 2018, Coast Guard Capt. K. Luttrell issued a new administrative order for Taylor Energy to design and install a new containment system to capture and remove the leaking crude oil. Taylor Energy has argued that performing more work at the leak site could be dangerous and cause more environmental harm than good. It also claims residual oil is oozing from sediment on the seafloor and not the wellbores.

On 20 Dec 2018, Taylor Energy filed three federal lawsuits, one against the Interior Board of Land Appeals that refused to excuse the company from well abandonment requirements, one against Couvillion Group LLC contracted by the Coast Guard to design and construct the new containment system, and one against the federal government to recover funds in a trust to pay for leak response work. Future work to be performed under the Trust Agreement will be determined based on the results of the lawsuits and the availability of applicable technology.

Fig. 14.9 Containment system deployed in 2009 above the MC-20 wellbores and removed in 2013. *(Source: USCG.)*

REFERENCES

Kaiser, M.J., Liu, M., 2018. A scenario-based deepwater decommissioning forecast in the U.S. Gulf of Mexico. J. Petrol. Sci. Eng. 165, 913–945.
Kunzelman, M., 2018. N.O. company sues to block order to contain oil leak. The Advocate, December 22.
Schleifstein, M., 2017. 12 years after Gulf oil platform destroyed, feds start investigating environmental damage. The Times-Picayune. July 28.

PART FOUR

Critical Infrastructure Issues

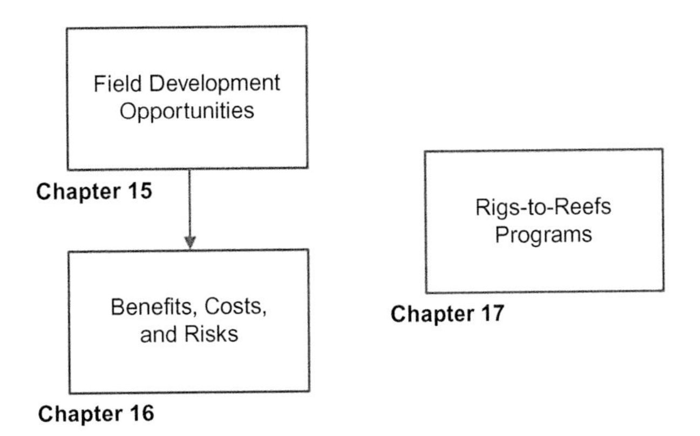

CHAPTER FIFTEEN

Field Development Opportunities

Contents

Abstract

High levels of decommissioning in the shallow-water GoM over the past decade have drawn attention to the important role of infrastructure in oil and gas production and raise questions regarding its impact on future development in the region. The purpose of this chapter is to identify the role of critical infrastructure, what's important and why, and the nature of the business decisions of operators. Four tiers of infrastructure organize the discussion, and opportunities within and outside field development are discussed using conceptual economic-risk models. Within field opportunities are more easily ranked by cost and risk compared with outside field opportunities that involve a larger number of factors and higher levels of uncertainty. Project cost and risk for within field opportunities normally increase from workovers to redevelopment to secondary and tertiary production, and return on investment is usually smaller at each stage. Progressive deployment of evolving technologies is considered key to successful development. For investment opportunities that lie outside the field or lease boundary, project cost and risk are more complicated to evaluate and depend on additional factors. Without defined problem boundaries, system constraints, and field-specific data, outside field opportunities cannot be performed quantitatively and compared. The most critical offshore infrastructures are large producers, structures containing a high concentration of pipeline linkages or high volume throughput, and large-diameter high-volume pipelines.

Decommissioning Forecasting and Operating Cost Estimation
https://doi.org/10.1016/B978-0-12-818113-3.00015-1

15.1 OVERVIEW

In 2017, over 80% of the oil production in the GoM was produced in water depth >400 ft from around 100 structures, and the remaining 20% of crude oil was produced in shallow water from 866 producing structures. Deepwater oil production has yet to peak, while shallow-water oil production continues to decline. Gas production in both shallow-water and deepwater regions is also in decline.

Estimates of undiscovered resources in the GoM are significant (74 Bboe total, 13 Bboe in shallow water) and exceed total cumulative production through 2017, but they are also hypothetical and speculative with unknown timing, location, and amount. Future exploration and development activity is expected to follow current trends and remain concentrated in deepwater where 80% of oil and gas reserves reside. The shallow-water GoM will continue to see new discoveries, but these will be small and marginal in nature and of decreasing frequency. Redeveloping fields with existing infrastructure is expected to provide the majority of incremental recovery in shallow water. Secondary production methods in shallow water were already being employed decades ago, and the opportunities for enhanced oil recovery methods (i.e., CO_2 injection) are unlikely to be implemented in large scale because of poor economics and the composition of the shelf players.

Pipelines are used to deliver almost all GoM oil production to shore, and the pipeline network is a recognized vital part of the supply chain. As product flows head northward, they are aggregated into larger pipelines and then split at platform hubs to deliver to different onshore destinations. The deepwater oil pipeline network is segregated from the shelf network, while the gas pipeline networks are more integrated. Currently, few pipeline constraints or capacity bottlenecks exist. If individual structures or pipelines are responsible for a significant portion of supply (e.g., say >10%), then such infrastructure would be considered critical in common sense usage, but in the GoM, concentration of this form has not (yet) arisen, and so, either all of the active deepwater structures and export pipelines that deliver crude to shore could be considered critical infrastructure or none at all because of its distributed nature. The deepwater GoM region is widely recognized as critical to the Gulf's continued success.

15.2 INFRASTRUCTURE TIERS

According to Merriam Webster, a system is defined as a "group of devices or objects forming a network for distributing something or serving a common purpose," and critical refers to an "indispensable, vital, or absolutely necessary component for the operation of a system." Following this definition, four tiers of infrastructure are identified with critical infrastructure defined as the most important and vital for system operations:

I. Critical
II. Active

III. Inactive—potentially useful
IV. Inactive—probably not useful

Active structures are producing or currently serving a useful function and include in-service pipelines. Critical infrastructure is a subset of active infrastructure. Inactive (or idle) infrastructure is grouped as potentially useful and probably not useful. The distinction between these two classes is difficult to determine without additional information from operators, and thus, the ambiguity of the inactive class is intentional and inherent in its classification.

15.3 INFRASTRUCTURE HIERARCHY

Wells, structures, flowlines, and export pipelines are the main components used in offshore oil and gas development (Fig. 15.1). The importance of system components increases at every stage along the value chain from well to onshore destination as flows are combined and increase in volume for processing and export. Large hydrocarbon production sources are important independent of the transport system. Structures are the first point of aggregation followed by hub and transportation platforms. For wet wells, manifolds are commonly used in subsea developments to aggregate production before delivery to the host. The absence, incapacity, damage, or destruction of aggregation points (hubs) would cause immediate cessation of those upstream operations that do not have alternative routes for transport. GoM pipeline systems are interconnected, but the degree of interconnection varies geographically.

Old wells have the potential for sidetracking and injection operations, but whether such activities are a good investment is site-, time-, and operator-specific. Maintaining inactive wells is widely recognized as potentially useful because of the cost savings associated with reusing wells to access reservoirs, but as fields mature, sidetrack operations become increasingly risky because success is uncertain and incremental production is typically small. Seeking untapped faulted reservoirs and bypassed targets for production will flatten out the decline curve, but strict cost management is necessary to be successful. For wet wells, operators have much less experience in sidetracking because of the high cost involved and the problems associated with delivering high water cut production.

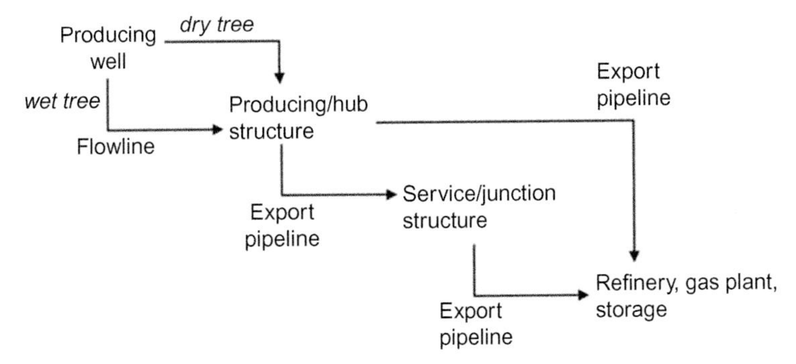

Fig. 15.1 Processing hierarchy from well to end consumer.

Structures installed to drill wells will usually also process production, but in shallow water, a broad mix of development options has been pursued. Nondrilling central processing facilities were common in early developments. Producing structures handle and process hydrocarbon fluids and if off-lease fields are involved or other export pipelines cross the structure are often classified as a hub. The processing capacity at the structure and the oil and gas export pipeline capacity, or the volumes processed or throughput, are the key parameters quantifying hub operations. Assets can be ranked by production and pipeline capacity, flow rate, throughput, or degree of connectivity. Large producing platforms and high-capacity export systems would generally be considered more critical than marginal producing structures and low-capacity export lines. Pipelines with high volume throughput are more critical than pipelines with low volume throughput.

Structures near the end of their producing life are usually not critical assets unless they are also serving as a hub, but may be useful for future development. The benefits of standing structures are well known to operators and regulators alike but outside specific assets and development plans are difficult to quantify. Marginal producing structures are not considered critical unless serving as a pipeline hub. Structures that are no longer producing may or may not be useful in the future. Potential benefits should be considered speculative unless supported by detailed operator spending and development plans tied to the assets in review.

All deepwater producing platforms and pipelines and hub platforms represent critical infrastructure. Shelf junction platforms are another important category of critical infrastructure. Several shelf platforms interconnect with deepwater operations and are owned and operated by the same operators or their affiliates.

15.4 LIFE CYCLE STAGES

The initial production phase of oil and gas is the easiest and most profitable phase because of the initial driving forces present. The best areas of the reservoir are drilled in primary development, and other less significant blocks are drilled later if the economics of the incremental barrels are favorable. Primary or free-flow production refers to the fluids produced by the original reservoir energy. As production proceeds, the pressure in the reservoir falls, reducing the natural flow rates of the hydrocarbons, especially oil. The rate of reduction depends on the drive mechanism and development methods.

Primary production uses the reservoir's natural energy that comes from fluid and rock expansion, solution gas, gravity drainage, aquifer influx, or a combination drive as described in every reservoir engineering textbook. This natural energy (pressure) is created by a pressure differential between the higher pressure in the rock formation and the lower pressure in the wellbore. In aquifer support, for example, reservoir pressure declines slowly—almost linearly—as long as the water replaces the oil in the pores during production. For gas-cap mechanisms, the gas cap expands with oil production, and production decline is more rapid than with aquifer support. In most oil reservoirs, only a small percentage of the original oil-in-place is recovered during the primary production period, and secondary methods may be used to produce a portion of the oil that remains. In gas reservoirs, recovery rates are usually quite high (>85%), and secondary recovery methods are not frequently used.

Primary recovery factors for oil range broadly from 5 to 40% of the oil-in-place depending on a number of factors such as wells drilled, reservoir complexity, fluid properties, development methods, and reservoir drive mechanisms (Fig. 15.2). This leaves a considerable amount of oil in the reservoir after the pressure has been depleted that may be economic to recover. Improved oil recovery refers to engineering techniques that are used to recover more oil from depleted fields, known variously as secondary and tertiary production, pressure maintenance, waterflooding, or enhanced oil recovery (Fig. 15.3). Primary techniques include artificial lift, and secondary techniques include injection wells to maintain pressure or sweep efficiency (Jahn et al., 2008).

Secondary recovery consists of replacing the natural reservoir drive or enhancing it with an artificial drive to maintain the pressure in the reservoir. The use of injected water or natural gas is the most common method, and when water is used, it is referred to as waterflooding. Waterfloods or gas injection may be implemented at the start of production or after initial production depending on the behavior of the reservoir pressure decline.

Tertiary recovery methods (also called enhanced oil recovery, EOR) seek to enhance the sweep efficiency and reduce the capillary forces by making the fluids miscible or improving their mobility. They are usually classified into thermal, chemical, and miscible displacement categories. Tertiary methods are expensive and risky, economics are usually

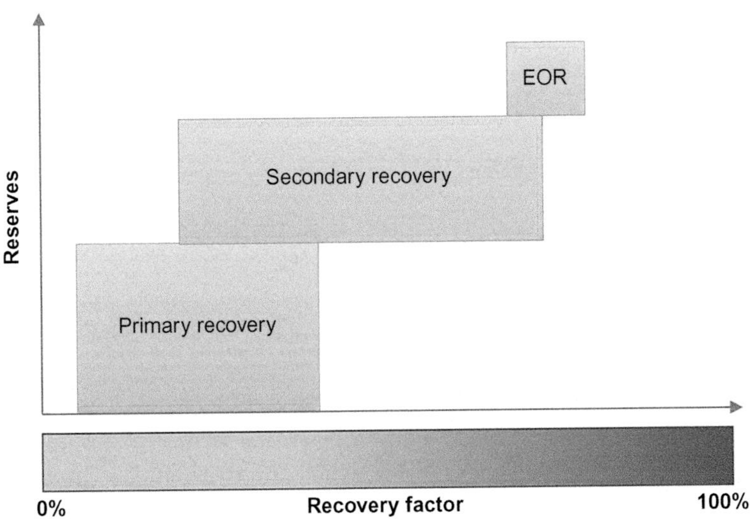

Fig. 15.2 Representative recovery factors and oil production across life-cycle stages of development.

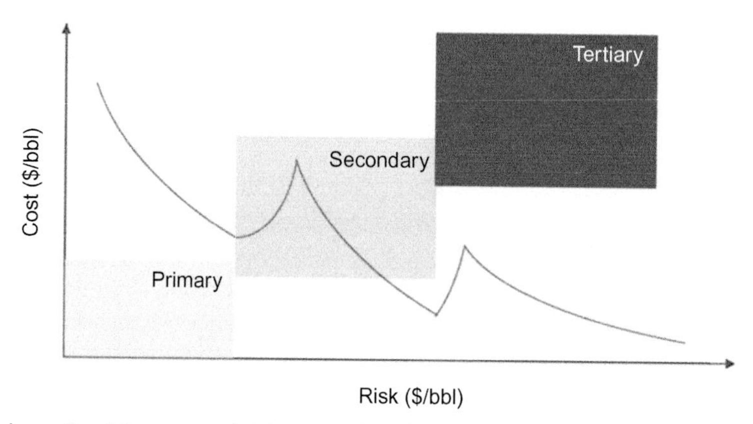

Fig. 15.3 Schematic of the cost and risk across the life-cycle stages of production.

difficult, and performance is uncertain. EOR methods are only pursued after primary and secondary methods are performed, and circa 2017, contributes about 5% of US oil production and is most frequently used in heavy oil and poor permeability fields. All EOR projects in the United States are onshore, and most projects require well-capitalized companies due to high investment requirements. Companies with significant tertiary activity include Occidental, Hess, Kinder Morgan, Chevron, Denbury Resources, and EOG Resources. In the shallow-water GoM where most players are juniors, EOR opportunities are unlikely to be pursued.

From an economic perspective, the cost per barrel of secondary methods is considerably higher than the cost of the primary recovery stage since material and energy inputs are greater and produced volumes are lower. Similarly, the cost of tertiary methods is higher than secondary methods and in most cases both the cost and outputs are more uncertain and therefore riskier.

15.5 WITHIN FIELD OPPORTUNITIES

Within field opportunities, project cost and risk described on a dollar-per-barrel ($/bbl) basis normally increase from workovers, to redevelopment, to secondary and tertiary production, and return on investment is usually smaller at each stage (Table 15.1). This economic characteristic causes larger companies to sell or abandon mature assets, while smaller companies may seek such opportunities to grow their production. Project economics in brownfield secondary recovery projects depends upon accurate risk assessment and mitigation strategies and an acceptable return on investment.

As a general rule and as illustrated in the schematic in Fig. 15.3, anytime there is a significant bump (increase) in the production profile of a field or well, one should assume/recognize that an investment was required that caused the change. The bump

Table 15.1 Opportunities Within and Outside Field Developments

Within Field Opportunities
Workover
Redevelopment
Secondary production
Tertiary production
Outside Field Opportunities
Satellite development—full
Satellite development—partial
Transport hub
Others

did not appear by magic nor was created for free—it required capital spending for information, technology, equipment, and material. This seemingly obvious observation is neither widely appreciated nor accounted for in many economic assessments. Whether the investment paid for itself and increased the value of the asset is difficult to estimate since work activities and project costs are proprietary.

15.5.1 Workover

Wells need to be maintained for maximum productivity, and in the event that a well stops producing (well failure), a decision has to be made on the merits of attempting to bring the well back online. In many instances, a simple back of the envelope economic justification is adequate, such as would occur with a new high-producing well with formation damage or other impairment, but for expensive or risky operations, a more detailed evaluation may be required. Workover decisions in old wells are risky since the operator may not get their investment back, and so once an older well stops producing, a workover may not even be contemplated if production rates are low. For wet wells, workovers are more expensive and uncertain than dry tree wells and introduce additional complications in evaluation.

Wells are "worked over" to increase production and accelerate reserves recovery, reduce operating cost, or reestablish borehole integrity. There are many different types of workovers that may be required, and workover activity is often budgeted as part of operating and maintenance expense and is not considered capital expenditures. Examples of workover activity include replacing corroded tubing or packer leaks or drilling out and plugging a section of casing to reduce water cut or cross flow in the well or behind casing. If the formation is damaged by pore-throat plugging or other impairment, it may be necessary to fracture or acidize the area or perform a solvent soak. If flows are restricted due to sand production or wax or scale deposition, remediation may be required. In a primary cement failure, such as water invading a producing zone behind the casing and between the bad cement, significant production decline may occur. Workovers are the most common activity performed at offshore structures to extract maximum value from field development.

The primary objective of stimulation is to restore impaired well/reservoir connectivity. The nature of impairment, treatment options, and posttreatment production issues often change over the life of the well, and the selection of wells and frequency of stimulation are site-specific, usually based on comparison between peer wells in analog reservoirs, well-flow modeling, and available budgets. The results of treatment are complex and variable, which makes decision-making difficult and risky. Stimulation can result in immediate improvement in production rate, but the impairment may come back rapidly (Fig. 15.4). In other cases, the initial improvement may not be dramatic, but the longer-term performance is improved. In some cases, a total loss of production may result and represent the downside risk of well intervention.

The causes of near-wellbore impairment are well documented. During completion operations, perforations often become plugged with rock fragments and debris and residue from fracturing (frac) fluids. During production, formation impairment is primarily due to movement and rearrangement of reservoir fines, clay-sized particles that cause a reduction in permeability, deposition of solids (e.g., asphaltenes) from produced crudes, and deposition of mineral scales from produced water. Three categories of well impairment are thus defined:

- Scale—inorganic minerals deposited from water
- Solids—organic deposits from crude, often a combination of asphaltenes and resins
- Fines—clay-sized particles from pore bodies that capture in pore throats

Scale formation may occur in the reservoir and inside the production tubing and may be removed chemically or mechanically (Jordan et al., 2001). A well producing at high

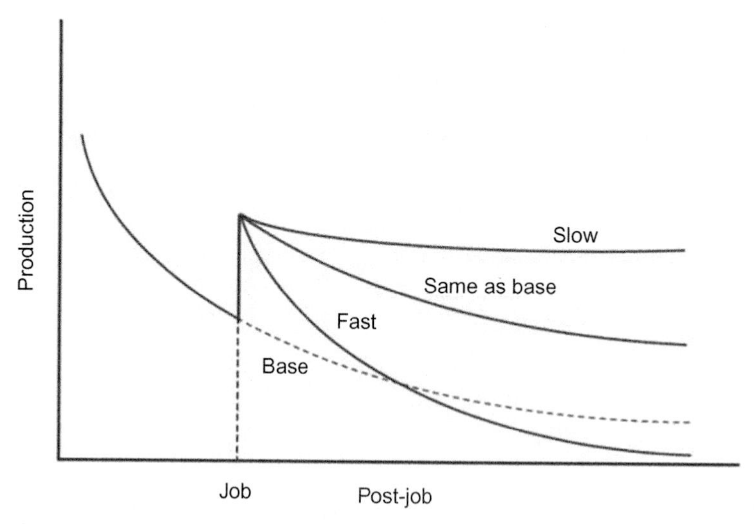

Fig. 15.4 Redevelopment requires investment and is accompanied by uncertain outcomes on the amount and timing of incremental production.

water cut may be choked back, or a change of perforation interval may be considered to shut off unwanted fluids. Skin problems may be resolved by acidizing or additional perforations. Solids and fines depend on the reservoir pressure and the change in pressure between the reservoir and the wellbore. Fine migration often occurs early in field life, solids and scale later.

The best way to assess the production benefit of stimulation is to evaluate the well production before and after treatment under similar conditions (e.g., choke setting and separator pressure). Unfortunately, few publications quantify the results of stimulation, but an important paper by Morgenthaler and Fry (2012) provides guidance on how such studies should be evaluated and the nature of outcomes.

Example. Stimulation production improvement

Morgenthaler and Fry (2012) quantified the impact of 128 stimulation treatments performed on deepwater GoM wells in water depth from 1000 to 7000ft from 2002 to 2011. The majority of the stimulations were on oil wells (80%) and direct vertical access wells, but interventions on subsea wells were also performed. Morgenthaler and Fry defined the ratio of the posttreatment production rate divided by the pretreatment rate as the production improvement factor (PIF) and used this indicator to quantify the distribution of PIFs for scale, solvent, and fine migration treatments (Fig. 15.5).

Although only 25% of scale treatments yield a PIF > 1.5, some scale treatments were performed to prevent tubing or subsurface safety valve plugging, benefits that are hard to quantify. Solvent and acid treatments have similar performance curves and entail some risk in that 2% to 5% of treatments lead to total production loss. However, there is also a 50% probability to achieve a PIF > 1.5 and a 10% to 20% chance that PIF > 2.

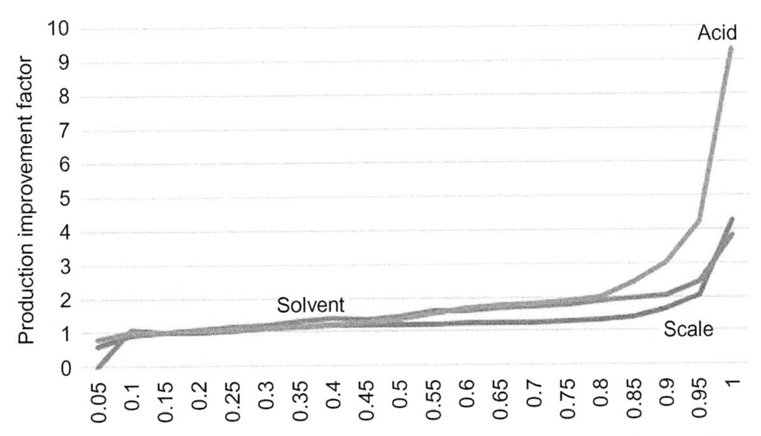

Fig. 15.5 Production improvement factor distribution for 128 deepwater well stimulations, 2002–11. *(Source: Data from Morgenthaler, L.N., Fry, L.A., 2012. A decade of deepwater Gulf of Mexico stimulation experience. In: SPE 159660. SPE Annual Technical Conference and Exhibition. San Antonio, TX, Oct 8–10.)*

Following solvent treatments, wells tended to return to their prestimulation performance rapidly and consistently (Fig. 15.6), and because operations can be conducted relatively cheaply and cause few problems during flowback, they are considered economically attractive and are performed regularly to maintain production.

Scale and acid treatments are more complex and varied as shown by the normalized production rates pre- and post treatment (Figs. 15.7 and 15.8).

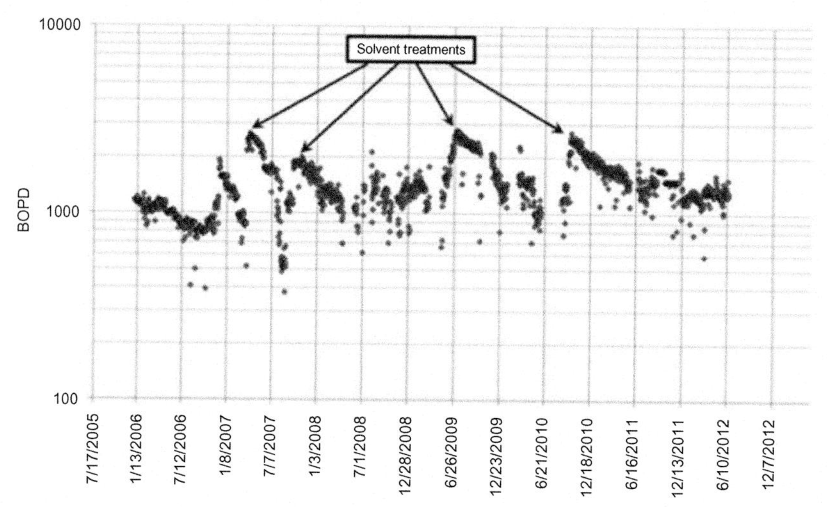

Fig. 15.6 Typical response for solvent treatments. *(Source: Morgenthaler, L.N., Fry, L.A., 2012. A decade of deepwater Gulf of Mexico stimulation experience. In: SPE 159660. SPE Annual Technical Conference and Exhibition. San Antonio, TX, Oct 8–10.)*

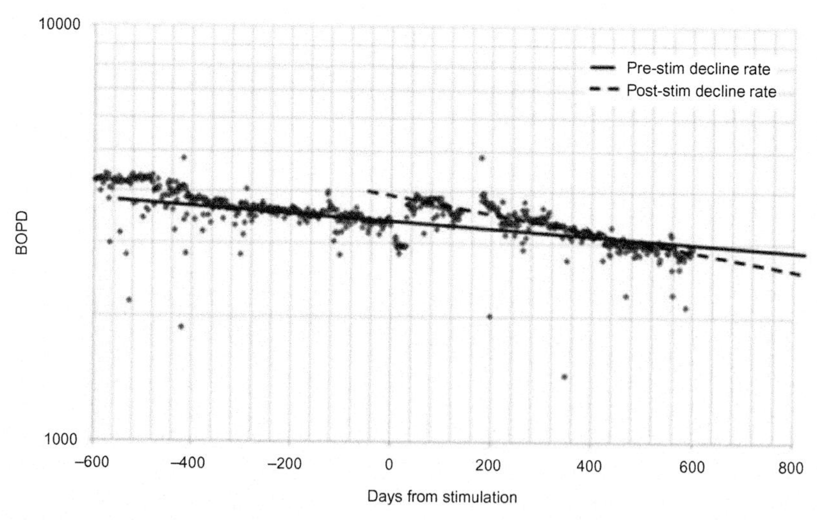

Fig. 15.7 Typical accelerated production response. *(Source: Morgenthaler, L.N., Fry, L.A., 2012. A decade of deepwater Gulf of Mexico stimulation experience. In: SPE 159660. SPE Annual Technical Conference and Exhibition. San Antonio, TX, Oct 8–10.)*

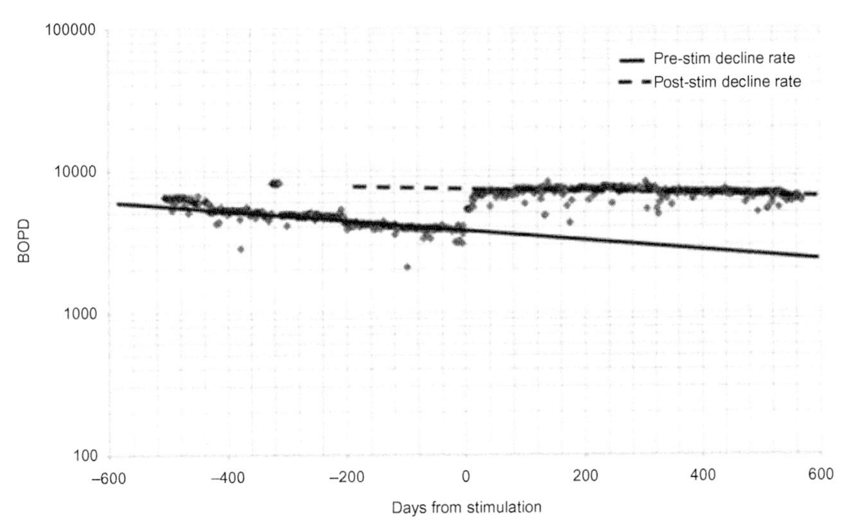

Fig. 15.8 Production response with shallower decline and increased volume recovery. *(Source: Morgenthaler, L.N., Fry, L.A., 2012. A decade of deepwater Gulf of Mexico stimulation experience. In: SPE 159660. SPE Annual Technical Conference and Exhibition. San Antonio, TX, Oct 8–10.)*

15.5.2 Redevelopment

Redevelopment can take many different forms and is distinct from workovers with different objectives. Normally, redevelopment activity involves capital expenditures that are depreciated, not expensed as workover cost. Redevelopment usually entails drilling new wells with the objective to accelerate production or increase reserves. Operators attempt to reuse existing wells to reduce cost, but this is not always possible. Reentry wells may cost one-third to one-half a new well due in large part to the cost savings from not having to drill and complete the top hole and setting casing.

Directional drilling can be used to reach most targets within GoM lease blocks from one location, and extended reach drilling can usually achieve a 3–5 mi horizontal distance. For wells that can be drilled from the platform, almost all sands with oil and gas deposits that the operator believes will yield a positive return on investment will be drilled. In other words, as long as the expected revenue from production is anticipated to exceed the drilling and completion cost, the target will be drilled. High risk targets may also be drilled.

Hydrocarbons in isolated fault blocks or cellar oil left behind below production intervals will require new wells, sidetracks, or extended reach drilling (Fig. 15.9). Attic oil left behind above production intervals is usually accessible with a new perforation, a considerably cheaper and faster operation than well deepening or sidetracking. These wells will increase reserves rather than accelerate recovery and are classified as capital expenditure.

As geologic models improve and become more detailed, it may be possible to identify reserves that are not being drained effectively or within the economic life of the asset.

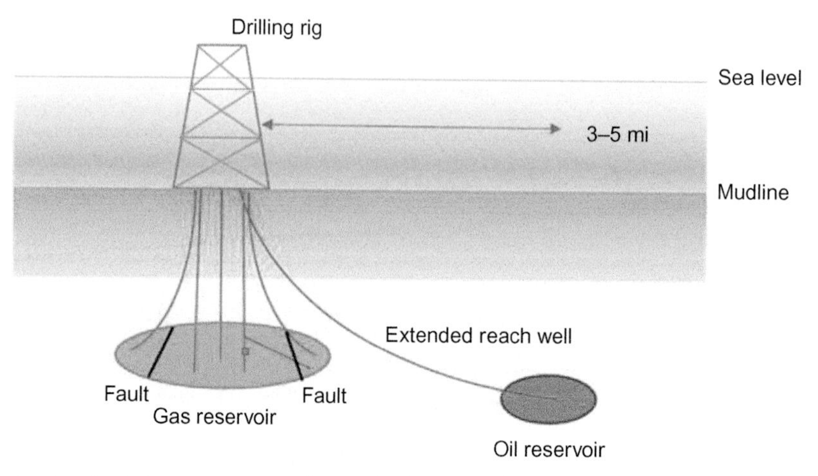

Fig. 15.9 Anchor field development and extended reach drilling.

These are usually referred to as infill wells. Infill drilling means drilling additional wells between the original development wells to accelerate recovery, not necessarily to increase reserves, although incremental reserves additions may be an outcome. If the original wells were too far apart, for example, additional wells will reduce the distance the oil and gas has to travel to the wellbore that accelerates production. Accelerated recovery projects are not necessarily economic.

In theory, the economics of redevelopment is straightforward. Compare the (discounted) income from the incremental recovery Δ expected with the investment cost C using probability estimates. If the activity is expected to yield production profile $Q'(t)$ as shown in Fig. 15.10 and the status quo (no investment) profile is $Q(t)$, then the production increment $\Delta(t)$ at time t is computed simply as the difference:

$$\Delta(t) = Q'(t) - Q'(t)$$

The net present value at assumed future commodity price $P(t)$ and corporate discount rate D from the time of the investment $t=0$ through the time of impact $t=u$ is

$$NPV(\Delta) = \sum_{t=0}^{u} \frac{P(t)\Delta(t)}{(1+D)^t}$$

If $\text{NPV}(\Delta) > C$, the investment satisfies economic criteria and would likely be pursued. The investment cost C can be estimated reasonably accurately, but the impact Δ is difficult to estimate reliably, both in magnitude and profile. Capital can be spent to reduce this uncertainty, say, by acquiring additional geophysical data, and this is often done, but there is always a trade-off between collecting more data and its value in decision-making. The risk of the investment is described by the probability $\text{NPV}(\Delta) < 0$ but in most cases cannot be computed reliably.

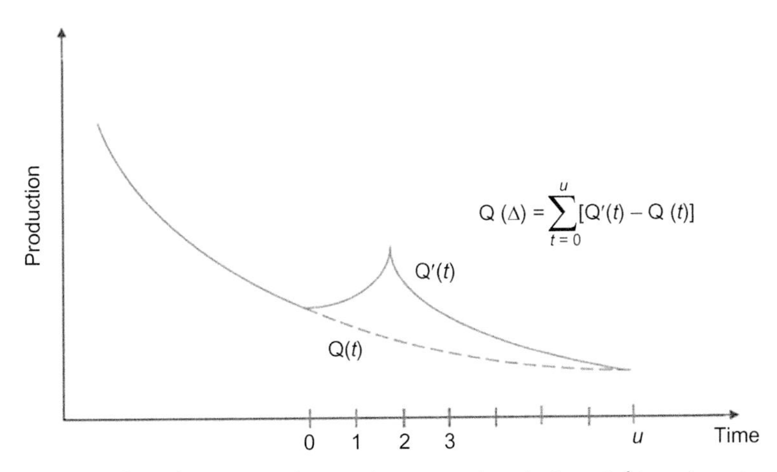

Fig. 15.10 Incremental production is values and compared against capital investment.

15.5.3 Secondary Production

A waterflood involves injecting water through injection wells into the depleted oil reservoir rock. It can be initiated either before or after the reservoir drive has been fully depleted, and both methods are common. The injection wells are either drilled or converted from producing wells, the latter obviously providing significant cost savings, and the water is sourced as seawater or from a water-producing zone. The water normally has to be treated so that it is compatible with the producing formation, which involves filtering to remove suspended solids, chemical treatment to reduce oxygen levels, and biocide treatment to control bacteria to prevent corrosion and fouling.

In pressure maintenance, water is used to replace the produced fluids to maintain high production rate and is normally performed when it is discovered that reservoir pressure is decreasing at a high rate relative to production volumes. When water is used as the primary source of reservoir energy, it is common to apply a voidage replacement policy, that is, produced hydrocarbon volumes are replaced by injected volumes in a specific (e.g., 1:1 or 1:2) ratio. In a waterflood, water is used to sweep out some of the remaining oil in the reservoir and in this form is usually performed near the end of primary production.

In gas injection, there is no need to control for hydrocarbon dew point, but it is necessary to dehydrate the gas to avoid water dropout. Injection gas pressures are usually much higher than gas-lift or gas export pipeline pressures and require compression (Jahn et al., 2008). In cycling operations, gas is reinjected into the reservoir to prevent condensate liquids from dropping out in the reservoir. Like black oil reservoirs where production engineers strive to stay above the bubble point of the reservoir, in gas fields, it is desirable to keep the reservoir pressure above the dew point to prevent condensate from forming since these liquid hydrocarbons block the flow path of gas in the reservoir rock and are essentially lost (i.e., not producible). After the gas rises to the surface and is collected topsides, the condensates are stripped out, and the dry gas is reinjected through injection wells.

15.5.4 Enhanced Oil Recovery

Enhanced oil recovery involves the injection of substances that are not naturally found in the reservoir, such as CO_2, steam, or chemicals. EOR can be initiated after primary production or waterflooding. Offshore, tertiary methods are generally not feasible unless several conditions prevail simultaneously; onshore, CO_2, steam injection, and miscible fluid displacement have been successfully and (presumably) profitably applied but only for a small number of projects. Less than a handful of offshore EOR projects have been implemented worldwide, and these were all short-term pilot projects (see Section 16.1.7). For some of the logistic considerations in offshore chemical EOR projects, see (Raney et al., 2011; Baginski et al., 2015). Useful reviews on carbon capture and storage issues can be found in (Kheshgi et al., 2011) and (Anderson, 2017).

15.6 OUTSIDE FIELD OPPORTUNITIES

For investment opportunities that lie outside the field or lease boundaries, project cost and risk are more complicated to evaluate and depend on additional factors. Outside field opportunities are difficult to assess without detailed data on all available options and development requirements. Without defined problem boundaries, system constraints, and field-specific data, evaluations cannot be performed quantitatively and compared.

15.6.1 Satellite Development

In most hydrocarbon basins, the largest fields are discovered and developed first because they are the easiest to find and provide the best economic returns. The structures, processing facilities, and export systems used in development often have a considerable working life that may prove useful to develop smaller fields in the region that would otherwise be uneconomic to develop on a stand-alone basis. Ownership of the reserves and surrounding infrastructure at the time of development will play a large role in decision-making. Operators that maintain a large presence will have more options to monetize their assets relative to smaller players with less infrastructure.

In a satellite development, a portion of the existing offshore infrastructure is used for development, which may prolong the economic life of the infrastructure and reduce decommissioning rates. The role of the host facility in a satellite development can vary from providing all production and processing support (e.g., as in a subsea tieback), to providing partial support (say for gas-lift or water injection to a nearby field), to providing minimal support but allowing access to one or more export pipeline routes (e.g., host receives pipeline-quality product for compression or pumping services).

Reservoirs/targets outside a 3–5 mi range will be drilled by a MODU, and the platform may serve as host if the development is economic (Fig. 15.11). Large satellite fields will support a greater tieback distance than smaller fields, and gaseous fluids can be transported farther than liquids and require less demanding flow assurance strategies. As distances

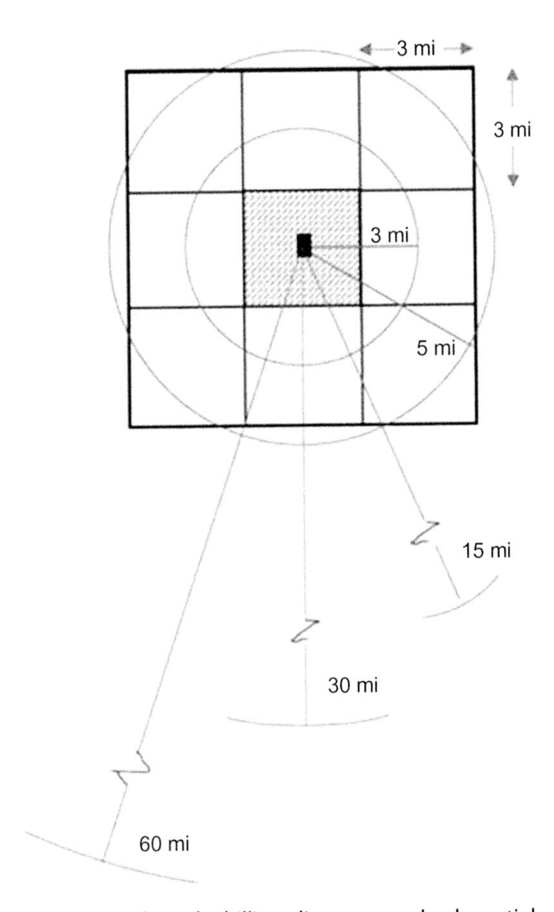

Fig. 15.11 Representative extended reach drilling distances and subsea tieback zones.

increase and product quality decreases, flow assurance requirements and capital spending increase. Commerciality depends on many factors, but reservoir size and production rate, fluid type, ownership, and development cost are usually the most important (Table 15.2).

Off the platform, drilling and development cost increase, and prospects must be larger to achieve the economic-risk criteria established by the company. Subsea wells (especially oil) are expected to strand more reserves than platform-drilled wells, and less redevelopment will occur because of the higher cost of access. For subsea gas wells, there is less concern of stranding resources, but if problems develop, premature abandonment may also arise. For third-party tiebacks, successful negotiation is required to bring product back to a host.

15.6.2 Transportation Hub

At service/junction structures (also called transportation platforms), export pipelines board the structure, and compression/pumping stations raise the pressure and then reinject the fluid into export systems to enable flow onward to another transportation platform or

Table 15.2 Hypothetical Economic Thresholds by Reserves Size and Tieback Distance

Size (MMboe)	Tieback Distance < 5 mi	5–30 mi	30–60 mi	> 60 mi
< 1	–	– –	– – –	– – –
1–5	?	–	– –	– – –
5–10	+	?	–	– – –
10–50	++	+	?	– –
50–100	+++	+++	++	++

Note: Plus sign indicates likely economic and minus sign indicates likely uneconomic. The number of signs translates into higher confidence/probability levels.

one or more onshore destinations. Transportation platforms serve as a connecting point for export pipelines to enter the offshore pipeline network and connect to their onshore destinations. In some cases, export pipelines may tap into a pipeline directly and do not need to board a platform. Operators often place a premium on having multiple routes to different destinations to increase the netback value of their production, and several of the larger GoM operators have invested in a robust pipeline network to maximize their options.

Strictly speaking, service/junction platforms do not have processing capacity available unless the structure previously served in field development and equipment is still operable, and so, the fluids boarding have already been processed to pipeline specification and only need to be pumped or compressed to reach their destination. Ancillary services offered normally include metering, liquid removal, pig catcher, heating, and cooling. Structures may be manned or unmanned. If the service/junction structure handles multiple deliveries and departures for different operators, it will likely have several lines entering and leaving the platform.

15.6.3 Future Development

When fields are no longer economic to produce, structures installed for drilling will no longer be viable. Within the 3–5 mi radius centered at the platform, viable targets presumably no longer exist. Processing capacity at the facility may still be useful if of sufficient size and in good working order, and because export pipelines are connected to the GoM network, if production from another structure or field can be tied into the structure at reasonable cost, then new routes for export do not need to be constructed. Although drilling activities at the facility are no longer viable, the work space created by the platform is potentially valuable since the structure can host compression, pumping, metering, and export equipment or serve as a fuel station for helicopters.

However, there are always a number of technical constraints associated with structure reuse, some of which are easy to remedy and others which are deal breakers. The production volumes from a tieback might be too small or large for the equipment at site, or the product fluids might not be the right type for the equipment or export pipeline (e.g., sour crude and high CO_2 gas). The structure has to have the capacity to accept new risers and the space/weight to store the subsea control systems and storage tanks, and if the structure is no longer active, it is significantly less viable as a potential host. If there is no significant drilling activity in the region, the value of the structure declines quickly, and since the operator of new discoveries will consider all potential options in development, they may simply prefer a new structure (especially if the field is of modest size).

Idle/marginal structures compete against other structures, both existing and new build, which reduces its potential value and likelihood of reuse. In shallow water, operators are unlikely to rank prospects for drilling based on trying to maintain low-cost infrastructure, whereas in deepwater, such considerations will play a larger role because of the greater capital investments and value of infrastructure (hundreds of millions of dollars vs. a few million dollars) along with the continuing value that such infrastructure holds for future development.

BOX 15.1 Location Location Location

In the initial Na Kika development, six fields were developed simultaneously as tiebacks to a semisubmersible located in 6340ft water depth located approximately centrally between the fields (Fig. 15.12). Subsea production systems were used with one to six wells per field (Rajasingam and Freckelton, 2004). The Galapagos project fields Isabella, Santa Cruz, and Santiago are the most recent tiebacks (Fig. 15.13), and Faunas, Cypress, and Teton are prospects (Salunke and Hamman, 2015).

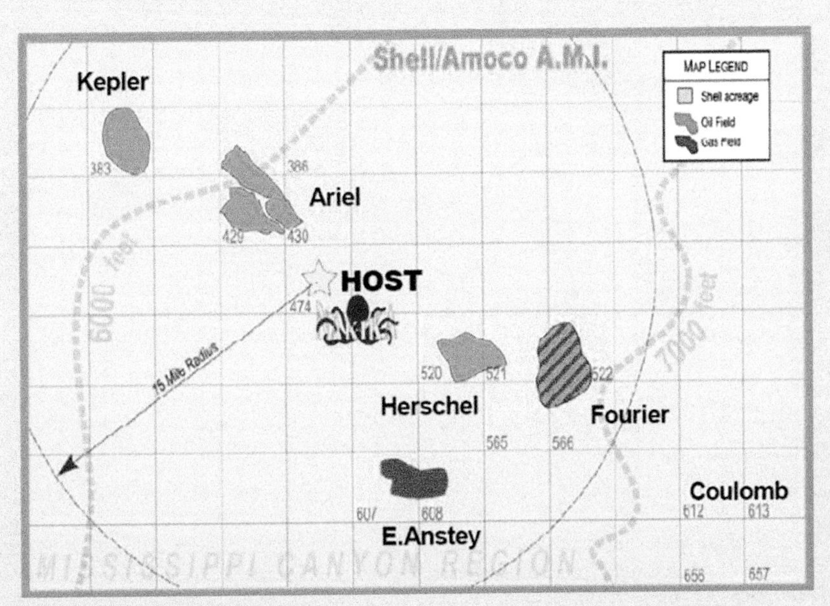

Fig. 15.12 Na Kika field development circa 2004. *(Source: Rajasingam, D.T., Freckelton, T.P., 2004. Subsurface development challenges in the ultra deepwater Na Kika development. In: OTC 16699. Offshore Technical Conference, Houston, TX, May 3–6.)*

Continued

BOX 15.1 Location Location Location—cont'd

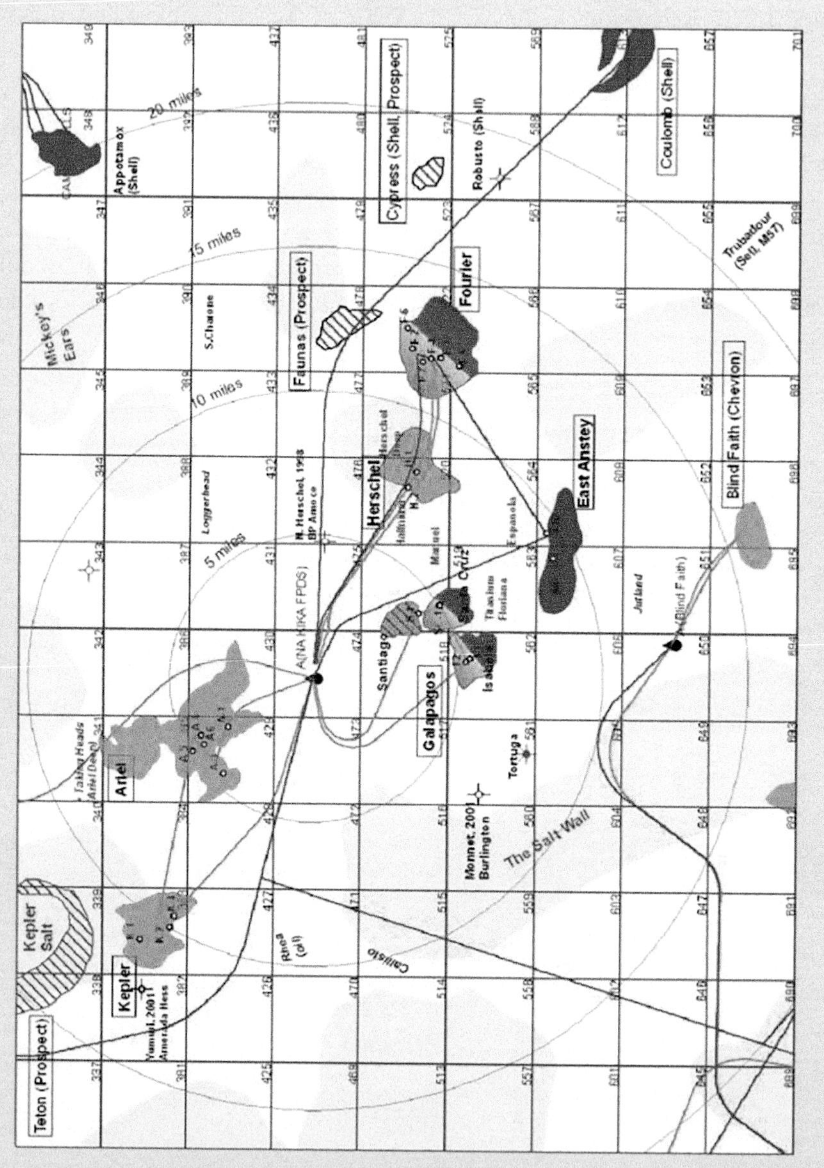

Fig. 15.13 Na Kika area map circa 2015. *(Source: Salunke, S., Hamman, J., 2015. Applying a consistent evaluation approach to thin-bedded sands in a Gulf of Mexico deepwater field. Soc. Petrophys. Well Log Anal. 56 (5), 511–520.)*

REFERENCES

Anderson, S.T., 2017. Risk, liability, and economic issues with long-term CO_2 storage – a review. Natl. Resour. Res. 26 (1), 89–112.

Baginski, V., Larsen, D., Waldman, T., Manrique, E., Ravikiran, R., 2015. Logistic considerations for safe execution of offshore chemical EOR projects. In: SPE 174698. SPE Enhanced Oil Recovery Conference. Kuala Lumpur, Malaysia, Aug 11–13.

Jahn, F., Cook, M., Graham, M., 2008. Hydrocarbon Exploration and Production, 2nd Ed. Elsevier, Amsterdam, the Netherlands.

Jordan, M.M., Sjuraether, K., Collins, I.R., Feasey, N.D., Emmons, D., 2001. Life cycle management of scale control within subsea fields and its impact on flow assurance, Gulf of Mexico and the North Sea basin. In: SPE 71557. SPE Technical Conference and Exhibition. New Orleans, LA, Sep 30–Oct 3.

Kheshgi, H.S., Thomann, H., Bhore, N.A., Hirsch, R.B., Parker, M.E., Teletzke, G.F., 2011. Perspectives on CCS cost and economics. In: SPE 139716. SPE International Conference on CO_2 Capture, Storage, and Utilization. New Orleans, LA, Nov 10–12.

Morgenthaler, L.N., Fry, L.A., 2012. A decade of deepwater Gulf of Mexico stimulation experience. In: SPE 159660. SPE Annual Technical Conference and Exhibition. San Antonio, TX, Oct 8–10.

Rajasingam, D.T., Freckelton, T.P., 2004. Subsurface development challenges in the ultra deepwater Na Kika development. In: OTC 16699. Offshore Technical Conference, Houston, TX, May 3–6.

Raney, K., Ayirala, S., Chin, R., Verbeek, P., 2011. Surface and subsurface requirements for successful implementation of offshore chemical enhanced oil recovery. In: OTC 21188. Offshore Technical Conference, Houston, TX, May 2–5.

Salunke, S., Hamman, J., 2015. Applying a consistent evaluation approach to thin-bedded sands in a Gulf of Mexico deepwater field. Soc. Petrophys. Well Log Anal. 56 (5), 511–520.

CHAPTER SIXTEEN

Benefits, Costs, and Risks

Contents

Abstract

The benefits, costs, and risks associated with maintaining offshore infrastructure are described and illustrated using several examples. The trade-offs and uncertainties are described in general terms rather than trying to resolve, quantify, or evaluate system-specific issues which are, of course, impossible to perform without additional detailed data. Valuing benefits for hypothetical scenarios without understanding system constraints, cost and risk requirements, and technological hurdles is of limited value. The role government policy plays in encouraging offshore activity and managing the diverse range of operators concludes the chapter.

Decommissioning Forecasting and Operating Cost Estimation
https://doi.org/10.1016/B978-0-12-818113-3.00016-3

16.1 POTENTIAL BENEFITS

There are many potential benefits associated with maintaining structures or extending their useful life, but each project and its potential benefits are site-, time-, and operator-specific. Unless supported with detailed project-specific data, benefits cannot be evaluated quantitatively. Potential benefits may arise in field redevelopments, life extension projects, integrated asset modeling, Rigs-to-Reefs programs, pipeline reuse, alternative marine applications, and enhanced oil recovery. Not all of these opportunities will be viable or economic, and each is associated with its own costs, risks, and limitations. The following examples are meant to be illustrative of the type of benefits that can arise.

16.1.1 Field Redevelopment

Oil and gas companies pursue investment opportunities that make economic sense and that satisfy their risk-reward criteria. In field redevelopment, the operator seeks to identify new targets to drill within the field where the incremental revenue from drilling is expected to exceed the cost of completion. Many opportunities commonly arise, and hundreds of examples have been reported of operators applying new technology to improve the interpretation and understanding of old producing fields. Subsea fields present additional challenges (Stright et al., 2012).

Cost, risk, and degree of difficulty for prospects vary significantly and are often ranked in drilling campaigns. In the early 1990s, horizontal drilling became widespread in the GoM and provided new profit opportunities in mature fields. Horizontal reentries drill new horizontal wells by sidetracking from existing production casings and enjoy lower drilling costs than setting conductor, surface, and production casings in new wellbores (Batchelor and Mayer, 1997). However, geology and target geometry cause horizontal drilling candidates to differ appreciably in risk and degree of difficulty. For example, drilling up dip prospects in thin dipping beds on a salt flank structure has greater risk than prospects contained in a low-relief faulted anticline.

The technologies to identify prospects and improve recovery have changed significantly over time, but the strategy is always the same: invest in those opportunities consistent with corporate risk–reward criteria and rates of return that add value to shareholders. Do not pursue opportunities where the costs and risks outweigh the benefits. Generate more discounted revenue than the cost of the investment.

Some redevelopment projects are successful and result in additional recovery at a suitable return, while other projects are not successful. What is considered successful will vary with the operator, but to create value, the investment must yield a return that satisfies company thresholds or a positive net present value. Eventually, new opportunities at a given field dwindle or do not satisfy company requirements. Operators may sell their interest before this time to a company interested in marginal production or redevelopment or may decide for liability reasons that it is better to decommission the facilities themselves. Assets commonly pass "down the food chain" as transfers are known colloquially.

Example. Eugene Island 188

The Eugene Island 188 field was discovered in 1956 as a shallow piercement-type salt diaper located in blocks EI 188, 189, 190, and 192 in 70 ft water depth (Mason et al., 1997). The field was developed in the 1960s, redeveloped in 1983–84 and again in 1992 as indicated by the production bumps (Fig. 16.1). From 1956 to 1975, cumulative production was 39.4 MMbbl oil and 53.7 Bcf gas, and through September 2017, the field had produced 60.8 MMbbl oil and 234.6 Bcf gas. Remaining reserves circa 2017 is estimated at 8.2 MMbbl and 30.3 Bcf gas.

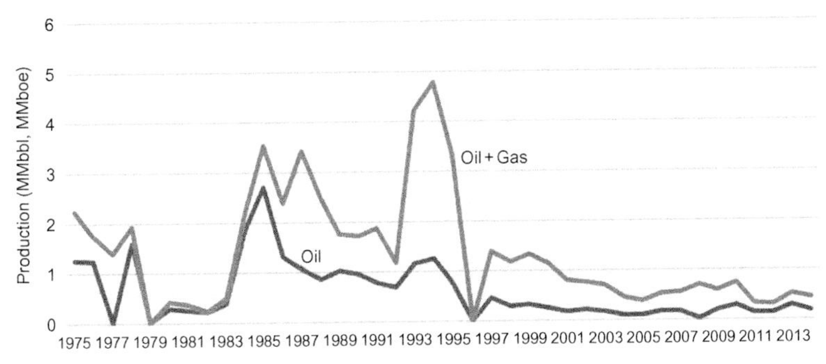

Fig. 16.1 Eugene Island 188 field oil and gas production plot, 1975–2015. *(Source: BOEM 2017.)*

In 1991–92, a field study was performed to identify investment opportunities, and all pay and potential pay zones were remapped using a new seismic survey. Five well locations were identified and drilled resulting in four successful wells and one dry hole and a large increase in gas production. Recoverable reserves from the drilling program were estimated at 10 MMboe, and the operator considered the investment a success.

Example. Main Pass 73

The Main Pass 73 field was discovered in 1974 in 135 ft water depth and started production in 1979. The field consists of hydrocarbon-bearing sands located in blocks MP 72, 73, and 79 truncated against steeply dipping salt domes. Circa September 2017, cumulative production was 52.8 MMbbl oil and 263 Bcf gas (Fig. 16.2), with remaining reserves approximately 4 MMbbl and 12.6 Bcf.

In 2007, Energy XXI acquired the field, and using ocean bottom node technology and sophisticated image processing algorithms, more accurate and detailed earth models were created to support prospect generation and new drilling (Ammer et al., 2015). Typically, the salt bodies defining the reservoir edges are not well imaged on seismic data, making the accurate mapping of the producing reservoir difficult (Fig. 16.3). The revised model indicated smaller detached salt bodies and led to a new and optimistic interpretation of the producing sands (Fig. 16.4). Two new prospects located updip to older well penetrations were drilled in 2011 and completed as producers. The drilling success resulted in the identification of several new prospects located near the boundary of the salt dome.

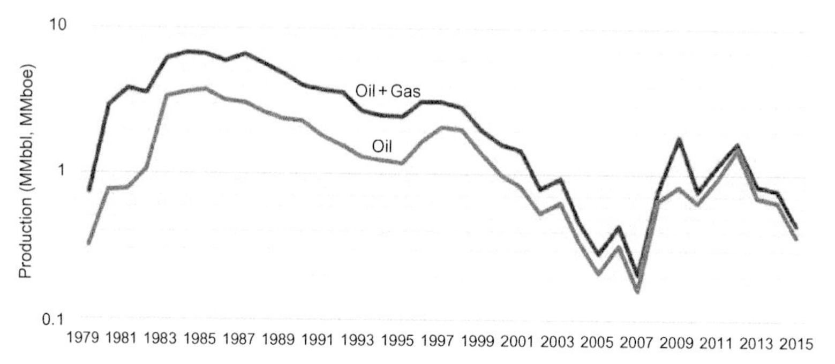

Fig. 16.2 Main Pass 73 field oil and gas production plot, 1979–2015. *(Source: BOEM 2017.)*

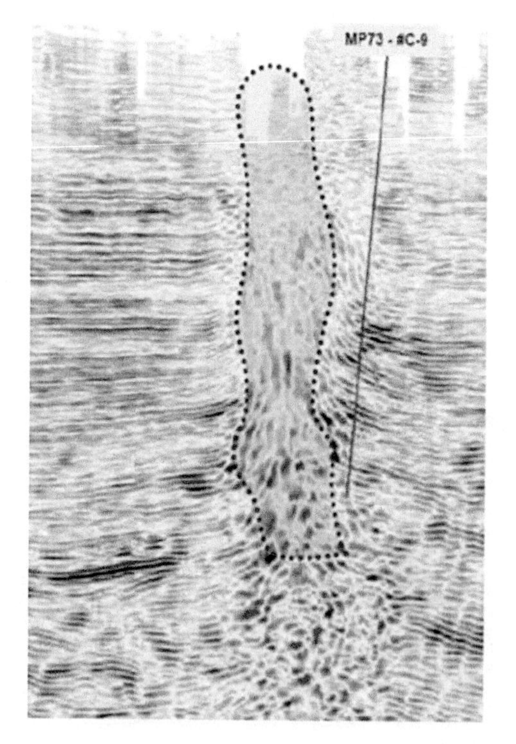

Fig. 16.3 Salt body image at Main Pass 73 field circa 2007. *(Source: Ammer, A., Saunders, R., Henry, C., Wilkinson, T., Frismanis, M., Bradshaw, M., Codd, J., Kessler, D., 2015. New life to an old field – Main Pass 73 Gulf of Mexico shelf. In: SEG New Orleans Annual Meeting, pp. 1807–1812.)*

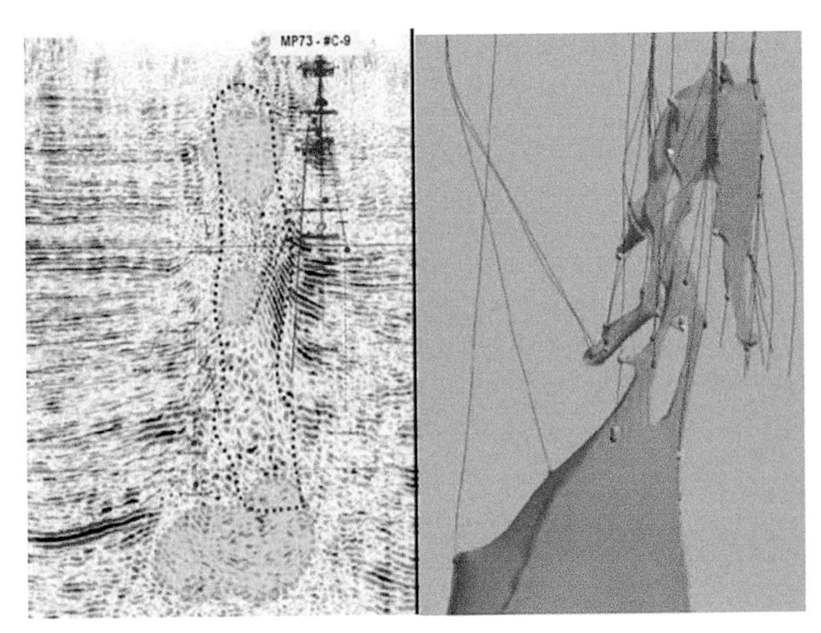

Fig. 16.4 Salt body image at Main Pass 73 field and model reconstruction circa 2011. *(Source: Ammer, A., Saunders, R., Henry, C., Wilkinson, T., Frismanis, M., Bradshaw, M., Codd, J., Kessler, D., 2015. New life to an old field – Main Pass 73 Gulf of Mexico shelf. In: SEG New Orleans Annual Meeting, pp. 1807–1812.)*

16.1.2 Life Extension

Structures that are located in an area of active drilling may extend their life by hosting subsea tiebacks if nearby discoveries arise and the outcomes of negotiation between the parties (if host and reserves owners are different) are successful. Tiebacks require flow lines and umbilicals to reach the host, and additional processing equipment. New export pipeline may or may not be required. If processing capacity is available, its use would result in savings for the operator, but the size of the savings will vary between projects. Platform owners are always interested in new ways to add revenue to existing assets, and third-party fees are simply another form of revenue, while pipeline owners are interested parties since they desire to keep their pipelines full to maintain a steady income. Resource owners are interested parties if the tieback is the best (or only) option for development.

Example. Thunder Hawk

The Thunder Hawk field was developed in 2009 using a semisubmersible production unit in Mississippi Canyon in 6060ft water depth with nameplate production capacity of 45 Mbopd and 70 MMcfpd (Yoshioka et al., 2016). Initially, one well was connected, and two flow lines were tied back with a single-control umbilical. After 5 years of production and without any new wells drilled in the field, available

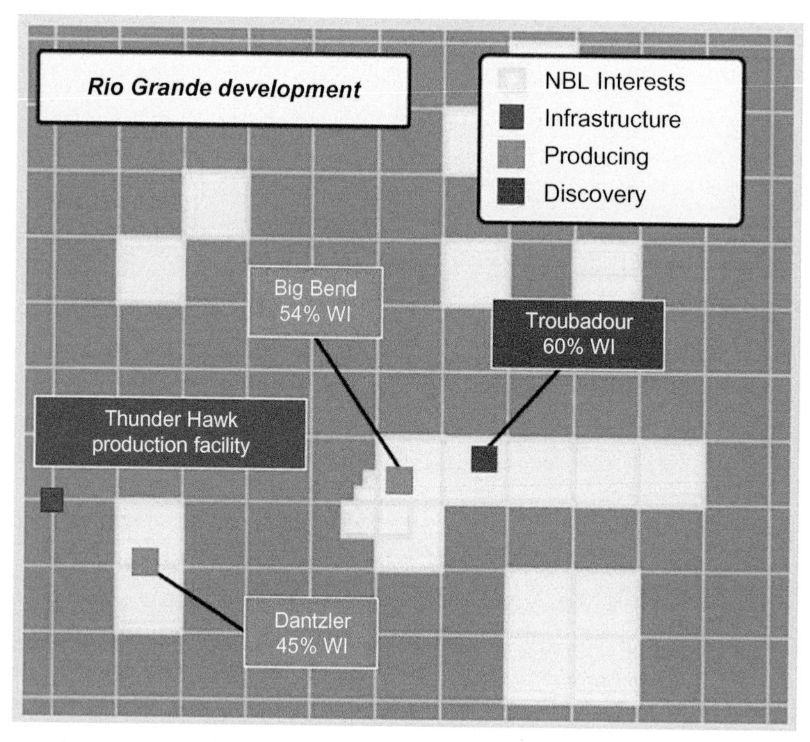

Fig. 16.5 Rio Grande fields are tied back to the Thunder Hawk production facility. *(Source: Noble Energy, BOEM 2016.)*

production and export capacity created an opportunity for a third-party tieback. In 2015, the Big Bend and Dantzler fields located about 12 and 6 mi away in 7200 ft water depth were tied back to Thunder Hawk using dual pipe-in-pipe flow lines and insulated steel catenary risers (Fig. 16.5). Production capacity at the host was expanded to 60 Mbopd to accommodate well production, and gas-lift capability was added.

16.1.3 Gas Lift

The most common types of artificial lift for offshore oil production in the GoM are gas-lift and downhole pumping. In gas-lift systems, gas is injected directly into the wellbore to lower the hydrostatic head. The fundamental mechanism behind gas lift is that the gas injected in the tubing reduces the density of the fluids that act to lower the flowing bottom-hole pressure. Lowering bottom-hole pressure increases the pressure difference between reservoir and wellbore and increases flow. Gas compression depends on the source pressure, and since gas lift is essentially a closed-loop system except at start-up where another source of gas (e.g., nitrogen) may be required, little gas is consumed in operations. Power generation is required to drive downhole electric pumps.

Gas lift is common throughout the shallow-water GoM and is also used in the deep-water GoM at wells and at the base of production and imports risers to help lift the fluids through the water column (Everitt, 1994; Stair, 1999; Borden et al., 2016). In gas lift, gas is delivered from surface compressors by way of the annulus between the casing and production tubing to a series of gas-lift mandrels (conduits), which are positioned within the production tubing string. When a predetermined gas pressure is reached, valves within the mandrels open, causing gas to be "injected" into the production tubing, thus reducing the density of produced fluid and enhancing/enabling flowback to the surface. If the platform has formation gas available in excess of utility fuel requirements and is in the volumes required, then changing the tubing strings and adding horsepower and a compressor package can usually be accommodated. Capital investments are modest, and gas-lift systems are reliable and can be maintained with wireline services that reduce maintenance cost. If adequate gas supply is not available at the platform, then a flow line supplying gas from a nearby platform will need to be laid and the gas purchased which will reduce profits from the investment. Automation is common to maintain performance on complex systems (Reeves et al., 2003; Borden et al., 2016).

Example. Amberjack gas lift redesign

The Amberjack oil field was developed in 1991 as a single jacket platform in a water depth of 1030 ft in Mississippi Canyon. In 2003, the field had 34 dry tree wells, six of which were dual completions and 27 of which employed gas lift (Hannah et al., 1993; Reeves et al., 2003). Wells drilled from 1991 to 1994 were all on gas lift by 1995, and in 2001, these wells were reported to be declining at a 36% per year average decline rate. In 2002, gas-lift valves were redesigned and installed in 10 wells, and in 2003, a gas-lift automation system was implemented to improve operational decision-making. After implementation, there was a 600 boe/day increase in production for several months, and the payback of the project was less than a year.

Electric submersible pumps are not nearly as popular as gas-lift applications in the shallow-water GoM because they are less reliable and cannot accommodate high-angle wellbores and the subsea safety equipment, but in deepwater applications, booster pumps that sit on the ocean floor are increasingly common for high viscous crudes, and in many recent Lower Tertiary developments (e.g., Perdido, Cascade/Chinook, Jack/St Malo, and Stones), they are an essential part of the development (Kondapi et al., 2017). Typically, ESPs have high failure rates of the electric cable and/or pump operating in a high-temperature environment, and have difficulty with the presence of solids or free gas in the produced fluid. ESPs do not require a gas supply which is an obvious advantage over gas-lift systems.

16.1.4 Integrated Asset Modeling

All producing oil and gas fields represent potential redevelopment opportunities since drive mechanisms are known, artificial lift methods are usually in place (if used), flow line networks are established, and equipment capacity is available. Integrated asset models provide the opportunity to quantify various improvement scenarios and give new life to production. Such options may include gas-lift allocations, surface back pressure optimization, and process facility adjustments to name a few (Ruiz et al., 2015). Reservoir models need to be coupled with oil production, water injection, and gas-lift distribution networks in order to evaluate viable investment opportunities.

Example. Ewing Bank 873 (Lobster)

The Lobster platform was installed in 1994 in Ewing Bank block 873 in 775 ft water depth. The platform was originally owned by Chevron Texaco and Marathon and is currently operated by EnVen Energy Ventures. The platform processes production from the Lobster field (EW 873, 824, 914) and South Timbalier 308 using platform wells and wet trees, as well as three other deepwater fields Seattle Slew (EW 914, 915, 871), Arnold (EW 963), and Manta Ray (EW 1006) as shown in Fig. 16.6. In 2009, a new drilling campaign based on 4-D seismic technology designed to target remaining oil and potentially untested targets reachable from the platform was implemented (Roende et al. 2009). Circa September 2017, the EW 873 field had produced 172 MMbbl oil and 158 Bcf gas (Table 16.1).

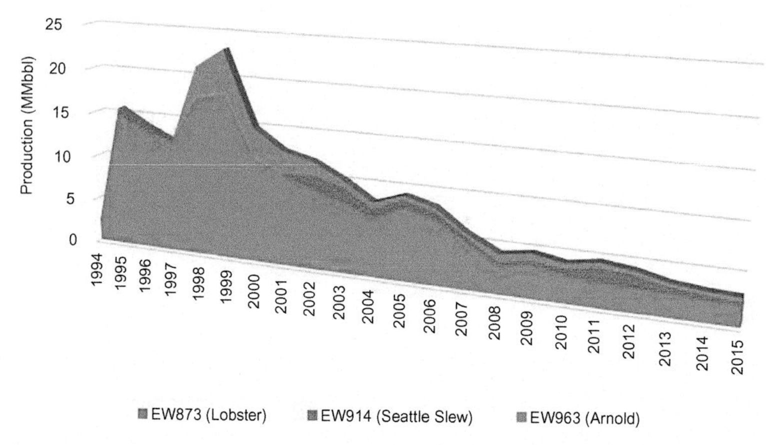

Fig. 16.6 Ewing Bank 873 field and subsea tieback oil and gas production plot, 1994–2015. *(Source: BOEM 2017.)*

Table 16.1 Lobster and Subsea Tieback Cumulative Production Circa September 2017

Name	Field	Operator	Oil (MMbbl)	Gas (Bcf)
Lobster	EW 873	EnVen Energy, Fieldwood	172	157.7
Seattle Slew	EW 914	Walter Oil & Gas	13.4	17.0
Arnold	EW 963	EnVen Energy	22.7	20.0
Manta Ray	EW 1006	Marathon Oil, Walter Oil & Gas	4.7	12.4

Source: BOEM 2017.

Another opportunity frequently requiring an integrated production model in decision-making is subsea well stimulations (Ma et al., 2016). In addition to accelerating recoveries, stimulation can improve recovery volumes and reduce decline rate following treatment. The cost of subsea well intervention is significantly greater than direct vertical access and dry tree well intervention, however, and will likely require upside production to proceed. In a multiwell recovery, additional production from one well may back out production from other wells, adding to the risk of the operation and requiring an integrated model to account for interwell interactions and longer-term production factors.

16.1.5 Rigs-to-Reefs

In the GoM, active Rigs-to-Reefs programs exist offshore Louisiana and Texas where the state accepts liability for reefed structures in designated areas (of federal waters) and the operator splits the cost savings with the state in lieu of transporting the platform to shore, which is set aside in a trust fund to support the administration and management of the reef programs (Kaiser and Pulsipher, 2005). Platforms are decommissioned by complete removal, partial removal, or toppling-in-place, and in Louisiana, Special Artificial Reef Sites have been established to accept "rigs of opportunity" such as destroyed by hurricane events. One of the largest artificial reefs in the world is at the site of a former sulfur mine (Box 13.1).

Louisiana's Rigs-to-Reefs program celebrated its 30th anniversary in 2017, and since its inception, 363 structures have been donated to the program, on average about 13 structures per year. Texas's Rigs-to-Reefs program started in 1990, and through 2017, about 154 structures have been donated or about six structures per year. In total, about 20% of all jacket platforms decommissioned since 1990 have been placed in the Texas and Louisiana reef programs.

Example. Eugene Island 331A platform reefing

In 2008, Hurricane Ike passed through the Eugene Island 330 field and caused significant damage to Shell's EI 331A platform (Box 12.1). Previously a 24-slot production platform, well plug and abandonment work was completed on EI 331A in 2005, and the facility was operated as a pipeline hub for Auger oil production at the time of Ike's arrival (Abadie, 2010).

There was no visible damage to the topside structure, but when the jacket was inspected by divers, three legs were found to be severely damaged, and numerous diagonal braces were buckled, broken, or missing. Shell engineers determined it was not feasible to repair the jacket, and the platform was removed to avoid risk of collapse in future storms (Abadie et al., 2011). Eleven pipelines crossing the structure had to be rerouted or abandoned.

The topsides and deck of the EI 331A platform were removed before the 2009 hurricane season, and in 2010, the platform was towed using the Versabar to an existing Special Artificial Reef Site about 4 mi away

Fig. 16.7 Eugene Island 331A jacket tow using Versabar. *(Source: Abadie, R., 2010. Eugene Island 331A decommissioning – successful execution of an unplanned project. In: Decommissioning and Abandonment Summit Gulf of Mexico. Houston, TX, Mar 28–29.)*

in EI 313 (Fig. 16.7). The jacket was toppled and reefed in one piece to provide the best habitat for marine life and joined several other jacket structures at the site (Fig. 16.8). Underwater remotely operated vehicle cameras documented fish and other marine life following the jacket during its tow operation and reefing.

Platforms act like artificial reefs (Reggio, 1989; Blaine, 2001) and are home to some of the most prolific ecosystems in the oceans (Claisse et al., 2014; Ajemian et al., 2015). They are frequently the preferred destination of recreational fishing and commercial diving (Stanley and Wilson, 1989) and are capable of harboring threatened species, providing reef habitat, boosting recruitment of overfished species, raising ornamental fish and invertebrates, and acting as foraging sites for top-order predators. For recent reviews of the habitat value of platforms and their importance to recreational fishing and diving, see Schuett et al. (2016) and Braddy et al. (2016). Platform reefs contain coral, algae, bacteria, and sponges (Kolian et al., 2017; Sammarco et al., 2004), and tantalizing evidence suggests they can produce antiviral, antibacterial, and anticancer compounds (Rouse, 2009).

16.1.6 Pipeline Reuse

The vast majority of decommissioned pipelines are left in place on the seafloor with their ends plugged and buried three feet below the mudline, although in near-shore areas with potential reserves of hard minerals such as sand, gravel, or shell resources, pipeline removal is now required. In principal, decommissioned or out-of-service pipelines can be reused in development. In practice, however, pipelines are rarely reused after decommissioning since several conditions need to be satisfied. Only under very special

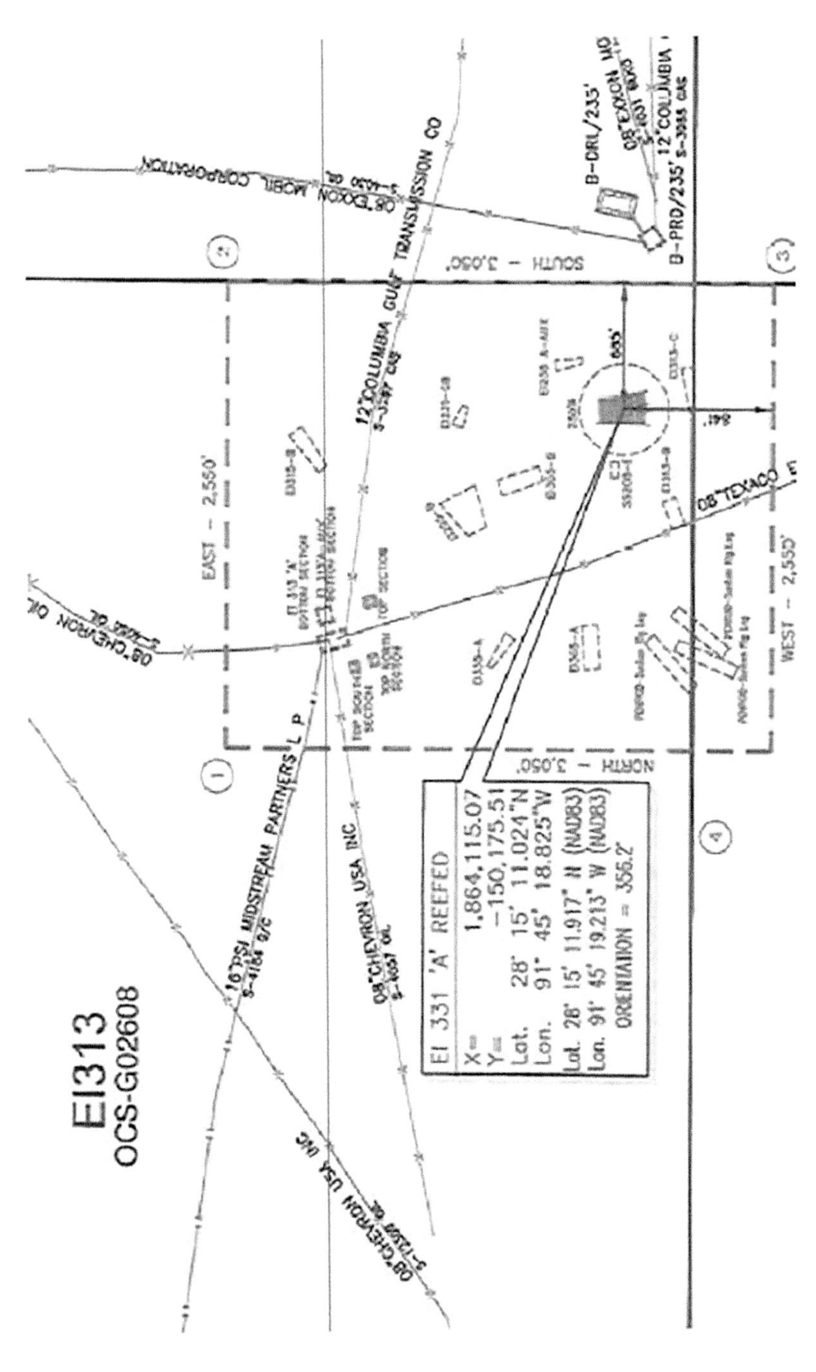

Fig. 16.8 Eugene Island block 313 Special Artificial Reef Site reefing survey. *(Source: LDWF.)*

circumstances, if the pipeline is of sufficient length and quality and located near the field development, and if the savings are significant, does the potential for reuse arise. Reuse projects are exceptional one-off events.

Example. Lucius and South Hadrian

The oil export pipeline for the Lucius field in Keathley Canyon is an 18 in, 147 mi pipeline consisting of three segments: (1) 74 mi of 18 in new build pipe from the Lucius spar to the Phoenix pipeline, (2) test and reuse of a 47 mi section of the out-of-service Phoenix gas pipeline converted to oil service, and (3) 26 mi of 18 in new build pipeline to South Marsh Island block 205 transportation platform (Fig. 16.9). About one-third of the oil

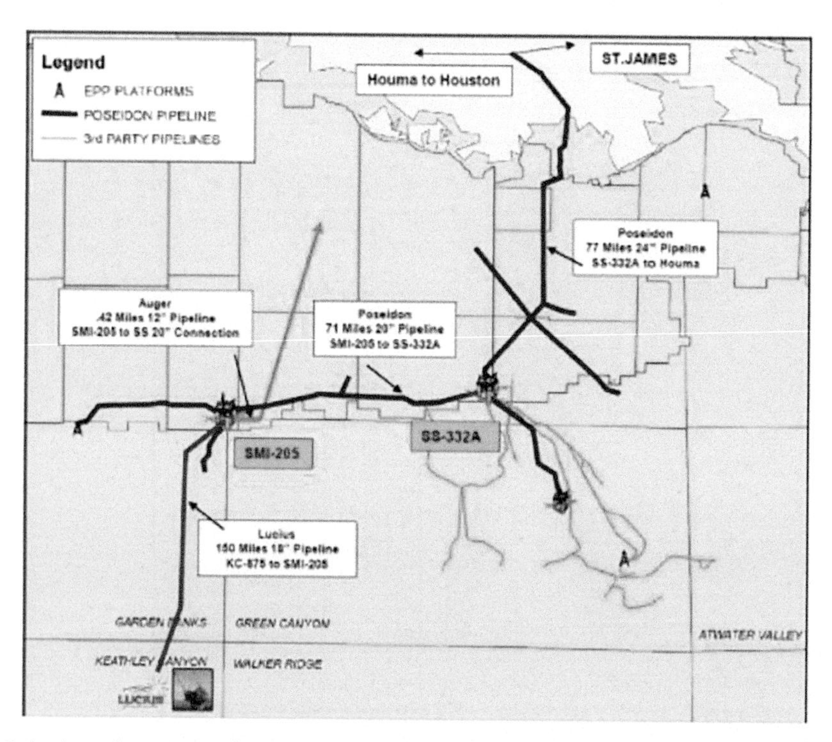

Fig. 16.9 Lucius oil export and Poseidon system. *(Source: Schronk, L., Biggerstaff, G., Langmaid, P., Aalders, A., Stark, M., 2015. Lucius project: export oil and gas pipeline development and execution. In: OTC 26006. Offshore Technology Conference. Houston, TX, 4–7 May.)*

pipeline reused abandoned gas pipeline from Anadarko's Red Hawk spar, which was taken out of service in 2008. Pipeline reuse required thorough cleaning of the pipeline, caliper runs, dehydrating, ROV inspection, cathodic protection checks, and span data analysis for fatigue and stress. First oil was delivered in 2015 (Schronk et al., 2015).

16.1.7 Alternative Uses

Section 388 of the Energy Policy Act of 2005 (Public Law 109-58) allows retired oil and gas platforms to be redeployed for alternative uses such as the production of wind, wave, and current energy; sustainable fisheries; or any other marine-related purpose (NOAA, 2018).

Mariculture and offshore wind have been discussed and proposed in the GoM for many years, but no ventures to date have assembled the capital required to test their business plans and establish commercial operations. Offshore developments require a high-value resource to exploit to cover the high-cost and high-risk operating environment. The reuse of platforms in the GoM for non-petroleum related activities is generally constrained by poor economics, difficult logistics, and high risk.

Fish are not that valuable (yet), offshore wind in the region is not very strong, and low regional population densities mean that markets are relatively distant and conspire to prevent ventures from gaining much traction (e.g., Kaiser et al., 2010). Hence, although there is always the potential to reuse GoM structures for alternative uses, economic and risk factors have prevented ventures progressing to the financing stage.

16.1.8 EOR via CO_2 Injection

Enhanced oil recovery (EOR) is the third stage of oil recovery after primary and secondary methods have been employed and involve the injection of various materials beyond water and miscible gas into a reservoir to extract additional oil. EOR is most frequently used in fields with heavy oil, poor permeability, and irregular fault lines, and CO_2 injection is the most popular method employed in the United States followed by thermal methods. In 2010, EOR projects were producing about 300,000 bopd and accounted for about 5% of total US production. About half of all EOR projects were CO_2 and the other half thermal recovery. In the Gulf Coast, the Jackson Dome in Mississippi supplies over 1 Bcf of CO_2 daily to 15 fields, and oil production associated with these projects was estimated at 38,000 bopd or about 14 MMbbl per year in 2011 (Davis et al., 2011).

Although offshore EOR volumes are intriguing to contemplate, they are speculative and highly uncertain. CO_2 project economics are difficult and require several conditions to be viable, including amenable reservoirs, reliable and long-term CO_2 supply, CO_2 pipeline, appropriate well patterns, waterflooded fields, and well-capitalized firms (Bondor et al., 2005; Koperna and Ferguson, 2011).

Offshore, CO_2 EOR projects are much more difficult and expensive than onshore projects, and circa 2018, there were no commercial projects implemented or under construction worldwide. Offshore, CO_2 EOR operations are expensive due to the expense of transporting CO_2 to location, the high cost of processing and recompressing produced gas in offshore settings (recycle), the high cost of drilling and reworking offshore wells, the corrosive nature of CO_2, and the high cost of offshore operations and maintenance.

The performance of CO_2 EOR projects is highly uncertain because reservoir behavior is difficult to model and predict, that is, after filling in the pore volume with CO_2, how much oil will be recovered and how fast, and what will the recycle rate be? Analogs and normalized production curves are usually applied to estimate operational results, but the uncertainties remain high, which will obviously limit interest from investors. Over the past half century, only a handful of offshore CO_2 pilot projects have been conducted worldwide.

Example. Carbon storage and EOR projects

The Sleipner and Snøhvit natural gas fields offshore Norway and the K12-B gas field offshore Netherlands have high concentrations of CO_2 (5–13 vol%) that must be processed to satisfy the 2.5 vol% limit for European natural gas pipeline specifications. The CO_2 is separated from the produced natural gas and reinjected into sandstone formations for storage. Statoil operates the Norway facilities, and Gaz de France Production operates the K12-B field. These are carbon storage operations and exceptional due to the high CO_2 content of the produced gas.

There have been nine offshore CO_2 EOR projects and one EGR project reported by operators, all short-term pilot projects considered technically successful but not implemented on a large scale after the pilot program was terminated and, thus, not economically viable (Sweatman et al., 2011). At Weeks Island, Bay St. Elaine, Quarantine Bay, and four oil fields along the Louisiana coast in the 1980s, CO_2 was transported inside refrigerated tanks on barges pushed by tugboats to site and then injected into wells. At Timbalier Bay, CO_2 was transported via tanker trucks to a pipeline station at Port Fourchon where it was pumped to a compressor and onto an injection well. One project offshore Malaysia (Dulang) employed water alternating gas injection.

In a 2014 study sponsored by the Department of Energy, DiPietro et al. (2014) identified oil reservoirs in the GoM amenable to CO_2 EOR and simulated a CO_2 flood for each reservoir using production and cost estimates to determine which oil fields were economic for CO_2 EOR. A total of 391 fields out of 531 oil fields were screened out based on size (applying original oil-in-place thresholds for reservoirs <10 MMbbl and fields <50 MMbbl as too small), residual oil saturation, and well spacing. For the remaining 140 oil fields and 696 reservoirs, CO_2 EOR simulations were performed with cash flow models to assess the economics at a reservoir level similar to an earlier study by Brashear et al. (1982). A minimum 20% rate of return was required to be considered economic.

To perform the economic calculations, the study assumed that groups of proximate fields will be served by a \$1.5 billion CO_2 pipeline originating at Baton Rouge, Louisiana, supplying 1 Bcf/year CO_2 at a levelized transportation cost \$1.06/Mcf CO_2 (\$20/mt), oil price of \$90/bbl, CO_2 price of \$1.6/Mcf, 18.75% royalty, and a 20% rate of return before taxes. Economically recoverable resources were estimated at 800 MMbbl with a 5.8 Tcf CO_2 demand, 390 MMbbl in shallow water, 80 MMbbl in deepwater, and 340 MMbbl at undiscovered fields. Under a scenario with next-generation performance assumptions, economically recoverable resource values are larger by an order of magnitude.

Because many preconditions are required for success, only a few offshore CO_2 EOR projects have been pursued worldwide, and they have all been short-term pilot (feasibility) projects that were not implemented because of economic reasons. Hypothetical studies do not translate into reality because of the difficult economics and complex technical issues associated with modeling outcome and reducing uncertainty. Projects are usually considered successful in a technical sense, but technical success and profitability are not the same. In short, there are too many complexities and difficulties associated with offshore CO_2 projects that need to be resolved, which will prevent commercial development in the midterm to long-term future.

16.2 COSTS AND RISKS

There are various costs associated with maintaining idle infrastructure, including inspection cost, lighting cost, maintenance and repair cost, and insurance cost. Inspections must be performed periodically above and below water, and maintenance cost will vary with the level of corrosion and size of the structure and business plan of the operator.

16.2.1 Inspection Requirements

The OCS Lands Act authorizes and requires the BSEE to provide for both an annual scheduled inspection and periodic unscheduled (unannounced) inspection of all oil and gas operations on the OCS to monitor the adequacy of the corrosion protection system and determine the condition of the platform. BSEE regulations require operators perform in-service inspection intervals for fixed platforms according to API Recommended Practice 2A-WSD (NTL 2009-G32).

The time interval between platform inspections depends upon exposure category (L-1, L-2, L-3), survey level (Level I, Level II, Level III), and manned status (Table 16.2). The three levels of exposure to life safety are manned and nonevacuated (L-1), manned and evacuated (L-2), and unmanned (L-3). Consequences of failure encompass damage to the environment, economic losses to the operator and the government, and public concerns. Economic losses to the operator can include the costs to replace, repair, and/or remove destroyed or damaged facilities; costs to mitigate environmental damages due to released oil; and lost revenue.

A Level I survey is required to be conducted for each platform at least annually and a grade assigned to the coating system. A Level II survey is required for each platform at the minimum survey interval, at least every 3 years for L-1 platforms and at least every 5 years for L-2 and L-3 platforms. A Level III survey is required at least every 6 years for L-1 platforms and at least every 11 years for L-2 platforms.

Table 16.2 In-Service Inspection Intervals for Fixed Manned and Unmanned Platforms

Level	Exposure Category L-1	L-2	L-3
I	1 year	1 year	1 year
II	3 years (5 years)	5 years (10 years)	5 years (10 years)
III	6 years (6 years)	11 years (11 years)	

Note: Unmanned platform inspection intervals are denoted in parenthesis.
Source: NTL 2009-G32.

16.2.2 Maintenance Cost

There is a wide variation in the maintenance programs operators perform to protect their assets. For new capital intensive facilities, operators are likely to invest in continuous, year-round, or seasonal maintenance programs, while for idle structures, painting will only be performed on an as-needed basis or as dictated by deficiencies found in BSEE audits. The costs to maintain structures that are not producing or serving an active/useful role in operations primarily include inspections (periodic and mandatory) and, if applicable, painting and blasting repair and insurance. Personnel must be transported to site to perform activities, and if the structure is audited and found to have safety or other hazards, additional cost will be incurred.

Painting and blasting operations on inactive structures are less likely to be prioritized relative to active structures, and so, as rust develops and steel degrades, structures may become hazardous to visiting personnel. If there is no maintenance painting, safety and hazardous conditions may arise in operations and decommissioning activity.

Operators allocate their "paint budget" across their fleet of structures. For deepwater platforms and floaters, maintenance painting may range from $500,000 to $1.5 million per year per structure after the service life of the coating is reached (perhaps 7–12 years after application). For shallow-water platforms, costs are significantly lower and depend on the extent and distribution of paint degradation and on exposure category. Activity-based methods to estimate painting and maintenance cost are described in Chapter 23.

16.2.3 Lighting Cost

All platforms must have navigation lighting at all times between sunset and sunrise from the time of installation according to federal regulations. Periodic inspections are made to ensure compliance with lighting requirements. Inactive structures must maintain their lighting systems or will be found in violation of regulations. Structures are designated into one of three classes as A, B, or C depending on the water depth and marine commerce traffic routes. The number of lights required depends on platform size. If one side of a platform is <30 ft wide, one light is required. If one side of the platform is >50 ft, four lights must be installed at each corner of the platform. If the platform is between 30 and

50 ft wide, two lights must be installed on diagonal ends. On class A structures, lighting must be visible for 5 nm.

16.2.4 Hurricane Exposure

Many of the damaged and destroyed structures in the paths of hurricanes that passed through the GoM from 2005 to 2008 were inactive and cleanup activities have taken nearly a decade (Box 16.1). Now only the most expensive and difficult destroyed

BOX 16.1 Hurricane Clean-Up, 2004–2008: Job (Almost) Done

From 2004 to 2008, five major hurricanes (Ivan, Katrina, Rita, Gustav, and Ike) destroyed 181 structures and 1670 wells in the Gulf of Mexico. Through 2016, 167 of the destroyed structures have been decommissioned, representing 92% of the destroyed structure inventory and 97% of the associated wellbores (Table 16.3). The decommissioning sector's ability and success in dealing with unprecedented levels of destruction is a notable accomplishment for the industry.

Table 16.3 Hurricane Destroyed Structures and Wells Remaining Circa 2017

		Destroyed		Remaining	
	Year	Structures	Wells	Structures	Wells
Ivan	2004	7	136	1	19
Katrina/Rita	2005	114	1000	7	19
Gustav/Ike	2008	60	534	6	9
All		181	1670	14	47

Source: BOEM.

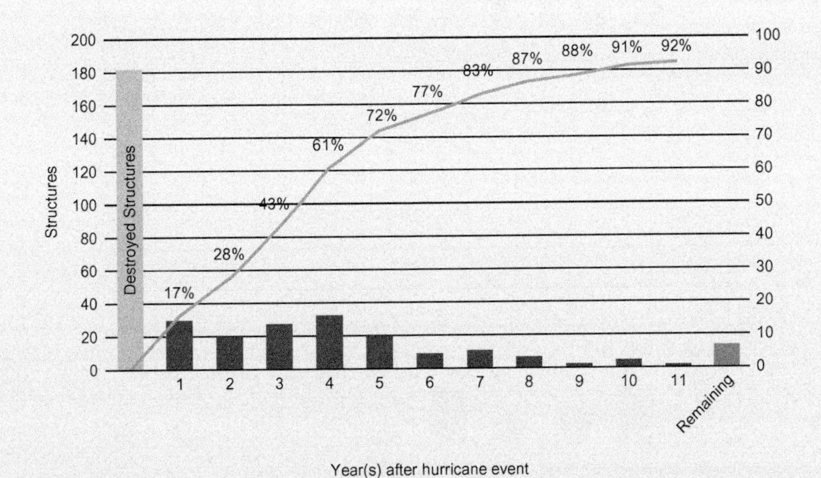

Fig. 16.10 2004–08 hurricane-destroyed structures and structure decommissioning activity, years after event circa 2017. *(Source: BOEM 2017.)*

Continued

BOX 16.1 Hurricane Clean-Up, 2004–2008: Job (Almost) Done—cont'd

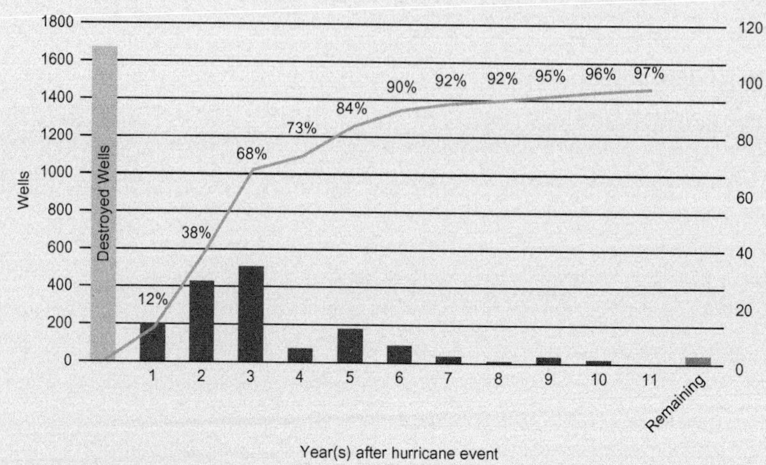

Fig. 16.11 2004–08 hurricane-destroyed structures and well abandonment activity, years after event circa 2017. *(Source: BOEM 2017.)*

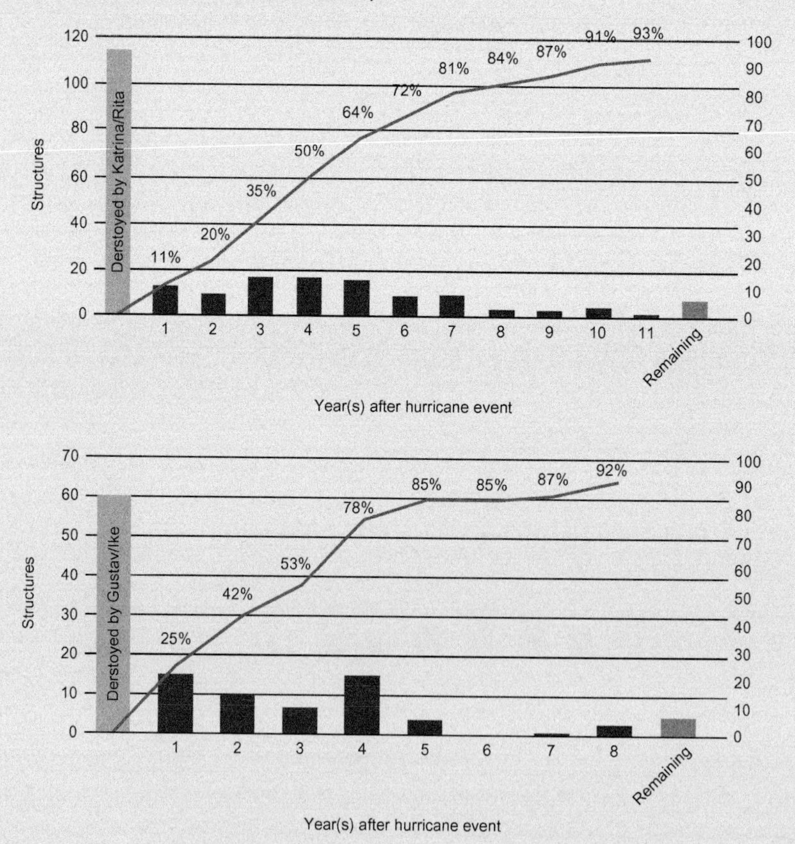

Fig. 16.12 Hurricane-destroyed structures and cleanup activity by hurricane, years after event circa 2017. *(Source: BOEM 2017.)*

Continued

BOX 16.1 Hurricane Clean-Up, 2004–2008: Job (Almost) Done—cont'd

Hurricane-destroyed structures are either toppled to the seafloor, leaning, or physically damaged beyond repair and are reported to the government a few months after each event (Kaiser, 2014; Kaiser and Narra, 2017). Structures that were significantly damaged and where operators later decided to perform early decommissioning are not included in the counts.

All the destroyed wells and structures from Hurricane Ivan have been decommissioned except Taylor Energy's Mississippi Canyon 20 structure and 19 of its remaining wells (Figs. 16.10–16.12). Hurricanes Katrina and Rita destroyed 114 structures and 1000 wellbores; seven structures remain circa March 2017. Hurricanes Gustav/Ike destroyed 60 structures and 534 wells; circa March 2017, six structures remain.

structures remain (Kaiser and Narra, 2017). In one exceptional case, Taylor Energy's Mississippi Canyon 20 toppled platform is unlikely to ever be fully decommissioned because of the difficulty, uncertainty, and hazards of operations (Box 14.1). Hurricane cleanup cost can be up to ten times greater than normal decommissioning activity, and several operators went bankrupt or left the region after being negatively impacted by hurricane events. Today, there are significantly less structures in the shallow-water GoM and fewer idle structures, so the impact of future destructive hurricanes will be smaller, but for those idle structures remaining, they will continue to pose a liability for the operator and potential environmental risk if wells are not permanently abandoned.

16.2.5 Environmental and Safety Risk

Structures that no longer serve a useful purpose expose operators to hurricane risk and present collision risk to vessels and tankers, which may lead to environmental issues from leaks, fires, and safety concerns. If a hurricane occurs and damages/topples the structure, additional cost will be incurred in cleanup and potential environmental damage if wells are not permanently abandoned.

In the 21st edition of API RP2A, a consequence-based design criterion was introduced based on two classes of risk, those associated with life safety and those associated with consequences of failure. If the platform is of modest likelihood of failure and low consequence (e.g., old, unmanned, and with most wells abandoned), much of the risk of holding the structure in inventory is an economic one rather than a hazard to human life or the public interest. If the platform is of modest likelihood of failure and high consequence (e.g., old, manned, no wells abandoned, and no subsurface safety valves), then risk to the operator and public increases.

16.2.6 Bankruptcy Risk

A large number of GoM operators declared bankruptcy from 2015 through 2018 (Table 16.4). In many cases, companies were successfully reorganized with stock holders incurring the majority of losses (Haynes and Boone, 2018), but in some cases could not continue operations under a reorganized company, presenting a risk to the government of taxpayer-funded decommissioning.

Bankruptcy risk among small independents is higher than large independents and majors and is reflected in their lower credit ratings that quantify on an aggregate (gross) basis the probability a company will default on its debt obligations. If operators go bankrupt or are otherwise unable to pay for their decommissioning obligations, there is a risk that the public will bear the cost of cleanup. The risk is relatively small, however, since the regulatory structure includes a number of safeguards that reduce the exposure of the government.

The risk to the taxpayer is a function of the number of jointly and severally liable lessees in the chain of title and the financial strength of the individual lessees. If a lease includes a major or a large independent, there is high financial strength in the chain of title, and the risk of default for these leases is near zero. If there is no major or large independent, the risk is tied to the financial condition of the other independents in the chain of title. All of the companies in the chain would have to default simultaneously for the government to be required to pay the liabilities. The greater the number of companies in the chain of title for the lease, the less likely it is that they would all default together and the less risky the lease. The riskiest leases are those with only one lessee in the chain of

Table 16.4 Corporate Bankruptcies With Significant Gulf of Mexico Presence, 2015–18

Filing Date	Debtor	Secured (MM $)	Unsecured (MM $)
08/11/2015	Black Elk Energy Offshore	69	76
10/26/2015	RAAM Global Energy Company	304	
03/24/2016	Whistler Energy II, LLC	132	60
04/14/2016	Energy XXI Ltd	1543	1206
05/16/2016	SandRidge Energy, Inc.	1830	6429
08/11/2016	Bennu Titan, LLC	188	4
08/12/2016	Northstar Offshore Group, LLC	98	34
11/30/2016	Bennu Oil & Gas, LLC	495	229
12/14/2016	Stone Energy Corporation	353	1092
06/02/2017	Rooster Energy, LLC	52	0.2
06/08/2017	Deep Operating, LLC	0.3	1.1
06/29/2017	King's Peak Energy, LLC	9	14
12/14/2017	Cobalt International Energy, LLC	1476	1429
02/15/2018	Fieldwood Energy, LLC	3287	57

Source: Haynes and Boone, LLP, 2018. Oil patch bankruptcy monitor. Mar 31.

title. In such case, the risk of default is tied to the individual lessee's financial strength. Sole lease ownership (no title chain) represents the highest-risk leases.

According to a 2016 Opportune study (Sherman et al., 2016), there were 2243 leases in the GoM with decommissioning liabilities estimated at about $24 billion, and of those leases with liabilities, 408 have no major or large independent in the chain of title with an estimated liability of about $1.4 billion. Decommissioning liabilities at risk represent <5% of the total exposure in the GoM.

16.2.7 Aging Infrastructure

Structures that are no longer producing or useful for operations are unlikely to receive the attention and maintenance of active fleets and may fall into a state of neglect and disrepair (Flint et al., 2011). Operators with large inventories of idle infrastructure may fall behind on maintenance from corrosion and struggle to keep up with audits from the BSEE. Safety and environmental risks may ensue from rusting structures and require additional precautions and cost in decommissioning.

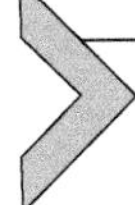

16.3 GOVERNMENT POLICY

16.3.1 End-of-Life Royalty Relief

Under 43 U.S.C. 1337(a)(3)(A), the US Department of Interior may reduce or eliminate the royalty or net profit share specified for producing OCS leases to promote increased production. The basic rule specifies that when royalties exceed 75% of net revenues generated, a lease is becoming uneconomic. Under these circumstances, BOEM may modify the royalty rate to extend the productive life of the lease.

The purpose of royalty relief is to allow operators reasonable financial returns to increase recovery and augment receipts to the Federal Treasury. To qualify for royalty relief, operators need to demonstrate and document the need for royalty relief via supporting information using engineering and economic data.

Royalty relief for end-of-life leases was enacted in March 1999 but has not been popular with operators, indicating that royalty relief near the end of production is not an adequate stimulant for operators to continue with marginal operation or that the program is cumbersome for applicants. End-of-life production is characterized by marginal cash flows and low margins, and the overall benefits of reducing costs via royalty rate reduction/elimination are expected to play a relatively small factor in the economics of operations and decisions of most operators.

16.3.2 Producing in Paying Quantities

In NTL 2008-N09, Extension of Lease and Unit Terms by Production in Paying Quantities, the Minerals Management Service (MMS) stated that they will periodically perform lease-holding reviews on leases with minimal and/or intermittent production to ensure that leases are "producing in paying quantities":

> *Under prudent operator standards and historical precedent, the MMS interprets production in paying quantities…to yield a positive stream of income after subtracting normal expenses (i.e., operating costs), which include the sum of minimum royalty or actual royalty payments, whichever is greater, and the direct lease operating costs.*

MMS stated that regional offices will perform an initial review using standard operating expenses and compare against revenue generation. If revenue generation is less than cost, the regional supervisor will ask operators to provide cost and production data to demonstrate that their leases are producing in paying quantities:

> *If the Regional Supervisor determines that your lease did not produce in paying quantities for a period that exceeded 180 days, MMS either may issue an order to show cause as to why your lease did not expire by its own terms at the end of the 180-day period or may issue a determination that your lease has expired.*

16.3.3 NTL 2010-G05

NTL 2010-G05 dissolved the lease boundary in determining decommissioning time lines and redefined the regulatory requirements at the individual wellbore and structure level by specifying the maximum number of years wells and structures are allowed to remain idle before they have to be abandoned. For wells that have not produced for 5 years or more, operators have 3 years to either permanently or temporarily abandon the well. For structures that have not produced for 5 years or more, operators have 5 years to remove the structure.

16.3.4 Financial Assurance and Bonding

BOEM regulations apply a multitiered financial assurance system to ensure that OCS obligations are met. There are three stages in the life of a lease when financial assurance is required by regulation: (1) lease issuance, (2) approval of an exploration plan, and (3) approval of a development and production plan or a development operations coordination document (30 CFR §§ 556.900(a) and 556.901(a) and (b)).

BOEM has established minimum thresholds for each of nine ratios:

- Cash flow from operations/total debt
- Current ratio
- Earnings before interest and taxes/interest expense
- Quick ratio
- Return on assets
- Return on equity
- Total debt/capital
- Total debt/earnings before interest, taxes, depreciation, and amortization

- Total debt/equity

The number of such thresholds that BOEM requires an operator to exceed to determine financial capacity exists in excess of lease and other obligations.

On 14 July 2016, BOEM issued NTL No. 2016-N01, Requiring Additional Security, to clarify the procedures and criteria used for sole liability properties in determining if and when additional security may be required. On 12 February 2017, BOEM withdrew its NTL No. 2016-N01 orders, and it appears that BOEM has put implementation on hold pending a review of the requirements.

16.3.5 OCS Sediment Resources Policy

In addition to oil and gas, OCS resources include sediment deposits such as clay, silt, sand, gravel, and shell found on or below the surface of the seabed. Public Law 103–426 allows BOEM to negotiate on a noncompetitive basis the right to use OCS sand, gravel, or shell resources for shore protection or beach or wetland restoration projects by federal, state, or local agencies or for use in construction projects funded or authorized by the federal government.

In 2009, the Minerals Management Service implemented the Significant OCS Sediment Resources policy and issued a Notice to Lessees and Operators (NTL 2009 G04) to clarify how it would implement the regulatory requirements aimed at preventing waste of OCS natural resources and identified use conflicts to prevent oil and gas infrastructure from unduly interfering with other uses of the OCS and its surface and shallow subsurface mineral deposits.

In 2016, BOEM and BSEE clarified that pipeline decommissioning on the OCS in areas with potential reserves of hard minerals must be removed. BOEM considers abandoned pipelines in these designated areas to unduly interfere with other uses of the OCS and takes the position that such pipelines must be decommissioned by removal.

REFERENCES

Abadie, R., 2010. Eugene Island 331A decommissioning – successful execution of an unplanned project. Decommissioning and Abandonment Summit Gulf of Mexico. Houston, TX, Mar 28–29.

Abadie, R.J., Trail, K.L., Riche, R.S., 2011. Success in the abandonment of a damaged offshore platform through stakeholder involvement. In: SPE 142174. SPE Americas E&P Health, Safety, Security and Environment Conference. Houston, TX, Mar 21–23.

Ajemian, M.J., Wetz, J.J., Shipley-Lozano, B., Shively, J.D., Stunz, G.W., 2015. An analysis of artificial reef fish community structure along the northwestern Gulf of Mexico shelf: potential impacts of 'rigs-to-reefs' programs. PLoS One 10 (5), 20126354.

Ammer, A., Saunders, R., Henry, C., Wilkinson, T., Frismanis, M., Bradshaw, M., Codd, J., Kessler, D., 2015. New life to an old field – Main Pass 73 Gulf of Mexico shelf. In: SEG New Orleans Annual Meeting, pp. 1807–1812.

Batchelor, B.J., Mayer, M.C., 1997. Selection and drilling of recent Gulf of Mexico horizontal wells. In: OTC 8462. Offshore Technology Conference. Houston, TX, May 5–8.

Blaine, M., 2001. Artificial reefs: a review of their design, application, management and performance. Ocean Coast Manag. 44 (3), 241–259.

Bondor, P.L., Hite, J.R., Avasthi, S.M., 2005. Planning EOR projects in offshore oil fields. In: SPE 94637. SPE Latin American and Caribbean Petroleum Engineering Conference. Rio de Janeiro, Brazil, June 20–23.

Borden, Z.H., El-Bakry, A., Xu, P., 2016. Workflow automation for gas lift surveillance and optimization, Gulf of Mexico. In: SPE 181094. SPE Intelligent Energy International Conference and Exhibition. Aberdeen, UK, Sep 6–9.

Braddy, S., Yoskowitz, D., Santos, C., Lee, J., Carollo, C., 2016. Socio-economic impact analysis of recreational scuba diving on Texas natural and artificial reefs. Prepared for the Texas Parks and Wildlife Department Artificial Reef Division, Austin, TX.

Brashear, J.P., Morra Jr., F., Hass, M.R., Payne, S., 1982. Enhanced oil recovery from offshore Gulf of Mexico reservoirs. In: SPE/DOE 10698. SPE/DOE Joint Symposium on Enhanced Oil Recovery. Tulsa, OK, Apr 4–7.

Claisse, J.T., Pondella, D.J., Love, M., Zahn, L.A., Williams, C.M., Williams, J.P., Bull, A.S., 2014. Oil platforms off California are among the most productive marine fish habitats globally. Proc. Natl. Acad. Sci. 111 (43), 15462–15467.

Davis, D., Scott, M., Roberson, K., Robinson, A., 2011. Large scale CO_2 flood begins along Texas gulf coast. In: SPE 144961. SPE Enhanced Oil Recovery Conference. Kuala Lumpur, Malaysia, Jul 19–21.

DiPietro, P., Kuuskraa, V., Malone, T., 2014. Taking CO_2-enhanced oil recovery to the offshore Gulf of Mexico. In: SPE 169103. SPE Improved Oil Recovery Symposium. Tulsa, OK, Apr 12–16.

Everitt, T.A., 1994. Gas-lift optimization in a large, mature GOM field. In: SPE 28466. SPE Annual Technical Conference and Exhibition. New Orleans, LA, Sep 25–28.

Flint, M., Angelsen, S., Amundsgard, O., 2011. Ageing infrastructure offshore – a risk-based approach to supporting investment/divestment. In: SPE 146198. SPE Offshore Europe Oil and Gas Conference and Exhibition. Aberdeen, UK Sep 6–8.

Hannah, R.R., Park, E.I., Poter, D.A., Black, J.W., 1993. Combination fracturing/gravel-packing completion technique on the Amberjack, Mississippi Canyon 109 field. In: SPE 26562. SPE Annual Technical Conference and Exhibition. Houston, TX, Oct 3–6.

Haynes and Boone, LLP, 2018. Oil patch bankruptcy monitor. Mar 31.

Kaiser, M.J., 2014. Hurricane clean-up activity in the Gulf of Mexico, 2004-2013. Mar. Policy 51, 512–526.

Kaiser, M.J., Narra, S., 2017. Decommissioning of storm-destroyed assets nears completion. Offshore 77 (6), 34–35.

Kaiser, M.J., Pulsipher, A.G., 2005. Rigs-to-reefs programs in the Gulf of México. Ocean Dev. Int. Law 36, 119–134.

Kaiser, M.J., Yu, Y., Snyder, B., 2010. Economic feasibility of using offshore oil and gas structures in the Gulf of Mexico for platform-based aquaculture. Mar. Policy 34 (3), 699–707.

Kolian, S.R., Sammarco, P.W., Porter, S.A., 2017. Abundance of corals on offshore oil and gas platforms in the Gulf of Mexico. Environ Manage. 60 (2), 357–366.

Kondapi, P.B., Chin, Y.D., Srivastava, A., Yang, Z.F., 2017. How will subsea processing and pumping technologies enable future deepwater field developments. In: OTC 27661. Offshore Technology Conference. Houston, TX, May 1–4.

Koperna, G.J., Ferguson, R., 2011. Linking CO_2-EOR and CO_2 storage in the offshore Gulf of Mexico. In: OTC 21986. Offshore Technology Conference. Houston, TX, May 2–5.

Ma, X., Borden, Z., Porto, P., Burch, D., Huang, N., Benkendorfer, P., Bouquet, L., Xu, P., Swanberg, C., Hoefer, L., Barber, D.F., Ryan, T.C., 2016. Real-time production surveillance and optimization at a mature subsea asset. In: SPE 181103. SPE Intelligent Energy International Conference and Exhibition. Aberdeen, UK Sep 6–8.

Mason, E.P., Beets, J.W., French, M.D., Johnston, H.E., Weaver, S.M., 1997. Salt morphology and hydrocarbon trapping at EI 188: detailed salt surface mapping yields new opportunities in an old field. In: OTC 8328. Offshore Technology Conference. Houston, TX, May 5–8.

National Oceanic and Atmospheric Administration, A Guide to the Application Process for Offshore Aquaculture in U.S. Federal Waters of the Gulf of Mexico. Washington, DC.

Reeves, D., Harvey, R., Smith, T., 2003. Gas lift automation: real time data to desktop for optimizing an offshore GOM platform. In: SPE 84116. SPE Annual Technical Conference and Exhibition. Denver, CO, Oct 5–8.

Reggio, V.C., 1989. Petroleum structures as artificial reefs: a compendium. Fourth International Conference on Artificial Habitats for Fisheries. Rigs to Reefs Special Session. Miami, FL. OCS Study/MMS 89-0021.

Rouse, L., 2009. Evaluation of oil and gas platforms on the Louisiana continental shelf for organisms with biotechnology potential. OCS Study MMS 2009-059 U.S. Dept. of Interior, MMS, Gulf of Mexico Region, New Orleans, LA.

Ruiz, F.G., Ponce, G., Silva, M.C.C., Silva, R., Mora, J.A.R., 2015. Monitoring and optimizing a brown offshore oilfield with an integrated asset modeling study. In: OTC 26148. Offshore Technology Conference. Rio de Janeiro, Brazil, Oct 27–29.

Sammarco, P.W., Atchison, A., Boland, G.S., 2004. Expansion of coral communities within the northern Gulf of Mexico via offshore oil and gas platforms. Mar. Ecol. Prog. Ser. 280, 129–143.

Schronk, L., Biggerstaff, G., Langmaid, P., Aalders, A., Stark, M., 2015. Lucius project: export oil and gas pipeline development and execution. In: OTC 26006. Offshore Technology Conference. Houston, TX, 4–7 May.

Schuett, M.A., Ding, C., Kyle, G., Shively, J.D., 2016. Examining the behavior, management preferences, and sociodemographics of artificial reef users in the Gulf of Mexico offshore from Texas. N. Am. J. Fish Manag. 36, 321–328.

Sherman, J., Steed, P., Price, D., Busch, D., Loucks, D., Dusek, R., Brannum, L.B., 2016. A Cost-Benefit Analysis of the BOEM NTL on OCS Bonding. Opportune.

Stair, C.D., 1999. Artificial lift design for the deepwater Gulf of Mexico. In: SPE 48933. SPE Annual Technical Conference and Exhibition. New Orleans, LA, Sep 27–30.

Stanley, D.R., Wilson, C.A., 1989. Utilization of offshore platforms by recreational fishermen and scuba divers off the Louisiana coast. Bull. Mar. Sci. 44 (2), 767–776.

Stright, S., Srivastava, V., King, G., Smith, D.W., Tears, N., 2012. Regeneration of first-generation subsea fields: the challenges of new wells in old infrastructure. In: IADC/SPE 151202 IADC/SPE Drilling Conference and Exhibition. San Diego, CA, Mar 6–8.

Sweatman, R., Crookshank, S., Edman, S., 2011. Outlook and technologies for offshore CO_2 EOR/CSS projects. In: OTC 21984. Offshore Technology Conference. Houston, TX, May 7–10.

Yoshioka, R., Salem, A., Swerdon, S., Loper, C., 2016. Turning FPU's into hubs: opportunities and constraints. In: OTC 27167. Offshore Technology Conference. Houston, TX, May 2–5.

CHAPTER SEVENTEEN

Rigs-to-Reefs Programs

Contents

Abstract

The Louisiana and Texas Rigs-to-Reefs programs enjoy widespread public, industry, and government support and have become models for similar programs, around the world. Louisiana's Rigs-to-Reefs program is the largest in the world, and since its inception in 1986, about 363 oil and gas platforms have been donated or on average about 12 structures per year. Texas's Rigs-to-Reefs program started in 1990, and since this time, about 154 structures have been donated or about six structures per year. In both programs, capture rates increase significantly with water depth. While Texas has several reefed platforms in water depths less than 100 ft, Louisiana has recently expanded its program in shallow waters to preserve marine habitat. A summary update of the Louisiana and Texas reef programs circa 2017 is provided along with recent changes in legislative activity. Donation trends and statistics are reviewed.

Decommissioning Forecasting and Operating Cost Estimation
https://doi.org/10.1016/B978-0-12-818113-3.00017-5

17.1 INTRODUCTION

Over the years, it was widely recognized by fishermen and the scientific community that oil and gas platforms were extremely valuable as marine habitat. The legs were covered in marine growth, and each platform created a marine reef microcosm increasing the acreage of reef habitat with each installation. It was no surprise that their removal and scrapping caused concern by scientists and citizens and stimulated the need to protect this existing marine habitat. In 1980, the Minerals Management Service (MMS), along with other agencies, academia, and petroleum industry, initiated an effort to develop a Rigs-to-Reefs program for the GoM through an interagency agreement with the National Marine Fisheries Service (NMFS). The agreement laid the framework to

- develop a national policy that recognized the artificial reef benefits of oil and gas platforms,
- prepare a Rigs-to-Reefs program plan for the Gulf of Mexico,
- establish standard procedures to ensure and facilitate timely conversion of obsolete platforms as reefs,
- identify research necessary to optimize the use of platforms as reefs, and
- identify legal restrictions that could prevent using obsolete platforms as reefs.

This goal was realized when the National Fishing Enhancement Act was signed into public law (Public Law 98-623, Title II) in 1984. The US Congress signed the Act because of increased interest and participation in fishing at offshore platforms and widespread support for effective artificial reef development by coastal states. The Act recognizes the social and economic values in developing artificial reefs, establishes national standards for artificial reef development, provides for creation of a National Artificial Reef Plan, and provides for establishment of a reef-permitting system.

In 1985, NMFS laid the foundation for federal endorsement of offshore artificial reef projects when it developed and published the National Artificial Reef Plan under Title II of the National Fishing Enhancement Act of 1984 (33 USC 2101). Since then, federal regulators have supported and encouraged the reuse of obsolete oil and gas platform jackets as artificial reef material and can grant a departure from removal requirements under 30 CFR §250.1725(a) and applicable lease obligations provided that

- the structure becomes part of a state reef program that complies with the National Artificial Reef Plan,
- the state agency acquires a permit from the US Army Corps of Engineers (CoE) and accepts title and liability for the reefed structure once removal/reefing operations are concluded,
- the operator satisfies any US Coast Guard (USCG) navigational requirements for the structure, and
- the reefing proposal complies with Bureau of Safety and Environmental Enforcement (BSEE) engineering and environmental reviewing standards.

Table 17.1 Platforms Deployed as Artificial Reefs in the Gulf of Mexico and Active Inventory Circa 2017

State	Platforms Reefed	Active Platforms	Rigs-to-Reefs Program
Florida	4		No
Alabama	7	39	No
Mississippi	16		Yes
Louisiana	363	1354	Yes
Texas	154	148	Yes
Total	517	1541	

Note: Federal waters only. Alabama and Mississippi active platform counts are combined, and in the last column yes signifies a legislatively mandated Rigs-to-Reefs program.
Source: BOEM 2018; LDWF 2018; TPWD 2018.

Establishment of the National Artificial Reef Plan provided guidelines for artificial reef development in state and federal waters and paved the way for the development of state artificial reef plans. In 1986, the Louisiana Artificial Reef Program (LARP) was created under the authorization of the Louisiana Fishing Enhancement Act, and in 1990, the Texas Artificial Reef Program (TARP) was created by the Texas legislature's authorization of the Artificial Reef Act of 1989. In 1999, Mississippi's legislature approved state statutes that gave the Commission on Marine Resources the authority to develop a Rigs-to-Reefs program (Table 17.1).

Dauterive (2000) discusses the development of the Rigs-to-Reefs policies through the 1990s, Kaiser and Pulsipher (2005) provides a status review circa 2005, and Kaiser et al. (2019) summarizes programs circa 2018.

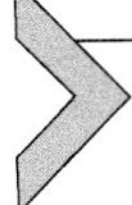

17.2 RIGS-TO-REEFS PROGRAM BASICS

17.2.1 General Process

When an Outer Continental Shelf (OCS) lease expires and/or structures are no longer useful for operations, companies are obligated to decommission and remove their facilities (30 CFR §250.1725(a)) and clear the seabed of all obstructions (30 CFR §250.1740). Wells are plugged and abandoned; platforms are decommissioned; and the structure, piling, and conductors are removed down to 15 ft below the mudline and taken to shore for scrapping or reuse per federal regulation.

In lieu of complete removal, a company can request a departure from BSEE per 30 CFR §250.1730 and enter into an agreement with a state that has an authorized Rigs-to-Reefs program to reef the structure. The state will direct the conditions of reefing through a material donation agreement (or similar) that will detail where the reefing will occur, how the structure is to be reefed, when the state will accept liability for the structure, and the negotiated donation amount to the state reef program for assuming future liability and maintenance.

State agencies coordinate applications for reefing platforms with the BSEE, and if a candidate is approved for reefing, a reefsite permit (Section 10 of the Rivers and Harbors Act of

1899) is required from the CoE. Before a platform can be reefed, a material donation agreement (or equivalent) is executed. This process forms the agreement between the state and the company for transfer of the platform and specifies the monetary donation the state will accept in exchange for taking ownership and liability for the platform. The income generated by these donations provides the revenue needed to operate state reefing programs. After reefing, legal title and liability for the jacket is transferred to the state.

In a typical platform donation, the petroleum company will save money by entering into an agreement with a state to reef the platform. The company reefs the platform at their expense while donating a percentage (normally 50%) of the realized savings to the state management agency. The realized savings (RS) is calculated as

$$RS = TC - RC$$

Total removal cost (TC) is the key variable in calculating realized savings. In many cases, reefing cost (RC) can be justified through contract data submitted by petroleum companies but is still considered proprietary information. However, total removal costs are always subject to speculation since contractors are not actually bidding on total removal, but instead create an estimate for the operator. Since TC is an estimate, it typically results in a negotiation of the final RS. Donation amounts are unique to the structure and vary due to the size and type of platform, its location, water depth, and other factors.

Not all platforms are good candidates for reefing, and factors that play a role in decision-making include the size of the structure, structural integrity, US Coast Guard clearance requirements, proximity to navigational safety fairways, water depth, and tow distance. Decks are not typically incorporated into reefing programs, since they essentially need to be stripped down to structural steel to meet environmental requirements, and caissons, tubular structures that surround wellbores, are not suitable reef material. Under special circumstances, decks may be considered for reefing provided they are free from all potentially hazardous and nonstructural items.

17.2.2 Deployment Options

After wells are plugged and abandoned, three reef deployment options are available for the jacket: topple-in-place, tow-and-place, and partial removal (Fig. 17.1). The topple-in-place and tow-and-place reefing options require the well conductors and jacket legs to be severed 15 ft below the mudline. The severance of the well conductors and jacket legs is performed via either explosives or mechanical methods. Mechanical removal is the preferred method by regulators with the least environmental impact, but is not always feasible or preferred by operators. The main advantage of mechanical severance is the decreased impacts and mortality to marine life associated with the platform.

Once the legs are severed, the entire platform jacket can be either toppled in place or towed to a designated deployment location as permitted. Toppling the jacket generally involves a derrick barge pulling it over so that the jacket lays on the seafloor in a

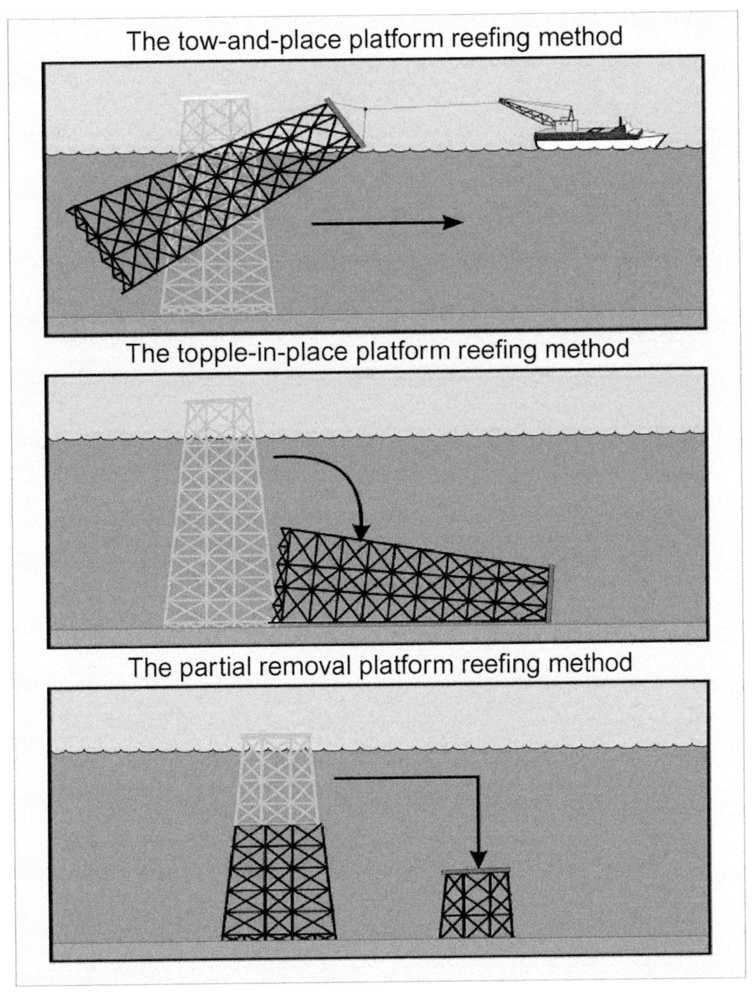

Fig. 17.1 Reefing options—tow, topple-in-place, and partial removal. *(Source: MMS.)*

horizontal position (Fig. 17.2). Towing the structure requires lifting the jacket from the seafloor to a safe towing height, usually 40 to 100 ft above the seafloor, using a derrick barge or other heavy lifting vessels. Depending on the depth at the deployment location and the jacket configuration, the towed jacket is either lowered and left in a vertical position standing upright or toppled horizontally.

Partial removals are often favored by state agencies. In a partial removal, the upper portion of the jacket is cut and placed on the seafloor next to the bottom portion of the jacket or towed to a shallower reef for deployment (Fig. 17.3). The bottom section of the jacket remains undisturbed from the seafloor up to the height where the upper portion is cut. The objective of a partial removal project is to preserve the marine habitat in its current state with minimal disturbance; therefore, mechanical methods are used to severe the

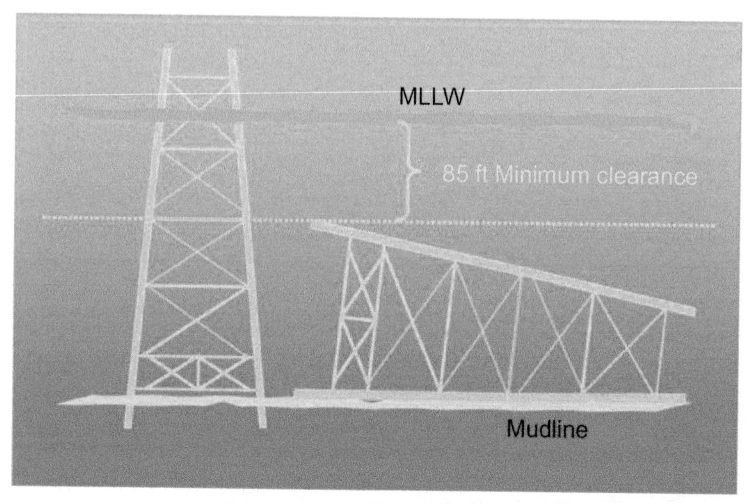

Fig. 17.2 Tow or topple-in-place reefing minimum clearance requirements. *(Source: LDWF.)*

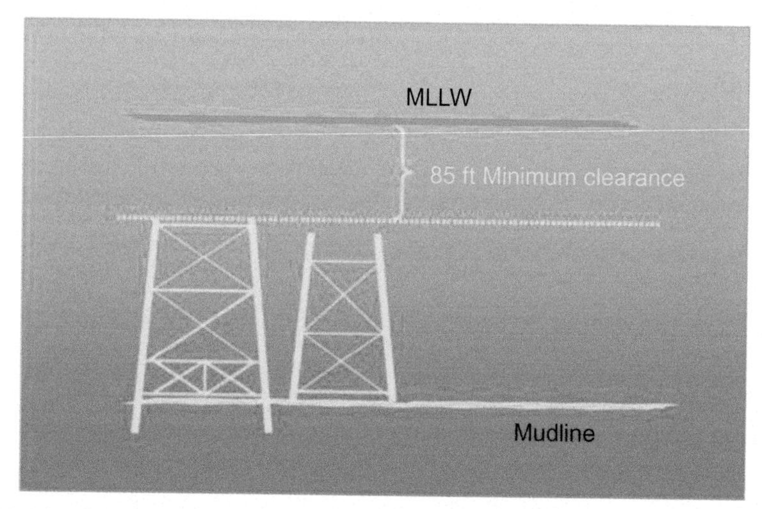

Fig. 17.3 Partial removal reefing minimum clearance requirements. *(Source: LDWF.)*

jacket piles at a depth that allows for safe navigational clearance. Conductors may be left in place and cut off at the same height as the jacket base or slightly lower, but this is dependent on state and BOEM approval.

The partial removal method is particularly beneficial with deepwater (>400 ft) structures that are converted into reefs. The relatively undisturbed base of the jacket remains in place and continues to provide habitat for a large number of pelagic and reef fish associated with the platform. The deployed upper jacket section provides an equal or slightly lower profile to compliment the standing jacket base and increases the overall surface area of the structure for habitat enhancement.

17.2.3 Clearance Requirements

The USCG is responsible for developing marking guidelines for obstructions to navigation under 33 CFR 64.30. In the Gulf of Mexico, the Eighth Coast Guard District (New Orleans, LA) has jurisdiction from western Florida to the Texas/Mexican border and determines when a Private Aids To Navigation (PATON, aka marker buoy) is required (USCG, 1990). The general "rule of thumb" is that if there is 85 ft of clearance above the platform, the USCG may waive the need for a PATON. If the clearance is shallower than 85 ft, a buoy with light may be required. Offshore reefs with less than 50 ft mean low low water (MLLW) are typically required to be marked permanently. Each PATON request is reviewed by the USCG on a case-by-case basis, and the need for a buoy is based on vessel traffic, fishing methods, and the potential for the platform to create a navigational hazard in the area.

To avoid the expense of installing and maintaining PATONs to mark Rigs-to-Reefs, most states target platforms in water depths greater than 100 ft, with 200 ft and deeper considered prime reefing depths. Along Louisiana and the upper Texas coast, the 100 ft contour exists between 30 and 75 mi offshore, making some reefs inaccessible to many fishermen. The 100 ft contour is significantly closer offshore the lower Texas coast at about 15 mi.

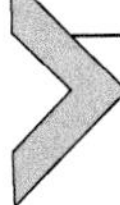

17.3 LOUISIANA ARTIFICIAL REEF PROGRAM

17.3.1 Establishment

In 1986, the Louisiana Artificial Reef Program (LARP) was created under the authorization of the Louisiana Fishing Enhancement Act of 1984 to enhance access to maritime resources and provide benefit to marine ecosystems. This act created LARP within the Louisiana Department of Wildlife and Fisheries (LDWF) and (1) required an artificial reef plan to be written; (2) formed the Artificial Reef Council (ARC) to oversee the program development and provide guidance on policy, site selection, and fund allocation; and (3) established the Artificial Reef Trust Fund to fund management and operational activities.

The LARP consists of three programs:

- Inshore—inside coastline to the intracoastal waterway (state waters only)
- Nearshore—coastline to 100 ft water depth contour (state and federal waters)
- Offshore—100 ft water depth contour to US Exclusive Economic Zone boundary (federal waters only)

The offshore program is the oldest and was established in 1986 (Wilson et al., 1987), while the inshore and nearshore programs are more recent additions established in 2014 and 2016, respectively (LDWF, 2018).

Louisiana offshore reefs are located entirely in federal waters between the 100 ft water depth contour and the US Exclusive Economic Zone boundary and consist of planning areas, Special Artificial Reef Sites, and deepwater reefs. Nearshore reefs are those artificial reefs developed in either state or federal waters between the coastline of Louisiana and the 100 ft depth contour. Inshore reefs are defined as those artificial reefs developed in Louisiana state waters between the Louisiana Intracoastal Waterway and the Louisiana coastline and within Lake Pontchartrain.

Each of the offshore, nearshore, and inshore areas were selected after an extensive and lengthy public consultation process to avoid conflicts with shipping, commercial trawling, fisheries, sand mining, and oil and gas exploration activities. The offshore and nearshore programs primarily use oil and gas structures in reef development, whereas the inshore program avoids the use of oil and gas infrastructure. Inshore and nearshore reefs may be in areas of high use by numerous user groups and subject to high-energy events requiring more extensive design and planning considerations.

17.3.2 Offshore Program

The LARP's offshore program consists of 48 planning areas, 18 SARS, and 10 deepwater reefs circa 2017 (McDonough, 2017).

Planning Areas

The original 1986 LARP program comprised eight planning areas, a ninth planning area in Ship Shoal was added at the request of fishermen, and several smaller areas have been incorporated over the years. Circa 2017, there are 48 planning areas (Fig. 17.4). Planning areas facilitate decommissioning planning by oil and gas companies and provide flexibility in site selection within the planning areas, encouraging industry participation. One highly populated planning area is the South Marsh Island 146 reef area (Fig. 17.5).

Special Artificial Reef Sites

In the early 1990s, there were opportunities to create reef sites arising from hurricane-damaged and hurricane-destroyed structures that were outside designated planning areas. In South Timbalier 86, for example, a jack-up rig collapsed in 1986 during Hurricane Juan. The location subsequently became a popular diving and fishing spot but was not located within one of the original planning areas. The program coordinators and council members felt that expansion of an existing planning area to encompass the downed

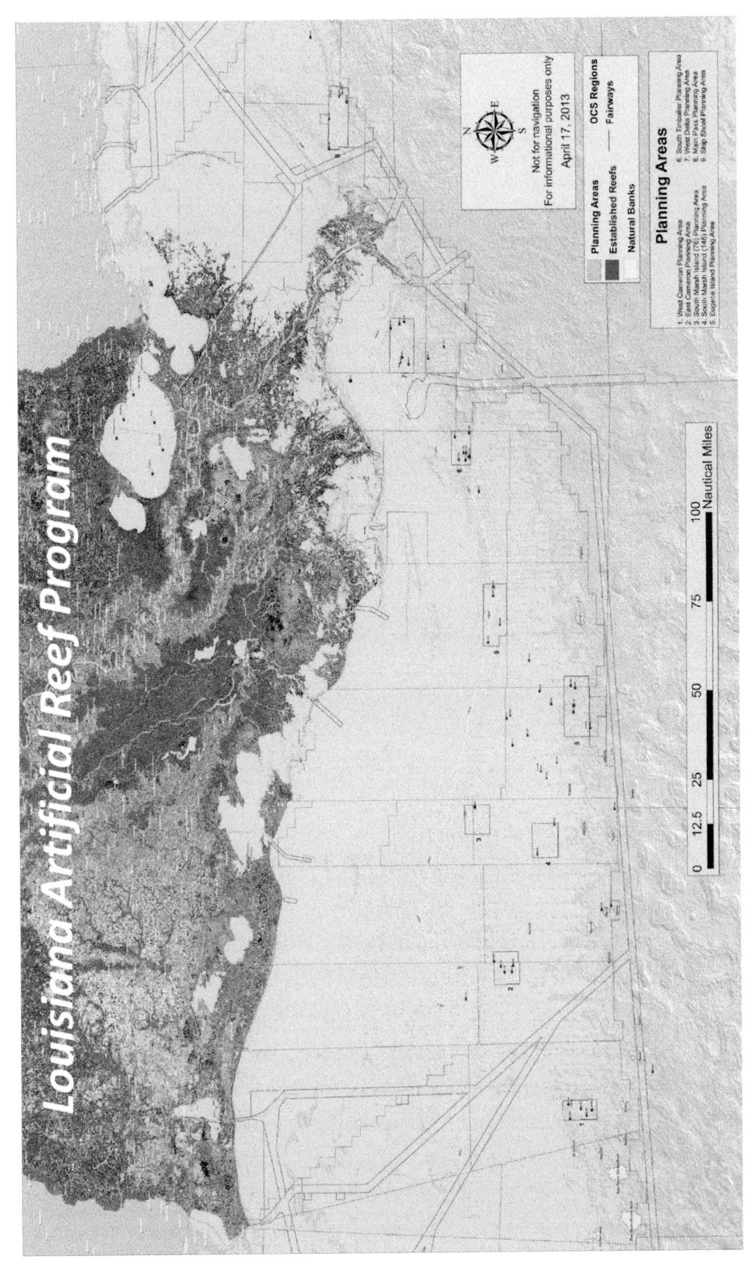

Fig. 17.4 Louisiana Artificial Reef Program planning areas circa 2017. *(Source: LDWF.)*

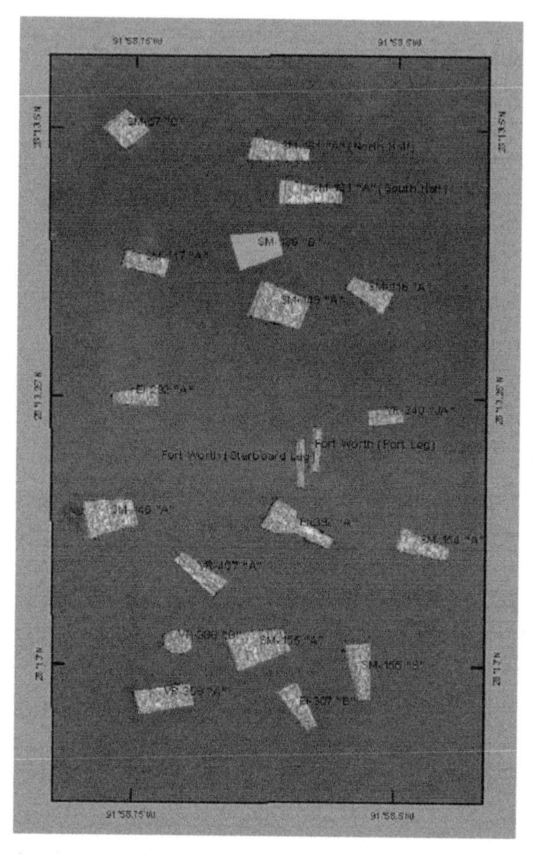

Fig. 17.5 South Marsh Island 146 reef area circa 2017. *(Source: LDWF.)*

structure would set a bad precedent and instead created a Special Artificial Reef Site (SARS) to establish selected reef sites under unusual circumstance.

In August 1992, Hurricane Andrew damaged or destroyed over one hundred platforms and caissons, five of which entered LARP as SARS (Kasprzak, 1998). From 2004 to 2008, five major hurricanes (Ivan, Katrina, Rita, Gustav, and Ike) destroyed 181 structures (Box 16.1), several of which entered LARP via the SARS program (Kaiser and Kaspzrak, 2008; Kaiser, 2014).

For hurricane-destroyed structures, reefing in place is usually the operator's preferred option if regulatory approval can be granted since the savings from having to remove the platform to shore or another reef site would likely be large. The problem from an environmental and regulatory perspective, however, is that the downed structure may not afford significant ecological benefit at its location, especially if the structure is in deepwater or has been buried under significant mud cover, and since the structure was not prepared for reefing, it may not be possible to satisfy the program requirements. Taylor

Energy's Mississippi Canyon 20 toppled platform is an example of a structure that provides limited habitat value (Box 14.1). In other cases, the requirements a toppled structure must satisfy as an artificial reef, such as waterline clearance and structure stability, may be difficult to achieve without extensive hazards to diving personnel or long complicated operations.

According to LARP guidelines, for every SARS created, an area of equal size (not containing permitted or established artificial reefs) is eliminated from an existing, adjacent planning area, and a reasonable effort is made to remove an area of higher shrimp-harvesting areas.

To be considered for SARS, qualifying criteria include (must meet one or more of the following):

- Historical or biological significance,
- Cooperative effort with LARP,
- Contains shipwrecks and other derelicts, and
- Integral part of LARP or demonstration project.

Mandatory criteria for SARS include (must meet all of the following):

- Benefit commercial or recreational fishing or provide fish habitat,
- Removing material would have a negative impact on fish,
- Does not pose a threat to navigation,
- Overall positive impact on other user groups,
- Free of hazardous material,
- Area does not occupy currently trawlable bottom, and those water bottoms are free of obstructions, and
- The structure must be standing and undamaged by storm or other disasters.

SARS became controversial in 2008 after large numbers of toppled structures from Hurricanes Katrina and Gustav were accepted into the program. Shrimpers complained that SARS were taking too much trawlable bottom and regulators were concerned about perceptions that the program was being used as a dumping ground.

In 2009, the MMS issued a moratorium on reef proposals outside of planning areas in response to public criticism and NGO assertions that the Rigs-to-Reefs program was becoming a *de facto* "Ocean Dumping Program" (Peter, 2013). The Rigs-to-Reefs Policy Addendum (MMS 2000-073) outlined enhanced reviewing and approval guidelines for hurricane-destroyed and toppled platforms (MMS, 2009).

In 2013, BSEE issued an Interim Policy Document that lifted restrictions on SARS but kept in place a restriction on reefing toppled structures (BOEM, 2013). The Interim Policy Document did not eliminate storm-toppled platforms from consideration as reefs, but it specifies that no departures from removal requirements will be granted for structures toppled due to structural failure.

In 2017, the Louisiana ARC lifted the SARS moratorium, and revisions to the SARS amendment were made that stipulated among other criteria that the structure must be standing and undamaged by storm or other disasters to be accepted into the program (McDonough, 2017).

Deepwater

There are approximately 100 active structures in the deepwater GoM circa 2017, most (~90%) in the central GoM, split approximately equally between fixed platforms and floating structures.

Established deepwater reef sites and deep planning area reefs circa 2017 include Eugene Island 371-372-384, South Pass 89, Green Canyon 6, Garden Banks 236, Mississippi Canyon 486, South Marsh Island 205, and Vermilion 395–412 (Fig. 17.4). In 2016, two new deepwater reefs were proposed, West Cameron 645 and Mississippi Canyon 148.

When new deepwater reefs are proposed, they are first evaluated and approved by the Louisiana ARC. LDWF then seeks approval and a permit from the federal government. Federal approvals are usually given, but sometimes, there is disagreement over science and engineering considerations that delay or prevent the necessary permit.

To date, two floating structures have been reefed at deepwater sites in the central GoM—Typhoon's mini tension leg platform in 2005 and the Red Hawk spar in 2014 (Box 17.1). In the future, as more deepwater structures reach the end of their useful life, there will be more deepwater reef applications, especially from spars and tension leg platforms that are difficult to bring back to shore. Semisubmersibles are readily towed back to port after demooring, and operators of these structures are unlikely to pursue reefing.

17.3.3 Nearshore Program

The LARP's nearshore planning areas were approved by the Louisiana ARC in 2016 and operate similar to the offshore Rigs-to-Reefs program. The program's stated aim is to preserve as much habitat as practicable by keeping the base of the structure, rockpiles, and unique profile features when they are present.

Twelve nearshore planning areas were selected in a public process focusing on high-density infrastructure and low shrimping/trawling activity locations (Fig. 17.9). From west to east, the nearshore planning areas are named Johnson Bayou, West Cameron 66 (WC 66), East Cameron 2 (EC 2), East Cameron 14 (EC 14), South Marsh Island 233 (SM 233), Ship Shoal, South Timbalier 34-51-52 (ST 34-51-52), Bay Marchand 3 (BM 3), West Delta/Sandy Point (WD/Sandy Point), Southwest Pass (SW Pass), East Bay, and South Pass. Bay Marchand 3, Southwest Pass East Bay, and South Pass at the east end overlap with state waters.

Clearance requirements in nearshore donations are determined by the USCG, and any material that cannot meet clearance must be removed. In cases where the top of the jacket will not meet clearance, the base of the structure would be suitable for reef

BOX 17.1 Typhoon and Red Hawk Deepwater Floater Reefs

Typhoon Mini Tension Leg Platform

Chevron installed the Typhoon mini tension leg platform (MTLP) in Green Canyon 236 in 2100 ft water depth, and from start-up in July 2001 to September 2005, the field produced about 50 MMboe. Damage sustained during Hurricane Rita caused Typhoon to break its tendons, flip upside down, and move in a northwesterly direction like an iceberg. It beached about 70 mi away in 186 ft water depth, the height of the structure (Fig. 17.6). TLPs are positively buoyant with columns serving as the principle source of buoyancy, so when Typhoon broke its tendons, the structure flipped because it was top-heavy but floated in the water like a fishing bob.

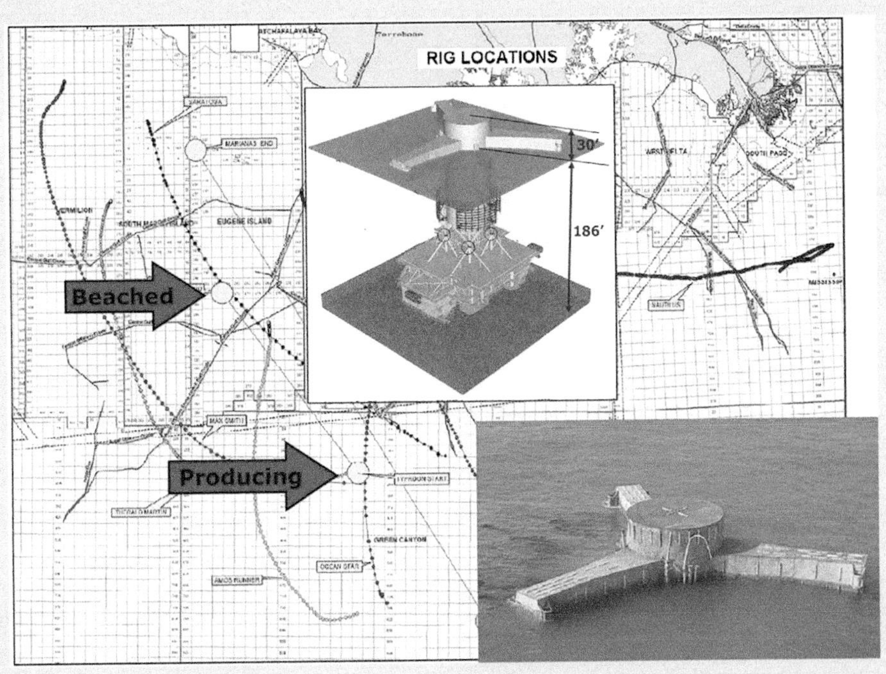

Fig. 17.6 Typhoon was beached in 186 ft water depth, the height of the platform. *(Source: Thornton, W., 2010. A hurricane story about a Typhoon – overview of the Typhoon TLP decommissioning project. In: Presentation at the Decommissioning and Abandonment Summit Gulf of Mexico. Houston, TX, March 28–29.)*

The management directive was to recover the hull, if practical, and remove the hazard by the start of the 2006 hurricane season (Thornton, 2010). Temporary mooring was installed at the beached location, and the hull tanks were hot tapped, a method of making a connection to pressure vessels without emptying the vessel. Hull and deck recovery were not considered feasible for technical and safety reasons, and the decision was made to float the hull and deck for reefing. Before reefing, production equipment was flushed, topside cleaning was performed by divers system by system, hazardous material was removed, and ballast systems were installed. The refloated hull was towed about 50 mi southeast to an existing reef site at Eugene Island 263 and was then toppled to provide topside clearance (Fig. 17.7).

Continued

BOX 17.1 Typhoon and Red Hawk Deepwater Floater Reefs—cont'd

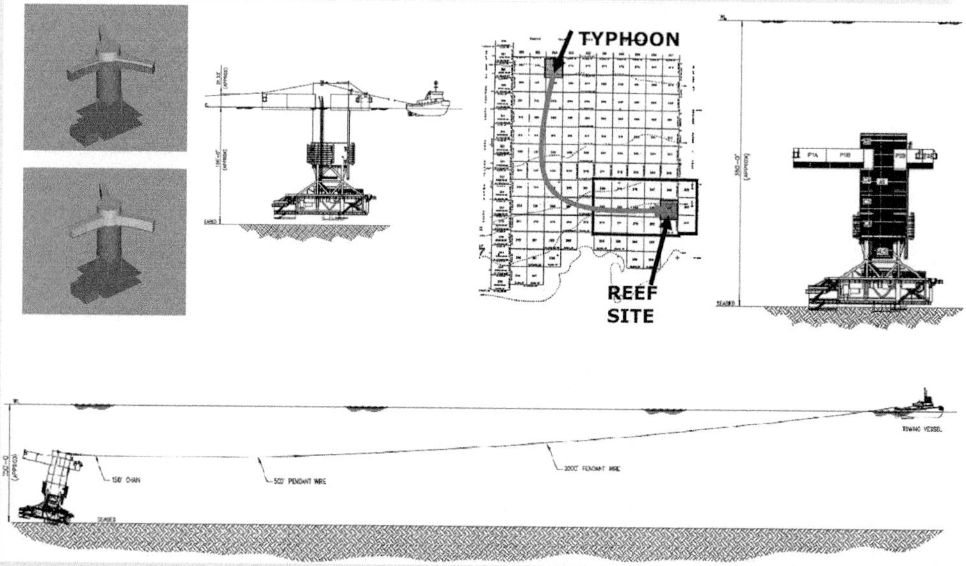

Fig. 17.7 Selected recovery plan – refloat, tow, sink, and topple. *(Source: Thornton, W., 2010. A hurricane story about a Typhoon – overview of the Typhoon TLP decommissioning project. In: Presentation at the Decommissioning and Abandonment Summit Gulf of Mexico. Houston, TX, March 28–29.)*

When Typhoon capsized, the flare tower, turbine skid, and water polishing units were dislodged from the deck and fell to the seabed. Over 800 sonar hits were examined and evaluated for recovery and potential environmental impact by the new owners of the lease (Sutherland and Macklin, 2011). The oil and gas risers were severed along with five flow lines and five umbilicals, and most of the debris field was confined to the GC 236 and GC 237 blocks. The integrity of the large pieces would not permit them to be lifted and safely loaded for onshore disposal, and a collection site for unrecoverable debris was laid out with regulatory agency approval. The flow lines were cut, flushed, and plugged for abandonment. A majority of all production deck equipment was recovered from the seabed and taken to onshore disposal sites.

Red Hawk Spar

The Red Hawk gas field in Garden Bank 876–877 in 5300 ft water depth was developed using cell spar technology (Lamey et al., 2005). The cell spar is a collection of several cylindrical sections, six tubes 20 ft in diameter all connected by structural steel and surrounded by a large tube 64 ft in diameter. Three of the tubes extend as legs to the keel that holds the permanent ballast (Fig. 17.8).

Continued

BOX 17.1 Typhoon and Red Hawk Deepwater Floater Reefs—cont'd

Fig. 17.8 Red Hawk spar. *(Source: Anadarko.)*

Red Hawk was installed in 2004 with extra deck space and weight capacity to be used in a hub-and-spoke operation, but regional opportunities did not arise. In 2008, the wells watered out after producing 125 Bcf natural gas, about half of the original reserve estimate. Anadarko Corporation spent the next several years "shopping the facility around," but with more competitive options available and no viable tieback opportunities, the decision to decommission was made (Furlow, 2015). A crew of seven manned the facility during this time performing routine maintenance, safety testing, and inspections.

The wells were permanently abandoned in 2011. In 2014, after the topsides were removed, the 564 ft long, 64 ft-diameter hull was towed vertically to the Eugene Island 384 reef site. The hull was deballasted to a draft of 400 ft and pulled by tugboat 70 mi to the 430 ft deep Eugene Island 384 location. At site, the spar was ballasted until it rested on the seabed, and then, holes were cut into the hull, and the spar pulled over and laid on its side. The spar lies 64 ft in the water column and is less than the two other jackets reefed in place at the site that extend 85 ft in the water column.

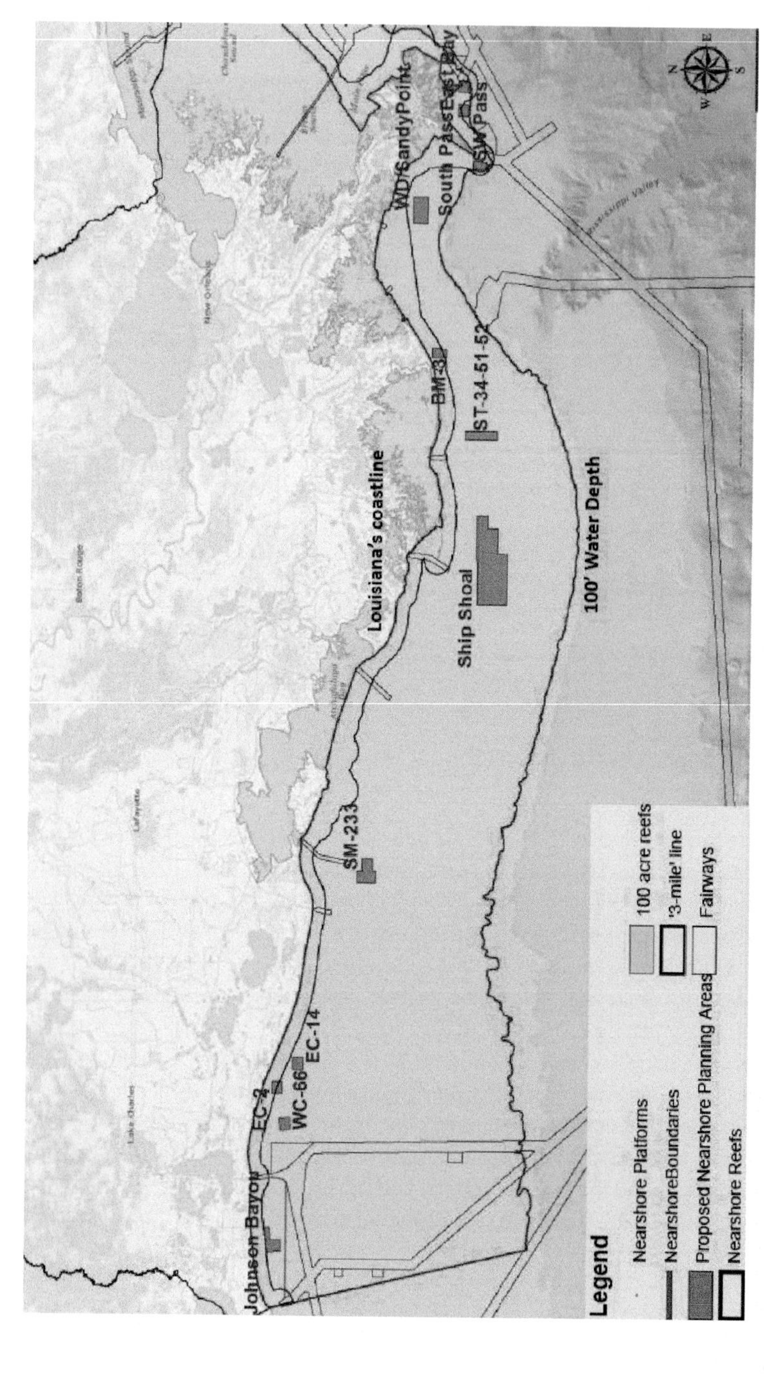

Fig. 17.9 Nearshore planning areas in the Louisiana Artificial Reef Program circa 2017. *(Source: LDWF 2018.)*

material if decommissioning operations employ mechanical cutting methods. Nearshore reefs with less than 50 ft MLLW are typically required to be marked permanently.

17.3.4 Inshore Program

Inshore and nearshore reefs can be utilized by a greater number of Louisiana's fishermen due to shorter travel distances and provide options to stay closer to shore during inclement weather conditions. Unlike the offshore and nearshore programs, however, the inshore program avoids the use of oil and gas infrastructure. Inshore reefs typically require a minimum 6 ft clearance at MLLW and are constructed of low-profile materials to meet navigation requirements. Circa 2017, there were 31 established reef sites using shell, limestone, concrete, and reef balls.

17.3.5 Donations

From 1987 to 2017, 363 platform jackets were donated to LARP, about 12 structures per year (Fig. 17.10). Eight drill rig legs, 40 armored personnel carriers, one jack-up barge, and one tugboat have also entered the program. Structures that are partially removed with tops moved to another site are counted as one structure, while complexes that contain multiple structures are counted per individual structure. Platforms are only counted once regardless of the number of reefs made from the platform, and once created, reef sites are equivalent from a programmatic perspective, and no distinction is made between planning area and SARS reefs.

In 2017, there were 10 oil and gas structures deployed, 15 in 2016, and 13 in 2015 (Table 17.2). Historically, about 70% of platforms have been towed to reef sites, 20% have been toppled in place, and 10% partially removed. Decks are almost always taken to shore, while jacket sections, if partially removed, may be used to create more than one reef site if the top section is moved to a location away from the base of the jacket. Caissons are not suitable reef material and are rarely reefed, but three-pile jackets are employed. In

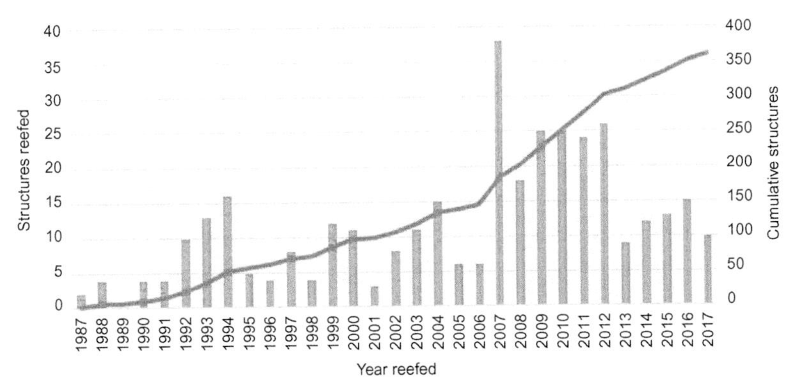

Fig. 17.10 Platforms donated to the Louisiana Artificial Reef Program. *(Source: LDWF 2018.)*

Table 17.2 LARP Donated Structures by Deployment Method, 2008–17

Year	Tow	Partial Removal	Topple	All
2008	13	4	1	18
2009	17	3	5	25
2010	16	2	5	21
2011	18	2	3	23
2012	15	2	9	26
2013	7	1	1	9
2014	9		3	12
2015	10	1	2	13
2016	14	1		15
2017	9	1		10

Source: LDWF 2018.

recent years, Apache Corporation, Chevron U.S.A. Inc., Fieldwood Energy, Stone Energy Corporation, and W&T Offshore have been leading donors.

From 2007 to 2012, 156 structures were donated to LARP, an average of 26 structures per year, twice the historic average donation levels due in large part to numerous SARS reefs. From 2007 to 2012, the number of annual donations ranged from 18 to 38 structures per year.

17.3.6 Capture Statistics

LARP offshore planning areas are all located in water depth greater than 100 ft, and only a few decommissioned structures in water depth <100 ft have been reefed to date. Through 2017, seven jacket structures have been reefed in water depth <100 ft out of 1367 total decommissioned platforms (excluding caissons) in the central GoM. As water depth increases, donations as a percentage of decommissioned jacket strucures increase significantly, from 31% (100–200 ft), to 63% (200–400 ft), to 93% (400–1000 ft). In water depths from 100 to 1000 ft, LARP has captured 42% of all available platforms since its inception (Table 17.3). LARP's overall capture rate is 16%.

17.3.7 Average Donations

The LARP has received about $100 million in donated funds since 1986. From 2004 to 2018, LARP received about $80 million from 204 structure donations or about $392,000 per structure (LDWF, 2018). From 2012 to 2018, the average annual donation was $562,000 per structure (Table 17.4). By comparison, in the first decade of the program, $8.8 million was donated from 67 platforms, an average donation of $132,000 per structure (Kasprzak, 1998).

Table 17.3 LARP Capture Probabilities in the Central Gulf of Mexico, 1986–2017

Water Depth (ft)	Donations (#)	Decommissioned (#)	Percentage (%)
<100	7	1367	1
101–200	174	553	31
201–400	161	256	63
401–1000	14	15	93
Total	356	2191	16

Note that decommissioned structure counts do not include caissons.
Source: BOEM 2018; LDWF 2018.

Table 17.4 LARP Revenue and Average Donation per Structure, 2004–18

Fiscal Year	Revenue ($1000)	Donations (# Structures)	Average Donation ($1000 per Structure)
2004	3300	11	300
2005	3100	9	344
2006	1400	15	93
2007	7800	26	300
2008	4900	15	327
2009	9100	22	414
2010	8700	32	272
2012	11,078	16	694
2013	9572	12	798
2014	7338	15	489
2015	6546	12	546
2016	5600	12	467
2017	1490	7	213
2004–18	79,900	204	392

Note: Fiscal year 2011–12 revenue is not reported in public records.
Source: LDWF 2018.

17.4 TEXAS ARTIFICIAL REEF PROGRAM

17.4.1 Establishment

In 1989, Texas passed legislation that directed Texas Parks and Wildlife Department (TPWD) to promote, develop, maintain, monitor, and enhance the artificial reef potential in state and federal waters adjacent to Texas. This legislation also directed TPWD to "actively pursue acquiring offshore platforms for use as artificial reefs in the Gulf of Mexico, in deference to other structures" (Stephan et al., 1990). The Texas Artificial Reef Management Plan was approved in 1990 formally establishing the Texas Artificial Reef Program. Texas applies an exclusion approach under which any area is assumed to be an appropriate reef site unless excluded because of more important alternative uses (Stephan et al., 1990).

When the TARP Rigs-to-Reefs program was first established, a large area of the High Island OCS was designated by the CoE, Galveston District, as a Regional General Permit Area. This included a total of 316 OCS lease blocks within the South Addition, East Addition, and South Extension. Reef sites were generally 40 ac, encompassed 1/16 of a square mile, with enough space to cluster at least nine jacket structures. If a specific proposed 40 ac reef site met all of the designated conditions of the general permit, the CoE could issue the permit without the standard 30-day public notice. If a proposed site did not meet all conditions, such as being too close to a safety fairway or existing reef site, then it could receive an individual reef permit after a 30-day public notice. The general permit covered all artificial reef sites within its boundary for 10 years, and 5- to 10-year extensions were granted upon expiration.

17.4.2 Recent Changes

In June 2016, the CoE restructured its Regional General Permit by eliminating specific areas of coverage for the permit and encompassing all areas off the Texas coast. As before, if a proposed site meets the conditions of the general permit, the CoE can issue a Section 10 of the Rivers and Harbors Act of 1899 permit without a 30-day public notice. Individual permits are still issued in cases where all conditions of the general permit are not met. The main difference between the new permitting system and the prior general permit system is the time of expiration. New permits expire within a 2-year period or sooner if reefing is completed. The complication with this new system is that TARP cannot leave reef sites open for future reefing unless it renews the permits through the CoE on a case-by-case basis (Kaiser et al., 2019).

17.4.3 Reef Sites

The TARP comprises three subprograms: Rigs-to-Reefs, nearshore reefs, and Ships-to-Reefs. Circa 2018, TARP has 92 reef sites ranging in size from 20 to 1650 ac (Fig. 17.11).

The Rigs-to-Reefs program is the major focus of TARP. To date, 157 petroleum structures and components have been reefed at 62 sites in water depths from 60 to 658 ft. Most platforms (~90%) have been reefed in water depths ranging from 100 to 300 ft.

The focus of the nearshore program is to create and enhance marine habitat in coastal waters offshore to the 9 nm state boundary. This provides habitat for adult and juvenile marine species using low-profile materials in waters from 50 to 80 ft in depth. These reefs are important to local fishermen and divers but are generally too shallow to allow for the reefing of oil and gas platforms.

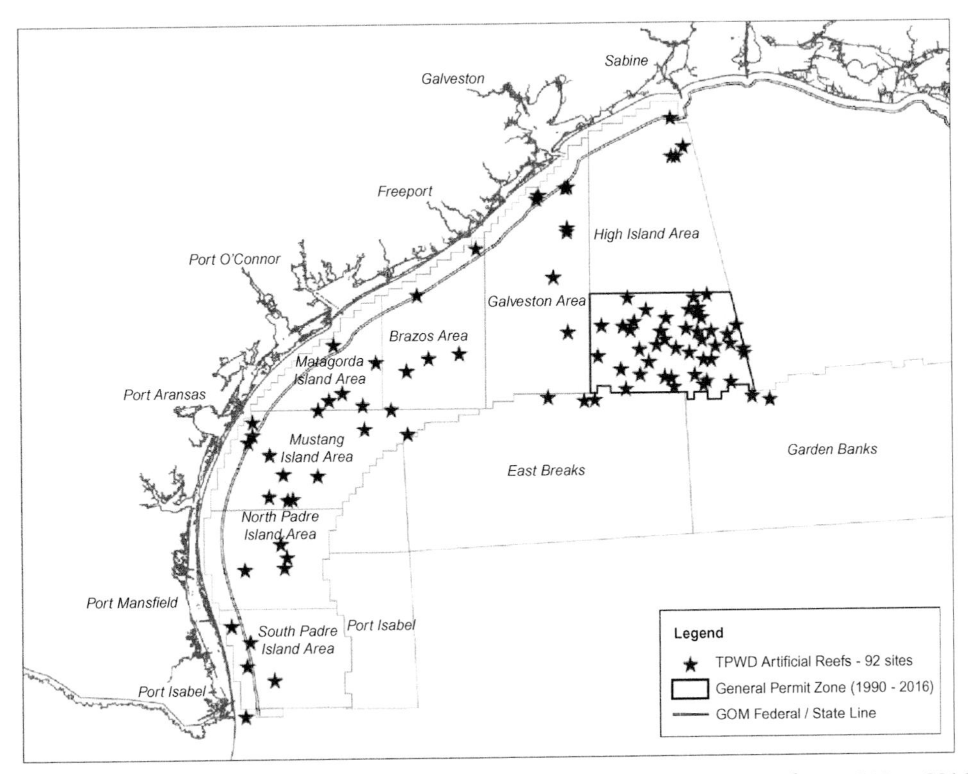

Fig. 17.11 Texas Artificial Reef Program reef sites and the general permit zone from 1990 to 2016. *(Source: TPWD 2018.)*

The Ships-to-Reefs program develops reef sites using derelict and obsolete vessels such as barges, shrimp boats, tugboats, tankers, WWII Liberty ships, and larger ships. These materials make excellent marine habitat while providing for additional fishing and recreational diving opportunities.

17.4.4 Donations

From 1990 to 2017, TARP accepted 154 platforms and various other components of structures (e.g., net guards, one deck, rig legs, and caissons) reefed in 62 permitted reef sites (Fig. 17.12). The vast majority of platforms are in the High Island area. Most structures were towed to the reef site (60%) or partially removed (30%). Toppling in place was the least common deployment, and most toppled structures were performed at the onset of the program before 1995. Over the past three years, 16 structures have been placed in TARP, seven towed to reef sites and nine partial removals (Table 17.5).

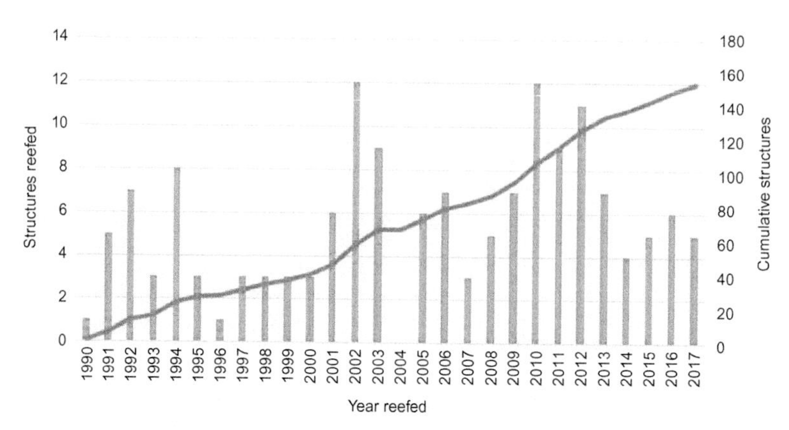

Fig. 17.12 Platforms donated to the Texas Artificial Reef Program. *(Source: TPWD 2018.)*

Table 17.5 TARP Donated Structures by Deployment Method, 2010–17

Year	Tow	Partial Removal	All
2010	10	2	12
2011	6	3	9
2012	11		11
2013	5	2	7
2014	4		4
2015	2	3	5
2016	4	2	6
2017	1	4	5

Source: TPWD 2018.

17.4.5 Capture Statistics

About one-quarter of all available jacket structures in the western GoM were reefed from 1990 to 2017. To date, TARP has reefed nine structures in water depth <100 ft, 75 platforms in 101–200 ft, 71 platforms in 201–400 ft, and two platforms in 401–1000 ft water depth (Table 17.6). Nine jacket structures have been reefed in water depth <100 ft out of 263 decommissioned platforms in the western GoM (again, excluding caissons), about 3% of the total. As water depth increases, the chance a platform is reefed increases significantly, from 36% (101–200 ft) to 70% (201–400 ft). In water depths from 100 to 1000 ft, TARP capture percentage is 47%.

17.4.6 Average Donations

Since 1990, TARP has generated over $30 million in revenue through the Rigs-to-Reefs program (Table 17.7). From 2005 to 2017, about $19 million was received from 88 donations, an average donation of $216,000 per structure. Annual average donations from 2005 to 2017 have ranged from $47,000 to $732,000.

Table 17.6 TARP Capture Probabilities in the Western Gulf of Mexico, 1990–2017

Water Depth (ft)	Donations (#)	Decommissioned (#)	Percentage (%)
<100	9	263	3
101–200	75	208	36
201–400	71	101	70
401–1000	2	3	67
Total	157	575	27

Note that decommissioned structure counts do not include caissons.
Source: BOEM 2018; TPWD 2018.

Table 17.7 TARP Revenue and Average Donation per Structure, 2005–17

Calendar Year	Revenue ($1000)	Donations (# Structures)	Average Donation ($1000 per Structure)
2005	685	6	114
2006	1332	7	190
2007	1436	3	154
2008	575	5	115
2009	1135	7	217
2010	2623	12	219
2011	2310	9	257
2012	1571	11	143
2013	1708	7	244
2014	895	4	224
2015	233	5	47
2016	121	6	202
2017	4390	6	732
2005–17	19,014	88	216

Source: TPWD 2018.

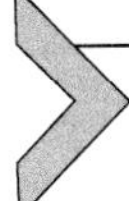

17.5 OTHER GULF COAST REEF PROGRAMS

17.5.1 Mississippi Artificial Reef Program

In 1999, Mississippi's legislature approved state statutes that gave the Commission on Marine Resources the authority to develop the Rigs-to-Reefs program. Legislation similar to that of Louisiana and Texas was developed to allow for the inclusion of platforms into its reef program and associated donations. In 2000, Mississippi received its first two platforms, donations reefed 25 mi southeast of the Chandeleur Islands in 185 ft of water. Through 2017, 16 platforms have been placed offshore Mississippi.

17.5.2 Alabama Artificial Reef Program

Alabama does not have specific legislation creating a Rigs-to-Reefs program, but the state has accepted platforms into its Marine Resources Division Artificial Reef Program.

Alabama saw its first Rigs-to-Reefs structure with the reefing of a Marathon Oil Corporation platform in 1983. Since that time, Alabama has incorporated six other platforms into its reef program. Alabama does not have specific reef areas designated for petroleum structures; they are reefed within its 10 mi^2 permitted reefing area.

17.5.3 Florida Artificial Reef Program

Florida does not have specific legislation authorizing a Rigs-to-Reefs program, but four structures have been reefed offshore. The first reefing took place in 1980 with the placement of an Exxon experimental subsea template that was transported from offshore Louisiana and placed in 106 ft of water 27 mi offshore Franklin County, Florida. Three platforms were deployed off Escambia and Dade/Broward counties between 1980 and 1993.

17.6 OUTLOOK

Decommissioning in the shallow-water GoM since 2000 have considerably reduced the number of structures in the region and available source material for Rigs-to-Reefs programs. Fewer decommissioned structures mean that fewer platforms will enter Rigs-to-Reefs programs, which may reduce funding for inshore and nearshore projects. Deepwater structures have the potential to add to program revenues, but greater effort and planning are required from both operators and regulators to create floater reefs.

The number of standing jacket structures in the central and western GoM circa 2018 in water depth less than 1000 ft is 1342 and 143, respectively, and represents the inventory from which future reefs will be sourced (Figs. 17.13 and 17.14). The Texas program has already reefed more structures than those available circa 2018, while Louisiana's program represents about one-third of 2018 inventory totals. Rigs-to-Reefs programs are unlikely to see donation activity exceed historic levels.

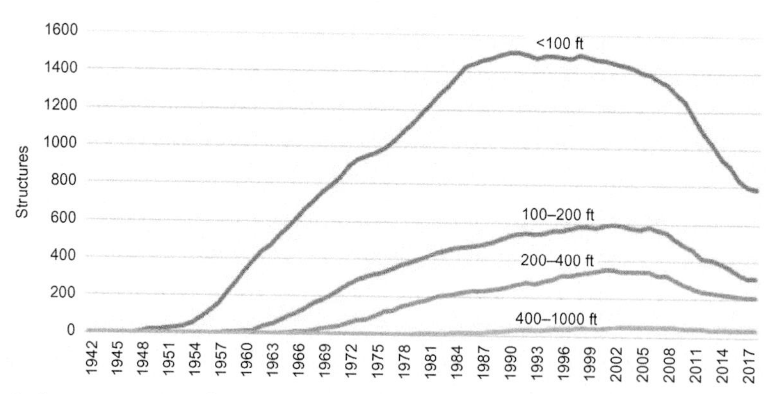

Fig. 17.13 Active structures in the central Gulf of Mexico by water depth. *(Source: BOEM 2018.)*

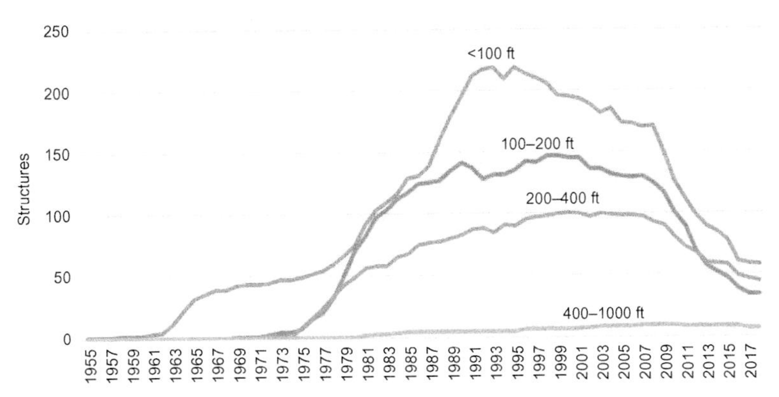

Fig. 17.14 Active structures in the western Gulf of Mexico by water depth. *(Source: BOEM 2018.)*

REFERENCES

Bureau of Ocean Energy Management, 2013. Rigs-to-Reefs Interim Policy Document. IPD No: 2013–07, Herndon, VA.

Dauterive, L., 2000. Rigs-to-Reefs policy, progress, and prospective. U.S. Department of the Interior, Minerals Management Service, Gulf of Mexico OCS Region, New Orleans, LA OCS. MMS 2000-073.

Furlow, W., 2015. Lessons learned as world's first cell spar laid to rest. SPE Oil Gas Facilit. 2, 16–19.

Kaiser, M.J., 2014. Hurricane clean-up activity in the Gulf of Mexico, 2004-2013. Mar. Policy 51, 512–526.

Kaiser, M.J., Kasprzak, R., 2008. The impact of the 2005 hurricane season on the Louisiana Artificial Reef Program. Mar. Policy 32, 956–967.

Kaiser, M.J., Pulspher, A.G., 2005. Rigs-to-Reefs programs in the Gulf of Mexico. Ocean Development International Law 36(2), 119–134.

Kaiser, M.J., Shively, J.D., Shipley, J.B., 2019. An update on the Louisiana and Texas Rigs-to-Reefs programs in the Gulf of Mexico. Ocean Development International Law 50(1), 1–33.

Kasprzak, R.A., 1998. Use of oil and gas platforms as habitat in Louisiana's artificial reef program. In: OTC 8786. Offshore Technology Conference. Houston, TX, May 4–7.

Lamey, M., Hawlay, P., Maher, J., 2005. Red Hawk project: overview and project management. In: OTC 17213. Offshore Technology Conference. Houston, TX, May 2–5.

Louisiana Department of Wildlife and Fisheries, 2018. Louisiana Artificial Reef Program: Strategic Plan 2018-19 through 2022-2023, Baton Rouge, LA.

Louisiana Inshore and Nearshore Artificial Reef Plan, 2014. Louisiana Artificial Reef Council, Baton Rouge, LA.

McDonough, M., 2017. The Louisiana Artificial Reef Program. In: Presentation at the Louisiana Artificial Reef Council Meeting, Baton Rouge, LA, April 10.

MMS (Minerals Management Service), 2009. Rigs-to-Reefs policy addendum: enhanced reviewing and approval guidelines in response to the post-Hurricane Katrina regulatory environment. U.S. Department of Interior, Gulf of Mexico OCS Region, New Orleans, LA. 31 December 2009.

Peter, D., 2013. BSEE Rigs-to-Reefs program policy overview. In: Presentation at the Louisiana Artificial Reef Council Meeting, Baton Rouge, LA, November 14.

Stephan, C.D., Dansby, B.G., Osburn, H.R., Matlock, G.C., Riechers, R.K., Rayburn, R., 1990. Texas artificial reef fishery management plan. Fishery Management Plan Series #3, Texas Parks and Wildlife Department, Coastal Fisheries Branch, Austin, TX.

Sutherland, C.A., Macklin, J.P., 2011. Phoenix subsea field development: challenges, lessons learned and successes. In: OTC 21424. Offshore Technology Conference. Houston, TX, May 2–5.

Thornton, W., 2010. A hurricane story about a Typhoon – overview of the Typhoon TLP decommissioning project. In: Presentation at the Decommissioning and Abandonment Summit Gulf of Mexico. Houston, TX, March 28–29.

U.S. Coast Guard, 1990. Policy for marking artificial fishing reefs. In: Signed 4 APR 1990, R. J. Heym, Captain, Chief, Aids to Navigation Branch. Department of Homeland Security, Eighth Coast Guard District, Aids to Navigation Branch, New Orleans, LA.

Wilson, C.A., Sickle, V.R., Pope, D.L., 1987. Louisiana artificial reef plan. Louisiana Department of Wildlife and Fisheries Technical Bulletin 41, Baton Rouge, LA.

PART FIVE

Operating Cost Review

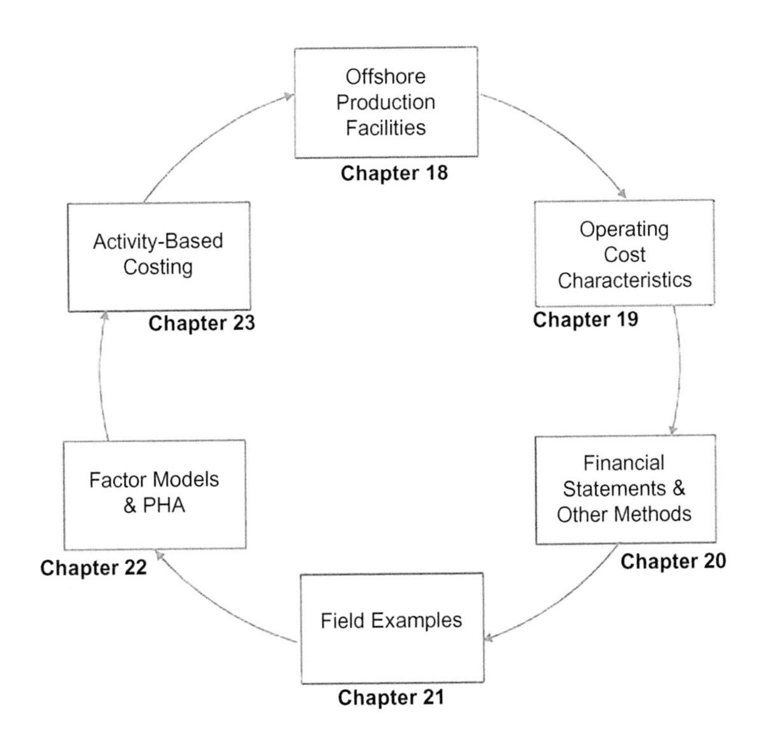

CHAPTER EIGHTEEN

Offshore Production Facilities

Contents

Abstract

The equipment between the wells and export pipeline or other transportation system is referred to variously as topsides and surface facility. The processes are for the most part easy to understand and reliable, consisting mostly of phase separation, temperature changes, and pressure changes that require various tanks and vessels to perform. There are no chemical reactions to make new molecules as in refining, and because natural gas is frequently available from processing, "fuel is free" and explains why fuel cost for the majority of a field's life is usually small. In this chapter, a summary description of the equipment that comprises an offshore production facility is described. The five chapters following this chapter provide the framework to evaluate and understand offshore operating cost models and bring together its many disparate branches.

18.1 GULF OF MEXICO OPERATIONS MARKET

The US GoM oil and gas operations market generates anywhere between $5 and $15 billion per year in spending, and as a percent of capital spending in the region, operations generally contribute between 15% and 30% of total expenditures. Market value estimates vary widely depending on the categories employed, and because methods and assumptions are rarely specified in commercial reports, comparisons are difficult or impossible to perform (Box 18.1). Estimates are usually accurate no better than 40%.

18.2 OFFSHORE PRODUCTION FACILITIES

Oil and gas wells produce a mixture of hydrocarbon gas, condensate, and oil; water with dissolved minerals such as salt; other gases, including nitrogen, carbon dioxide, and possibly hydrogen sulfide; and solids, including sand from the reservoir, scale, and

Decommissioning Forecasting and Operating Cost Estimation
https://doi.org/10.1016/B978-0-12-818113-3.00018-7

BOX 18.1 Operations Market Estimates

Operations are typically defined to include services, maintenance, logistics, transportation, engineering, and decommissioning, and may or may not include production crew salary, workover expense, ancillary spending, and other categories. Manned platforms need to be self-sufficient in terms of their utility requirements, and crew has to be transported via helicopter or crew boat, and all materials have to be supplied via marine vessels. When wells need to be serviced and equipment replaced, there are significant costs involved with the vessels and crew required to perform the operations. In remote and harsh environments, costs increase further.

Unlike the capital expenditures required to drill a well, build a platform, equip facilities, and construct and install pipeline, operating cost is much less transparent with no readily available and reliable primary data sources. The data that are available come in different forms and quality and vary widely across applications. Inferences are required, and site-specific attributes need to be accounted for, most of which are not observable and can only be extrapolated with a high degree of uncertainty.

Companies maintain detailed production cost for their properties, of course, but this information is not shared or reported unless required by regulation. Thus, the amount and quality of operating cost information available for analysis are tightly constrained and misconceptions frequently arise, both in the data and in the limitations of models built from the data. As a corollary, it is fair to say that many companies do not understand the value of their operating cost data nor the manner in which it can be utilized to improve operations because of its opaque connection with accounting and the difficulty to organize and interpret the information in a useful manner.

corrosion products from the tubing. For the oil or gas to be sold, they must be separated from the water and solids, treated, measured, and transported to their sales point.

18.2.1 Typical Oil Facility

The first processing step in an oil well is separating the gas from the liquid and the water from the oil (Fig. 18.1). Because reservoir rock is largely of sedimentary origin, water was present at the time of rock genesis and therefore is trapped in the pores of the rock. Rocks deposited in lakes, rivers, or estuaries have fresher water than rocks that originated in a sea or ocean. Separation is achieved in a pressure vessel using gravity and may be two-phase, separating gas from liquids, or three-phase, separating gas, oil, and water (Bothamley, 2004; Deneby, 2011; O'Connor et al., 1997; Thro, 2007).

Since gas takes up a much larger volume than its equivalent mass of liquid, crude oil needs to reduce its gas content before transportation, and oil pipeline owners specify a maximum vapor pressure to prevent the lighter components in the oil from flashing into gas. Reid vapor pressure (RVP) is a measure of volatility and is defined as the pressure at which a hydrocarbon liquid will begin to flash to vapor under specific conditions. RVP specification is typically less than 12 lb per square inch (psi) at 100°F but for some pipelines

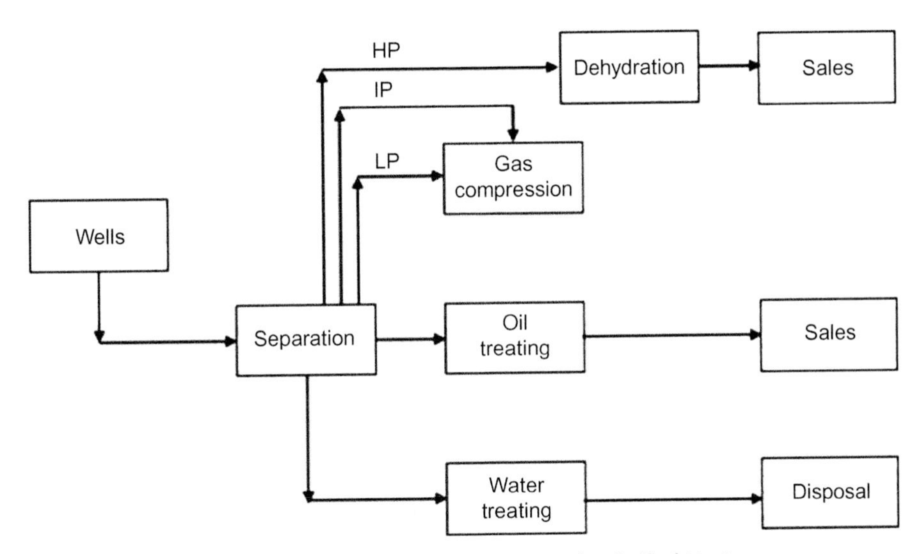

Fig. 18.1 Schematic representation of a typical oil facility in the Gulf of Mexico.

may be as low as 8.6 psi. The process of reducing the vapor pressure in the oil to meet pipeline specifications is called stabilization, and stabilized oil is also called "dead" crude.

Offshore, there are two or three stages of separation per train where the gas is flashed from the liquid, reducing the vapor pressure of the oil and releasing gas, which must then be compressed back to a higher pressure and combined with the separator gas (Bothamley, 2004). As additional stages are added, the horsepower required to compress the gas is lower, and more stabilized oil will be produced. Adding additional stages of separation and compression will increase liquid volumes and reduce compression horsepower, but the capital cost of equipment increases, and the incremental benefits to hydrocarbon value are less, so rarely are more than three stages applied (Arnold and Stewart, 2008).

No separation is perfect, and there is always some water left in the oil and oil left in the water (Bommer, 2016; Deneby, 2011). The purpose of oil treating is to get the water out of the oil for pipeline specifications, and the purpose of water treatment is to remove the oil-in-water content for overboard discharge or reinjection. The lower the API gravity, the less efficient the separation and the more energy requirements for treatment. Heat and an electrostatic grid are normally used to promote the coalescing of water droplets. Usually, acceptable water content for pipeline specification is 0.3% to 0.5% by volume, and basic sediment and water (BS&W) and other impurities are also specified, typically at 1% or less by volume. Restrictions on salt content may also be specified. High water and salt content can increase corrosion problems in production equipment and export pipeline if present.

Heat is used to aid oil-water separation and glycol regeneration. Facilities that produce a stabilized crude or receive cool production from remote wellheads require heating to attain export vapor pressure specifications. Glycol regeneration typically requires the highest temperature at the facility at 400°F. The choices for process cooling are air, direct seawater, and indirect cooling medium. Air cooling is typical on GoM platforms, and some facilities lift seawater for cooling purposes.

The amount of oil left in the water from a separator is normally between 100 and 2000 ppm by mass (Bothamley, 2004). This oil must be removed to acceptable levels before the water can be disposed of in the sea. In the GoM, producers are limited to a maximum measurement of 42 ppm for a single sample and no greater than 29 ppm average for a given month. Equipment used for produced water treating include water skimmers, plate coalesces, gas flotation devices, and hydrocyclones. A general rule of thumb is to use two types of water-treating equipment for a gas facility and three types for an oil facility (Bothamley, 2004).

18.2.2 Typical Gas Facility

Gas treating involves separation from the liquids and compression, dehydration, removing hydrogen sulfide (H_2S) and CO_2, and additional processes to control hydrocarbon dew point (Fig. 18.2).

Gas wells are often high pressure at the beginning of production that must be reduced at the point the gas flows through a wellhead choke. When gas pressure is reduced, the gas cools and liquids can condense, and hydrates can form that plug the choke and flow lines. High-pressure gas wells often require a line heater to keep the well from freezing. For

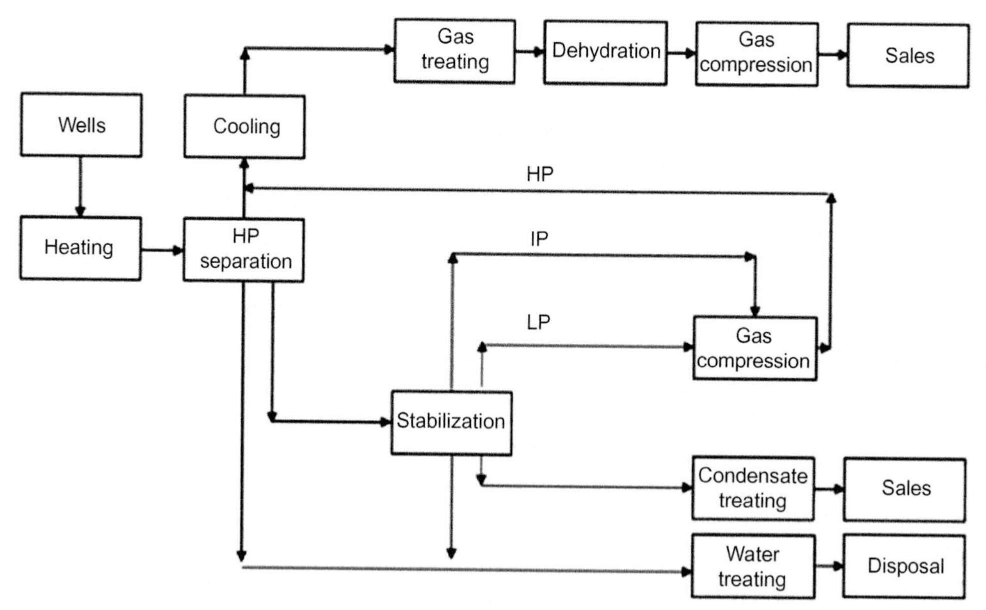

Fig. 18.2 Schematic representation of a typical gas facility in the Gulf of Mexico.

subsea wells, hydrate formation is often inhibited by injecting solvent such as methanol (MeOH) or monoethylene glycol (MEG).

A separator is used to settle liquid out from the gas. The separator pressure is set higher than the pipeline pressure so that as the gas goes through subsequent processes each with some pressure drop, it arrives at the required pipeline pressure. Hot gas leaving the high-pressure (HP) separator can cause process and corrosion problems with the downstream treating system, and because hot gas will carry more water vapor, dehydration systems would need to be larger and more expensive than if the gas were cooled first (Thro, 2007). Thus, it is sometimes necessary to install a gas cooler downstream of the first-stage separation. The cooler may be an aerial cooler or a shell-and-tube exchanger that uses seawater.

When natural gas is produced from a reservoir, it is saturated with water vapor and might contain heavy hydrocarbon compounds and impurities. Liquids and solids are easy to remove by screens and filters. Water vapor is reduced to prevent hydrate formation and corrosion in the pipeline. Gas that contains too much CO_2, H_2S, or N_2 must be treated. H_2S gas is highly toxic and CO_2 forms a strong acid in the pressure of water, and when combined, they are corrosive and possibly deadly even at low concentrations. If the quantity of H_2S is significant, it must be converted to a solid or liquid sulfur compound for sale or disposal. Nitrogen is neither corrosive nor hazardous, but it takes up space in the pipeline, increases handling cost, and reduces heating value. If the $CO_2 + N_2$ content is less than 3% to 4% by volume, most pipelines will accept the gas, and individual limits may also be specified (e.g., $CO_2 < 2\%$ and $N_2 < 3\%$). Heavy hydrocarbons are removed in significant concentration because they lead to operational problems if they drop out in the line as liquid. For small streams of rich gas remote from processing facilities, hydrocarbon dew-point control is necessary.

All offshore gas facilities require gas dehydration to avoid water condensing in the export pipeline. A standard specification for offshore pipelines is 4–7 lb per million cubic feet. Water is often removed with a glycol dehydration system that uses triethylene glycol (TEG) to absorb the water vapor from the gas. Wet TEG is then heated up to 400°F to release the water and is recycled through a loop system. An alternative is to flow wet gas into the pipeline and to add a sufficient amount of inhibitor (e.g., methanol) to prevent hydrate formation. Typical temperature requirements for gas export in the GoM range from 120 to 140°F.

Stabilization removes the light hydrocarbons from the liquid stream, either by reducing the pressure and letting the lighter components flash or by a combination of pressure reduction and heating. The resulting condensate has a low vapor pressure that can be stored in tanks without excessive vapor venting or reinjected back into the liquids or gas export line.

The vast majority of gas production in the GoM is processed at onshore gas plants and fractionators, and recovery of liquid hydrocarbons at offshore facilities in the form of natural gas liquids (NGLs) is not common and only performed if necessary to meet pipeline specifications on heat content or vapor pressure or if a C4/C5 fuel is required at site (Bothamley, 2004; Thro, 2007). If required, lean-oil absorption, refrigeration, or turboexpander plants are employed in these processes.

Gas streams that exit stabilization and other processes are often at lower pressure than the main gas stream and must be compressed so they can be processed with the rest of the gas and exported. Compression may also be required to inject into the export line at the pipeline pressure, and export gas from other facilities that cross the platform often needs compression to make it to shore.

18.3 EXPORT PIPELINE SPECIFICATIONS

18.3.1 Oil Export Pipeline

Oil export pipelines in the GoM generally transport spec crude, that is, crude with RVP < 12 psi, BS&W < 1 vol%, and temperature < 140°F, directly to refineries along the Gulf Coast region (Table 18.1). Because of the low vapor pressure crude spec, only very small amounts of butane or lighter material can exist in the stabilized crude product.

Sulfur content refers to the amount of sulfur and sulfur compounds present expressed as a percentage by weight. Oil pipelines usually specify sulfur not greater than 0.5% by weight. Oil high in sulfur and resin are usually of low viscosity and vice versa. To facilitate the flow of crude oil in pipelines, heating, chemical treatment, and mixing with lighter oil and condensate are common methods to improve viscosity. Light oil with high volatility is more likely to gasify during transportation and cause damage to the pipeline system via gas etching.

API gravity is defined as °API = 141.5–131.5/sg, where sg is the specific gravity of the crude. Since for most crudes sg < 1, °API > 10, and its value serves as a broad indicator for viscosity, sulfur content, chemical composition, and other characteristics (Riazi, 2005).

Table 18.1 Typical Crude Oil Pipeline Specifications in the Gulf of Mexico

Owner	Cypress	Marathon	Williams	ExxonMobil	Chevron	Shell
Origin	MP 69	EI 312	AC 25, EB 602, EB 643	SM, EI and SS Area	SP 78	GB 426 (Auger)
Destination	Empire, LA	Caillou Station, LA	Jones Creek Station, TX	Caillou Station, LA	SP 55A	SS 28
BS&W (%vol)	1		1	1	1	1
Water (%vol)		1				
Sulfur (%wt)	0.5				0.5	
RVP (psi at 100°F)	12	11		12	12	8.6
°API	≥ 20				≥ 20	

Note that impurities excluded in pipeline systems include chlorinated, oxygenated hydrocarbons, arsenic, and lead.
Source: Industry tariffs.

Table 18.2 Typical Natural Gas Pipeline Specifications in the Gulf of Mexico

	Garden Banks		Enbridge		Nautilus		Miss. Canyon		Destin	
Specifications	**Max**	**Min**	**Max**	**Min**	**Max**	**Min**	**Max**	**Min**	**Max**	**Min**
Oxygen (%vol)	0.2		1		0.2		0.2		0.2	
Hydrogen (%vol)									0.1	
Water (lb/MMcf)	7		7		7		7		7	
Condensate (bbl/MMcf)							40			
H_2S (grain/Ccf)	0.25		1		0.25		0.25		0.25	
Temperature (°F)	120	40	120	65	120	40	120	40	120	
Heating value (Btu/scf)					1400	980			1075	1000
Sulfur (grain/Ccf)	20		20		5		20		10	
N_2 (%vol)			3							
CO_2+N_2 (%vol)	3				4		3		3	
CO_2 (%vol)	2		3		2		2		2	
Dew point									0°F at 800 psi	

Note that impurities excluded in pipeline systems include PCBs, dust, gums, sand, oil, free water, sulfate-reducing bacteria, acid-producing bacteria, and any other microbiological agents that could have an adverse effect on pipeline system, iron oxides, salts, arsenic, mercury, lead, and oxides of nitrogen.
Source: Industry tariffs.

API limits may be specified to provide redundancy on other components, and for offshore GoM pipelines, °API > 20 is common.

18.3.2 Gas Export Pipeline

Gas export pipelines may operate in low-pressure multiphase, moderate-pressure single phase, or high-pressure dense phase. Gas export pipeline systems in the GoM generally transport dehydrated but hydrocarbon wet gas to onshore gas plants for additional processing and NGL recovery.

Gas pipelines generally require water content <7 lb/MMscf (shelf) and 2–4 lb/MMscf (deepwater) depending upon contract requirements or hydrate avoidance requirements (Table 18.2).

CO_2 specification is usually limited to at most 3% by volume. Hydrogen sulfide specifications typically range between ¼ and 1 grain per 100 scf, where 100 scf is sometimes abbreviated Ccf, and ¼ grain corresponds to 4 ppm, so the H_2S specification is equivalent to 4–16 ppm/Ccf. Heating value may be bounded by loose (e.g., 980–1400 Btu/cf) or tight constraints (e.g., 1000–1075 Btu/cf) depending on system configuration and operating environment.

REFERENCES

Arnold, K., Stewart, M., 2008. Surface Production Operations – Design of Oil Handling Systems and Facilities, 3rd ed. vol. 1. Gulf Professional Publishing, Elsevier, Burlington, MA.

Bommer, P.M., 2016. Chapter 9: An introduction to primary production facilities. In: Riazi, M.R. (Ed.), Exploration and Production of Petroleum and Natural Gas. ASTM International Manual, Mayfield, PA.

Bothamley, M., 2004. Offshore processing options for oil platforms. In: SPE 90325. SPE Annual Technical Conference and Exhibition. Houston, TX, Sep 26–29.
Deneby, D., 2011. Fundamentals of Petroleum, 5th ed. The University of Texas at Austin – Petroleum Extension Service, Austin, TX.
O'Connor, P., Bucknell, J., Lalani, M., Lake, L.W., 1997. Chapter 14: Offshore and subsea facilities. In: Arnold, K.E. (Ed.), Facilities and Construction Engineering. Petroleum Engineering Handbook, vol. III. SPE Richardson, TX.
Riazi, M.R., 2005. Characterization and Properties of Petroleum Fractions. ASME International, West Conshohocken, PA.
Thro, M., 2007. Chapter 1: Oil and gas processing. In: Arnold, K.E., Lake, L.W. (Eds.), Facilities and Construction Engineering. Petroleum Engineering Handbook, vol. III. SPE Richardson, TX.

CHAPTER NINETEEN

Operating Cost Characteristics

Contents

Abstract

There are many factors that impact operating cost, and they frequently change over time with different factors important at different times in a field's life. Unlike capital expenditures, operating cost is much less transparent and the identification of factors and their relative combination makes evaluation difficult. This combination of multiple interacting, dynamic, and unobservable conditions is the main reason that simple extrapolations are rarely successful in prediction. The chapter begins with basic terminology and important cost categories, and concludes with a general description of operating cost factors.

Decommissioning Forecasting and Operating Cost Estimation
https://doi.org/10.1016/B978-0-12-818113-3.00019-9

19.1 COST CHARACTERISTICS

19.1.1 Fixed Costs vs. Variable Costs

One way to view costs is to categorize them based on their behavior and how a given cost will respond as a level of activity changes. In the oil and gas industry, the most common measures of operating activity are production volumes and number of producing wells. As the level of production changes, some costs in total will not change (fixed costs) or not change much, while other costs will change proportionally or nearly so as the level of activity changes (variable costs). Costs that remain fixed with respect to activity levels are called fixed, and costs that change are referred to as variable cost.

An offshore worker's salary is an example of a fixed cost. Assuming the production worker salary is \$100,000 per year, the company will pay the \$100,000 regardless of how much the platform where the worker is assigned is producing. For a four-man production crew, the total cost for the crew will be fixed at \$400,000 per year ignoring inflationary changes and potential raises. As production declines, the crew cost per barrel oil produced will increase. For example, if the platform produces 200,000 bbl during the year, then the crew cost is \$2/bbl, but when production declines to 50,000 bbl, the crew cost is \$8/bbl.

The behavior of variable cost differs from that of a fixed cost. Whereas the total cost of a fixed cost will remain constant regardless of production, the total cost of a variable cost will change as the level of activity changes, increasing with increasing activity and vice versa. In other words, the total cost varies depending upon the level of activity. Chemical costs for treating produced water and hydrate control are examples of a variable cost. If a structure does not produce, the total chemical cost will be zero, but as the production rate increases, the total chemical cost will increase because treatment is volume-based. Black oil reservoirs are typically supported by water drive, and water cuts typically increase as production declines, necessitating greater chemical usage with total fluids produced as a field matures. Not all chemical costs are variable, however, and to complicate matters further, chemical costs depend upon market conditions and consumption rates depend upon factors such as operator preferences and design considerations.

19.1.2 Direct vs. Indirect Cost

Another way to categorize cost is based upon their connection to an activity, product, or department within an organization. Two categories are used to describe costs based on traceability, direct and indirect (Seba, 2003). A cost that can be traced directly to an activity, product, or department is referred to as a direct cost. A cost that requires some method of assigning it to an activity, product, or department is referred to as an indirect cost.

For example, consider a production crew responsible for operations not only at the complex where they are housed but also at three other remote (unmanned) platforms in the region. Various members of the crew visit and work on each of the platforms on a semiregular and as-needed basis. At the end of the year, the operator wants to know the labor and transport cost it took to maintain the unmanned facilities. Some method of allocation is required.

If the worker and flight hours at each platform are recorded, then it is an easy matter to assign labor and transport cost to the individual platforms, but if these data were not recorded or transmitted to the accounting department, then the operator has a choice on how it assigns crew salary and logistics. Production levels at each platform or a simple uniform method could be applied. For the transport cost, distance is a good proxy for flight hours, and the total number of visits per platform could be used in allocation. The main point is that regardless of the method, the crew salary and transport cost at the unmanned platforms are an indirect cost because the cost is assigned to each structure.

19.1.3 Operating Cost vs. Capital Expenditures

Capital expenditures (CAPEX) represent the investment required to design, construct, and commission the hardware for field development and include the wells, platforms, facilities, equipment, pipelines, and everything else with a lifetime greater than one year. CAPEX is typically defined as those items whose useful life exceeds one year, and as such, US tax regulations require that each item be depreciated on a specific schedule when computing net income (Gallun et al., 2001). Operating expenditures (OPEX), also referred to as lease operating expenses (LOE), lifting cost, or production cost, represent items whose useful life is one year or less and cost are expensed for accounts.

Unlike the capital expenditures required to drill a well, build a platform, equip facilities, construct and install pipeline, etc., operating cost is much less transparent with no readily available and reliable public data sources. The data that are available come in different forms and quality and vary widely across field applications. Site-specific attributes need to be accounted for, most of which are not observable or known to outside analyst, and can only be inferred with a high degree of uncertainty. Various organizations provide operating cost estimates, but these are of variable quality and usually do not describe the source data or methods in detail. The UK North Sea is a notable exception and provides the best and most transparent offshore operating cost data in the world.

For offshore development, the majority of capital expenditures occur upfront in the exploration and development stage, whereas operating costs start at first production and run through the life cycle of the field. It has often been suggested that over the lifetime of an offshore field, the total undiscounted OPEX will exceed the total undiscounted CAPEX, and while this certainly seems reasonable, empirical evidence to support the claim has never been presented. More importantly, because OPEX occurs over a long period of time compared with CAPEX, its impact to profitability is usually less significant than the schedule and cost overruns that impact CAPEX and the changes in commodity prices that occur over the life of the asset.

Example. UK North Sea cost expenditures, 1970–2015

The most reliable offshore expenditure description in the world is the UK North Sea because the regulator and producers cooperate on annual surveys of capital and operating expenditures (UK Oil and Gas Authority, 2017). In Fig. 19.1, the total expenditures for exploration and appraisal cost, development cost, operating costs, and decommissioning cost are presented in 2015 pounds. A notable characteristic of the data is that it is an actual record of expenditures and does not involve significant estimation procedures or other algorithms to infer cost.

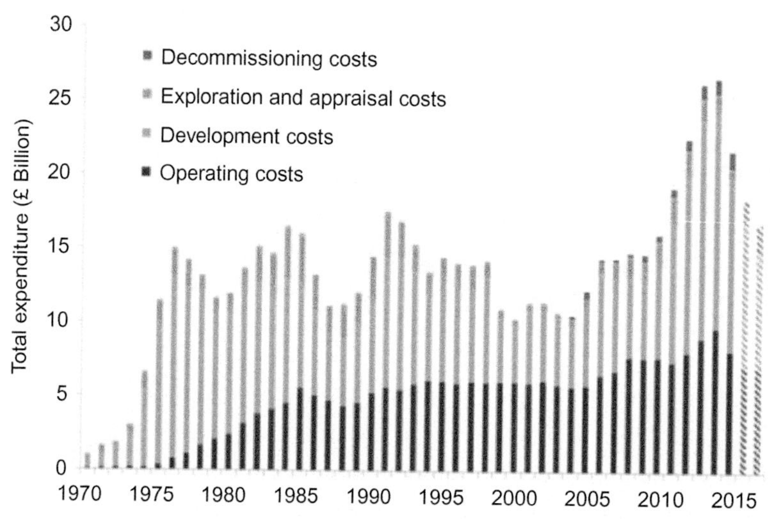

Fig. 19.1 Total expenditure in the UK North Sea by sector. *(Source: UK Oil and Gas Authority.)*

19.1.4 Unit Operating Cost vs. Life-Cycle Cost

Unit cost and life-cycle cost are normalized measures that provide complementary information about production operations. Unit operating cost (UOC) is usually calculated as the ratio of annual operating cost divided by annual production. If the entire life cycle of the field is considered, then the unit cost is referred to as life-cycle cost (LCC) or average cost. Crude oil, natural gas liquids, and natural gas are often added arithmetically to provide an oil-equivalent or gas-equivalent volume in normalization.

The UOC formula is simply

$$\text{UOC} = \frac{\text{Annual OPEX (\$)}}{\text{Annual Production (boe)}}$$

LCC aggregates and consolidates annual operating expenses and smooths out the variations and time trends present in UOC. Life-cycle cost refers to the average operating cost (undiscounted) over the life of the structure or field, computed as total operating cost divided by total production:

$$\text{LCC} = \frac{\text{Total OPEX (\$)}}{\text{Total Production (boe)}}$$

Example. GoM deepwater average production cost

The life-cycle operating cost for three deepwater fields in the GoM was estimated by consulting firm IHS using its Questar software. Average field life operating cost for Lucius (spar), Big Foot (ETLP), and Jack/St Malo (semi) was estimated at $9/boe, $11/boe, and $17/boe, respectively (IHS Global, 2015). Costs include labor, inspection and maintenance, logistics, chemicals, wells, insurance, and transportation.

The older an offshore asset, the more expensive it is to run due to increased maintenance and changing reservoir conditions. However, if regional operations or operating cost can be shared with other companies or if new wells add to production, operating cost may stabilize for a period of time. Operators actively look for ways to reduce expenses during high-cost and low-price commodity periods. There is no clear-cut relationship between the age of infrastructure and its unit operating cost because of these and other factors.

Example. Unit and life-cycle cost comparison

In 1960, a shallow-water lease block in the GoM was acquired for $2.5 million, and after spending about $5 million on exploratory drilling and assessment, the operator believed that a commercial deposit existed (Dickens and Lohrenz, 1996). The operator spent $44.5 million to develop the field, and in 1965 first production was achieved. In 1969 production peaked at 6.7 MMbbl (Table 19.1 and Fig. 19.2). Production began to decline in 1974, and the operator spent $4.1 million in 1980 and 1981 on capital improvements in an attempt to slow the decline, but abandonment came in 1987 and cost $5.2 million to plug the wells and decommission the structure.

The field produced 69 MMbbl oil and generated $417 million (undiscounted) gross revenue over its lifetime. Unit operating cost ranged from less than $2/bbl to $15/bbl. Total capital cost was $57 million, and total (undiscounted) operating cost was $83 million, yielding an average life-cycle development and production cost of $0.83/bbl and $1.21/bbl, respectively.

Table 19.1 Production and Cost Profile for a Gulf of Mexico Shallow-Water Oil Field

Year	Oil Price ($/bbl)	Production (MMbbl)	Capital Expenditures ($ million)	Operating Cost ($ million)
1960			2.5	4.0
1961				0.1
1962				0.2
1963			0.5	0.1
1964			13.0	0.6
1965	2.86	0.9	15.5	0.5
1966	2.86	3.3	16.0	1.8
1967	2.86	5.0	3.0	2.8
1968	2.86	5.6	0.8	3.2
1969	2.86	6.7	1.0	3.6
1970	3.18	5.6		3.5
1971	3.39	5.0		3.2
1972	3.39	5.5	0.4	3.6
1973	3.96	4.8	0.2	3.8
1974	6.88	6.0		3.9
1975	7.67	4.1		6.2
1976	8.19	3.9		5.8
1977	8.57	3.2		5.4
1978	9.00	3.1		2.8
1979	13.99	2.4		3.6
1980	22.49	1.8	2.0	8.0
1981	31.13	1.0	2.1	6.0
1982	28.52	0.2		3.0
1983	26.19	0.3		1.5
1984	25.88	0.1		0.4
1985	23.95	0.1		0.3
1986	11.36			0.3
1987	15.40			5.2

Source: Dickens, R.N., Lohrenz, J., 1996. Evaluating oil and gas assets: option pricing methods prove no panacea. J. Financ. Strateg. Decis. 9 (2), 11–19.

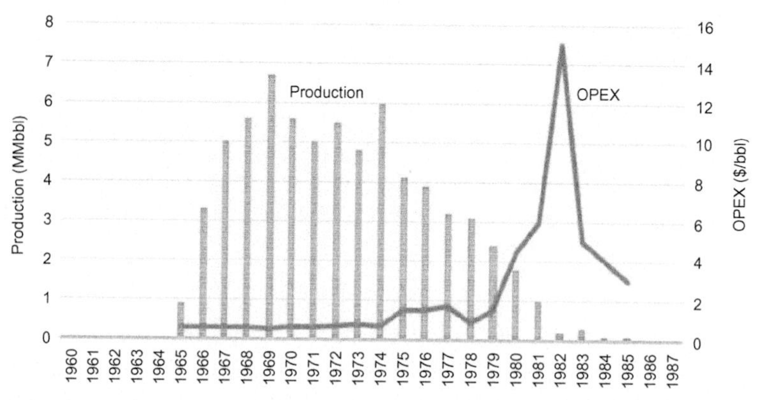

Fig. 19.2 Production and unit operating cost for a shallow-water Gulf of Mexico oil development. *(Source: Data of Table 19.1.)*

19.1.5 Marginal Revenue vs. Marginal Cost

The free cash flows of an offshore development are composed of the revenues derived from all the individual wells and structures in the field and any third-party fees, minus the total operating expenditures associated with production. To understand field operations and to improve performance, it is necessary to account for cost at the well and structure level and to identify the types of cost involved (Doering, 1993). For example, a well with a high variable cost might be a candidate for closure, but closing a well with high fixed cost can decrease the profitability of other wells in the field that must absorb the additional fixed cost. Granular and data intensive analysis is required to evaluate trade-offs.

Maximum production does not guarantee maximum profit. High-water-cut wells frequently need to be shut in to reduce treatment and compression cost to improve financial performance. It is the marginal revenue and marginal cost that are important in decision-making, but these are often difficult for operators to compute. For mature properties (e.g., when water cuts approach 90% or more), a sustained focus on cost is necessary to stay profitable, and when producing from an integrated system in subsea developments, success depends on precise and clear knowledge of cost components.

19.2 COST COMPARISON

19.2.1 Rules-of-Thumb

The first rule of thumb is that rules of thumb should be avoided. Although rules of thumb are useful to guide design and decision-making, technology and best practices change, and markets evolve over time. Site- and region-specific conditions are usually highly variable and need to be understood to inform discussion. Within companies, rules of thumb develop over years of experience, and these are often supported with empirical evidence which lends credibility to their application. Unlike development and equipment costs that can be decomposed and estimated accurately, operating costs depend upon site- and time-specific conditions that can vary widely across many dimensions. Individual variation across properties is wide, and general/generic bounds provided by rules of thumb should only be used as a last resort and if no other information is available.

19.2.2 Cost Component Range

For offshore GoM operations, personnel and logistics typically represent 40%–50% of operating costs, with repairs and maintenance 10%–40% of total cost, and chemicals and workovers 10%–20% (Fig. 19.3). Insurance, if an operator opts for coverage, usually will not exceed 5%–10% of direct LOE, and gathering and transportation expenses contribute another 5%–10%. Third-party processing fees apply to well owners that do not

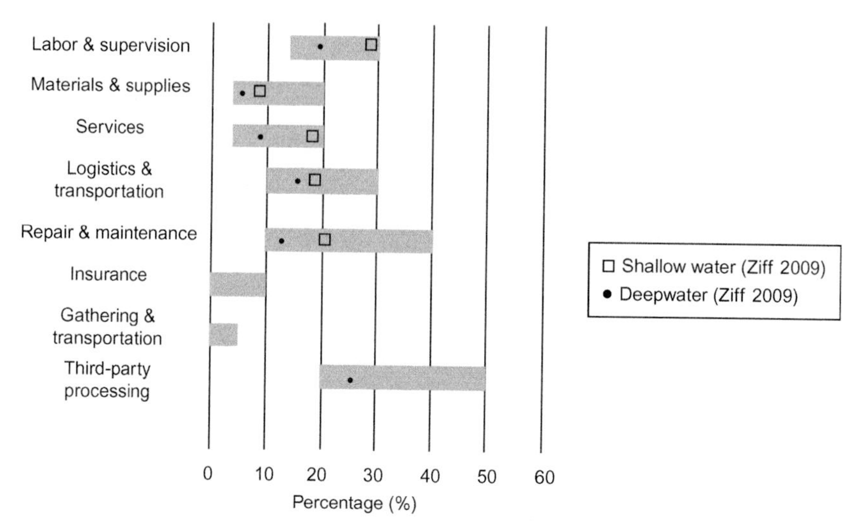

Fig. 19.3 Typical operating cost range in the Gulf of Mexico.

have a working interest in a processing platform and pay fees for access, often between 20% and 50% of their total operating cost. These ranges are typical, and variation over time will occur.

As properties age, they generally require more workovers, and chemical cost may increase, but labor cost is mostly fixed, while transportation cost and insurance will fluctuate with market conditions and can increase or decrease. Workovers are largely discretionary, and operators typically plan for a workover when they wish to accelerate or enhance production, which usually occurs when commodity prices rise. Insurance is discretionary for operators who self-insure and mid- to small-size operators are the primary players paying for coverage. Gathering and transportation fees are volume-based and may increase or decrease with time.

19.2.3 Regional Cost Comparison

Regional comparisons give a sense of the different ranges of cost involved per basin/region but need to be interpreted with an understanding of the source data and the uncertainty and variability involved since most organizations do not specify their data sources, methods, uncertainty, or limitations. The best feature of graphs is their ability to consolidate a large amount of information, but the reader should understand that the representations should not be considered at the same level of accuracy as a cost accounting survey or operator survey.

19.2.4 Relative Price Comparison

Crude oil and natural gas are fluids that flow from the reservoir through piping and are processed using equipment driven in large part by the temperature and pressure of the hydrocarbon streams generated at site. On a relative basis, operating costs per barrel or cubic foot are much less than the price of the commodity being sold, usually ranging from <5% to at most 20% of the commodity price throughout its lifetime. By comparison, in coal and oil sands, operating costs usually range from 40% to 50% or more of the commodity sales price and increase after the cutoff reserves is achieved, similar to mineral ore deposits (Box 19.1).

19.2.5 Cost Indices

How does production cost for an offshore facility or region change over time? For a facility, the answer needs to be gathered from operator data, while on a regional basis, cost indexes are a common approach. Can an inflationary index reliably capture

BOX 19.1 Mining Operations, Movin' Dirt

Mining operations are labor and equipment-intensive as anyone visiting a mining operation observes immediately. Large amounts of dirt/rock must be moved, crushed, and processed that is energy-intensive, and all mechanical things and moving equipment (trucks, conveyor belts, grinders, etc.) require large numbers of operators and breakdown with use and require frequent repair and/or replacement because of the abrasive nature of solids and the requirements of processing (Hustrulid and Kuchta, 2006; Darling, 2011).

Each truck requires an operator and a repair crew for the fleet, and additional personnel are required for surveying, drilling, blasting, processing, and refining. Overburden must be removed to access the reserves unless underground entry is made that is even more expensive to provide safe access and operating conditions, ventilation, equipment, and material transfer (Hartman and Mutmansky, 2002). Rock is physically extracted and crushed, and minerals separated using energy-intensive and chemical methods with waste streams collected in pits that often require remediation. All energy for operational activity must be imported and water-sourced.

There are usually lots of different chemical forms of minerals and separation techniques from a benefaction and extraction standpoint, along with different benefits and risks in development. Operators first concentrate the mineral and then extract it in a form salable in the market. Investment risk is related to processing technology and market access. Not all deposits are amenable to serve all markets, and each market has its own product specifications. For example, in the lithium market, there are lithium carbonates, lithium hydroxide, and lithium metal foil that each serves different consumers (Walther, 2013).

Traditional oil and gas operations use wells to make contact with reservoir sands, and rock is only removed in drilling. Flowing fluids are significantly easier to process, handle, and transport than solids. Changes in temperature and pressure between the well and surface facilities are used for processing and provide a significant percentage of the energy needed, and only near the end of primary production where the reservoir energy has been used up does operating cost start to increase. BP reported GoM shelf operations consumed 2% to 3% of the energy exported, and deepwater GoM operations consumed 1% to 2% of the energy exported (Edwards, 2004).

changes in operating cost at a property/field or regional level? Caution should always be exercised when applying cost indexes to adjust or otherwise normalize operating cost data because an unknown uncertainty is being introduced in the assessment.

No evidence has ever been reported that shows inflationary indexes are useful for LOE adjustment or that they accurately reflect changes that occur. The manner in which specific categories of cost change provides important clues to what cost indexes can and cannot accomplish. For chemicals, inflationary indexes work well, but these costs usually only contribute a small proportion of the total cost of LOE. LOE may increase, decrease, or stay the same from year to year depending as much on the decisions of the operator as the behavior of cost components. The age of infrastructure, market changes, and operator reaction impact the individual components of operating cost in different ways confounding the ability of a cost index to reliably reflect.

19.3 OPERATING COST FACTORS

Offshore structures were categorized in Part 1 according to structure type, manned status, number of wells, water depth, complex, production type, and well type. All of these factors play a role in the economic limits computed in Chapters 8 and 10 and are also important in operating cost considerations. In this section, additional factors are described to provide a more detailed and nuanced view. The direction in which factors impact cost is intuitive and easy to understand qualitatively (recall Fig. 4.1), but the relative impacts and actual changes that occur are much more complicated to evaluate.

19.3.1 Labor

The number of personnel required for offshore operations represents a significant fixed cost associated with offshore production and influences many other cost components. Once staffing requirements are defined, other associated costs such as personnel logistics and catering can be estimated. To determine labor costs, staffing and job classification levels need to be defined. Personnel may be organized into different groups such as field management, production crew, multiskill personnel such as mechanics and electricians, roustabouts, housekeeping, and catering (Steube and Albaugh, 1999). In the GoM, catering services that provide housekeeping and food are usually quoted on a dollar per person per day basis. Contracts on a cost-plus basis may also be used.

All deepwater producing facilities in the GoM are manned 24 h a day, while about one-third of shallow-water structures circa 2018 are continuously manned. In shallow water, a manned facility is often responsible for several facilities in the area, but in deepwater, crews are generally responsible for just one facility and its subsea tiebacks, if any. If a manned platform and crew are responsible for other facilities, the labor costs for the other platforms that are serviced are allocated according to man-hours or similar metric.

Example. Deepwater bed count

Bed counts for the largest deepwater floaters in the GoM range from 150 to 200, while for smaller floaters without drilling capacity, bed counts range from 20 to 40. Thunder Horse, Perdido, and Mad Dog report quarters of 186, 170, and 150, respectively, whereas Boomvang, Devils Tower, and Lucius report bed counts of 20, 26, and 44 (Fig. 19.4).

Fig. 19.4 Mad Dog *(top left)*, Thunder Horse *(top right)*, Boomvang *(bottom left)*, and Lucius *(bottom right)*. *(Source: BP, Anadarko.)*

19.3.2 Dry vs. Wet Trees

Dry tree and wet tree wells have significantly different operating costs. Dry tree wells are produced from lines tied back to the rig floor, whereas subsea wells are tied into a manifold on the ocean floor and connected to the production facility by a riser. Subsea wells are more expensive to operate, more expensive to maintain, more expensive to repair, and more expensive to abandon than dry tree wells.

For platform (dry tree) wells, well intervention is normally straightforward and relatively cheap. If the intervention can be performed without a rig (rigless), costs are reduced further and interruptions to drilling schedules avoided. Subsea well interventions

Table 19.2 Record Subsea Development Water Depths and Tieback Distances Circa 2017

Operator	Project	Field	Water Depth (ft)	Tieback Distance (mi)
Shell	Tobago	Oil	9627	6
Anadarko	Cheyenne	Gas	9014	44.7
Shell	Penguin	Oil	574	43.4
Noble	Tamar	Gas	5446	93

on the other hand are rarely easy or cheap and carry significant risk. They require mobilization of a rig or multipurpose service vessel that has high day rates, and significant lead time for equipment and planning is required. If a subsea well fails, it may be shut in for a long period of time, and as a result, there tend to be fewer well interventions on subsea wells.

19.3.3 Flow Assurance

Flow assurance refers to keeping flow paths open and are generally not a concern in shallow water and with dry tree or direct vertical access wells, but are primary problems in deepwater development and subsea systems (Bai and Bai, 2012; Jamaluddin et al., 2002). Wax, asphaltenes, and hydrates in the hydrocarbon streams have the potential to disrupt production due to deposition in the production system. Hydrates are the most common problem facing developers (Cochran, 2003), and differences in tieback distances arise in oil and gas wells (Table 19.2). Flow assurance requires energy or chemicals or both and regular pigging operations.

Example. Gemini hydrate management strategy

The Gemini field is located in Mississippi Canyon 292 in 3400 ft water depth and is tied back 27.5 mi to the Viosca Knoll 900 platform. The initial development consists of three subsea wells tied into a four-slot cluster manifold connected together via a pigging loop (Fig. 19.5). Dual 12 in uninsulated flow lines transport the produced fluids to the host platform. Gemini's fluid consists of 96% methane with 3.5% light ends (C_2–C_5) and 1.5% C_{6+} fraction. The condensate-gas ratio was predicted to be about 16 bbl/MMcf, and peak production rates of 80 MMcf/day were expected per well.

Reservoir modeling predicted limited water production throughout field life, and continuous methanol injection was selected to control hydrate formation (Kashou et al., 2001). Both flow lines enter the hydrate formation region <1000 ft from the manifold and remain in the hydrate region for the next 27.5 mi (Fig. 19.6). Low operating temperatures, high operating pressures, long tieback, and the presence of water create a high-risk environment for hydrate plug formation.

Fig. 19.5 Gemini subsea development.

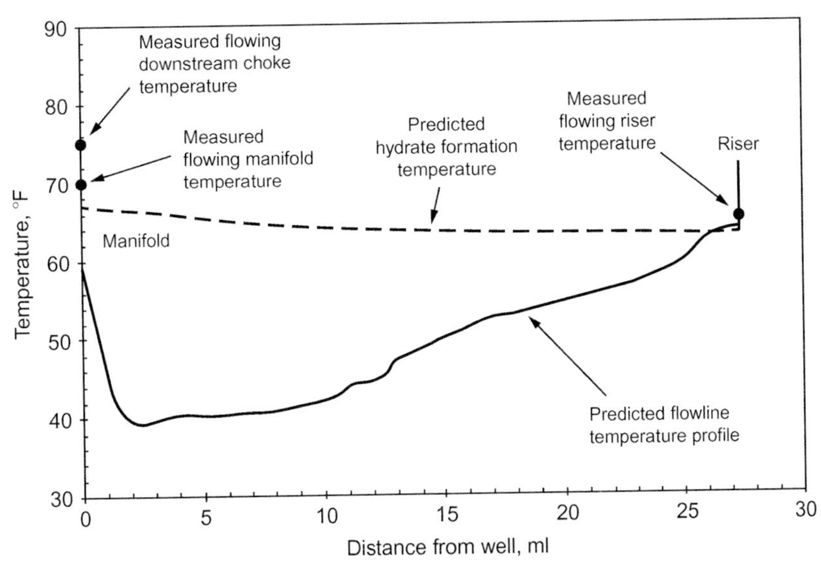

Fig. 19.6 Gemini flow line and hydrate formation temperature. *(Source: Kashou, S., Matthews, P., Song, S., Peterson, B., Descant, F., 2001. Gas condensate subsea systems, transient modeling: a necessity for flow assurance and operability. In: OTC 13078. Offshore Technology Conference. Houston, TX, Apr 30–May 3.)*

Subsea processing is an emerging technology that treats produced fluids at or below the seabed to improve recovery rates and include multiphase pumping, subsea separation, gas compression, and seawater injection (Kondapi et al., 2017). Circa 2017, there were only 25 subsea boosting systems and six subsea separation systems installed worldwide. In the GoM, Perdido employs a subsea separation system, and Cascade and Chinook, Jack/St Malo, Stones, and Julia host subsea boosting.

Example. Wax-on, wax-off strategy at Coulomb

The Coulomb field is a gas/condensate development in 7500 ft water depth tied back to the Na Kika semisubmersible via a single 27 mi, 8 in flow line (Manfield et al., 2007). Monoethylene glycol is injected at the tubing head of each well to provide continuous hydrate inhibition so that solid accumulation would not constrain production. MEG usage requirements for the two wells are defined by the MEG/condensate ratios 1:55 and 1:230.

The two wells produce fluids with significantly different condensate-gas ratios, 65 bbl/MMcf and 200 bbl/MMcf, and fluids from one well were significantly waxier than the design basis and caused the well to be temporarily shut in shortly after first production. The deterioration in flow line performance was believed to be due to accumulation of a highly viscous material, either a wax/glycol/condensate emulsion or a wax slurry. Engineers determined that burying the flow line 6–8 ft with backfill would improve heat retention and mitigate the phenomenon.

The flow line burial operation was executed over several days while one well was flowing. After completion of the first burial pass, operators reported a rapid change in the measured pressure drop across the line, and within hours, a large slug of liquid and wax began to arrive at the platform. Approximately 3000 bbl of liquid inventory was unloaded from the flow line after burial, about 40% of the total volume of the line. Prior to burial, well C-2 was producing at 68 MMcf/day, and the flow line pressure drop ΔP was 3900 psi. After burial, well C-2 production reached 102 MMcf/day with ΔP of 2700 psi.

Example. Who Dat flow assurance

The Who Dat field is located in Mississippi Canyon blocks 503, 504, and 547 (Fig. 19.7). The field is primarily oil and consists of 10 stacked reservoirs in a salt withdrawal minibasin. There is a wide variation in the physical properties of Who Dat fluids in each reservoir, and a substantial amount of asphaltenes and waxes is present in all oils (Simms et al., 2012). The high wax content in some of the oils raised concern of potential wax deposition problems, but the wax crystal formation rates showed that wax deposition tendencies were small.

Initial field development consisted of 12 subsea wells flowing to three four-slot manifolds in two drill centers. Each drill center manifold is connected to the production facility via dual 6 in wet insulated flow lines and flexible risers with roundtrip pigging capability. The infield flow lines are about 3 mi and follow a downslope seafloor to the host (Fig. 19.8).

Steady State Production. Wet insulation, maximum production rates, and/or commingling production are used to prevent hydrate formation and wax deposition during steady-state operations. Although production field temperatures were expected above the wax appearance temperature for the vast majority of the producing life and only required paraffin inhibitor during cold well start-up operations, paraffin inhibitor is injected continuously at subsea wellheads to provide inhibition for the oil export pipeline.

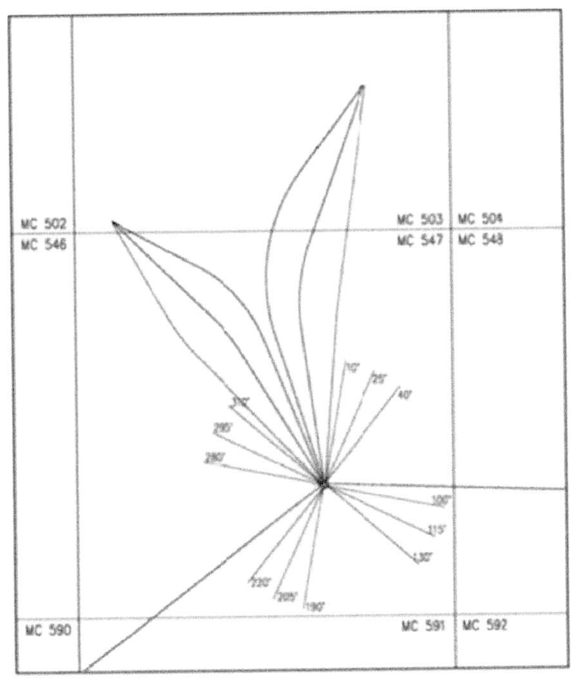

Fig. 19.7 Subsea layout for the Who Dat development. *(Source: Simms, G.J., Cooley, B., Fowler, R., Leontaritis, K.J., Krishnathasan, K., 2012. Subsea Who Dat project – producing light and heavy oil in a deepwater subsea development. In: OTC 23378. Offshore Technology Conference. Houston, TX, Apr 3–May 3.)*

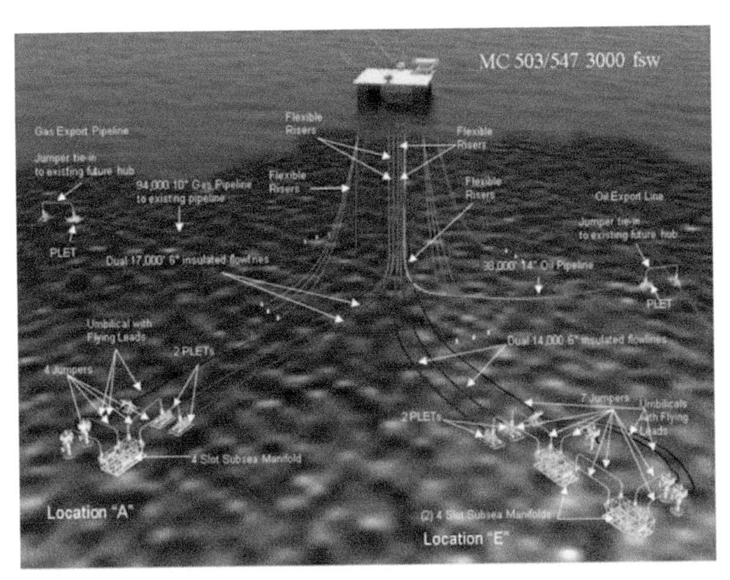

Fig. 19.8 Manifolds, flow lines, and platform for the Who Dat development. *(Source: Simms, G.J., Cooley, B., Fowler, R., Leontaritis, K.J., Krishnathasan, K., 2012. Subsea Who Dat project – producing light and heavy oil in a deepwater subsea development. In: OTC 23378. Offshore Technology Conference. Houston, TX, Apr 3–May 3.)*

Transient Production. For hydrate management, methanol injection into the subsea system is used during start-up and shutdown. For long-term unplanned shutdowns, flow lines and risers are depressurized and displaced if necessary.

Remediation. Although pigging capabilities are available, plans are not to employ an aggressive pigging program and instead to use surveillance and a chemical pill if wax buildup is suspected.

19.3.4 Well Configuration

GoM wells on the shelf are predominately low-angle directional wells that are cased and perforated. Deepwater wells are a mixture of moderate- to high-angle wells, most with open-hole gravel packs and open-hole frac packs, a completion technique that merges hydraulic fracturing and gravel packing. Unconsolidated sands require sand control, and in the deepwater GoM, frac packing is the most common completion technique followed by high-rate water pack and open-hole gravel pack (Weirich et al., 2013).

A larger percentage of dry tree wells are high-angle wells with longer completion intervals than subsea wells that are likely to be near vertical with shorter step-outs. Single completion shallow-depth wells that are near vertical allow for relatively simple wire line operations, but deep boreholes with multiple completions and deviated or complex boreholes are more complicated to reenter and perform remediation efforts. Workover and repair cost on complex well configurations are expected to be more expensive than on simple wells for all other things equal.

19.3.5 Old vs. New

Generally speaking, new wells have higher production and lower unit cost, less problems after start-up, and fewer repairs and maintenance compared with old wells. For all things equal, older structures will require greater repair and maintenance cost than younger structures. Offshore, platform space and load capacity represent restrictions, and low-cost operating and maintenance options for mature production may not be as viable as onshore. For example, in mature oil fields, increasing volumes of produced water may restrict production because of equipment capacity limitations. In some cases, the operating cost of large processing facilities may simply be too much to be carried by production from a smaller field.

19.3.6 Materials and Supplies

Materials and supplies are an important component of operating cost and depend on product type (oil, gas, condensate) and quality (e.g., sour, paraffinic, corrosive, water cut) of production. The cost to handle produced water and the scale and corrosion that results typically increases over field life (Cavaliaro et al., 2016). When seawater for injection is mixed with fresher aquifer water, scale may develop that will require inhibitor to manage. Gas compression is typically required on mature properties and require a prime driver (e.g., diesel engine) and a fuel source, and if adequate space is not available, a new platform may need to be installed. The prime driver being equipment and long lasting is depreciated, while the fuel source, if gas is produced at site, is essentially free, although sales volume will be reduced. If fuel has to be transported and stored at site or on a nearby lease, operating cost will increase.

Example. Eugene Island 11 complex upgrade

The Eugene Island 11 platform was designed with a capacity of 500 MMcfd and 6000 bopd to service production from five wells in federal waters and five wells in Louisiana state waters (Fig. 19.9). In September 2010, Contango installed a companion platform and two pipelines to access alternative markets. In July 2014, Contango reported investing $12 million to build and install a turbine-type compressor to serve all wells.

Fig. 19.9 Eugene Island 11 complex. *(Source: Contango.)*

19.3.7 Development Tradeoffs

Development trade-offs are important to recognize since they have a significant impact on operating cost. There are always trade-offs between capital and operating expense that engineers make in design, but these are unusually hard to disentangle, and for the inexperienced may not even be recognized. At the design stage, these trade-offs can have a large impact on where and how costs are allocated over the life cycle of production, and so, comparisons between developments need to be made carefully.

The selection of dry trees versus wet trees, for example, is readily identified and will have a significant impact on future operating costs, but the manner flow assurance is designed and managed is much less obvious and might have to be inferred. Prevention of hydrates can be undertaken by using insulation or heating to maintain temperature high enough to operate within the hydrate-free zone, a high-capital low-operating-cost option, or MEG dosing may be continuously employed, a low-capital high-operating-cost option.

Example. Ukpokiti operating strategies

Conoco's first offshore development in Nigeria was the Ukpokiti field that employed a converted FPSO designed to process 20,000 bopd, 40,000 bbl/day water injection, 25 MMcfd produced gas, and 14,000 bbl/day of produced water (Steube, 2000). Four operating strategies were considered in development selection (Table 19.3). Each strategy had a specific objective and involved different trade-offs. For example, strategy B's objective was to balance downtime risk and volume while managing market value. Strategy C was eliminated due to the risk of increased downtime, and strategy D, in which Conoco shared logistic support services for personnel and materials with another operator located 16 mi away (CanOxy at the Ejulebe field), was eventually selected.

Table 19.3 Operating Strategies in Nigerian Offshore Field Development

Strategy	Objective	Direct OPEX ($/day)	Lifting Cost ($/bbl)
A	Full operational control	32,000	1.64
B	Control through min cost	22,000	1.13
C	Minimize cost	18,000	0.96
D	Control through sharing	26,000	1.34

Source: Steube, C., 2000. Offshore operating costs for marginal fields: selecting the best operating strategy. In: SPE 66098. SPE Technical Conference and Exhibition. Abuja, Nigeria, Aug 7–9.

19.3.8 Produced Water

As reservoirs mature, especially if secondary recovery methods are used, the quantity of water increases and often exceeds the volume of the hydrocarbons before wells are shut in. Initially, water represents a small percentage of produced fluids, but over the life of the well, the water-to-hydrocarbon ratio increases. The cost of producing, handling, and disposing of the produced water defines the economic life of some fields.

Most offshore platforms dispose produced water directly into the ocean but have to meet stringent regulations on the entrained and dissolved oil and other chemicals in the produced water. In the GoM, produced water discharge specifications are based on monthly average oil and grease content of 29 ppm. Some deepwater GoM operators self-impose more stringent discharge requirements of 10 ppm to safeguard against oil sheen in the overboard discharge (Wiggett, 2014).

The wide variation in the concentration and type of constituents sometimes make produced water challenging to treat and discharge (Usher et al., 2015). The physical and chemical properties of produced water vary depending on the geographic location of the field, the geologic formation, and the type of hydrocarbon produced (Veil and Clark, 2011). The major constituents of concern are salt content (expressed as salinity, conductivity, or total dissolved solids), oil and grease (organic compounds captured through an n-hexane extraction procedure), inorganic and organic compounds introduced as chemical additives to improve drilling and production operations, and naturally occurring radioactive material.

Changes in produced water due to pressure and temperature changes from production can have serious impacts through precipitation of scales and corrosion that may lead to leaks and costly repairs if not inhibited and monitored (Jordan and Feasey, 2008). Inhibition of most scales is through the application of organic compounds that act to poison (prevent) the growth sites of the crystals. Corrosion mitigation typically takes investment in corrosion-resistant alloys and/or a chemical corrosion-inhibition/monitoring program.

19.3.9 Water and Gas Injection

Water and gas may be injected into reservoirs to supplement recovery. Water will generally require treatment, and the type of treatment and cost depends on the source and issues identified. If operators inject water into reservoirs to maintain pressure, they typically use seawater with some chemicals since this is the lowest-cost option (McClure, 1982). In some cases, subsurface water may be processed if seawater causes injection problems (Ogletree and Overly, 1977). To inject produced water, suspended solids and oil must be removed to an appropriate degree to avoid plugging and fouling the reservoir. About 5% to 10% of the produced water in the GoM is injected for pressure maintenance (Veil and Clark, 2011).

The operational requirements for seawater injection generally require filtration, deoxygenation, and corrosion control. The details of the treatment steps are specific to each project. For example, some projects may require injected water to be filtered to 1 μm, while other systems may require 10 μm. Deoxygenation in some systems may be achieved by chemical addition, other systems may require gas stripping and chemical treatment, and each process will have its own capital and operating cost requirement.

Example. Deepwater GoM water injection projects

Water injection has not been used in most deepwater GoM developments because in the majority of Neogene reservoirs there is good primary recovery, while drilling cost and facility limitations are significant. In over 80 fields and 450 reservoirs circa 2015, water injection was implemented in only 18 reservoirs in 13 fields (Li et al., 2013). Fields that have used water injection for one or more sands/reservoirs include Amberjack, Bullwinkle, Holstein, Horn Mountain, Lena, Lobster, Mars, Morpeth, Petronius, Pompano, Ram-Powell, and Ursa/Princess.

In five of these fields (Holstein, Horn Mountain, Lena, Lobster, and Petronius), water injection commenced near the start of production and has been extensive. Four fields (Amberjack, Bullwinkle, Morpeth, and Ram-Powell) have had only short periods of water injection. The Mars and Ursa/Princess fields are examples where primary production was followed by water injection after the reservoir pressure was reduced in late secondary recovery projects.

In Middle to Lower Miocene age reservoirs such as Atlantis, Mad Dog, K2, Shenzi, Neptune, Tahiti, Blind Faith, and Thunder Horse, the structural style of the reservoirs and reservoir properties indicate that water injection will be beneficial at later stages of recovery, and the facilities for many of these fields were designed with water injection capacity. Since 2001, when the Paleogene horizon was first encountered, there are about two dozen fields including Great White, Jack, St Malo, and Cascade and Chinook that will benefit from water injection if the technical hurdles can be overcome.

Gas can be injected into reservoirs to supplement recovery by maintaining reservoir pressure or as a means of disposing of gas that cannot be flared or transported. Generally, there is no need to control hydrocarbon dew point as in export gas since injected gas will get hotter not cooler, but it may be attractive to remove heavy hydrocarbons for economic reasons (Jahn et al., 2008). Dehydration is always required to avoid water dropout and corrosion problems. Gas injection is rarely used in the GoM except to enhance oil recovery since an extensive pipeline network exists and offshore gas has always been in demand as a fuel source and for gas-lift operations.

19.3.10 Artificial Lift

The most common types of artificial lift for offshore oil production in the GoM are gas lift and downhole pumping. In gas-lift systems, gas is injected directly into the wellbore to lower the hydrostatic head. Gas compression depends on the source pressure, and since gas lift is essentially a closed-loop system except at start-up where another source of gas (e.g., nitrogen) may be required, little gas is consumed in operations. For downhole pumping, power generation is required to drive the electric pumps.

19.3.11 Well Intervention

Well interventions (aka workovers) are performed throughout the life of a well to protect value (e.g., by preventing corrosion and maintaining gas-lift systems) or to create value (e.g., by shutting off water or adding gas lift to accelerate production) and are a primary means of protecting or increasing reserves and production. There is a wide variation among operators on their production surveillance, and the frequency of intervention is dependent on many factors—well vintage, production type, well type, operations policies, production level, oil and gas prices, corporate budget, etc. Some operations apply sophisticated methods in evaluation; others use simple well reviews. The best stimulation candidates with the greatest business value are usually high-production, high-recovery wells.

19.3.12 Operating and Maintenance Budget

During production, the surface facilities are managed to maximize system capacity and availability. There are many pieces of equipment involved in separation and treatment, and the equipment is monitored for performance and periodically inspected and tested. The production crew is responsible to monitor well constraints that may limit the reservoir potential and remediation options. If the cost of action is significant an economic evaluation may be performed; otherwise, these activities are part of the annual operating budget and are prioritized. If the operating budget for the year does not allow all activities to be performed, they are postponed.

Maintenance refers to how the equipment is maintained to ensure that it is capable of performing the tasks for which it was designed. Since mechanical performance deteriorates with use due to normal wear and tear, corrosion, vibration, contamination, etc. that may lead to failure and safety issues, the maintenance department plays an important role to achieve production objectives. Maintenance strategy may be proactive or reactive, and maintenance budgets will vary with the annual operating budget. Different equipment will be maintained in different ways depending on their criticality and failure mode (Logan et al., 1984). Preventative maintenance work is usually based on a schedule of planning cycles within the year.

19.3.13 Market Conditions

Offshore markets in the GoM are highly competitive and described by supply and demand conditions that often lag changes in oil and gas prices. When oil and gas prices are high, demand for services, chemicals, service boats, etc. is usually high, which places upward pressure on day rates and chemical prices. In weak markets, when oil and gas prices are low, prices for services and chemicals are reduced. Operators continually make operational decisions on workover and maintenance schedules in response to market conditions.

REFERENCES

Bai, Y., Bai, Q., 2012. Subsea Engineering Handbook. Gulf Professional Publishing, Waltham, MA.
Cavaliaro, B., Clayton, R., Campos, M., 2016. Cost-conscious corrosion control. In: SPE 179949. SPE International Oilfield Corrosion Conference and Exhibition. Aberdeen, Scotland, May 9–10.

Cochran, S., 2003. Hydrate control and remediation best practices in deepwater oil developments. In: OTC 15255. Offshore Technology Conference. Houston, TX, May 5–8.
Darling, P., 2011. SME Mining Engineering Handbook, 3rd ed. vols. 1 & 2. Society Mining Metallurgy & Exploration, Denver, CO.
Dickens, R.N., Lohrenz, J., 1996. Evaluating oil and gas assets: option pricing methods prove no panacea. J. Financ. Strateg. Decis. 9 (2), 11–19.
Doering, M.A., 1993. Acquisition pitfalls: operating-cost forecasts. In: SPE 25841. SPE Hydrocarbon Economics and Evaluation Symposium. Dallas, TX, Mar 29–30.
Edwards, J., 2004. Improving energy efficiency in E&P operations. In: SPE 86604. SPE International Conference on Health, Safety, and Environment in Oil and Gas Exploration and Production. Calgary, Alberta, Canada, Mar 29–31.
Gallun, R.A., Wright, C.J., Nichols, L.M., Stevenson, J.W., 2001. Fundamentals of Oil & Gas Accounting, 4th ed. PennWell Books, Tulsa, OK.
Hartman, H., Mutmansky, J.M., 2002. Introductory Mining Engineering, 2nd ed. John Wiley & Sons, Hoboken, NJ.
Hustrulid, W., Kuchta, M., 2006. Open Pit Mine Planning & Design. Volume 1. Fundamentals, 2nd ed. CRC Press, Boca Raton, FL.
IHS Global Inc, 2015. Oil and gas upstream cost study. In: Energy Information Administration – Upstream Cost Study Final Report, Washington, DC.
Jahn, F., Cook, M., Graham, M., 2008. Hydrocarbon Exploration and Production, 2nd ed. Elsevier, Amsterdam, the Netherlands.
Jamaluddin, A.K.M., Nighswander, J., Joshi, N., Calder, D., Ross, B., 2002. Asphaltenes characterization: a key to deepwater development. In: SPE 77936. SPE Asia Pacific Oil and Gas Conference and Exhibition. Melbourne, Australia, Oct 8–10.
Jordan, M.M., Feasey, N.D., 2008. Meeting the flow assurance challenges of deepwater developments: from capex development to field start up. In: SPE 112472. SPE North Africa Technical Conference and Exhibition. Marrakech, Morocco, Mar 12–14.
Kashou, S., Matthews, P., Song, S., Peterson, B., Descant, F., 2001. Gas condensate subsea systems, transient modeling: a necessity for flow assurance and operability. In: OTC 13078. Offshore Technology Conference. Houston, TX, Apr 30–May 3.
Kondapi, P.B., Chin, Y.D., Srivastava, A., Yang, Z.F., 2017. How will subsea processing and pumping technologies enable future deepwater field developments. In: OTC 27661. Offshore Technology Conference. Houston, TX, May 1–4.
Li, X., Beadall, K.K., Duan, S., Lach, J.R., 2013. Water injection in deepwater, over-pressured turbidities in the Gulf of Mexico: past, present, and future. In: OTC 24111. Offshore Technology Conference. Houston, TX, May 6–9.
Logan, W.A., Hasell, B.H., Trussel, P., 1984. An integrated computerized maintenance system for offshore operations. In: SPE 12983. European Petroleum Conference. London UK, Oct 25–28.
Manfield, P., Nisbet, W., Balius, J., Broze, G., Vreenegoor, L., 2007. Wax-on, wax-off: understanding and mitigating wax deposition in a deepwater subsea gas/condensate flowline. In: OTC 18834. Offshore Technology Conference. Houston, TX, Apr 30–May 3.
McClure, C.C., 1982. Seawater injection experience – an overview. In: SPE 9630. SPE Middle East Oil Technical Conference and Exhibition. Manama, Bahrain, Mar 9–12.
Ogletree, J.O., Overly, R.J., 1977. Sea-water and subsurface-water injection in West Delta block 73 waterflood operations. In: SPE 6229. SPE Annual Offshore Technical Conference. Houston, TX, May 3–5.
UK Oil & Gas Authority, 2017. Analysis of UKCS operating costs in 2016, London, UK.
Seba, R.D., 2003. Economics of Worldwide Petroleum Production. OGCI and Petroskills Publications, Tulsa, OK.
Simms, G.J., Cooley, B., Fowler, R., Leontaritis, K.J., Krishnathasan, K., 2012. Subsea Who Dat project – producing light and heavy oil in a deepwater subsea development. In: OTC 23378. Offshore Technology Conference. Houston, TX, Apr 3–May 3.
Steube, C., 2000. Offshore operating costs for marginal fields: selecting the best operating strategy. In: SPE 66098. SPE Technical Conference and Exhibition. Abuja, Nigeria, Aug 7–9.

Steube, C., Albaugh, E.K., 1999. Evaluating OPEX for marginal fields in international operations. Offshore Mag. 59(3).
Usher, M., Herrington, D., Kozikowski, M., Scott, E., 2015. Offshore waste water management system reduces operating costs. In: OTC 26277. Offshore Technology Conference. Rio de Janeiro, Brazil, Oct 27–29.
Veil, J.A., Clark, C.E., 2011. Produced water volume estimates and management practices. In: SPE 125999. SPE International Conference on Health, Safety and Environment in Oil and Gas Exploration and Production, Rio de Janeiro, Brazil, Apr 12–14.
Walther, J.V., 2013. Earth's Natural Resources. Jones and Bartlett Learning, Burlington, MA.
Weirich, J., Li, J., Abdelfattah, T., Pedroso, C., 2013. Frac packing: Best practices and lessons learned from more than 600 operations. SPE Drill. Complet. 28 (2), 119–124.
Wiggett, A.J., 2014. Water – chemical treatment, management and cost. In: SPE 172191. SPE Saudi Arabia Annual Technical Symposium and Exhibition. Al-Khobar, Saudi Arabia, Apr 21–24.

CHAPTER TWENTY

Financial Statements and Other Methods

Contents

Abstract

Public oil and gas companies listed on US stock exchanges are required to disclose operating expense in their financial statements, but because data are consolidated over many different assets and areas, its ability to inform operating cost in a specific region, field, or asset is limited except in special circumstances. In this chapter, financial data for 10 public oil and gas companies with the majority of their production and reserves in the GoM are reviewed. The strengths and weaknesses of lease operating statements, surveys, software tools, and computer methods in determining operating cost are also reviewed. The UK North Sea operating cost survey is the best publicly available offshore data in the world, and results are highlighted.

Decommissioning Forecasting and Operating Cost Estimation
https://doi.org/10.1016/B978-0-12-818113-3.00020-5

20.1 FINANCIAL STATEMENTS

Companies maintain detailed production cost for their properties, but this information is not reported unless required by regulation. The amount and quality of operating cost data available for analysis are therefore tightly constrained, and misconceptions often arise due to the lack of transparency and misunderstanding operational requirements (Table 20.1).

Table 20.1 Lease Operating Cost Data Sources, Strengths, and Limitations

Source	Limitations	Strengths
Financial statements	Consolidated statements across multiple properties and geographies restrict comparisons, regional players diverse	Publicly disclosed according to standardized categories, regularly audited, high-quality field data available in some cases
Lease operating statements	Company- and asset-specific, confidential, not representative	High-quality, detailed information asset and time specific
Benchmarking studies	Small number of participants, difficult to normalize across diverse assets, confidential to participants	High-quality, detailed information based on lease operating statements and normalized according to common cost categories
Survey methods	Narrow focus, variable quality, lack of transparency unless assumptions clearly specified	Snapshot of industry conditions, direct linkage with service company cost, commonly reported
Computer methods	Significant time and resource for development requiring experts, narrow application focus	Advanced quantitative approach, measurable benefits
Factor models	Often hypothetical and not supported by empirical data or historical activity, does not reflect site-specific conditions	Easy to apply, may capture aggregate cost if based on historical data, reasonable bounds can usually be inferred
Production Handling Agreements	Contract terms are only partially disclosed and complex terms and options usually not incorporated in analysis	Factor model application defensible and reasonably easy to develop and bound with transparent assumptions
Activity-based costing	Quality of evaluation depends on user experience and assumptions, requires detailed evaluation and data collection, potentially time-consuming	Detailed, bottom-up cost model can capture cost elements in direct, transparent fashion, includes engineering and market conditions

20.1.1 Regulatory Requirements

Exploration and production companies are required to present certain information about their oil- and gas-producing activities specified in Financial Accounting Standards Board Accounting Standards Certification (FASB ASC) Topic 932—Extractive Industries—Oil and Gas (Gallun et al., 2001).

According to Item 1204 of Regulation S-K, public companies are required to disclose operating expense according to direct, other, and indirect cost categories. Direct expenses primarily include labor cost, chemicals, services, repair and maintenance, and rentals. Other lease operating expenses include production and severance taxes, insurance, and gathering and transportation fees. Indirect operating expenses normally refer to overhead, general, and administrative expenses.

20.1.2 Direct Lease Operating Expenses

Direct lease operating expenses (LOE) are the costs associated with the recovery of produced hydrocarbons from wells and usually include the following:

1. Labor to operate equipment and facilities and to provide service for production
2. Materials, supplies, and fuel consumed and services utilized in operating the wells and related equipment and facilities
3. Services used in daily operations, storage, handling, transportation, processing, and measurement
4. Repair and maintenance
5. Rental of special and heavy tools and equipment
6. Equipment and facilities with a life of less than one year
7. All technical costs other than those specifically classified as capital costs

Labor

Labor cost is one of the more significant offshore operating expenses. Labor costs include the salary of employees who are directly involved in production activities, services (such as general repairs and maintenance performance), and supervision. Payroll and benefits for corporate staff are not included. Employee benefits, such as insurance and medical service, may be included. Some benefits are required by local laws, and employee benefit packages usually vary by company.

Example. Delta House

Delta House is a semisubmersible facility located in Mississippi Canyon 254 that has a 48-person living quarters with permanent crew requirements for 24 person working on a 2-week on/off basis (Fig. 20.1). Supply boat trips to Fourchon shore base near Grand Isle, Louisiana, take about 6 h, and a helicopter ride to the Galliano bases takes about an hour.

Fig. 20.1 Delta House semisubmersible in Mississippi Canyon block 254. *(Source: LLOG.)*

Materials and Supplies

Materials and supplies refer to equipment, tools, and other supplies that are used in production activity and repair and maintenance. Materials and supplies are usually held in inventory and charged to costs at the time they are sent to operations. The costs of materials and supplies purchased or furnished include not only the cost of the materials and supplies themselves but also all costs associated with acquiring and transporting the materials and supplies. These associated costs may include broker fees, loading and unloading fees, license fees, and in-transit losses not covered by insurance. Fuel, power, and water consumed in operating the wells and related equipment are sometimes considered a subcategory of materials and supplies and reported in a separate fuel, power, and water account.

Services, Logistics, and Transportation

Services refer to activities required for daily operations and to maintain production and include catering and personnel transportation and workover operations. Logistics refers to the organization of people and supply and storage of materials. The transport of people is related to the mode of manning the operation. For a typical GoM shelf operation, the transport of personnel to and from facilities is by helicopter or crew boat and is commonly shared among several operations or operators, while for a deepwater facility, crew transport is always by helicopter because of the greater distance, and sharing is not common. Material transport is by supply boat. Transportation cost include all expense involved in shipping materials and supplies and staff to site.

Repair and Maintenance

Repair and maintenance refer to normal activities to maintain safe and continued production and include maintenance of wells (e.g., when corrosion has made it necessary to replace downhole production equipment), related equipment, and facilities (e.g., repairing a generator and lubricating a pump) and the maintenance and repair of the structure (e.g., inspecting and replacing corroded or pitted braces and joints). When wells are involved, repairs and maintenance are usually referred to as workovers, and if the purpose is to restore or increase production, such costs are expensed. If the workover adds new proved reserves, costs are capitalized.

The performance of equipment deteriorates with use, and maintenance is required to ensure that the equipment and offshore structure are capable of safely performing the tasks for which it was designed. Wells and equipment break down and require intervention, and all offshore structures require periodic inspection and maintenance to provide a safe working environment. The cost to repair system failures includes the additional service company personnel and logistics expenses incurred, while the cost of the replacement equipment will be depreciated. Different maintenance strategies are employed by operators depending on the design of the system and criticality of the equipment. For example, an operator may employ a spare export pump, run to failure, and then switch to the spare pump during repair.

Repair and maintenance are often divided into ordinary repair and maintenance and major repair. Ordinary repair and maintenance, which repairs or maintains the asset to its original operating condition, are generally expensed when incurred and included in direct lease operating expenses. In contrast, major repair (e.g., overhaul), which materially increases the useful life or the productivity of an asset, should be capitalized and amortized in subsequent fiscal periods. The duration and the distribution of the amortization are regulated by US GAAP based on the type of asset. Amortization tends to smooth out the spending incurred through major repair.

Example. Hurricane damage at Mars

In 1996, Shell brought the giant Mars field into production using a tension leg platform (Fig. 20.2), and through 2017, the field has produced over 750 MMboe and is estimated to hold about 4 Bboe hydrocarbons in place. In 2005, the Mars TLP was directly in the path of Hurricane Katrina and suffered extensive damage when a drilling rig on the platform was toppled by the storm and shattered the upper decks and living quarters (Fig. 20.3). Two export pipelines were also damaged by a dragged anchor from a semisubmersible

Fig. 20.2 Mars tension leg platform. *(Source: Shell.)*

drilling rig (Paganie and Buschee, 2005). Shell chartered a six-story flotel (floating hotel) from the North Sea and linked it to Mars via a pontoon during repairs that required 8 months to complete and 1 million man-hours estimated to cost between $250 and $300 million (Paganie, 2006).

The Mars field is so large[1] that any loss—even complete destruction—would have been repaired/replaced, and there was never any question that the field would return to production. At capacity, Mars can produce 140,000 bopd, and on a full production basis, hurricane repairs and maintenance are estimated as $5.4 per barrel capacity:

$$\frac{\$275\,\text{million}}{(140{,}000\,\text{bopd})(365\,\text{d})} = \$5.4/\text{bbl}$$

The operator also incurred approximately $10 million loss per day business interruption since all production ceased during the repair.

It is not unusual for a company to build up a separate account outside lease operating expenses for amortizing major repair cost, such as would occur with significant and long-lasting hurricane damage. Separated accounts will carve out all expenses related to major repair from direct operating expense, including labor, transportation, material and service, rental tools, and equipment. For companies that build separate accounts, the major repair cost does not affect lease operating expense but will increase the final total expense.

[1] In 2015, a second TLP (Olympus or Mars B) was installed on an adjacent lease about a mile southwest of Mars at a cost of $7.5 billion to continue to develop the field (Newberry, 2014).

Fig. 20.3 Hurricane Katrina damaged the Mars platform and required significant repairs. *(Source: Shell.)*

20.1.3 Other Lease Operating Expenses

Production and Severance Taxes

Property tax for producing properties depends on location, and ad valorem taxes are levied and administrated by local tax districts. In the United States, oil and gas production books for tax purposes must be kept on an individual property basis. Production and severance taxes can be revenue- or volume-based and are normally a small part of production cost. In federal waters, there are no production taxes, while onshore and state water operations are subject to production and severance taxes.

Gathering and Transportation Fees

Gathering and transportation refer to the cost to gather and transport the raw fluids to the processing facility and the cost to transport the processed oil and gas to shore. These costs are usually not broken out, and they may or may not be reported separate from direct LOE. Export fees are often reported separately when the export pipeline owner is a third party, but large variation on reporting practices exists. If the export pipeline owner is a subsidiary or affiliate of the structure owner, export fees may be included within other LOE or allocated and accounted for differently. Gathering and transportation costs depend on the cost and age of the system and the volume and type of fluid that flows through the system. Most pipeline tariffs in the GoM are proportional rates and on a volume basis are usually no more than 3% to 5% commodity prices.

Example. Energy XXI Pipeline, LLC South Timbalier tariffs

The cost to transport crude from South Timbalier block 27 to Fourchon Terminal, Lafourche Parish, Louisiana, is 77.79 cents per barrel and from South Timbalier block 63 to Fourchon Terminal is 155.56 cents per barrel effective 1 July 2016.

Insurance

Companies maintain insurance for some, but not all, of the potential risks and liabilities associated with production. Most larger companies are self-insured, but many smaller and midsize companies rely on coverage to access capital markets. Offshore, the major risks are property damage due to hurricanes and severe weather, and companies may purchase coverage for removal of wreck, control of well, and business interruption. The oil and gas industry suffered significant damage from Hurricanes Ivan, Katrina, and Rita in 2004 and 2005 and Gustav and Ike in 2008, and as a result, insurance costs increased in the immediate aftermath of the storms. From 2009 to 2018, there has been no major hurricane to hit the GoM oil patch, and the insurance market has considerably weakened. Due to market conditions, premiums and deductibles for certain insurance policies may be economically unavailable or available only for reduced amounts of coverage. If a company cannot obtain insurance or believe the cost is excessive relative to the risks presented, insurance coverage may decrease.

Technically, the basic function of offshore property/casualty insurance is the transfer of risk, to reduce financial uncertainty, and to make loss manageable. It does this by substituting payment of a small fee—an insurance premium—to a professional insurer (underwriter[2]) in exchange for the assumption of the risk of a large loss (property damage limit) with a small and uncertain probability and the promise to pay in the event of such loss. Insurance companies collect premiums from payers and invest the funds, so that as events arise in the future, they have the funds to pay policyholders. To remain a going concern, payouts have to be less than funds collected and invested.

There are several types of coverage in the offshore oil and gas industry, the most common types being (Sharp, 2000) the following: business interruption, wind damage, physical damage, removal of wreck, control of well, operator extra expense, and pollution liabilities. Generally, the larger the number of premium payers (premium pool) in the market, the lower the premiums per policyholder, but as the premium pool shrinks or if the loss incidence increases above expected values, insurers will charge more premium and provide less coverage to remain a viable business enterprise. In theory, the more accurate insurers are able to estimate probable losses, the lower the amount of premium that will need to be collected, but this is most valid when sample sets are large and events are common (e.g., car accidents) as opposed to the infrequent events (e.g., blowouts and hurricane destruction) that occur in the oil and gas industry.

In the GoM, there were no major hurricanes that impacted the offshore industry from 2009 to 2018, and since significant decommissioning activity has occurred during the same period, there is much less exposure, and one would expect a soft insurance market to develop for windstorm and removal of wreck coverage, price reductions, and broadening terms and conditions offered by the industry to attract business due to a reduced premium pool. If premiums pools are reduced too much, however, coverage may become increasingly difficult for operators to obtain at affordable rates. The supply and demand dynamics are complicated.

It is difficult to give any sort of pricing guidelines or premium estimates for insurance products in the GoM due to the unique makeup of each operator's business, dynamic market conditions, and the proprietary nature of the business. Historically, most property damage and removal of wreck include a deductible, usually a few percent (e.g., 5%) of the structure replacement value or expected cost to remove, and premiums are usually priced at a few percent of the replacement value. The limit of liability equals the total coverage purchased; for example, the insured pays a $10 million premium for $150 million

[2] Underwriters got their name from the practice in seventeenth century England of investors signing their names as guarantors, for a fee, under posted listings of marine voyages and cargoes.

windstorm damage limit. In soft markets, the insurer may reduce the damage limit, increase deductibles, increase premiums, or aggregate events.

Example. Stone Energy windstorm insurance, 2005–13

In 2005, Stone Energy's windstorm insurance for all of their GoM platforms was covered under $150 million property damage limit per named storm and a $1 million deductible per storm (Siems, 2014). The annual premium for this coverage was reported as $6.5 million, or about 4% of the damage limit. In 2005, Hurricane Rita toppled eight Stone Energy platforms with 29 wells, and the following year, insurance cost in the GoM soared as underwriters adjusted loss calculations and operators dropped out of the premium pool.

In 2006, Stone Energy held insurance for windstorm damage for one-third of their platform inventory under a $100 million property damage for the season and a $25 million deductible per storm. The annual premium for coverage was $55 million, an astonishing 55% of the damage limit. Deductibles increased 25 times from the previous year, premiums increased 8.5 times, and the coverage limit was broadened to storm season rather than storm event.

In 2008, after six structures and 34 wells were toppled by Hurricane Ike, Stone Energy accelerated decommissioning activity to reduce exposure and potential future liability. From 2008 to 2011, they plugged 419 wells and from 2012 to 2014 decommissioned 102 idle structures. In 2013, Stone Energy completely eliminated their windstorm insurance.

20.1.4 Indirect Operating Expenses

The general rule for charging costs directly to an operation is that those charges must be for work physically performed at the project site, or if not on-site, they must be performed specifically and exclusively for that operation (Seba, 2003). All other costs that may be incurred at a distant location for a number of different operations are considered indirect operating costs or overhead and need to be allocated. Supervisory and administrative expenses that are shared with other properties are often prorationed.

Indirect operating expenses include the expenses that are not directly related to production activities but provide a benefit to operations. General and administrative expenses are costs incurred for overhead, including payroll and benefits for corporate staff, costs of maintaining headquarters, costs of managing production and development operations, audit and other fees for professional services, and legal compliance. The major component (sometimes the only component) of indirect operating expenses is operating overhead expense. Operating overhead expense refers to expense that, although incurred in a firm's day to day operations, cannot be directly assigned to a specific department or product.

The specific method for allocating indirect expenses is arbitrary, but specific rules should be developed to maintain uniformity. A common rule is to charge a percentage on top of certain direct costs. The percentage is often determined as a percent of direct operating costs plus a percent of capital expenditures required for the project when capital is spent.

Table 20.2 Lifting and Total Expense Metrics

$$\text{Lifting cost per Mcfe} = \frac{\text{Total LOE} + \text{Indirect operating expense}}{\text{Mcfe production}}$$

$$\text{Lifting cost per Mcfe} = \frac{\text{Total expense} - \text{Other LOE} - \text{Total major repair}}{\text{Mcfe production}}$$

$$\text{Total expense per Mcfe} = \frac{\text{Total expense}}{\text{Mcfe production}}$$

20.2 LIFTING AND PRODUCTION COST

Lifting cost, production cost, and total expense per unit of production are ratios that measure how much a company spends to produce (lift) one volume unit of hydrocarbon out of the ground and to make ready for transport. Generally speaking, the smaller the lifting and production cost, the greater the ability of the company to profit from its production and the more efficient its extraction processes. Production cost and total expense do not have the same meaning as lifting cost (in an accounting sense) and include more terms. Lifting cost is the most visible metric directly related to LOE, and it is important to understand what the metric includes, what it doesn't include, and how it is computed since there is no universal definition and regulatory agencies do not require reporting.

Lifting cost usually includes the cost to operate and maintain wells and related equipment and facilities and may or may not include indirect operating expense (Table 20.2). Production cost is usually defined as lifting cost plus gathering and transportation costs and may include production taxes, or these may be reported separately. Since gathering and transportation costs are often small on a unit basis, their inclusion in lifting costs will usually not significantly impact the metric. If lifting cost is computed based on total expense, it will be greater than a ratio that does not include other expense and major repairs.

Example. Energy XXI GOM operating expense, 2012–14

Lifting cost, production cost, and total operating expenses per boe as reported by Energy XXI GOM from 2012 to 2014 are shown in Table 20.3. Total lease operating expense includes direct LOE, insurance, and workover and maintenance that are broken out separately. Production taxes and gathering and transportation are also reported separately. Depreciation, depletion, and amortization (DD&A) of capital expenditures and overhead and management of salary company-wide via general and administrative (G&A) expenses contribute to total operating expenses. Note that DD&A accounts for more than half of total operating expense.

Table 20.3 Energy XXI GOM Operating Expenses per Barrel Oil Equivalent, 2012–14

	2014	2013	2012
Lease operating expense			
Insurance expense	1.90	2.08	1.77
Workover and maintenance	4.04	4.15	3.49
Direct lease operating expense	16.31	15.23	13.99
Total lease operating expense per boe	22.25	21.46	19.25
Production taxes	0.33	0.33	0.45
Gathering and transportation	1.43	1.54	1.01
Depreciation, depletion, and amortization	25.75	23.95	22.76
General and administrative	5.87	4.56	5.34
Other	2.19	2.08	1.98
Total operating expenses per boe	57.82	53.92	50.79

Source: Energy XXI GOM.

Companies normally identify their production as "primarily oil" or "primarily gas" using total aggregate oil and gas production volumes via the gas-oil ratio (GOR), expressed in cubic feet natural gas per barrel crude oil. Condensate from natural gas production is often included in the liquid reporting unless it exceeds 10%–15% crude volumes in which case it is broken out separately as required by SEC regulations. If hydrocarbon GOR < 10,000 cf/bbl, the company can be considered an "oil" company; otherwise, the company is classified as a "gas" company. An operator will usually report lifting cost as $/boe if production is primarily oil and $/Mcfe if primarily gas, but this is not a requirement, and operators may report production in units different than their primary product.

Lifting cost is a popular measure, used internally within companies as a means of measuring operating efficiency, and applied by business analyst and bond-rating agencies to compare operating performance across companies (e.g., Moody's, 2009). However, since lifting cost is not required to be released in financial statements, the ratios are often calculated based on different accounting treatments and will vary from company to company. Comparisons need to be made cautiously. Except in special situations, lifting cost is measured over all properties within a company's reporting divisions and fiscal period, and so, the ability to extract useful information for a specific region or property is rare.

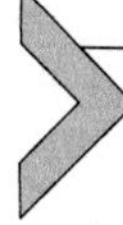

20.3 COMPANY FINANCIAL STATEMENTS

20.3.1 Sample

Ten public oil and gas companies with the majority of their production and reserves in the federal waters of the GoM were identified and LOE data tabulated from 2008 to 2015 (Table 20.4). All the companies are independents, primarily regional players, except Freeport-McMoRan Oil & Gas, which is a subsidiary of Freeport-McMoRan, a mining

Table 20.4 Unit Production Cost Reported by Gulf of Mexico Companies, 2008–15

Company	Item	2015	2014	2013	2012	2011	2010	2009	2008
ATP	Direct LOE				11.57	10.34	11.54	11.31	
($/boe)	Workover				1.87	3.26	5.76	3.15	
	All				13.44	13.60	17.30	14.47	9.60
Callon	Direct LOE			2.33	2.43	1.60	1.60	1.42	1.53
($/Mcfe)	Transportation				0.04	0.05	0.08	0.10	0.09
	All				2.47	1.65	1.68	1.52	1.62
Contango	Direct LOE		1.17	1.30	0.49	0.56	0.41	0.40	0.49
($/Mcfe)	Others				0.33	0.41	0.23	0.27	0.33
	All				0.83	0.97	0.63	0.67	0.81
Energy XXI	Direct LOE		16.31	15.23	14.00	14.99	12.54	11.15	13.40
($/boe)	Insurance		1.90	2.08	1.77	1.99	2.77	3.39	2.26
	Workover		4.04	4.15	3.49	2.72	2.67	1.59	2.84
	Transportation		1.43	1.54	1.01	1.41			
	All		23.68	23.00	20.27	21.11	17.98	16.13	18.50
EPL	Direct LOE				17.88	15.73	10.64	10.98	14.98
($/boe)	Transportation				0.12	0.17	0.27	0.19	0.23
	All				18.00	15.90	10.91	11.17	15.21
Energy Res. Tech.	Direct LOE				17.75	14.73	10.70	10.74	10.38
($/boe)	Workover				3.17	1.90	2.94	1.32	1.38
	Transportation				1.02	0.99	0.88	1.14	0.72
	Maintenance				1.29	1.36	0.98	1.86	2.64
	All				13.44	13.60	17.30	14.47	9.60
McMoRan	Direct LOE				2.00	1.66	1.79	1.57	1.49
($/Mcfe)	Workover				0.42	0.79	0.39	0.25	0.44
	Hurricane-repairs				0.03	0.00	0.12	0.19	0.26
	Insurance				0.23	0.21	0.45	0.32	0.25
	Transportation				0.41	0.36	0.36	0.29	0.44
	All	3.10	3.35	2.86	3.09	3.02	3.11	2.62	2.88
RAAM	Direct LOE				1.42	1.43	1.40	1.05	
($/Mcfe)	Workover				0.11	0.32	0.46	0.34	
	All				1.53	1.76	1.86	1.39	
Stone Energy	Direct LOE	1.15	1.89	1.99	2.33	2.20	1.90		
($/Mcfe)	Transportation	0.68	0.69	0.42	0.24	0.11	0.09		
	All	1.83	2.58	2.41	2.57	2.31	1.99	2.00	2.68
W&T Offshore	Direct LOE	1.88	2.50	2.51	2.26	2.16	1.95	2.15	2.35
($/Mcfe)	Transportation	0.17	0.19	0.16	0.14	0.17	0.19	0.14	0.16
	All	2.05	2.69	2.67	2.40	2.33	2.14	2.29	2.51

Source: Company annual reports.

company, and Energy Resources Technology, a subsidiary of Helix Energy Solutions Group, Inc., an international offshore service company. Four companies of the sample went bankrupt during the evaluation period (ATP Oil and Gas Corporation, Energy XXI GOM, RAAM Global Energy, and Stone Energy), two companies sold off their assets and left the region (Callon and Freeport-McMoRan Oil & Gas), and two companies were purchased outright (EPL by Energy XXI and Energy Resource Technology by Talos). Energy XXI GOM is entirely a shelf gulf player, whereas Stone Energy, W&T Offshore, and ATP are primarily deepwater operators.

The sample was roughly split between oil and gas producers, and all company fiscal years ended December 31 except Energy XXI GOM, which ended June 30. Companies report consolidated information on their operations, net production, average sales prices, impact of derivatives, and average production cost. Average production costs are usually broken out by category for direct LOE, DD&A, accretion, taxes, and G&A expenses. LOE may include maintenance expenses, or these may be reported separately but exclude transportation tariffs. Four companies report direct LOE as a single cost category as per the US SEC Regulation S-X Section 210.4-10(a), and all other companies present some variation. Costs are presented in nominal (unadjusted) money-of-the-day dollars.

Companies with a significant portion of their production and/or reserves onshore cannot be employed because of the significant differences between onshore and offshore regions. Companies with a significant presence in the GoM and internationally such as Anadarko, Apache, Chevron, Newfield, and Shell do not break out LOE for the GoM and could not be used in the evaluation. Private companies with a significant presence in the GoM such as Fieldwood and Bennu Oil and Gas do not report data.

20.3.2 Direct LOE

Direct LOE is the largest and most important cost category for all companies all years and in most cases includes chemicals, fuel, insurance (if purchased), labor, logistics, repairs, and maintenance (workovers). In some cases, companies break out one or more subcategories such as workovers, repair and maintenance, and insurance. Transportation and gathering are usually considered a separate category but in a few cases are included within the direct LOE category.

Direct LOE may decrease from one year to the next if high-cost fields are divested, operating efficiencies are improved, and/or low commodity prices allow cost reductions during contract negotiation, say for transportation services. Conversely, older less prolific properties with greater water cuts and maintenance expense, sour crude service, and an increasing commodity price environment will usually act to increase service cost and unit operating cost from one year to the next. Declining production from suspension of drilling and intervention activities will act to increase unit operating cost because of reduced production relative to fixed costs. Property acquisitions may act to increase or decrease direct LOE.

Combinations of factors determines operating expense. For example, old, low-rate oil wells with high water cuts that require pressure maintenance and are isolated and far from port may cost five times or more on a unit basis relative to new, high-producing, high-quality wells that do not require pressure support.

20.3.3 Workover and Maintenance

Workover and maintenance expense are reported separately for half of the sample and range from $1.9 to $5.8/boe and $0.1 to $0.8/Mcfe from 2010 to 2012. When commodity prices rise, workover activity usually increases since it accelerates production, but no inferences can be inferred from the data. Direct LOE will increase if the marginal cost exceeds the incremental production. In low-price environments, workovers may not return the investment and are therefore postponed. For wet wells, workover costs are significantly greater than dry tree wells because of the need to mobilize a MODU.

Workover and maintenance cost depends on the nature and age of operations, level of production, well type, and oil and gas prices. Workovers often require installing smaller tubing, recompletion into a higher zone, restimulation, and replacing corroded tubing. Workover and maintenance expense are discretionary, meaning the operator decides if it wants to perform the activities and incur the cost. High-producing wells in high-price environments have a greater chance of returning the investment, and workovers are more likely to be performed relative to marginal producers.

Repairs and maintenance costs will be higher for remote mature properties and in low-price environments will be postponed, except for emergencies and as long as it does not involve a safety hazard. Repair and maintenance work is expected to be positively related to commodity prices to the extent that increasing prices provide budgets that allow work to be performed.

20.3.4 Hurricane Impacts

Significant hurricane impacts to infrastructure can reduce and/or defer future production and revenues, increase lease operating expenses for evacuations and repairs, and increase and/or accelerate plugging and abandonment costs (Kaiser, 2014). Assets damaged or destroyed by hurricanes will require greater budgets for longer durations to repair or decommission than undamaged or standing structures. McMoRan was the only company in the sample that reported hurricane-related repairs as a separate category, averaging $0.14/Mcfe from 2006 to 2012.

20.3.5 Insurance

Insurance cost depends on a company's structure and well inventory, type of coverage (property damage, repair of wreck, and business interruption), and market conditions. To understand coverage, it is necessary to enumerate the number, location, type, and replacement value of infrastructure and the type and nature of coverage. Affordable coverage in the years after a significant event may be difficult to obtain, and companies may choose to eliminate or reduce hurricane insurance coverage, which will reduce short-term operating expense but may increase expenses later. Energy XXI and McMoRan Oil & Gas reported insurance cost as a separate cost category that averaged $2.36/boe at Energy XXI and \$0.30/Mcfe at McMoRan over the reporting period. For companies that purchase insurance coverage, insurance cost often ranges from 5% to 10% LOE but in some cases may be greater.

20.3.6 Transportation and Gathering

Transportation and gathering costs depend on the capital investment and age of the pipeline(s), number of subscribers, volumes and distance transported, pipeline ownership and regulation, and product transported. The reservation rate is paid to reserve firm capacity regardless of usage. The commodity rate is paid based on volumes committed and shipped. Interruptible service is sometimes offered on a discounted basis.

Old, partially full pipelines generally require more maintenance than new full lines, and if line volumes are low, tariff rates will often be higher than on packed lines unless the capital cost is fully depreciated. Generally, transportation and gathering costs to deliver oil and gas to onshore markets are a small part of overall operating cost (<\$2/bbl and < \$0.50/Mcf), but exceptions exist. Typically, because transportation and gathering costs are volume-based, fees will change in proportion to production changes and will usually vary less than other cost categories and in smaller proportion to the total expense.

20.4 LEASE OPERATING STATEMENTS

Lease operating statements represent the accounting records of operating expenses on a property or lease basis for tax purposes and are used for internal management review and cost control. They are by far the best and most accurate source of data on the cost to operate oil and gas properties since they are a complete and faithful record of all operational expenses. Lease operating statements are not part of financial statements that are regulated and required to be disclosed to the public periodically, and hence, they are not available for review except to property owners.

Companies typically employ accrual-based accounting that recognizes expense and revenue as economic events occur, regardless of when cash transactions are recorded (Gallun et al., 2001). Accrual accounting is the standard accounting practice for most oil and gas companies, but the expenses and revenue depicted on the lease operating statement may not exactly match the (actual) cash flows in the same fiscal period.

20.5 SURVEY METHODS

Survey methods are a common means to evaluate lease operating cost, and over the years, various operators, consulting companies, and the federal government have performed studies. The UK Oil and Gas Authority is currently the gold standard and is the source of the best publicly available offshore operating cost data in the world (Box 20.1).

20.5.1 Oryx Energy GoM Survey

In 1990, Oryx Energy Company conducted surveys with a dozen shallow-water GoM operators to quantify the various practices, philosophies, and techniques with a view to benchmark operations (Bolin et al., 1990). Operating cost per boe was the most widely used measure of operational performance and for the sample ranged from $0.7 to $2.8/boe and $110,000 to $430,000 per well per year in 1990 dollars. There were 6.1 employees per platform and 14.3 employees per manned platform for structures with full

BOX 20.1 UK North Sea Operating Cost, The Gold Standard

The UK North Sea has the best publicly available offshore operating cost data in the world because the UK Oil and Gas Authority collects cost data on an annual basis using survey instruments that all operators are required to complete (UK Oil and Gas Authority, 2017). In 2016, average operating cost was reported at $16/boe. Large producers generally have the lowest unit cost, and the northern North Sea—with its late life fields, harsh weather conditions, large infrastructure, and more onerous logistic requirements—has the highest regional cost.

Operators producing >50 MMboe during 2016 had operating cost ranging between $4 and $15/boe, while for operators producing between 15 and 50 MMboe, operating cost ranged from $8 to $35/boe (Fig. 20.4). There is a 12-fold difference between the highest and lowest unit production cost across operators in the region (Fig. 20.5). Operating cost by field for the south North Sea, central North Sea, and north North Sea shows wide variation between developments and within regions (Fig. 20.6).

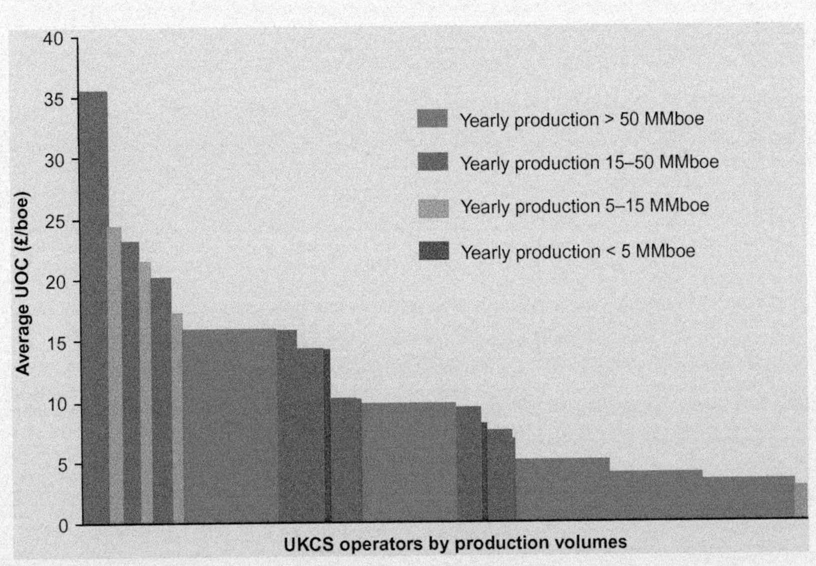

Fig. 20.4 Unit operating cost and hydrocarbon production by operator in 2016. *(Source: UK Oil and Gas Authority.)*

Continued

BOX 20.1 UK North Sea Operating Cost, The Gold Standard—cont'd

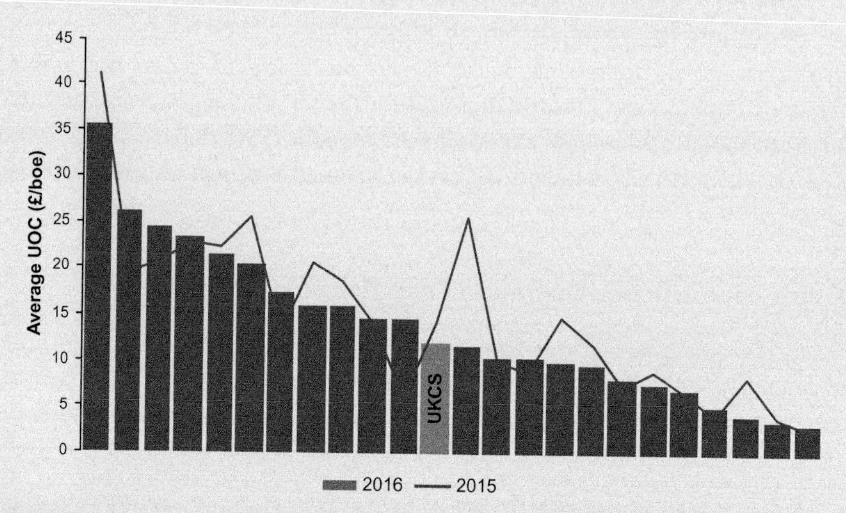

Fig. 20.5 Unit operating cost by operator for 2015 and 2016. *(Source: UK Oil and Gas Authority.)*

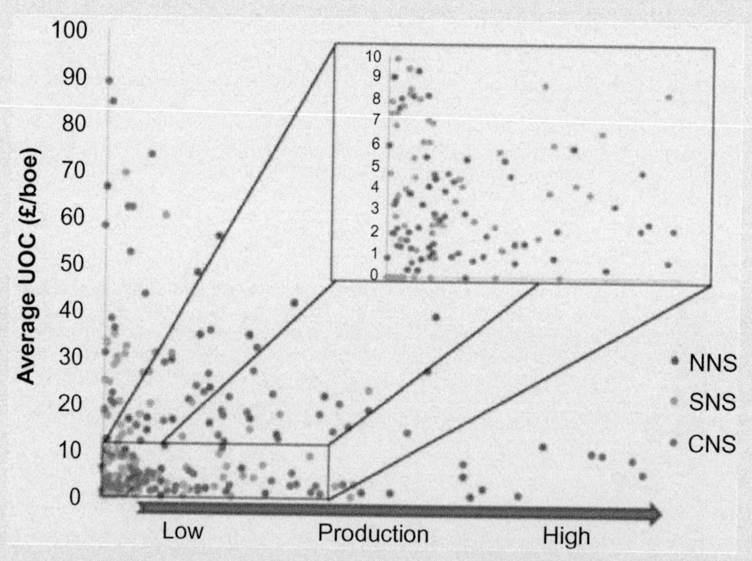

Fig. 20.6 Field operating cost for the south North Sea, central North Sea and north North Sea in 2016. *(Source: UK Oil and Gas Authority.)*

The average operating cost for manned platforms and floating facilities with 100,000 bopd processing capacity was $61 million per year and $80 million per year circa 2016, respectively, and on a production basis was $22/boe and $14/boe reflecting the higher production rates of floating facilities. Topside weight had the strongest correlative index for operating cost, and for manned platforms, the top quartile performance had an operating cost less than $3.5 million per 1000 tons of topside weight.

production capacity. Personnel and transportation accounted for slightly more than half of field expenses, excluding workovers. Cost variation was attributed to the nature of production and density of operations, number of manned/unmanned platforms, and degree of automation.

20.5.2 Energy Information Administration

From 1994 to 2009, the US Energy Information Administration (EIA) conducted a survey of domestic oil and gas well equipment in several regions of the United States (EIA, 2009). Individual items of equipment were priced by communicating with the manufacturers of the items in each region. The survey data on equipment and usage were combined with a factor model to reflect the direct cost incident to assumed production levels. The EIA survey was discontinued in 2010.

Operating costs for the Gulf of Mexico were estimated for 12- and 18-well slot fixed platforms assumed to be 50, 100, and 125 mi from shore corresponding approximately to water depths of 100, 300, and 600 ft. Crude oil production was assumed to total 11,000 bbl per day (4 MMbbl per year), and associated gas production was assumed to be 40 MMcf per day (14.6 Bcf per year). The GOR indicates a primarily oil-producing structure.

Meals, platform and well maintenance, helicopter and boat transportation of personnel and supplies, communication costs, insurance, and administrative expenses were included in the cost estimate. Export cost to shore and water disposal cost were not considered. Year 2009 estimates are shown in Table 20.5. The implied LOE for a 15-slot platform in 300 ft water depth producing 11,000 bopd crude and 40 MMcfpd natural gas would be \$1.5/boe:

$$\frac{\$9.6\text{ million}}{(17{,}667\text{ boed})(365\text{ d})} = \$1.5/\text{boe}.$$

20.5.3 IHS Upstream Operating Cost Index

The IHS Upstream Operating Cost Index measures quarterly changes in the cost of oil and gas upstream operations on a regional basis (Fig. 20.7). Introduced in 2000, the index is calculated by examining equipment and supply and service cost by vendors via survey for specific types of properties and operations in different regions in the world. The index is meant to be similar to the consumer price index (CPI) in providing a transparent

Table 20.5 EIA Offshore Gulf of Mexico Operating Cost Survey Results (Million 2009 $)

Platform	100 ft	300 ft	600 ft
12 slot	8.67	8.95	
18 slot	10.1	10.4	10.9

Source: EIA 2009.

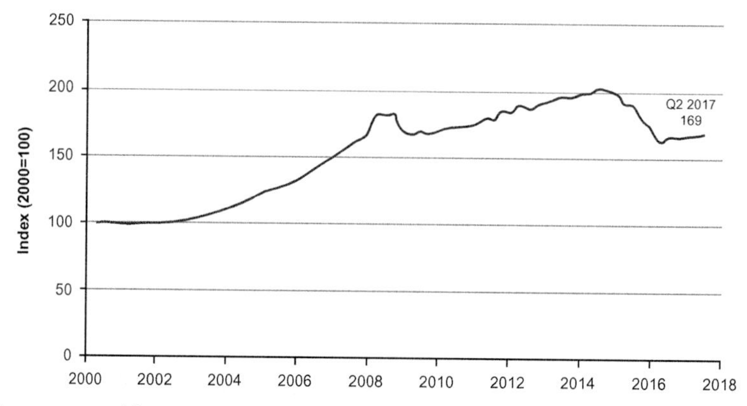

Fig. 20.7 Upstream world operating cost index, 2000–17. *(Source: IHS Markit.)*

benchmark for forecasting and as a commercial product provides greater levels of detail to subscribers.

20.5.4 Benchmarking

Benchmarking studies collect and analyze lease operating statements for participating companies and summarize the results and best practices using a common set of categories. Since oil and gas assets are unique and companies report and aggregate cost differently, the first step in all benchmarking studies is to normalize and account for the different factors to make comparisons. Wide participation helps to ensure surveys are representative. Benchmarking studies are not released to the public or otherwise reported except in the occasional press release. Participation is the means to gain access in client studies. HSB Solomon Associates LLC, Independent Project Analysis Inc., and Turner and Townsend Inc. (Performance Forum) are leading project benchmarking companies.

In 2002, Ziff Energy Group surveyed 312 GoM shelf fields for 24 companies representing over \$1 billion in annual operating expenses. The field weighted average operating cost for gas assets was reported to be \$0.29/Mcfe and for oil assets \$2.3/boe. In deepwater, 10 companies producing from 30 fields and representing about 80% of deepwater production had an average operating cost below \$2/boe.

In 2009, 125 GoM fields in water depth <1000 ft were evaluated (66 gas fields producing over 1 Bcfpd and 52 oil fields producing over 200,000 boepd). Unit operating cost nearly tripled to \$0.84/Mcfe for gas fields and \$7.83/boe for oil fields (Alba, 2008; Ziff Energy Group, 2009). The single largest cost component for shelf operations was reported as labor and field supervision accounting for about a quarter of operating expenses, followed by transportation for gas fields and repairs and maintenance for oil fields. Well servicing amounted to about a fifth of offshore expenses.

20.6 SOFTWARE TOOLS

There are three main software tools used by industry to estimate capital and operating costs: Aries Petroleum (Field Plan) from Halliburton, Questar from IHS, and Merak Peep from Schlumberger. In each, data are collected from industry, media reports, direct contact with operators, and other sources, and (proprietary) algorithms are applied using field characteristics and other information to normalize and estimate cost at various levels of granularity. These tools have been developed over a long period of time (25–30 years) and are generally considered accurate to ±20% in conceptual development studies. The application of SAP in operating cost management has also been employed in offshore operating cost applications (Taylor, 2016).

20.7 COMPUTER METHODS

Four computer/software technologies have been used in the petroleum industry for operating cost estimation—expert systems, fuzzy logic systems, neural networks, and genetic algorithms. Applications have been produced with each technique but expert systems and neural networks are the most common in operating cost applications, as much to organize data and for processing and prediction. Only a handful of papers have been published indicating the specialized nature and limited application of the procedures (Harvey et al., 2000).

An expert system is a computer program that provides expert advice (MacAllister et al., 1996). The program usually contains a data structure representing expert knowledge, algorithms to manipulate the data structure, and an interface for inputs and results. An example of an expert system developed specifically for operating expense modeling was described in Greffioz et al. (1993), but such systems have not gained much interest. Examples of operations that are amenable to sophisticated computer analysis include pressure maintenance via waterflood or gas injection, gas-lift operations, and subsea gas developments from multiple fields.

Neural networks are used to mimic the pattern-recognition capabilities that humans use in analysis and interpretation problems (Saputelli et al. 2002). Neural networks develop solutions implicitly through exposure to information about the problem domain. Usually, the exposure takes the form of solutions to several different examples to the problem class being solved. The neural network uses these examples to create a solution in a manner analogous to learning via trial and error. In Boomer (1995), a neural network model of lease operating cost was built.

Fuzzy set theory deals with the generalization of binary logic to include imprecise, vague, and ambiguous concepts to include a spectrum of possible states. Genetic algorithms provide a technology for solving optimization problems that cannot be handled using conventional linear programming approaches. Using the idea of fitness of each offspring generation, a population of solutions is generated, and only the strongest members survive in a nod to evolutionary processes.

REFERENCES

Alba, J.C., 2008. Benchmarking operating efficiency to enhance production from mature fields. In: 19th World Petroleum Council Proceedings.
Bolin, W.D., Brienen, J.W., Morgan, R.L., Pyles, S.R., 1990. Analysis of operational practices and performance of selected Gulf of Mexico operators. In: SPE 20664. SPE Annual Technical Conference and Exhibition. New Orleans, LA, Sep 23–26.
Boomer, R.J., 1995. Modeling lease operating expenses. In: SPE 30055. SPE Hydrocarbon Economics and Evaluation Symposium. Dallas, TX, Mar 26–28.
Energy Information Administration, 2009. Oil and Gas Lease Equipment and Operating Costs 1998 through 2006. Released: June 18, 2010. Washington, DC.
Gallun, R.A., Wright, C.J., Nichols, L.M., Stevenson, J.W., 2001. Fundamentals of Oil & Gas Accounting, 4th ed. PennWell, Tulsa, OK.
Greffioz, J., Oliver, A.J., Schirmer, P., 1993. TOPEX: an expert system for estimating and analyzing the operating costs of oil and gas production facilities. In: SPE 25315. SPE Asia Pacific Oil and Gas Conference and Exhibition. Singapore, Feb 8–10.
Harvey, H.E., Gration, D.P., Martin, S., 2000. Managing operating costs with a risk- and economic-evaluation tool. In: SPE 64448. SPE Asia Pacific Oil and Gas Conference and Exhibition. Brisbane, Australia, Oct 16–18.
Kaiser, M.J., 2014. Hurricane clean-up activity in the Gulf of Mexico, 2004-2013. Mar. Policy 51, 512–526.
MacAllister, D.J., Day, R., McCormack, M.D., 1996. Expert systems: a 5-year perspective. SPE Comput. Appl. 10–14. February.
Moody's Global Corporate Finance, 2009. Global oilfield services rating methodology. Moody's Investors Service. Dec 2009.
Newberry, D., 2014. 50 reservoirs, 48 well slots and two TLP's – maximizing recovery from the deepwater prolific Mars field. In: OTC 26578. Offshore Technology Conference. Houston, TX, May 5–8.
Paganie, D., 2006. Shell restarts Mars TLP, includes industry firsts. Offshore 66 (6), 2006.
Paganie, D., Buschee, P., 2005. Operators begin cleanup, repair from Katrina, Rita. Offshore 65 (10), 24–35.
Saputelli, L., Malki, H., Canelon, J., Nikolaou, M., 2002. A critical overview of artificial neural network applications in the context of continuous oil field optimization. In: SPE 77703. SPE Annual Technical Conference and Exhibition. San Antonio, TX, Sep 29–Oct 2.
Seba, R.D., 2003. Economics of Worldwide Petroleum Production. OGCI and Petroskills Publications, Tulsa, OK.
Sharp, D.W., 2000. Offshore Oil and Gas Insurance. Witherby & Co. Ltd, London, UK.
Siems, G., 2014. A practice risk mitigation program. In: Houston Marine Insurance Seminar, Houston, TX.
Taylor, J.A., 2016. Improving operational cost with SAP. In: SPE 181005. SPE Intelligent Energy International Conference and Exhibition. Aberdeen, Scotland, Sep 6–9.
UK Oil & Gas Authority, Analysis of UKCS operating costs in 2016, London UK.
Ziff Energy Group, 2009. Operating cost study to assist Gulf of Mexico shelf operators manage rising costs & enhance production operating efficiency. Tucker RM, VP Marketing. Press release. July 15.

CHAPTER TWENTY ONE

Field Examples

Contents

Abstract

Public companies listed on US stock exchanges are required to provide separate disclosure of reserves, production, and production cost for each country or field that controls more than 15% of its total proved reserves or production. Field information is useful because it is the lowest aggregation unit for production cost that is publicly available and provides insight into operational characteristics. In this chapter, seven offshore GoM fields are described, five on the shelf and two in deepwater, all major fields with reserves greater than 50 MMboe. One gas field in state waters (Mobile Bay 113) is represented along with a former oil and sulfur mine operation (Main Pass 299).

21.1 WEST DELTA 73

The West Delta 73 field is an oil-producing sandstone with bottom water drive and is one of the top 10 producing oil fields on the Gulf shelf circa 2017. The field was discovered in 1962 about 28 mi offshore of Grand Isle, Louisiana, in approximately 175 ft of water. The field is a large low-relief faulted anticline that produces from Pleistocene to Upper Miocene aged sands trapped structurally from 1500 to 13,000 ft below the mudline (Ogletree and Overly, 1977).

Energy XXI is the operator and 100% working interest owner in the West Delta 73 field, which covers seven lease blocks WD 73–75 and 89–92 (Fig. 21.1). At the end of 1975, cumulative oil production was 128 MMbbl oil and 170 Bcf natural gas. Through September 2017, the field had produced 277 MMbbl oil and 686 Bcf gas from 305 wells, and remaining reserves are estimated at 5.4 MMbbl oil and 17.3 Bcf gas (Fig. 21.2).

Decommissioning Forecasting and Operating Cost Estimation
https://doi.org/10.1016/B978-0-12-818113-3.00021-7

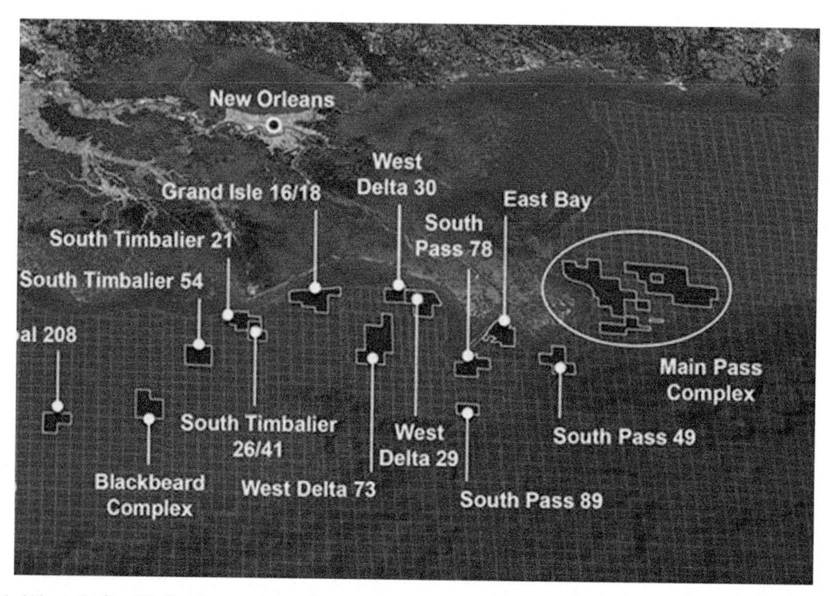

Fig. 21.1 West Delta 73 field location. *(Source: Energy XXI.)*

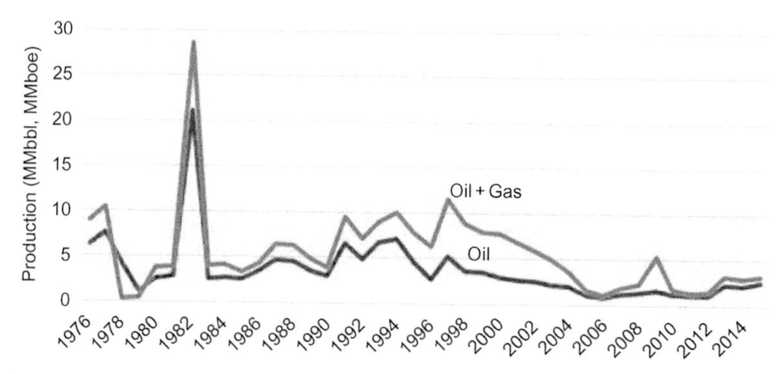

Fig. 21.2 West Delta 73 field oil and gas production profile, 1976–2015. *(Source: BOEM 2017.)*

In 2011, over $100 million was spent to identify, plan, and drill 21 horizontal wells and improve infrastructure, and by 2016, production expanded from 1800 to 7000 boepd (Iqbal et al., 2016). Circa 2017, there were six production platforms with 27 active wells and 46 shut-in wells (Fig. 21.3). From 2012 to 2014, production costs for the field ranged between $18.5 and $21.3 per boe (Table 21.1). Lease operating expense was about one fourth the average sales price indicating a late life field.

Fig. 21.3 West Delta 73 complex. *(Source: Energy XXI.)*

Table 21.1 West Delta 73 Production Cost, 2012–14

	2014	2013	2012
Net sales			
Oil (Mbbl)	1496	1278	840
NGLs (Mbbl)	3.7	3.7	3.7
Natural gas (MMcf)	2737	3285	2190
Oil equivalent (Mboe)	2007	1862	1241
Average sales prices			
Oil per bbl	$105.06	$109.11	$111.33
NGLs per bbl	$40.74	$33.50	$61.18
Natural gas per Mcf	$4.22	$3.4	$1.67
Production costs			
Oil equivalent ($/boe)	$19.76	$18.54	$21.30

Source: Energy XXI.

21.2 MAIN PASS 61 COMPLEX

The Main Pass 61 complex is located in approximately 100 ft of water near the mouth of the Mississippi River and consolidates the Main Pass 61, 73, and 311 fields and portions thereof. Energy XXI is the operator.

The Main Pass 61 field was discovered by POGO in 2000 and has been producing from the Upper Miocene sands since 2002. The Main Pass 73 field was discovered in

1976 by Mobil and began producing in 1979. In 2012, the last year in which data are reported, daily production from the complex totaled 6.5 Mbopd crude and 4.1 MMcfpd natural gas with production cost of $9.8/boe (Table 21.2).

There were 23 producing wells and three production platforms in operation circa 2016, one of which is shown in Fig. 21.4. Cumulative production through September 2017 from the three fields totaled 224 MMbbl oil and 953 Bcf gas.

Table 21.2 Main Pass 61 Production Cost, 2010–12

	2012	2011	2010
Net sales per day			
Oil (Mbopd)	6.5	7.2	5.8
Natural gas (MMcfpd)	4.1	3.0	3.5
Oil equivalent (Mboepd)	7.2	7.7	6.4
Average sales prices			
Oil per bbl	$105.94	$88.62	$75.37
Natural gas per Mcf	$2.91	$4.44	$4.99
Production costs			
Oil equivalent ($/boe)	$9.77	$10.30	$11.37

Source: Energy XXI.

Fig. 21.4 Main Pass 61 platform. *(Source: Energy XXI.)*

21.3 SOUTH TIMBALIER 21

The South Timbalier 21 field was discovered by Gulf Oil in 1957 on the south flank of the giant Timbalier Bay field 6 mi south of Lafourche Parish, Louisiana, in approximately 50 ft water depth (Lipari, 1962). Circa 2017, the field includes acreage in South Timbalier blocks 21, 22, 27, and 28 and two state leases (Fig. 21.1). The field is bounded on the north by a major Miocene expansion fault. Miocene sands are trapped structurally and stratigraphically from 7000 to 15,000 ft in depth.

There are 10 major production platforms, one of which is shown in Fig. 21.5, and 61 smaller structures located throughout the field. In 2010, the last year of reported data, crude production was 3.8 Mbopd, and natural gas was 4.6 MMcfpd with production cost of $27.2/boe (Table 21.3).

Through 1975, the field produced 150 MMbbl oil and 231 Bcf natural gas. Cumulative production through September 2017 from federal leases totaled 259 MMbbl oil and 426 Bcf natural gas (Fig. 21.6). Remaining reserves circa September 2017 are estimated at 1.5 MMbbl oil and 3.1 Bcf gas.

21.4 SHIP SHOAL 349 (MAHOGANY)

The Ship Shoal 349 field known as Mahogany covers Ship Shoal blocks 349 and 359 in 375 ft of water depth and was the first shelf subsalt development in the GoM (Montgomery and Moore, 1997). The field was discovered in 1993, and W&T Offshore, Inc. holds 100% working interest circa 2017.

Fig. 21.5 South Timbalier 21 platform. *(Source: Energy XXI.)*

Table 21.3 South Timbalier 21 Production Cost, 2008–10

	2010	2009	2008
Net sales per day			
Oil (Mbopd)	3.8	4.2	6.1
Natural gas (MMcfpd)	4.6	9.1	10.1
Oil equivalent (Mboepd)	4.6	5.7	7.8
Average sales prices			
Oil per bbl	$72.92	$65.96	$97.91
Natural gas per Mcf	$4.23	$6.14	$9.47
Production costs			
Oil equivalent ($/boe)	$27.21	$26.22	$14.76

Source: Energy XXI.

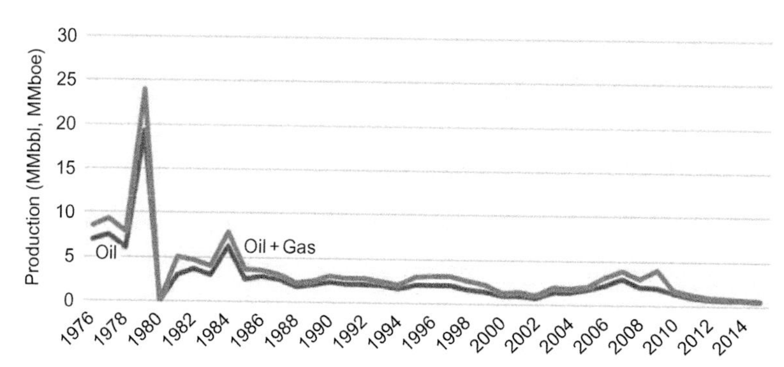

Fig. 21.6 South Timbalier 21 field oil and gas production profile, 1976–2015. *(Source: BOEM 2017.)*

The discovery was positioned beneath a salt sheet displaying high variable thickness and geometry (Fig. 21.7). The discovery well penetrated 3825 ft of salt and logged as many as 14 sandstone zones between depths of 12,300 and 16,300 ft (Voss et al. 2010). Below the salt, there was a noncompetent zone ("gumbo") approximately 1000 ft thick before entering the flank of the anticline with three-way dip closure. A production complex was installed above the field (Figs. 21.8 and 21.9).

Total reserves at the time of discovery were estimated at >100 MMbbl, but production has been slow to achieve these levels, and through September 2017, cumulative production was 34.4 MMbbl oil and 69.2 Bcf gas. Since 2011, the operator has completed several new challenging wells resulting in a significant production increase (Fig. 21.10). From 2013 to 2015, production costs ranged between $3.30 and $4.12 per boe (Table 21.4). Remaining reserves circa 2017 are estimated at 7.4 MMbbl oil and 14.4 Bcf gas.

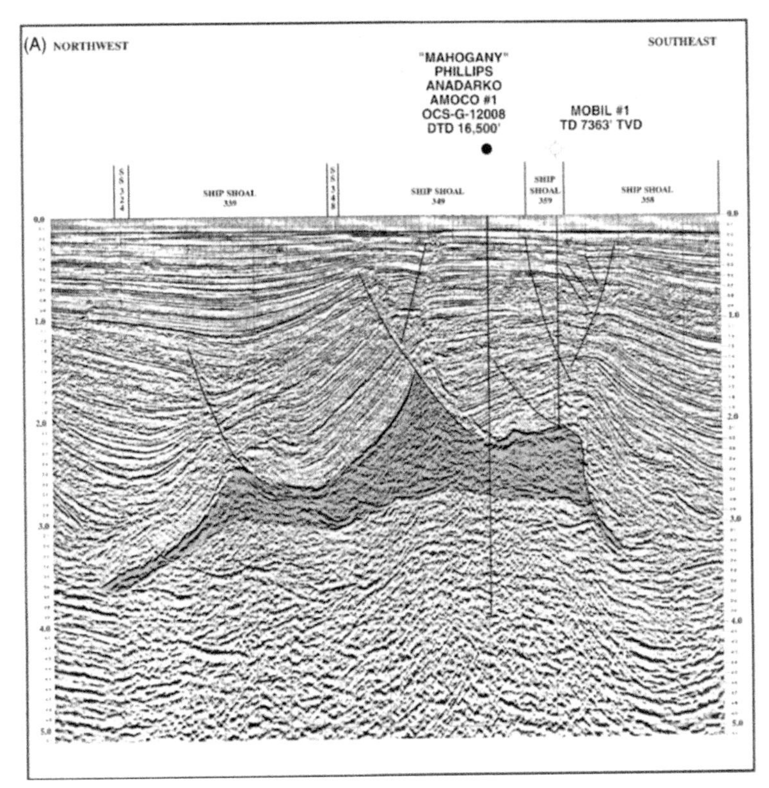

Fig. 21.7 Seismic profile from the Ship Shoal 349 #1 exploration well. *(Source: Montgomery, S.I., Moore, D.C., 1997. Subsalt play, Gulf of Mexico: a review. AAPG Bull. 81 (6), 871–890. AAPG © 1997. Reprinted by permission of AAPG whose permission is required for further use.)*

Fig. 21.8 Ship Shoal 349 (Mahogany) production complex. *(Source: W&T Offshore, Inc.)*

21.5 VIOSCA KNOLL 990 (POMPANO)

The Pompano field was developed by BP with a fixed platform in 1994 in 1290 ft water depth in the southeast corner of Viosca Knoll block 989 (Fig. 21.11). The development included a 10-well subsea tieback located about 4.5 mi away from the platform

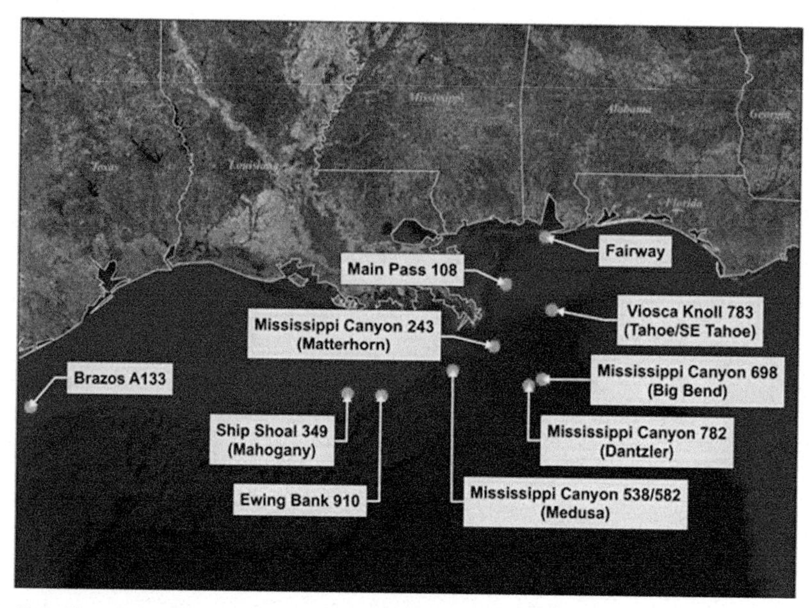

Fig. 21.9 Ship Shoal 349 (Mahogany) field location. *(Source: W&T Offshore, Inc.)*

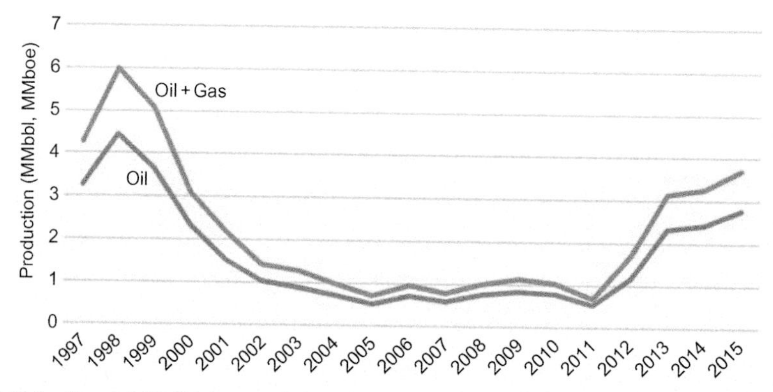

Fig. 21.10 Ship Shoal 349 (Mahogany) field oil and gas production profile, 1997–2015. *(Source: BOEM 2017.)*

in Mississippi Canyon 28 in 1850 ft water depth producing from Pliocene and Miocene reserves (Fig. 21.12). Cumulative field production through September 2017 was 143.9 MMbbl oil and 268.6 Bcf natural gas, and remaining reserves are estimated at 15.1 MMbbl oil and 17.1 Bcf gas (Fig. 21.13). Several tiebacks are hosted at the platform which serves as a regional hub.

Table 21.4 Ship Shoal 349 (Mahogany) Production Cost, 2013–15

	2015	2014	2013
Net sales			
Oil (Mbbl)	2313	2020	1943
NGLs (Mbbl)	97	104	90
Natural gas (MMcf)	3764	3433	3328
Oil equivalent (Mboe)	3037	2697	2589
Average sales prices			
Oil per bbl	$42.73	87.21	98.69
NGLs per bbl	$21.27	46.46	43.24
Natural gas per Mcf	$2.86	4.40	3.72
Oil equivalent per boe	$36.77	72.73	80.39
Production costs			
Oil equivalent ($/boe)	$3.30	$4.12	$3.68

Source: W&T Offshore, Inc.

Fig. 21.11 Viosca Knoll 990 (Pompano) production complex. *(Source: Stone Energy.)*

Reservoir geometry at Pompano is a series of stacked and connected composite sand bodies of moderate to high quality, which were deposited in varying thickness (Willson et al., 2003). Pliocene reserves are all located within reach of the platform, and some of the offset Miocene reserves can also be reached from the platform using extended reach drilling (Fig. 21.14). Miocene reserves that can't be efficiently drilled from the platform were developed with subsea wells. One of the first long distance tiebacks was the Mica field in 4350 ft water depth 29 mi away in Mississippi Canyon block 211 (Ballard, 2006).

In 1998, Pompano's production peaked at about 65,000 boepd, but by mid-2007, field production was approximately 11,500 bopd. Stone Energy acquired a working

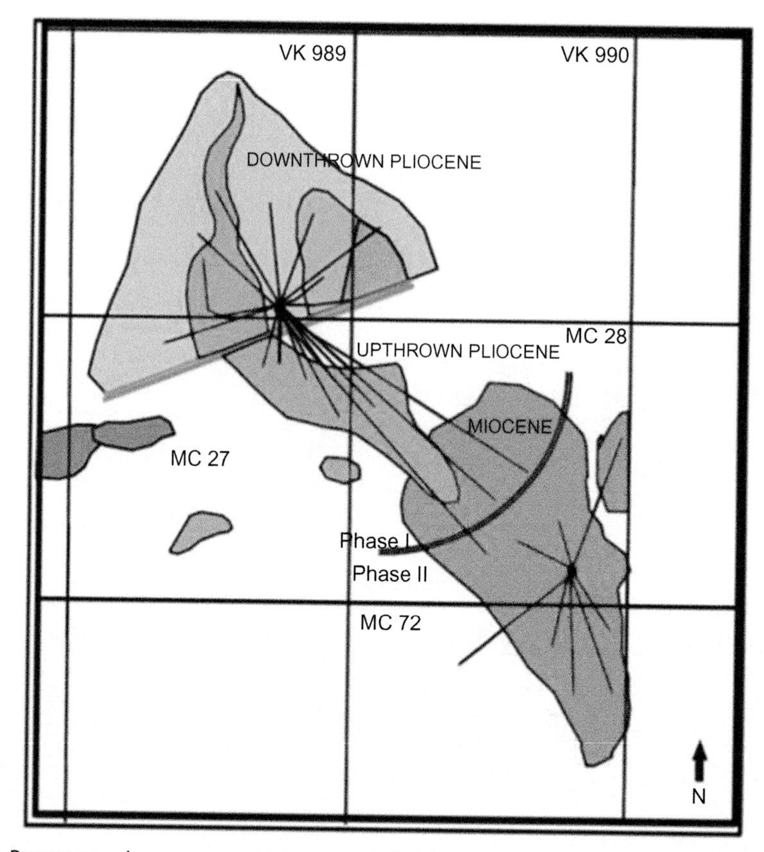

Fig. 21.12 Pompano phase-one and phase-two development. *(Source: Willson, S.M., Edwards, S., Heppard, P.D., Li, X., Coltrin, G., Chester, D.K., Harrison, H.L., Cocales, B.W., 2003. Wellbore stability challenges in the deep water, Gulf of Mexico: case history examples from the Pompano field. In: SPE 84266. SPE Annual Technical Conference and Exhibition. Denver, CO, Oct 5–8.)*

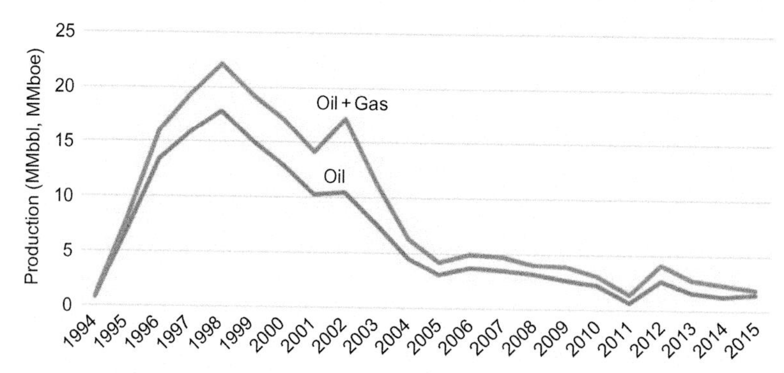

Fig. 21.13 Pompano field oil and gas production profile, 1994–2015. *(Source: BOEM 2017.)*

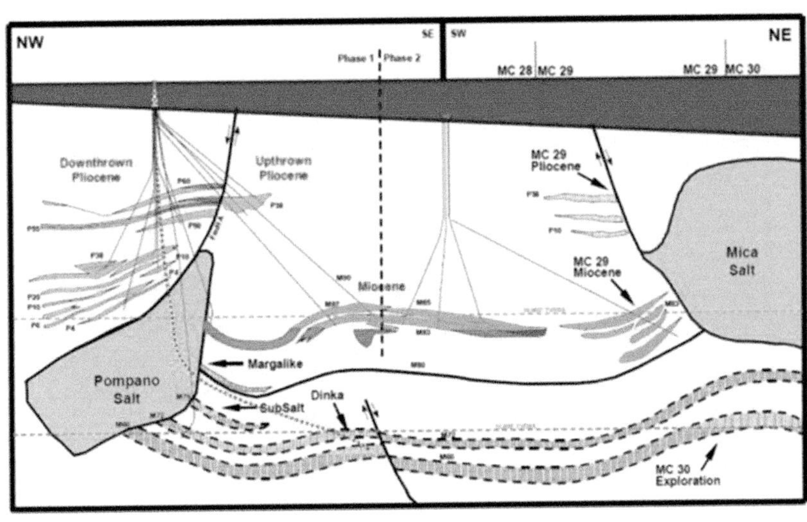

Fig. 21.14 Cross section through the Pompano field. *(Source: Willson, S.M., Edwards, S., Heppard, P.D., Li, X., Coltrin, G., Chester, D.K., Harrison, H.L., Cocales, B.W., 2003. Wellbore stability challenges in the deep water, Gulf of Mexico: case history examples from the Pompano field. In: SPE 84266. SPE Annual Technical Conference and Exhibition. Denver, CO, Oct 5–8.)*

interest in 2006 for $168 million and later became the sole working interest operator. In 2016, the Cardona and Amethyst subsea tiebacks were installed, and production has nearly doubled since 2013 (Fig. 21.15).

In 2013 and 2014, lease operating cost including major maintenance expense ranged from $1.98 to $2.75/Mcfe, but in 2015 with the doubling in production at the facility, unit cost was effectively cut in half to $0.96/Mcf (Table 21.5). Transportation, processing, and gathering expenses were reported at $0.1/Mcfe. In 2016, LOE was reported to be $4.85/boe = $0.81/Mcfe.

21.6 MISSISSIPPI CANYON 109 (AMBERJACK)

The Amberjack field is located on Mississippi Canyon blocks 108, 109, and 110 in water depths from 850 to 1050 ft (Fig. 21.16). Field development started in 1984, and 38 wells had been completed by 1992 and were producing oil and gas from various sand packages from a platform in a water depth of 1030 ft (Fig. 21.17).

The majority of the reserves in the Amberjack field are contained in the Green G-sand, an unconsolidated Pliocene reservoir deposited by a shelf edge delta system, which laid down the sediments as a series of shingled mouth bar sands separated by dipping silt beds, referred to as clinoforms (Johnston et al., 1993). Reservoir compartmentalization required the use of a large number of wells (Fig. 21.18).

Early in production, there were persistent problems with gravel packs to control sand production (Hannah et al., 1993). Horizontal wells were drilled in the early 1990s to

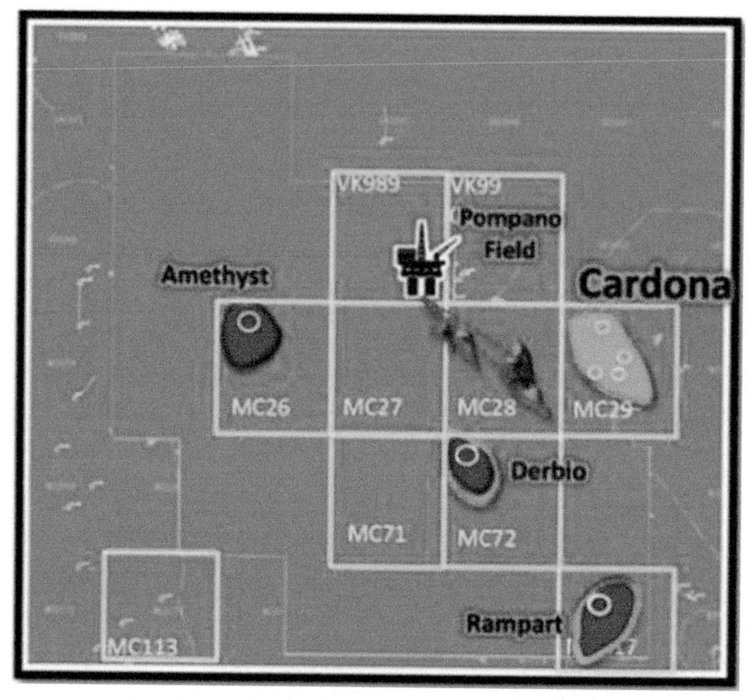

Fig. 21.15 Pompano field location and subsea tiebacks planned circa 2016.

Table 21.5 Viosca Knoll 990 (Pompano) Production Cost, 2013–15

	2015	2014	2013
Net sales			
Oil (Mbbl)	2839	1311	1255
Natural gas (MMcf)	3331	2894	2546
NGLs (Mbbl)	239	151	144
Natural gas equivalent (MMcfe)	21,799	11,666	10,940
Average sales prices			
Oil per bbl	$49.19	$92.53	$107.99
Natural gas per MMcf	$2.22	$ 3.10	$2.49
NGLs per bbl	$15.49	$41.27	$40.65
Natural gas equivalent per MMcfe	$6.91	$11.70	$13.50
Production costs			
Lease operating expenses ($/Mcfe)	$0.96	$2.75	$1.98
Transportation and gathering expenses ($/Mcfe)	$0.08	$0.13	$0.14

Source: Stone Energy.

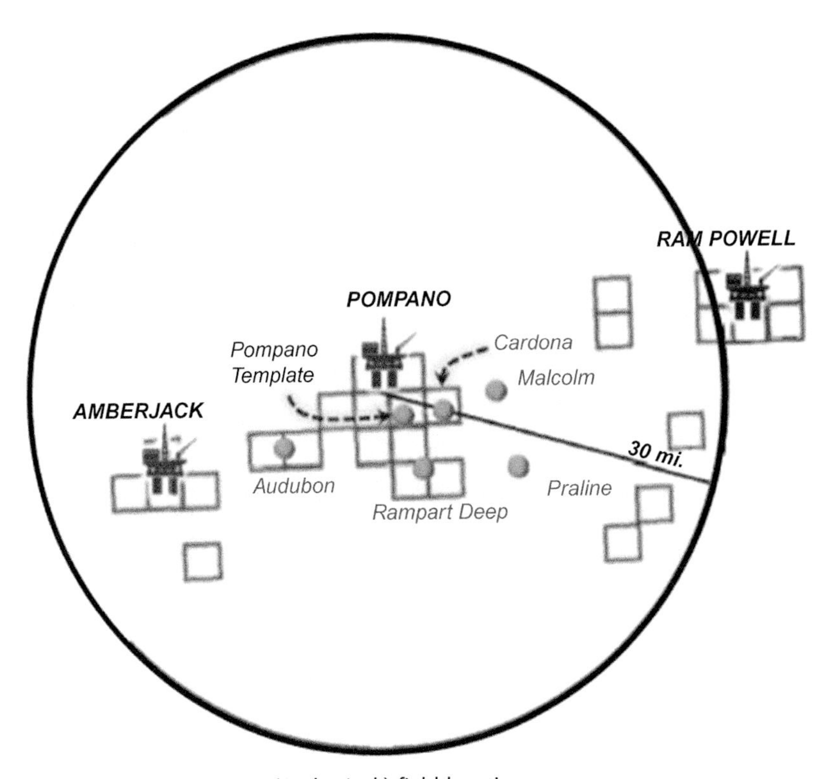

Fig. 21.16 Mississippi Canyon 109 (Amberjack) field location.

Fig. 21.17 Mississippi Canyon 109 (Amberjack) production platform. *(Source: Stone Energy.)*

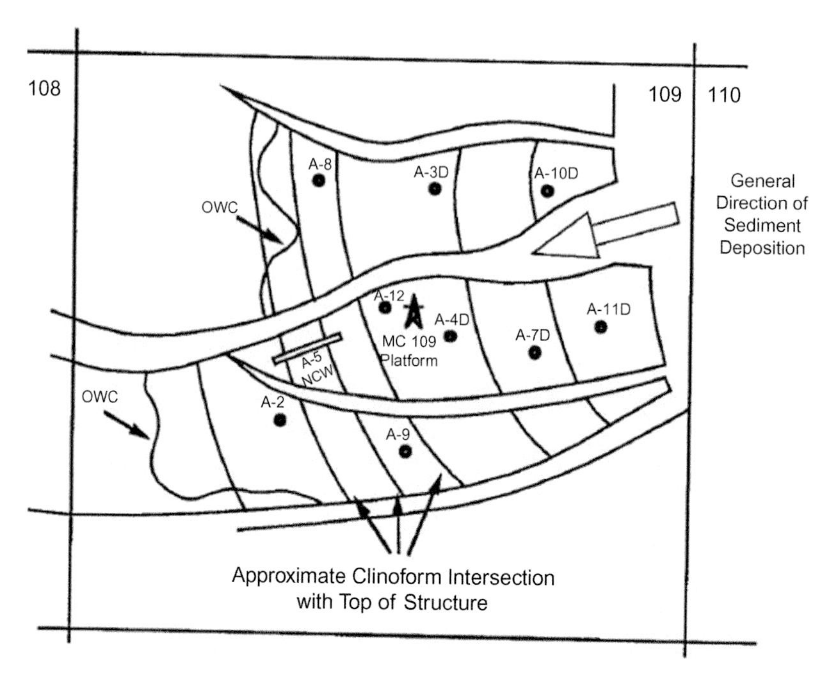

Fig. 21.18 Mississippi Canyon 109 clinoform intersections create reservoir compartmentalization. *(Source: Johnston, R.A., Porter, D.A., Hill, P.L., Smolen, B.C., 1993. Planning, drilling, and completing a record horizontal well in the Gulf of Mexico. In: OTC 7351. Offshore Technology Conference. Houston, TX, May 3–6.)*

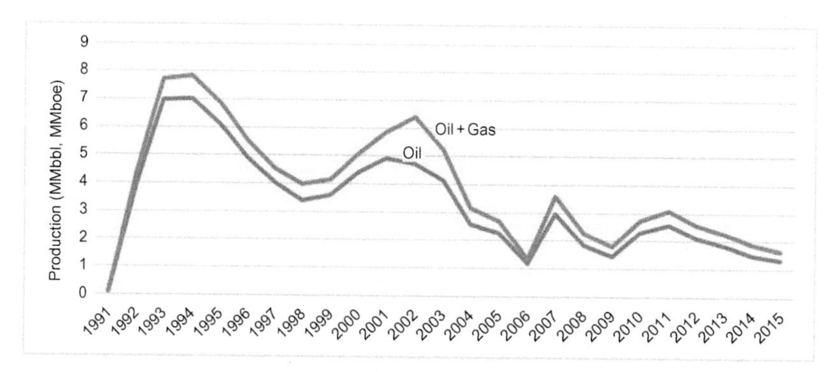

Fig. 21.19 Amberjack field oil and gas production profile, 1976–2015. *(Source: BOEM 2017.)*

improve productivity and drain several clinoforms. For example, well A-5 was oriented perpendicular to the clinoform depositional axes with the completion interval draining all three packages saving on well cost (Fig. 21.18).

In 1999, a redevelopment program was initiated with two sidetrack wells. Both wells experienced problems to reach the planned target depth but contributed to production increase (Fig. 21.19). One lost hole, one unplanned sidetrack, and persistent nondrilling time caused significant cost overruns. In 2003, one of the sidetracks was

Table 21.6 Mississippi Canyon 109 (Amberjack) Production Cost, 2008–10

	2010	2009	2008
Net sales			
Oil (Mbbl)	1389	861	1035
Natural gas (MMcf)	1689	1092	1700
Natural gas equivalent (MMcfe)	10,019	6256	7913
Average sales prices			
Oil per bbl	$77.29	$66.68	$107.99
Natural gas per Mcf	$4.89	$3.81	$9.55
Natural gas equivalent per Mcfe	$11.54	$9.86	$16.18
Production costs			
Natural gas equivalent ($/Mcfe)	$1.27	$2.19	$0.86

Source: Stone Energy.

reentered, but problems arose due to depletion, reduced fracture gradient, and wellbore instability.

In 2001, a gas-lift automation and optimization project was initiated, which resulted in production peaking at around 5 MMbbl oil. Of the 34 wells, 27 utilized gas-lift injection to facilitate flow, but daily operations were challenging because the system was not automated or optimized (Reeves et al., 2003). Following Hurricane Katrina in August 2005, the Amberjack platform required repairs and rerouting of a damaged oil pipeline, and from 2007 to 2010, Stone invested in a seven well drill program, which brought new production and flattened the decline curve (Fredericks et al., 2011).

From 2008 to 2010, Amberjack field production expense ranged between $0.86 and $2.19/Mcfe (Table 21.6). Through September 2017, the field has produced 84.7 MMbbl oil and 83.6 Bcf gas. Remaining reserves are estimated at 3.4 MMbbl oil and 6.1 Bcf gas.

21.7 MOBILE BAY 113 (FAIRWAY)

The Fairway field is composed of Mobile Bay blocks 113 and 132 and is located in 25 ft of water, approximately 35 mi south of Mobile, Alabama, in state waters (Fig. 21.9). The field was discovered by Shell in 1985 and is a Norphlet sand dune trend with one production horizon at about 21,300 ft. Development drilling began in 1990 and included four wells drilled from separate surface locations. W&T Offshore, Inc. acquired a majority working interest in 2011 and the remaining interest in 2014.

In 2016, cumulative field production was approximately 776 Bcfe (129 MMboe) with an average production rate of 4.7 Mboepd. From 2013 to 2015, unit production costs ranged between $1.49 and $2.08 per Mcfe (Table 21.7). As of 31 December 2015, the

Table 21.7 Fairway Field Production Cost, 2013–15

	2015	2014	2013
Net sales			
Oil (Mbbl)	10	7	5
NGLs (Mbbl)	319	415	268
Natural gas (MMcf)	8277	6899	4614
Natural gas equivalent (MMcfe)	10,250	9428	6373
Average sales prices			
Oil per bbl	$47.22	$101.94	$104.75
NGLs per bbl	$18.97	$27.41	$28.34
Natural gas per Mcf	$2.60	$4.07	$3.63
Natural gas equivalent per Mcfe	$2.73	$4.26	$3.99
Production costs			
Natural gas equivalent ($/Mcfe)	$1.49	$1.79	$2.08

Source: W&T Offshore, Inc.

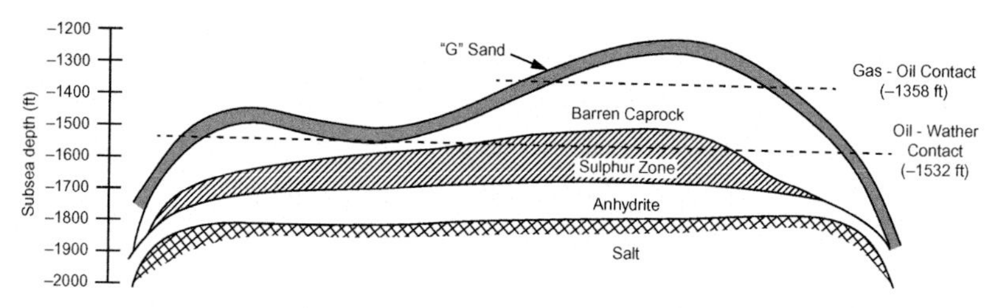

Fig. 21.20 Northwest to southeast cross section across Main Pass 299 block. *(Source: Lewis, L.R., Taylor, B.P., 1992. Early performance and planning for a simultaneous thermal enhanced oil recovery project and sulfur mining operation at Main Pass 299. In: SPE 24629. SPE Technical Conference and Exhibition. Washington, DC, Oct 4–7.)*

operator reported 84 Bcfe reserves. In 2015, the Fairway field generated about $28 million on net sales of 10,250 MMcfe and had production cost of $15 million, leaving about $13 million net revenue. Four platforms and numerous wells are used in production.

21.8 MAIN PASS 299

The Main Pass 299 field was discovered in 1962 on the periphery of a salt dome, and production began in 1967 from four lease blocks at Main Pass 142, 298, 299, and 300. In 1988, Freeport-McMoRan Oil & Gas Co. acquired a portion of block MP 299 on top of the salt dome and discovered a large accumulation of sulfur (67 million tons) and recoverable oil (39 MMbbl out of 99 MMbbl oil in place) in the caprock 2000 ft below sea level (Fig. 21.20).

Freeport-McMoRan's development plan called for a simultaneous thermally enhanced oil recovery project and sulfur mining operation (Lewis and Taylor, 1992). Pressurized hot water heated to 325°F injected into the sulfur zone would liquefy the sulfur where it was to be lifted to the surface using compressed air and off-loaded using barges (Box 21.1). The hot water injection for mining was expected to add 18 MMbbl to recoverable oil.

BOX 21.1 Offshore Sulfur Mining

Brief History. Sulfur is abundant on Earth's surface as native sulfur, in metal sulfides (especially pyrite, FeS_2, and galena, PbS), in mineral sulfates such as gypsum ($CaSO_4 \cdot 2H_2O$), as hydrogen sulfide in natural gas, and as organic sulfur in petroleum and coal (Craig et al., 2011). Sulfur is also associated with the cap rocks of particular salt domes (Fig. 21.21). Sulfur was known in the ancient world as brimstone, the stone that burns,[1] but it was the growth of the chemical industry

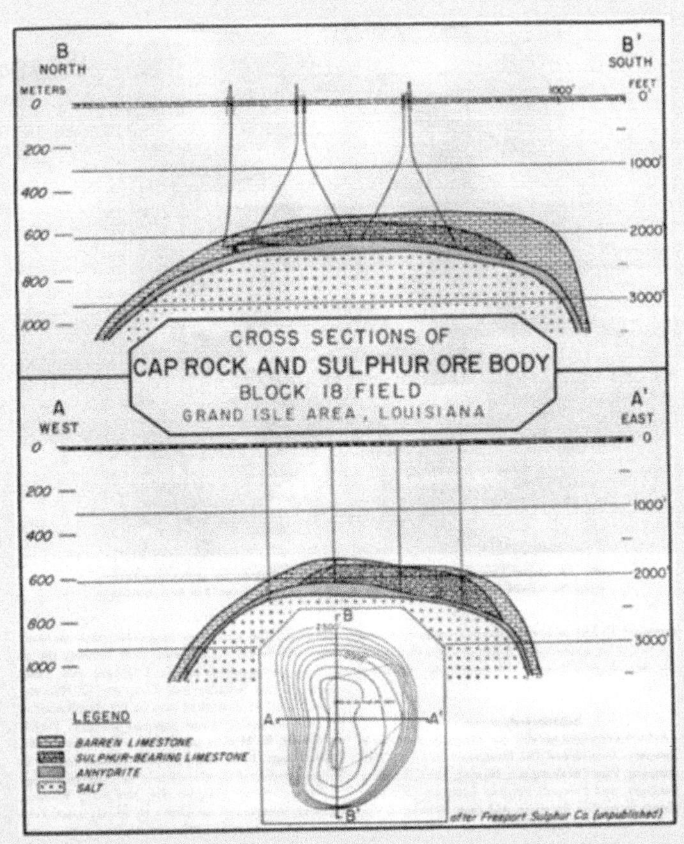

Fig. 21.21 Grand Isle block 18 cross section of cap rock for sulfur ore body. *(Source: Atwater, G.I., 1959. Geology and petroleum development of the continental shelf of the Gulf of Mexico. Gulf Coast Assoc. Geol. Soc. Trans. 9, 131–145. Reprinted by permission of Gulf Coast Assoc. Geological Societies, 2018.)*

[1] During the Peloponnesian War in the fifth century B.C.E., for example, Thucydides reports mixtures of burning sulfur and pitch used to produce suffocating gases to incapacitate soldiers.

Continued

BOX 21.1 Offshore Sulfur Mining—cont'd

in the 1800s that brought sulfur to its position of prominence. The primary use of sulfur for the past 250 years has been in the production of sulfuric acid, known as the "king of chemicals" and the first industrial chemical. In the early days, Sicily held a monopoly on the production of elemental sulfur, which was picked by hand from volcanic deposits (Haynes, 1942). A rise in demand and price increase resulted in a shift to pyrite, and in 1894, the Frasch process for mining subsurface sulfur deposits associated with Gulf Coast salt domes was introduced. Today, sulfur is primarily recovered as a by-product from oil and natural gas production.

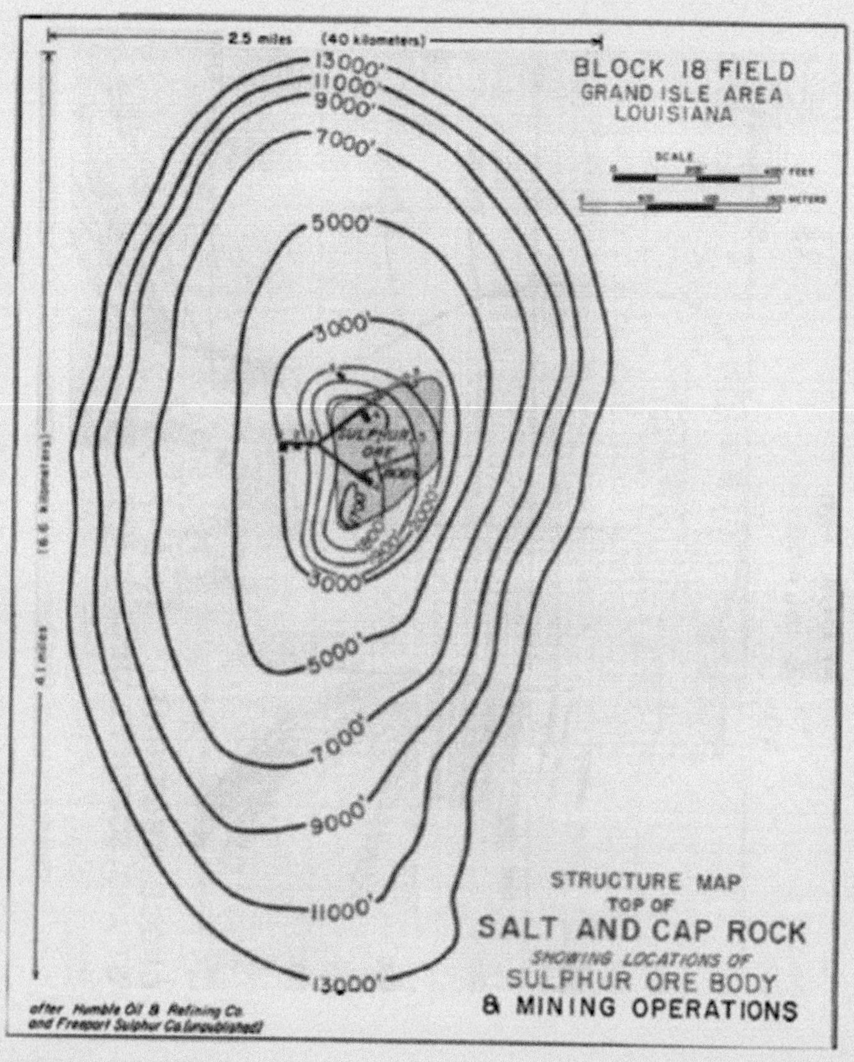

Fig. 21.22 Grand Isle block 18 structure map top of salt and cap rock for sulfur ore body mining operations. *(Source: Atwater, G.I., 1959. Geology and petroleum development of the continental shelf of the Gulf of Mexico. Gulf Coast Assoc. Geol. Soc. Trans. 9, 131–145. Reprinted by permission of Gulf Coast Assoc. Geological Societies, 2018.)*

Continued

BOX 21.1 Offshore Sulfur Mining—cont'd

Frasch Process. In the late 1880s, Herman Frasch developed a process to melt sulfur in host limestone and gypsum beds with superheated steam and pump it to the surface with compressed air where the molten sulfur cools and solidifies in vats. Sulfur melts at 246°F and solidifies at 238°F, so processes that melt sulfur must achieve temperatures >246°F. Using a system of three concentric pipes, superheated water pumped down the outside pipe melts the sulfur in situ; air forced down the smallest pipe lifts the liquid sulfur up the middle pipe, which flows to a holding vat and solidifies. After Frasch's Union Sulfur Company patents expired in 1912, new players entered the market, and by the 1930s, Freeport Sulfur Company of New Orleans dominated the industry (Peterson, 2000).

Freeport's Offshore Mines. Three offshore sulfur mines were built and operated in the GoM. The Grand Isle 18 mine was the world's first offshore sulfur mine constructed in 1956 and decommissioned in 1999 (Fig. 21.22). Caminada, also located in the Grand Isle area, was built in 1966 and shut down in 1994, and Main Pass 299 was constructed in 1991 and stopped sulfur production in the mid-2000s (Fig. 21.23). Oil and gas operations at Main Pass 299 terminated in 2016. Each of the facilities were large complexes using multiple platforms connected by bridges and support towers. Caminada was the smallest of the three mines and produced 5.7 million tons sulfur over its lifetime. The Main Pass mine was built to recover 67 million tons of sulfur, but only a small portion was recovered as operations ceased to be economic (Table 21.9).

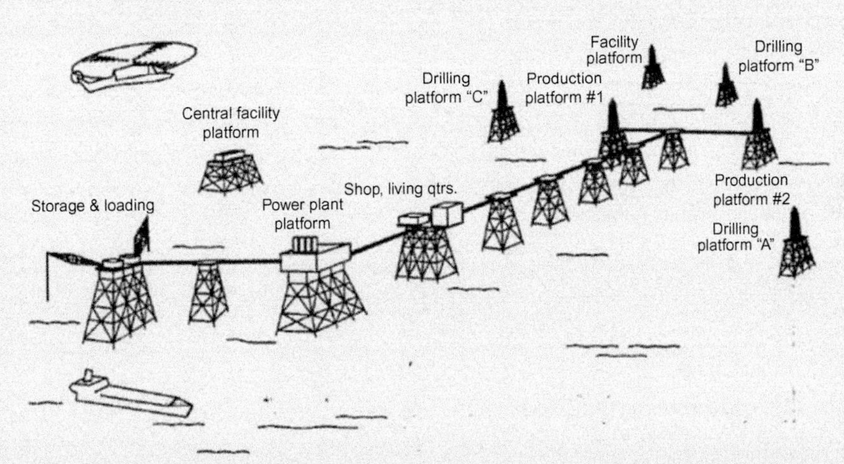

Fig. 21.23 Main Pass 299 sulfur mining complex. *(Source: Schmaltz, M., 1991. The automation of a Sulphur mine. In: OTC 6679. Offshore Technology Conference. Houston, TX, May 6–9.)*

Table 21.9 Sulfur Mining Operations in the Gulf of Mexico

Field	Status	Reserve/Production (Million Tons)	Oil (MMbbl)	Gas (Bcf)	Oil+Gas (MMboe)
Caminada	Terminated 1981	6/5.7			
Grand Isle 18	Reefed 1999	30/?	10.7	16.6	13.7
Main Pass 299	Terminated 2016	67/?	51.7	8.4	53.2

Source: Press releases (sulfur) and BOEM (oil and gas).

Continued

BOX 21.1 Offshore Sulfur Mining—cont'd

Operations. Platforms are positioned over the ore body, and vertical wells are drilled into the formation to deliver hot water and compressed air and receive molten sulfur. When a production area is depleted, a new platform is required with new bridgework. The bridge system carries the pipelines necessary to provide the required services from the plant, hot water under high pressure and compressed air to the production platform, and to deliver the molten sulfur to storage tanks (Schmaltz, 1991). Main Pass 299 was designed with 13.6 MW electric generation and consumed 24 MMcfpd natural gas to produce 5500 t sulfur per day, 2 million tons a year (McKelroy, 1991). For comparison, the Grand Isle and Caminada power plants had capacity of 5.7 and 4.3 MW, respectively (Nelson and Moody, 1991). Living quarters at Main Pass had accommodations for 188 beds.

Transportation. After the wells pump the liquid sulfur into collection tanks, it must be delivered to shore. At the Grand Isle mine, the sulfur was pumped through a 7 mi-long heated and insulated pipe-in-pipe system consisting of a 6 in sulfur line inside a 7 in hot water jacket with insulation inside a 14 in outer casing. The pipeline was in service from 1960 to 1990 when external corrosion allowed seawater to enter the casing. At the Caminada and Main Pass mines, sea going barges were used to take the sulfur to shore where it was off-loaded into an inland barge. At the Main Pass mine, two 12,000 t storage tanks were used with a 7500 t capacity ship to deliver product to shore. The Caminada mine had 4000 t storage.

Sulfur as By-Product. Crude oil and natural gas are processed in refineries and gas plants to remove sulfur to produce cleaner petroleum products and odorless gas. Today, >95% of sulfur is recovered as a by-product of processing, and the Frasch process is no longer used. Crude oil and refinery products containing sulfur compounds are treated with hydrogen and converted to a gas hydrogen sulfide. A process unit called a Claus unit burns this hydrogen sulfide with a catalytic converter into liquid sulfur. At gas plants, sour natural gas is fed to an amine solvent extraction unit to produce sweet natural gas and hydrogen sulfide. The hydrogen sulfide is then fed to a Claus unit as in refining operations.

Produced crude is heavy and sour (22° API and 2.5 wt% sulfur), and the solution gas and gas-cap gas have high H_2S content of 14.9 and 2.5 mol%, respectively, which is not surprising considering its association with sulfur bearing ore formation. Electric submersible pumps were utilized to lift the crude, and of the 18 initial development wells, 11 were drilled horizontal and placed high above the oil-water contact to eliminate water coning (Fig. 21.24).

Fifteen platforms make up the Main Pass complex and include one power plant, one living quarters, warehouse, and storage facilities (Fig. 21.25). Five of the six major platforms are connected via bridges spanning about a mile and supported by nine bridge platforms (Fig. 21.26). There are two drilling and production platforms. Eight wells were drilled on the "A" platform, five horizontal and two deviated oil wells and one gas-

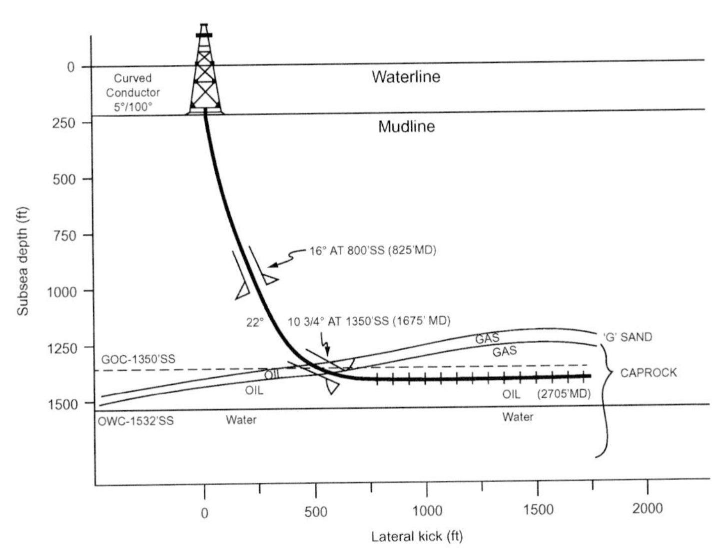

Fig. 21.24 Horizontal well completions at Main Pass block 299 were placed structurally high. *(Source: Lewis, L.R., Taylor, B.P., 1992. Early performance and planning for a simultaneous thermal enhanced oil recovery project and sulfur mining operation at Main Pass 299. In: SPE 24629. SPE Technical Conference and Exhibition. Washington, DC, Oct 4–7.)*

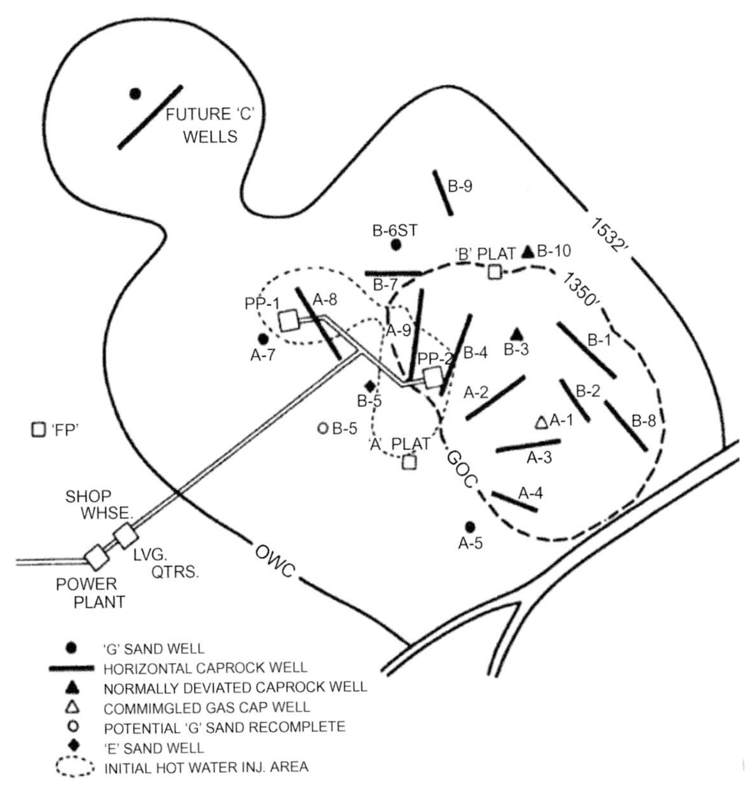

Fig. 21.25 Schematic of production complex and horizontal wells at Main Pass 299 circa 1992. *(Source: Lewis, L.R., Taylor, B.P., 1992. Early performance and planning for a simultaneous thermal enhanced oil recovery project and sulfur mining operation at Main Pass 299. In: SPE 24629. SPE Technical Conference and Exhibition. Washington, DC, Oct 4–7.)*

Fig. 21.26 Main Pass block 299 production and mining complex. *(Source: Freeport-McMoRan.)*

cap depletion well. Ten wells were drilled on the "B" platform, six horizontal and three deviated oil wells and one gas well.

Through September 2017, the MP 299 field has produced about 169 MMbbl of oil and 116 Bcf of gas (Fig. 21.27). Freeport-McMoRan's MP 299 lease block above the salt dome was highly prolific and produced about a third of the field's oil production (51.7 MMbbl of oil and 8.4 Bcf gas).

From 2010 to 2012, McMoRan Exploration Co. reported MP 299 lease oil production at an average annual rate of 361 Mbbl, average sales prices from \$73 to \$104/bbl,

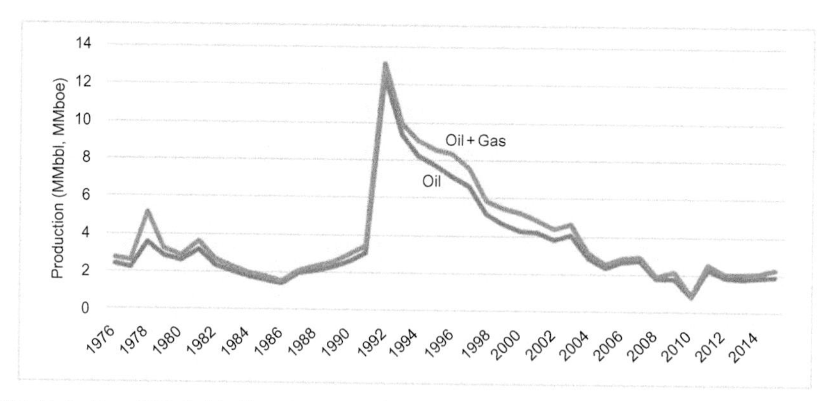

Fig. 21.27 Main Pass 299 field oil and gas production profile, 1976–2015. *(Source: BOEM 2018.)*

Table 21.8 Main Pass 299 Field Production Cost, 2010–12

	2012	2011	2010
Net sales			
Oil (Mbbl)	360	348	376
Average sales prices			
Oil and condensate per bbl	104.03	101.75	73.41
Production costs			
Oil equivalent ($/bbl)	$63.38	$97.83	$51.94

Source: McMoRan Oil & Gas, Inc.

and production costs from $52 to $98/bbl (Table 21.8). High production cost can be attributed to the difficulty of producing sour crude reservoirs. Workover expenses were reported as $1.9 million in 2010 (or $5.2/bbl of the $52/bbl total production cost), $16.2 million in 2011 ($46.6/bbl on a unit production basis), and $3.2 million in 2012 ($9/bbl). From 2005 to 2009, production cost ranged between $31/bbl and $69/bbl.

Freeport-McMoRan's MP 299 lease was terminated in 2016, but leases covering Main Pass 142, 298, 299, and 300 are still productive. Remaining reserves are estimated at 5.5 MMbbl oil and 4.3 Bcf gas.

21.9 DISCUSSION

In 2014, production from the West Delta 73 field was 5500 boe per day, 75% from crude oil and NGLs (4100 bbl crude oil and 100 bbl NGLs per day) and 7.5 MMcf gas per day, which generated approximately $170 million in gross revenue during the year:

$$4100\,\text{bopd}\left(\frac{365\,\text{d}}{\text{yr}}\right)\left(\frac{\$105}{\text{bbl}}\right)+7.5\,\text{MMcfpd}\left(\frac{365\,\text{d}}{\text{yr}}\right)\left(\frac{\$4.22}{\text{Mcf}}\right)=\$168.8\,\text{million}.$$

Crude oil contributed about 91% of revenue, 7% is due to gas sales, and 1% is due to NGLs. Production costs in 2014 totaled about $40 million:

$$5500\,\text{boepd}\left(\frac{365\,\text{d}}{\text{yr}}\right)\left(\frac{\$19.8}{\text{boe}}\right)=\$39.7\,\text{million},$$

about one-quarter of the total revenue, indicating the mature, high-cost nature of field production. In 2014, production weighted average sales price was $85/boe, and LOE as a percent of the sales price was 23%.

Main Pass 61 and South Timbalier 21 fields no longer contribute >15% of Energy XXI's reserves and have not been reported in recent years. In 2010, the sales prices received for crude oil and natural gas were about the same in the two fields, with Main

Pass crude and natural gas receiving slightly higher prices than from South Timbalier. These differences are attributed to the quality of the crude oil and amount of hydrocarbon liquids present in the gas streams. On the cost side, there is a large difference in production cost, $11.4/boe at MP 61 versus $27.2/boe at ST 21. South Timbalier 21 is an older field and has many more platforms that partially accounts for the cost difference.

Unit production cost at Mahogany is about one-third the level of MP 61 and one-fourth the level of WD 73 and ST 21 because production volumes are much greater and are centralized at one platform, well counts are smaller, and facilities are of younger vintage. Mahogany is classified as an oil field, and the NGL content of its associated gas is 25 bbl/MMcf, a typical medium-rich gas stream due to its association with crude production, but does not make a material contribution to revenue as the reader can readily verify.

Pompano produces from both wet and dry trees, and in 2015, gross revenue from crude oil was $139 million and $7.4 million from natural gas, while production cost totaled about $22.7 million. At 16% gross revenue, production cost is higher on a relative basis than at Mahogany where production cost is 9% gross revenue because the field is in deepwater and employs a large number of subsea wells that are expensive to operate and maintain. As new subsea wells bring additional production back to the host, lifting cost would be expected to stabilize or decline.

REFERENCES

Ballard, A.L., 2006. Flow-assurance lessons: the Mica tieback. In: OTC 18384. Offshore Technology Conference. Houston, TX, May 1–4.

Craig, J.R., Vaughan, D.J., Skinner, B.J., 2011. Earth Resources and the Environment. Prentice Hall, Upper Saddle River, NJ.

Fredericks, P.D., Smith, L., Moreau, K.J., 2011. ECD management and pore pressure determination with MPD improves efficiency in GOM well. In: SPE/IADC 140289. SPE/IADC Drilling Conference and Exhibition. Amsterdam, Netherlands, Mar 1–3.

Hannah, R.R., Park, E.I., Poter, D.A., Black, J.W., 1993. Combination fracturing/gravel-packing completion technique on the Amberjack, Mississippi Canyon 109 field. In: SPE 26562. SPE Annual Technical Conference and Exhibition. Houston, TX, Oct 3–6.

Haynes, W., 1942. The Stone that Burns. Plimpton Press, Norwood, CT.

Iqbal, S.A., Martinez, P., Tarazona, J., Chakalis, J., Sims, A., Taylor, M., Henry, C., Gaudet, B., 2016. A horizontal program utilizing geosteering collaboration is responsible for nearly 70% production growth over the original vertical program in the Gulf of Mexico. In: Society of Petrophysicists and Well Log Analysts Annual Logging Symposium. Reykjavik, Iceland, Jun 25–29.

Johnston, R.A., Porter, D.A., Hill, P.L., Smolen, B.C., 1993. Planning, drilling, and completing a record horizontal well in the Gulf of Mexico. In: OTC 7351. Offshore Technology Conference. Houston, TX, May 3–6.

Lewis, L.R., Taylor, B.P., 1992. Early performance and planning for a simultaneous thermal enhanced oil recovery project and sulfur mining operation at Main Pass 299. In: SPE 24629. SPE Technical Conference and Exhibition. Washington, DC, Oct 4–7.

Lipari, C.A., 1962. An engineering challenge – development of south Louisiana's giant Timbalier Bay field. In: SPE 407. SPE Annual Fall Meeting. Los Angeles, CA, Oct 7–10.

McKelroy, R.S., 1991. Offshore Sulphur production. In: OTC 6677. Offshore Technology Conference. Houston, TX, May 6–9.

Montgomery, S.I., Moore, D.C., 1997. Subsalt play, Gulf of Mexico: a review. AAPG Bull. 81 (6), 871–890.
Nelson, C.W., Moody, A.G., 1991. Design and construction of 6000-ton facilities module for Main Pass 299 Sulphur mine. In: OTC 6678. Offshore Technology Conference. Houston, TX, May 6–9.
Ogletree, J.O., Overly, R.J., 1977. Sea-water and subsurface-water injection in West Delta block 73 waterflood operations. In: SPE 6229. SPE Annual Offshore Technical Conference. Houston, TX, May 3–5.
Peterson, R.W., 2000. Giants on the River. Homesite Company, Baton Rouge, LA.
Reeves, D., Harvey, R., Smith, T., 2003. Gas lift automation: real time data to desktop for optimizing an offshore GOM platform. In: SPE 84116. SPE Annual Technical Conference and Exhibition. Denver, CO, Oct 5–8.
Schmaltz, M., 1991. The automation of a Sulphur mine. In: OTC 6679. Offshore Technology Conference. Houston, TX, May 6–9.
Willson, S.M., Edwards, S., Heppard, P.D., Li, X., Coltrin, G., Chester, D.K., Harrison, H.L., Cocales, B.W., 2003. SPE 84266. Wellbore stability challenges in the deep water, Gulf of Mexico: case history examples from the Pompano field. In: SPE 84266. SPE Annual Technical Conference and Exhibition. Denver, CO, Oct 5–8.

CHAPTER TWENTY TWO

Factor Models and Production Handling Agreements

Contents

Abstract

Factor models are popular in operating cost modeling because they are easy to implement in spreadsheets and allow for diverse applications. Unfortunately, the reliability of factor models is often poor because users typically assume model parameters without calibrating with empirical data. In this chapter, factor models for lease operating cost, workovers, and gathering and transportation services are illustrated, and their limitations are discussed. Production Handling Agreements are a special type of factor model negotiated between infrastructure owners and third-party producers who wish to use the facility for production operations. The terms and conditions of production handling agreements are reviewed for the Thunder Hawk and Independence projects, and the manner risk that is allocated between parties is highlighted using a hypothetical example concerning alternative futures for the Big Bend and Dantzler tiebacks.

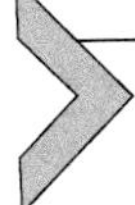

22.1 LEASE OPERATING EXPENSE

22.1.1 Model Specification

Operating cost models are typically described in terms of a fixed and variable component at a structure, lease, or field level. Fixed cost arises from the workforce salary, level of maintenance required to sustain production, overhead and administrative cost, and

Decommissioning Forecasting and Operating Cost Estimation
https://doi.org/10.1016/B978-0-12-818113-3.00022-9

related factors. Few properties can be operated without overhead costs because of accounting and reporting requirements, and occasionally, overhead costs may exceed direct costs. In modeling, fixed operating cost is frequently assumed proportional to the capital costs of the items to be operated and based on a percentage of initial capital expenditures or cumulative CAPEX over time.

Variable cost may be reported for each production and injection stream or components may be consolidated. Variable cost depends primarily upon chemicals, energy/electricity used in production and water handling, disposal of waste, and maintenance requirements. In modeling, variable operating cost is usually assumed proportional to the throughput of the primary (oil, gas, and water) or composite production fluids (oil and water, oil and gas, or oil gas and water).

One of the simplest OPEX factor models is specified as a combination of fixed and variable cost components:

$$OPEX_t = A \cdot CAPEX_t + B \cdot Production_t$$

where $CAPEX_t$ and $Production_t$ in year t are known or estimated and A and B are user-defined model parameters. $CAPEX_t$ and $Production_t$ are described in units of dollars and volume, respectively, usually on an annual basis, but for capital expenditures, a cumulative (running) total may be employed. Capital spending usually includes topsides and structure (i.e., the "facility") or the facility and subsea system, if applicable, and in some cases may include only the topside cost. For offshore GoM platforms, the value of A is typically assumed to range between 1 and 5%, and B between \$3 and \$15/boe or between \$0.5 and \$2.5/Mcfe. Ideally, the value of B would be obtained from an evaluation of the production cost or direct LOE of the operator or properties in the region (Kaiser, 2019).

An OPEX model that breaks production cost into individual streams would be written as

$$OPEX_t = C_{oil}\, NP_t + C_{gas}\, GP_t + C_{water}\, WP_t + C_{water\_inj}\, WI_t$$

where NP_t, GP_t, and WP_t represent the total production of oil, gas, and water for each year; WI_t is the cumulative injection of water during the year; and C_{oil}, C_{gas}, C_{water}, and C_{water_inj} are the variable operating expense for each unit of oil, gas, water produced, and water injected, respectively. This model requires detailed input from the user but is an improvement over the two-factor version if suitable data are available. For the deepwater GoM, C_{oil}=\$10/bbl, C_{gas}=\$1/Mcf, C_{water}=\$10/bbl, and C_{water_inj}=\$8/bbl might apply (Saputelli et al., 2009).

Example. Average production cost at Energy XXI GOM, 2012–14

Energy XXI reported average production cost from 2012 to 2014 for their shallow-water GoM properties as \$15.2/boe direct LOE, \$3.8/boe for workovers and maintenance, \$2/boe for insurance, and \$1.3/boe for

Table 22.1 Energy XXI GOM LLC Corporate Production Cost, 2012–14

	Cost ($/boe)	Percentage (%)
Direct LOE	15.18	68
Workover and maintenance	3.79	17
Insurance	2.01	9
Transportation and gathering	1.34	6
Total	22.32	100

Source: Energy XXI GOM.

transportation and gathering services. Total all-in production cost is the sum of the individual components and averaged $22.3/boe (Table 22.1).

22.1.2 Limitations

There are obvious structural flaws in these models since they predict operating expense will decline over time as production declines, which is the opposite of what is commonly experienced! Applying an inflationary factor to increase cost with time is not an appropriate strategy to resolve the issue since costs are increasing because of wear and tear and reservoir issues, which are independent of inflation and not the result of inflation. In the real world, LOE are not linearly related to anything and depend on more than the number of producing wells or the volume of hydrocarbons. It takes energy to produce water or operate a water/gas injection plant at a given pressure. It takes money for gas-lift operations and for production crew and workover expenses. As wells age, operating cost normally increases significantly.

To understand the cost to produce a barrel of oil and a cubic foot of gas, it is necessary to understand the primary operational variables and their interrelationships over time. These relationships are generally complex, poorly understood, and difficult to model. In many companies, it is laborious to establish a clear and effective link between cost drivers and the bottom line because of the complexity of financial reporting and transactional processes.

The most popular operating cost models typically relate operating expenses to (producing) well count, hydrocarbon volume or rate, water volume or rate, or time. Each summary statistic provides a different perspective. Dollars per oil volume, for example, assumes that changes in well counts or produced fluids do not impact cost (Boomer, 1995). The dollars per well metric assumes that increasing production volume does not increase operating expense. The dollars per month model assumes expenses will remain the same regardless of well count, fluid volumes, injection volumes, or injection pressure. Each metric is suitable under specific conditions.

22.1.3 Model Parameters

Oil is inherently more expensive to produce than natural gas, and dry tree wells are significantly cheaper to operate than wet tree wells, for all other things equal. Model parameters should depend on the number and type of wells and type of production at the facility. Fields that require multiple platforms or platforms developing multiple fields, platforms with more wells, or wells with more complex configurations will usually have larger production cost than fields with one platform or platforms with fewer and simpler wells, for all other things equal. Unmanned and automated facilities are cheaper to operate than manned platforms, but unit production cost may be higher or lower depending on circumstances.

Trends with commodity price are difficult to establish empirically since they impact operating cost components both directly and indirectly. Changes in commodity prices directly impact costs such as fuel and chemicals, which tend to be contractually tied to prices. Other items such as labor, boats, helicopters, and materials are indirectly impacted since as prices increase, industry activity and demand increase and, thus, costs, which enter the negotiation stage at contract renewal. Workover expenses are variable, and insurance expense depends on operator preferences.

Example. Two-factor model parameterization

If the operating cost for the GoM asset described in Chapter 19 (Table 19.1 and Fig. 19.2) was specified using the factor model $OPEX_t = A + B \cdot Production_t$, the model parameters determined from regression will yield A = \$2.6 million and B = \$0.26/bbl. The average production cost is smaller than the \$1.21/bbl life-cycle cost previously computed because of the inclusion of the fixed term in the model. Dividing the fixed term by the \$57 million capital expenditures of development yields 4.6% (= 2.6/57). Typically, A is assumed to range from 1% to 5% capital expenditures in cost models, but the empirical basis for this selection has never been demonstrated.

If the operating cost model is specified as $OPEX_t = A \cdot CAPEX_t + B \cdot Production_t$, the parameters range over a larger solution space. For example, if A = 3%, the fixed term is reduced to 1.7 (= 0.03*57), while B increases to \$0.46/bbl. For A = 5%, the fixed term increases to 2.9 (= 0.05*57), while the variable term decreases to B = \$0.21/bbl. User preference plays a large role in model specification.

22.2 WORKOVER EXPENSE

Workovers are performed to maintain production and produce at the highest possible rates. When commodity prices rise, operators seek to increase production, and the number of workovers performed typically increases. Workovers are a discretionary budget item, and a portion of the annual operating budget is often allocated toward workovers. Assuming an average well workover cost C_t and annual frequency of operation f_t, the workover cost model is described as the product of the unit cost, frequency of intervention, and well inventory:

$$WO_t = C_t \cdot f_t \cdot NW_t,$$

where the number of producing wells NW_t in year t is known, the workover cost is estimated, and the frequency of workovers is user-defined. A common assumption in the shallow-water GoM is to require a workover on producing wells every 3 years, but this will obviously vary widely. Workover cost also ranges broadly from tens of thousands of dollars per operation to several hundred thousand dollars to a million dollars or more. Companies maintain records of their well activities and cost but rarely publicize results.

Example. Shell's GoM deepwater portfolio

Shell's deepwater GoM portfolio circa 2012 consisted of about 120 wells, two-thirds direct vertical access and one-third subsea, and 80% oil wells and 20% gas wells. Morgenthaler and Fry (2012) reported that 128 stimulations were performed from 2002 to 2011, on average about one stimulation per well per decade, while the total number of stimulations per year varied from 6 to 22 (5% to 20% well portfolio per year). During the early, high-rate production period, depletion fine migration was the primary damage mechanism, and acid treatments accounted for more than 80% jobs. As reservoirs depleted closer to bubble-point, asphaltenes deposition became more prevalent, and as wells started to produce water, scale deposition became a problem. In 2011, half of treatments were for fine migration, 40% for asphaltene deposits and 10% for scale. Stimulation treatments for scale, solvent, and acid have distinct workover expense and outcomes.

22.3 TRANSPORTATION AND GATHERING

After a pipeline is installed, the primary purpose of the tariff is to recover the investment, operating and maintenance expense, and various other secondary costs. The investment and the manner in which pipeline is regulated are the most important factors that determine tariffs in the GoM, but the age of the line, the number of customers, and volume throughput are also important in determining rates at different times in the life cycle of the pipeline. The underlying principal of cost recovery is the universal theme in rate making.

Age-related factors can act in favor or against the producer. After the pipeline is fully depreciated, for example, tariffs may decline after investment costs are fully depreciated, but if volumes in the line decline, corrosion usually increases, and maintenance becomes more difficult and expensive. If the amount of product transported through the pipeline is low, the pipeline operator may have to increase rates on remaining subscribers to generate adequate revenue to cover operations and maintenance. The conditions governing each system are unique, and therefore, many factors potentially impact tariffs.

Historically, ownership of export pipelines in the GoM, especially gas trunk lines, has seen significant activity from gas transmission companies. If gas pipelines are regulated by the Federal Energy Regulatory Commission (FERC), the transportation agreements which set forth the terms and schedules are in the public domain; otherwise, the agreements are generally confidential. In many natural gas pipelines, energy content is measured in decatherms (Dth). A decatherm equals ten therms or 1 MMBtu.

The monthly bill for deliveries usually includes a reservation and commodity charge and other charges as applicable:

- Reservation charge—equal to the product of the reservation rate multiplied by the maximum daily quantity specified in the service agreement and multiplied by the shipper specific heating value and the number of days in the month.
- Commodity charge—equal to the commodity rate multiplied by the quantity of gas allocated to the delivery point in the month.
- Other charges—any applicable surcharges, new facility charges, repair charges, and any incidental expenses.

The sum of the charges is equal to the transportation fee for a particular segment of line. If product is delivered through multiple segments, a separate tariff may apply to each segment.

Example. Destin Pipeline firm transportation rate FT-1

A traditional firm service with fixed maximum daily quantity and reservation charge (FT-1 service) is depicted in Table 22.2 for the Destin Pipeline (Fig. 22.1). For a shipper reserving 50 MMcf/day capacity with a heat content of 1000 Btu/cf, the monthly reservation charge is computed as the reservation rate $7.19/Dth/month multiplied by the capacity and heat value:

$$\frac{\$7.19}{\text{Dth/mo}} \cdot \frac{50\ \text{MMcf}}{\text{d}} \cdot \frac{1000\ \text{Btu}}{\text{cf}} \cdot 30\ \text{d/mo} = \$10.785\ \text{million}$$

If the actual average amount of gas delivered to the pipeline was 48 MMcf/day for the month, the commodity rate is computed as

$$\frac{\$0.237}{\text{Dth}} \cdot \frac{48\ \text{MMcf}}{\text{d}} \cdot \frac{1000\ \text{Btu}}{\text{cf}} \cdot 30\ \text{d} = \$341,280$$

The transportation rate 0.3¢/Dth = $0.03/Dth is computed based on volumes delivered and being an order of magnitude smaller than the commodity rate is negligible in this case. The fuel retention percentage 0.3% represents a percentage of the quantity of gas delivered for transportation and used by the pipeline

Table 22.2 Destin Pipeline Rate Schedule Firm Transportation Rate

	Monthly Reservation Rate ($/Dth/mo)	**Daily Reservation Rate ($/Dth)**	**Transportation Rate (¢/Dth)**
Maximum rate	$7.19	$0.237	0.3¢
Minimum rate	0.00	0.00	0.3¢
Fuel retention	0.3%		

Source: FERC.

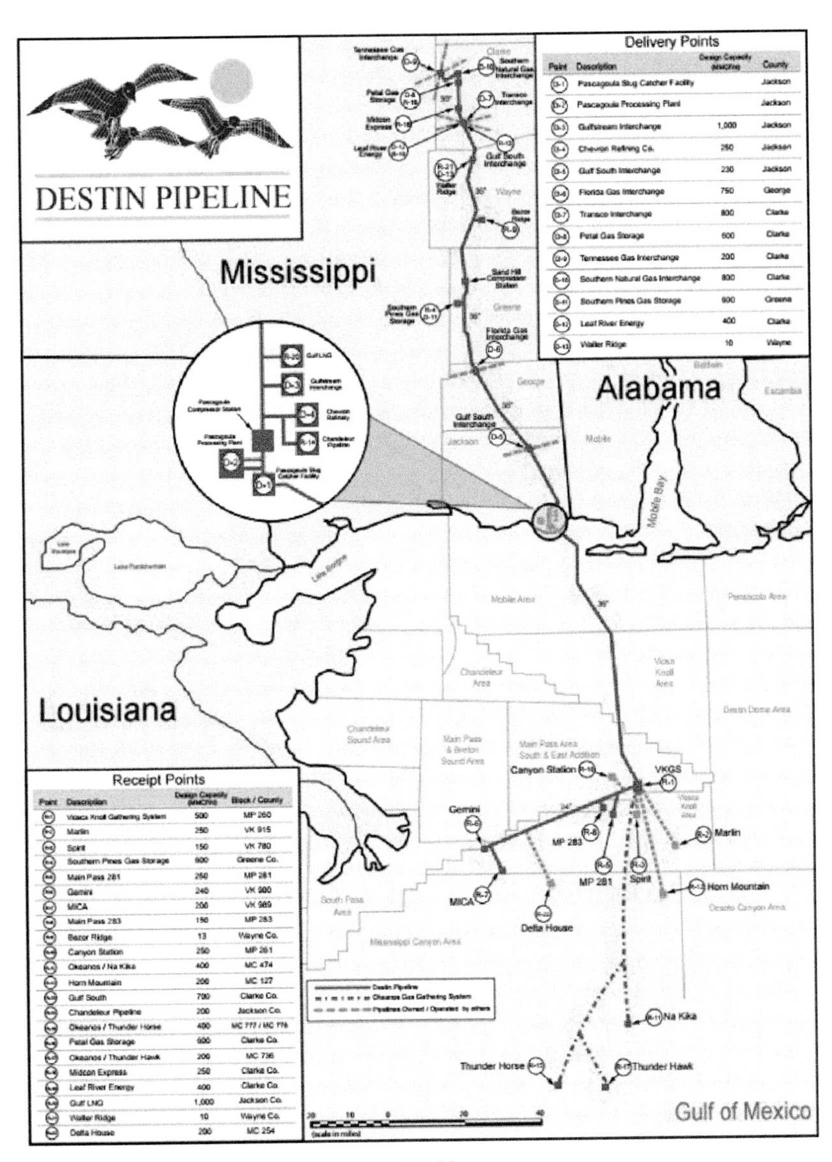

Fig. 22.1 Destin Pipeline system map. *(Source: FERC.)*

company for compressor fuel and gas otherwise used, lost, or unaccounted for and is also negligible. The total cost for the service for the month is $11.126 million, 3% of which is due to the variable cost and 97% for the reservation charge.

Example. Mars pipeline tariffs

The Mars platform in Mississippi Canyon block 807 was one of the first regional hubs in the GoM and circa 2019 serves as a central processing facility for several fields. The Mars pipeline is a 163 mi line originating approximately 130 mi offshore and delivers crude to salt-dome caverns in Clovelly, Louisiana.

Crude production from Mars A is transported to a service platform at West Delta 143 at the rate of $2.61/bbl and then onward to Bay Marchand block 4 at $1.16/bbl if <30,000 bbl/month was delivered and $0.70/bbl if >30,000 bbl/month was transported (Table 22.3).

Table 22.3 Mars Pipeline Tariff Firm Transportation Rate

From	To	Rate
MC 807 (Mars A)	WD 143	$2.61/bbl
WD 143	Bay Marchand 4	$1.16/bbl, if <30,000 bbl $0.70/bbl, if >30,000 bbl
Bay Marchand 4	Fourchon Terminal	$0.15/bbl
Fourchon Terminal	Clovelly/Caverns	$0.43/bbl

Source: Shell Midstream Partners.

A discounted rate applies because greater volumes for the pipeline mean greater revenue for the transporter to cover fixed cost. From Bay Marchand, product is delivered to Fourchon and into storage at Clovelly/Caverns.

22.4 PRODUCTION HANDLING AGREEMENTS

22.4.1 General Considerations

In the deepwater GoM, there are several developments where an unaffiliated third party owns the processing structure rather than the owner of the reserves. Canyon Express, Prince, Marco Polo, Independence Hub, Devils Tower, Tubular Bells (Gulfstar), Cascade/Chinook, Thunder Hawk, and, more recently, Who Dat, Delta House, and Stones are examples of deepwater projects where third parties (e.g., gas transmission companies, private equity, and conglomerates) without interest in the hydrocarbons own the host structure and lease it to the producers under negotiated terms referred to as Production Handling Agreements (PHAs).

Majors typically have the capital budgets and preference to own and operate all/most of the infrastructure in field development and believe they derive value from ownership (except for FPSOs), while independents in the region have sought various means to reduce capital outlays via third-party ownership to free up capital for exploration and production (Huff and Heijermans, 2003). In these transactions, the infrastructure owner provides the upfront capital and then collects monthly fixed fees and additional fees for demand and capacity based on the amount of production processed through the infrastructure. The downside for the reserves owner is less development freedom and higher production cost.

Infrastructure owners will seek recompense from third-party satellite (tieback) owners for operating expenses relating to processing production and a return on capital expenditures at a minimum, as well as a reservation of capacity for existing and other upstream projects in which they hold an interest. Generally speaking, the price parameters of PHAs are constrained on the upper end by the next best alternative available to the producer and on the lower end by the cost to upgrade equipment and process additional production (Thompson, 2009).

Conversely, an infrastructure owner might not allow access for third-party production for any number of reasons, including the lack of capacity, disagreement on the legal and economic terms on which access is granted, inability to handle the type of production for which access is proposed, and desire not to enrich a competitor (Sweeney, 2016). The use or increased use of a facility may also bring greater risks of liability and damage and faster rates of depreciation and mechanical breakdown, which need to be considered by the infrastructure owner.

Example. FPSO operating cost, rent vs. own

For a $600 million FPSO with daily process operational cost of $150,000, the difference between renting the FPSO on a 20-year lease and owning/depreciating the capital expenditures will be about $100,000 per day. A quick back-of-the-envelope calculation illustrates the numbers. The owner of the FPSO will amortize the capital investment over the 20-year lease period, which using straightline depreciation amounts to

$$\$600 \text{ million for } 20 \text{ years} = \$30 \text{ million per year} = \$82,200 \text{ per day}$$

The FPSO owner is responsible for all costs associated with vessel and process operations, maintenance, and repairs. A production crew of 15 and a marine crew of 10 on a 2-week on/off schedule and annual average salary of $100,000 per person equate to about $5 million per year or $13,700 per day. Combined, the owner of the FPSO will need to charge producers at least $100,000 per day to cover crew and depreciation cost.

22.4.2 Model Contracts

Several standard contracts for PHAs have been used for many years in the GoM, and the American Association of Professional Landmen (AAPL) was one of the first organizations to develop model contracts in the 1950s. Model contracts such as the AAPL Deepwater Model Form PHA and Shelf Model Form PHA are available through the AAPL website, and a few contracts are publicly accessible with the negotiated terms redacted. The contract arrangements and agreements are complex and consume time and resources in evaluation and negotiations. Contracts may exceed 100+ pages in length and take a year or longer to negotiate.

Table 22.4 Typical Fees and Expenses in Production Handling Agreements

Infrastructure access fee	Monthly \$/unit – Minimum monthly – Annual adjustment
Operation and maintenance expenses	Reimbursement of host operating expenses for oil, gas, and produced water – Monthly \$/unit – Shared expenses via throughput ratio – Sole expenses (host and satellite)
Deferred compensation	Compensation for deferment of host production during hookup of satellite production system
Oil quality bank	Monetary adjustments based on API gravity and sulfur value differentials

22.4.3 Contract Sections

The main sections of PHAs cover Infrastructure and Facilities, Services, Fees and Expenses, Capacity, Metering and Allocation, Gathering and Transportation, Suspension of Operations, Term, Liability, and Indemnification (Table 22.4). The Fees and Expenses section of contracts is the most important to understand for economic evaluations.

In the Infrastructure and Facilities section, the host and receiving facilities and the satellite production system are defined, including the host entry point (e.g., boarding valve) where satellite production enters host and the delivery point (e.g., export pipeline) where satellite production departs host. In the Services section, the production handling and operating functions are described.

In the Fees and Expenses section, an infrastructure access fee is charged on a monthly basis for utilizing the host and its associated facilities. In consideration for access to the host and utilization of facilities, including risers, porches, and boarding facilities; utilization of deck and riser space for satellite components; utilization of deck space for receiving facilities; and for the services provided, the producer pays a monthly infrastructure access fee. Minimum monthly fees and suspension of fees in the event the host is incapable of processing and handling satellite production are common features.

The producer pays for its pro rata share of certain host operating and maintenance expenses based on the quantity of oil, gas, and water delivered on a monthly basis. These expenses are subject to allocation charges if mechanical or operational problems occur with satellite or nonsatellite production. Deferred production compensates the host for deferment of host production during hookup of the satellite system. A typical formula applies the average daily oil and gas volume of the host during the first 14 days of the 21 days immediately preceding the shutdown, multiplied by the duration of the shutdown, the average prevailing oil and gas price during the duration of the shutdown, and a negotiated discount factor that multiples the expression (such as 0.22 or 0.35).

In the Capacity section, the host capacity defines baseline processing and handling capacity, while the satellite capacity establishes the maximum production rates by product (oil, gas, and produced water) on host. Production prioritization establishes constraint types and priority utilization principles in the event of curtailment due to operational issues of host and export pipelines. Fluid limits and operating parameters of satellite provide baseline specifications to determine conformance and establish rights, obligations, and responsibilities.

22.5 THUNDER HAWK PHA

Netherlands-based SBM Offshore, owner of the Thunder Hawk semisubmersible (Fig. 22.2), signed a PHA with Noble Energy to produce the Big Bend and Dantzler fields as subsea tiebacks. Big Bend is located about 18 mi from the Thunder Hawk facility in Mississippi Canyon 698 in 7200 ft water depth and Dantzler is about 7 mi away in 6580 ft in MC 782 (Fig. 16.5). In a 2016 press release, SBM Offshore announced production fees from the PHA were projected to generate $400 million in revenue over the 10-year primary contract period.

Denote the fixed and variable cost components of the PHA by A (\$/year) and B (\$/bbl) and assume the total fixed and variable cost over the duration of the contract are allocated according to the percentage p and $1-p$, respectively. For field reserves of X MMbbl

Fig. 22.2 Thunder Hawk semisubmersible. *(Source: SBM Offshore.)*

and contract duration D years, the total revenue R generated by the contract expressed in dollars is given by $R = D \cdot A + X \cdot B$. The assumption of the cost recovery split allows the total revenue to be decoupled into fixed and variable cost components:

$$\text{Fixed cost} = pR = A \cdot D$$
$$\text{Variable cost} = (1 - p)R = B \cdot X$$

For values of R, D, X, and p given or estimated, there are two equations and two unknowns to solve for A and B. For public companies, R is required to be reported if it is material to its operations, and D is usually also indicated in an accompanying press release. Reserves X may or may not be reported by the producer and is readily estimated. The cost split p usually varies between 40% and 60%, and although it is a negotiated (confidential) term, it is unlikely to deviate too far outside this range. In some cases, A and B may be structured to be time-dependent with an inflationary adjustment. Other variations are possible.

In the Thunder Hawk PHA, the revenue and duration of the contract are reported as $R =$ \$400 million and $D = 10$ years, while the field reserves $X = 50$ MMbbl and the cost split $p = 50\%$ are assumed. Using these input data, the cost equations are solved to yield $A =$ \$20 million per year and $B =$ \$4/bbl, the inferred terms of the PHA. The access cost to board the host for the producer and the revenue to the infrastructure owners is described as

$$ACCESS_t = \$20\,\text{million} + \$4/\text{bbl} \cdot Production_t$$

Access cost is determined in a form structurally equivalent to OPEX factor models. The full operating cost of the reserves owners requires estimates of workover expense, chemicals, engineers assigned to the field, logistics, and related items. The operating cost components not covered by the PHA can be represented by the variable C_t and total operating expense is then described by

$$OPEX_t = C_t + ACCESS_t$$

Platform owners are responsible for all costs associated with platform operations, maintenance, and repairs. Reserves owners are responsible for well cost and associated expenses.

22.6 INDEPENDENCE PROJECT PHA

The Independence Project consists of 10 natural gas fields in water depths from 7800 to 9000 ft in the Atwater Valley, DeSoto Canyon, and Lloyd Ridge areas (Fig. 22.3). No field by itself was large enough to support the capital investment, but combined and with third-party platform and export pipeline ownership, a commercial solution was found (Burman et al., 2007; Holley and Abendschein, 2007). The

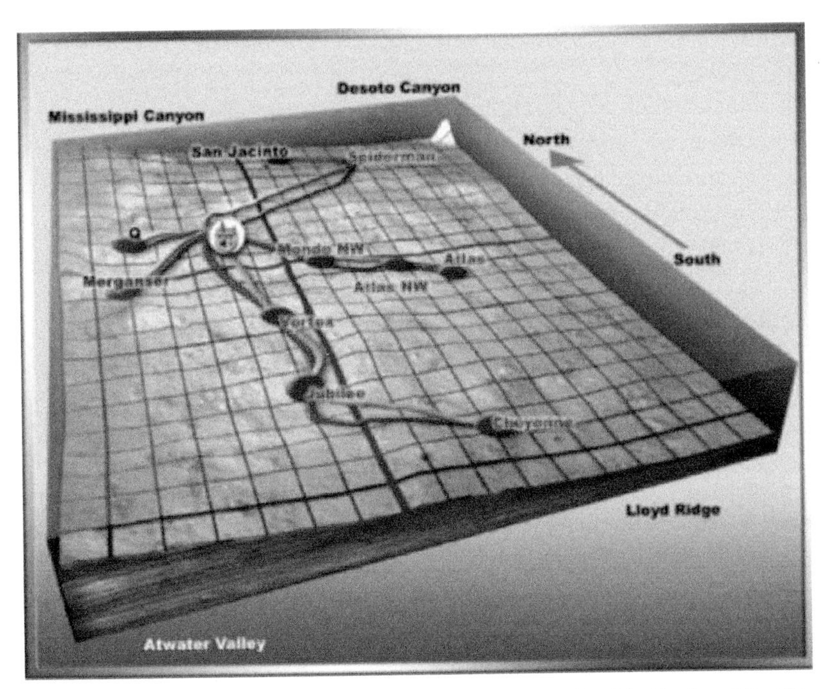

Fig. 22.3 Independence project layout. *(Source: Burman, J.W., Kelly, G.F., Renfro, K.D., Shipper, D.W., 2007. Independence project completion campaign: executing the plan. In: SPE 110110. SPE Technology Conference and Exhibition. Anaheim, CA, Nov 11–14.)*

Independence Hub semisubmersible was reported to cost $385 million, and the cost of the Independence Trail pipeline was reported as $280 million.

Under the terms of the PHA, each producer pays a monthly demand charge to the owners of the platform for a portion of its investment, and the remainder is recovered through a processing charge based on actual production. Each producer's capacity commitment determines its share of the total demand payment and has an allocated firm reserved capacity on the processing facility and pipeline for the first 5 years and a reduced reserve capacity in subsequent years.

The platform owners are responsible for all costs associated with platform operations, maintenance, and repairs. The demand charge is assumed to be recovered over 5 years, and the commodity charge is assumed to be recovered using proved reserves of 1 Tcf. The $385 million investment is split according to a 60% demand charge ($231 million) over 5 years, and a 40% commodity charge ($154 million) is spread over 1 Tcf reserves. The fixed cost for the first 5 years is therefore $46.1 million per year, and variable cost throughout production is $154/MMcf.

Processing capacity at the facility was designed for 1 Bcf/day, so if the first year of production achieved design capacity, then (1000 MMcf/day)(365 days/year) = 365,000 MMcf natural gas would be processed at a cost of (365,000 MMcf)($154/MMcf) = $56.2 million. As production declines, the variable cost component will decline, and after the entire reserves base has been recovered, the full investment will have been returned.

The pipeline owner is responsible for all costs associated with export pipeline operations, maintenance, and repairs (Al-Sharif, 2007). The export tariff recovers the investment cost on an annualized basis similar to the platform access fees. Assuming 2P reserves of 2 Tcf was the negotiated term of the contract, the $280 million would be recovered over 2 Tcf gas resulting in a smaller commodity charge payment: ($280 million)/(2 Tcf) = $140 MMcf.

In February 2016, production at Independence Hub ceased, and all the wells were permanently abandoned. Total production was 1.26 Tcf and 326 Mbbl condensate, meaning that an export tariff based on 2P reserves would not have recovered the full cost of the investment. Who Dat gas export currently flows through the Independence Trail pipeline, which provides some relief to the pipeline owners.

22.7 PHA RISK ALLOCATION

There are risks to the infrastructure owner if the reserves are not achieved in full since this is the cost basis used to recover their investment. If the commodity fee was based on 2P (proved and probable) reserves, for example, the infrastructure owner would have a greater exposure since the larger reserves both reduce and delay the cash flows received and 2P reserves may or may not be achieved over the lifetime of the field. For problems that may occur during drilling, see Box 22.1. If 2P reserves are used to determine the commodity charge and only P1 reserves are recovered, the infrastructure owners will probably not have reached their investment goals.

To secure the investment, the infrastructure owner will prefer discounting P1 reserves but may have to agree to a larger reserves base (to lower the tariffs) when negotiating with the producer group. Many factors are involved in negotiation, including the members of the producer group and their affiliation with the infrastructure owners, the time of development and outlook for the future, the nature of the geologic prospectivity in the region, and the reserves committed to the asset.

Allocation of the fixed and variable cost attempts to account for the degree of risk involved between the two parties, but the remedy is far from perfect. Two hypothetical scenarios for Thunder Hawk illustrate how risk enters PHAs for the different parties and why parties may seek to renegotiate contract terms.

BOX 22.1 Drilling Through Tar... Carefully

The Mad Dog field is located in the southern Green Canyon in 4400 to 6800 ft water depth near the Sigsbee Escarpment (Fig. 22.4). Solid circles indicate surface location of wells, and the lines emanating from the circles denote the horizontal offsets. Mad Dog wells are subsalt and are drilled to the Lower Miocene at approximately 19,000 ft TVD (Fig. 22.5).

BOX 22.1 Drilling Through Tar… Carefully—cont'd

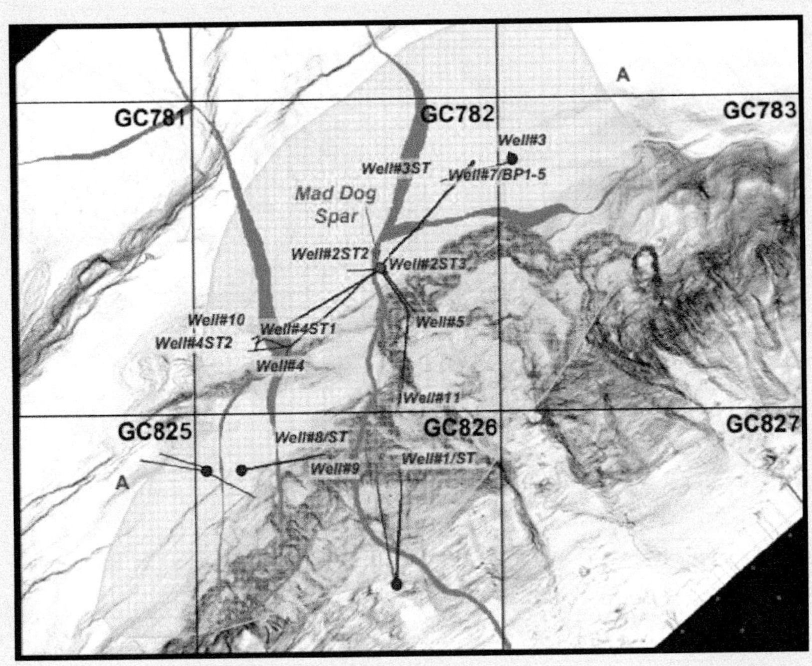

Fig. 22.4 Ocean bottom topography of Mad Dog field. *(Source: Romo, L.A., Prewett, H., Shaughnessy, J., Lisle, E., Banerjee, S., Wilson, S., 2007. Challenges associated with subsalt tar in the Mad Dog field. In: SPE 110493. SPE Annual Technical Conference and Exhibition. Anaheim, CA, Nov 11–14.)*

On rare occasions, tar has been encountered drilling in the deepwater GoM, and if it occurs, problems usually result. Tar is a highly viscous hydrocarbon rich in asphaltenes (high molecular weight components of oil), which may flow in the downhole temperatures encountered. At Mad Dog, tar was encountered at the base of salt. Well #7 drilled from the spar platform in 2005 encountered more extensive and more active tar in the wellbore than any previous well drilled in the field, and problems were reported to extend the drilling time by approximately 2 months (Romo et al., 2007).

GC782 #7 original hole encountered approximately 100 ft of tar at the base of salt (Fig. 22.6). Upon reaching the 10 3/4 in casing point at 18,084 ft MD, the drill pipe twisted off in a tar zone, and the decision was taken to sidetrack the original hole. Well #7 BP01 encountered a 60 ft tar zone thinning to 25 ft, but at 17,868 ft MD, the liner became stuck in tar, and at 19,330 ft MD, the hole started to take losses, and when drilling out the shoe, the hole was unintentionally sidetracked at 18,974 ft (BP02). The decision was taken to cement BP01 and BP02, kick off, and drill BP03. Well #7 BP03 was successfully drilled to total depth through Middle Miocene oil pay (Romo et al., 2007).

Continued

BOX 22.1 Drilling Through Tar... Carefully—cont'd

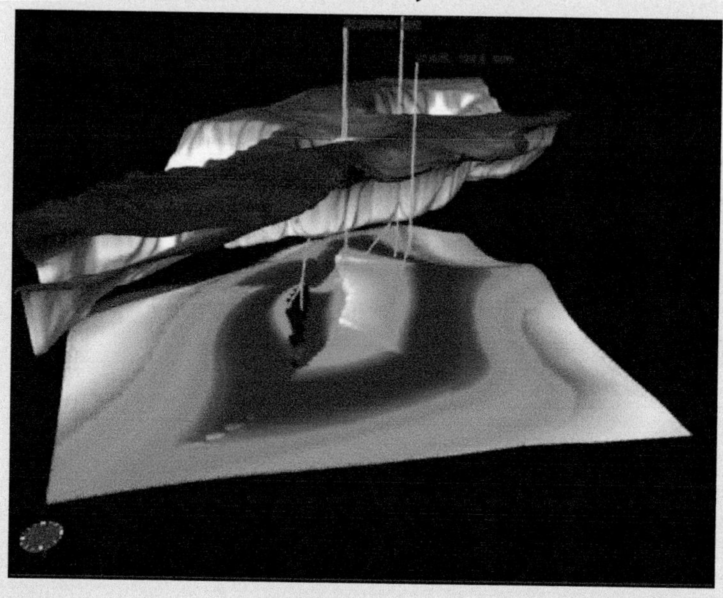

Fig. 22.5 Mad Dog 3-D model with well location. *(Source: Romo, L.A., Prewett, H., Shaughnessy, J., Lisle, E., Banerjee, S., Wilson, S., 2007. Challenges associated with subsalt tar in the Mad Dog field. In: SPE 110493. SPE Annual Technical Conference and Exhibition. Anaheim, CA, Nov 11–14.)*

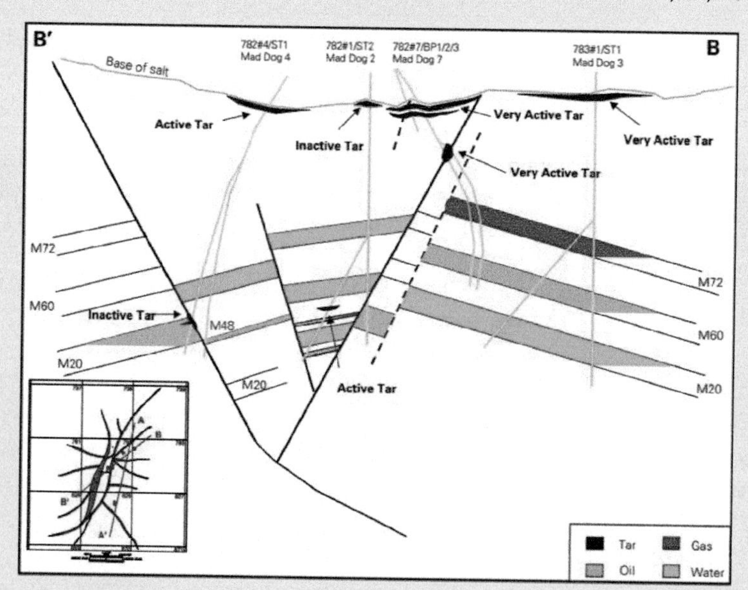

Fig. 22.6 Cross section across Mad Dog showing the geologic habitat of tar. *(Source: Romo, L.A., Prewett, H., Shaughnessy, J., Lisle, E., Banerjee, S., Wilson, S., 2007. Challenges associated with subsalt tar in the Mad Dog field. In: SPE 110493. SPE Annual Technical Conference and Exhibition. Anaheim, CA, Nov 11–14.)*

Example. Big Bend and Dantzler hypothetical futures

Scenario A: Fully depreciated asset by 2030

The year is 2030, and Big Bend and Dantzler have exceeded their 50 MMbbl initial reserves estimates with 65 MMbbl cumulative production. In 2026, the terms of the PHA were renegotiated. The fixed cost term *A* was set to zero since the capital expenditures of the investor were fully depreciated and the variable cost term was raised to $B = \$10$/bbl to handle SBM Offshore's higher operating and maintenance cost and aging infrastructure. In 2026, Big Bend and Dantzler production was 3 MMbbl oil, and commodity prices were relatively high, but by 2030, crude prices fell to $50/bbl, and production was 0.5 MMbbl. With a royalty rate of 18.75%, the producer net revenue in 2030 was about $20 million.

The processing fee to board/use Thunder Hawk is $5 million, and Big Bend and Dantzler's operating cost is about $15 million per year, so the producer will either need to shut in the field, attempt to renegotiate the terms of the PHA (say aiming to reduce *B* in the range $5 to $7/bbl), or try to sell the asset to another company. SBM Offshore may be willing to lower the terms of the PHA to keep the asset profitable for the operator, but they will not reduce the contract terms below their direct operating expense and insurance cost unless they have a clear strategic reason for doing so, say to keep the structure in place as nearby well results from other companies become known.

Scenario B: Reservoir problems in 2020

The year is 2020, just 4 years after the PHA was signed and first production was delivered. The operator has experienced numerous well failures, sand control problems, and compartmentalization far worse than anticipated, and the cost to remedy will yield much lower returns than originally expected. The operator does not anticipate recompletion or sidetracking additional wells and will run the existing wells until they no longer produce commercial quantities, which is expected within 2 years. Low oil prices have made the decision to drill new wells and workover existing wells uneconomic, and the operator will exit the field.

The production profile from Big Bend and Dantzler under this scenario yields revenue for SBM Offshore shown in Table 22.5. After 5 years, SBM Offshore has received $240 million of its $400 million investment. There have been no announced discoveries, and only a handful of exploration wells have been drilled within 40 mi of the asset, and it is considered unlikely that new regional production activities through 2025 would be supported at Thunder Hawk. A contingency fee on the fixed cost may require Noble Energy to pay some portion of the remaining $100 million to get out of the contract, but the variable costs are not included in the contingency, and SBM Offshore will incur at least a $60 million loss on its investment but thereafter is free to relocate and contract out its asset to other companies.

Table 22.5 Big Bend and Dantzler Hypothetical Production Profile and PHA Cost Terms

	Production (MMboe)	Fixed Cost ($ million)	Variable Cost ($ million)	Total Cost ($ million)
2016	10	20	40	60
2017	9	20	36	66
2018	8.1	20	32.4	5204
2019	5.6	20	22.4	42.4
2020	2.2	20	8.8	28.8
Total		100	140	

Note: Assumes fixed cost of $20 million per year and $4/bbl variable cost.

REFERENCES

Al-Sharif, M., 2007. Independence Trail – pipeline design considerations. In: OTC 19056. Offshore Technology Conference. Houston, TX, Apr 30–May 3.

Boomer, R.J., 1995. Modeling lease operating expenses. In: SPE 30055. SPE Hydrocarbon Economics and Evaluation Symposium. Dallas, TX, Mar 26–28.

Burman, J.W., Kelly, G.F., Renfro, K.D., Shipper, D.W., 2007. Independence project completion campaign: executing the plan. In: SPE 110110. SPE Technology Conference and Exhibition. Anaheim, CA, Nov 11–14.

Holley, S.M., Abendschein, R.D., 2007. Independence project overview – a producer's perspective. In: OTC 18721. Offshore Technology Conference. Houston, TX, Apr 30–May 3.

Huff, S.S., Heijermans, B., 2003. Cost-effective design and development of hub platforms. In: OTC 15107. Offshore Technology Conference. Houston, TX, May 5–8.

Kaiser, M.J., 2019. The role of factor and activity-based models in offshore operating cost estimation. J. Petrol. Sci. Eng. 174, 1062–1092.

Morgenthaler, L.N., Fry, L.A., 2012. A decade of deepwater Gulf of Mexico stimulation experience. In: SPE 159660. SPE Annual Technical Conference and Exhibition. San Antonio, TX, Oct 8–10.

Romo, L.A., Prewett, H., Shaughnessy, J., Lisle, E., Banerjee, S., Wilson, S., 2007. Challenges associated with subsalt tar in the Mad Dog field. In: SPE 110493. SPE Annual Technical Conference and Exhibition. Anaheim, CA, Nov 11–14.

Saputelli, L., Ramirez, K., Chegin, J., Cullick, S. 2009. Waterflood recovery optimization using intelligent wells and decision analysis. In: SPE 120509. SPE Latin American and Caribbean Petroleum Engineering Conference, Cartagena, Colombia, May 31–Jun 3.

Sweeney, D.H., 2016. Introduction to access to third party infrastructure in offshore projects: a comparative approach. LSU J. Energy Law Resour. 4 (2), 9.

Thompson, M., 2009. Production handling concepts and objectives. Rocky Mountain Mineral Law Foundation, Special Institute on Federal Offshore Oil and Gas Leasing and Development, Houston, TX, Jan 28.

CHAPTER TWENTY THREE

Activity-Based Costing

Contents

Abstract

Activity-based cost models apply work decomposition methods and knowledge of engineering and market conditions to estimate cost. The tasks required to be performed are identified, and the time and duration to perform each task are estimated. Activity-based costing is the most detailed and transparent method that can be applied in operating cost estimation and require a higher level of expertise to successfully apply. In this final chapter, work decomposition methods are described to illustrate how the operating cost components of labor, logistics and transportation, materials and supplies, and repairs and maintenance are estimated. Examples are provided to develop analytic skills and intuition regarding the relative importance of cost components. Regional diving, helicopter, and marine vessel contracts and service markets are reviewed as part of this discussion.

Decommissioning Forecasting and Operating Cost Estimation
https://doi.org/10.1016/B978-0-12-818113-3.00023-0

23.1 OPERATING COST CATEGORIES

The primary cost categories for offshore oil and gas operations include the following:

- Salaries of operating personnel
- Transportation of products and people
- Materials and supply services
- Repair and maintenance of wells and flow lines
- Repair, maintenance, and inspection of equipment and structure

Cost may be incurred hourly, daily, monthly, or annually; be volume- or capacity-based; or be per person (Table 23.1). In activity-based costing, the tasks required to be performed are identified, and the time and duration to perform each task are estimated. If only a small number of facilities are evaluated, a detailed approach is feasible, but for more than a few structures, the resources required to complete an activity-based cost study are significant (Kaiser, 2019).

Table 23.1 Offshore Operating Cost Categories and Cost Basis

Element	Basis
Personnel	
Process operators	Annual
Process maintenance	Annual
Supervision	Annual
Crew transportation	
Helicopters	Monthly fixed, hourly flight rate
OSV	Monthly fixed, day rate
Logistics	
OSV supply boats	Day rate
Standby vessels	Monthly
Docking charges	Monthly
Warehouse	Monthly
Chemicals	Volume
Fuel, water	Volume
Repairs and maintenance	
Service company personnel	Rate + schedule
Service company equipment	Rate + schedule
Contractor services	Rate + schedule
Equipment standby	Monthly
Pipeline tariffs	Volume, capacity, distance, age
Communications, data transmission	Annual
Catering	Per person per day
Insurance	Annual

23.2 LABOR

In the GoM, facilities that are manned 24h report a bed count to BOEM for the number of individuals that can be accommodated overnight (Fig. 23.1). Beds are required for production crew as well as service personnel, drilling and workover crew, supervisors, and visitors. The number of permanent crew required for operations typically range between one-third to one-half the number of permanent beds available.

Hourly wage rates for offshore production and drilling crew are about the same as onshore, but for offshore operations, the crews live onboard the platform which requires logistics planning, transportation, catering, safety process planning, and support staff, and makes the total expenses far greater than for onshore operations. Direct salary expense for permanent crew usually ranges between \$100 and \$150 thousand per year with many different personnel rates for different grades and occupations.

Example. Production crew salary

Offshore crews in the GoM typically work 2 weeks on and 2 weeks off, so a facility that requires a permanent crew of 24 such as Delta House would require an annual labor cost of 2 × 24 personnel × \$125,000 per person per year = \$6 million per year assuming a base salary of \$125,000 per person. A shallow-water facility such as a gas receiving platform may only require a four-man crew, which amounts to a crew salary of \$1 million per year.

Catering requirements are usually contracted on a dollar per person per day basis and include food service, cleaning, and laundry. Vendors can be contacted for the most up-to-date quotes, but values are easy to bound. It would be difficult for a GoM contractor to stay in business charging less than say \$10 per day per person to feed personnel three meals a day, while \$50 per day per person is an approximate upper bound. Actual values will depend on market competition and other conditions. Services such as cleaning linen and waste disposal may be added separately or included as part of the per person cost. Transportation to-from an onshore service base is provided by a logistic firm paid for by the operator unless the caterer contracts and charges for this service separately.

23.3 LOGISTICS AND TRANSPORTATION

Crews, supplies, and equipment must be transported to the platform, and so, the further the operation from the shore base, the greater the cost for fuel and vessel/helicopter rentals, and the more frequent the visits, the greater the expense. Larger vessels and helicopters usually charge a premium relative to smaller vehicles, and services with a shorter notice period and contract duration will also be more expensive, for all other things equal. All manned platforms in the GoM have helidecks, and all deepwater facilities are serviced using helicopters for crew, while closer to shore, both marine vessels and helicopters are utilized for crew change. Material transport is via supply vessel.

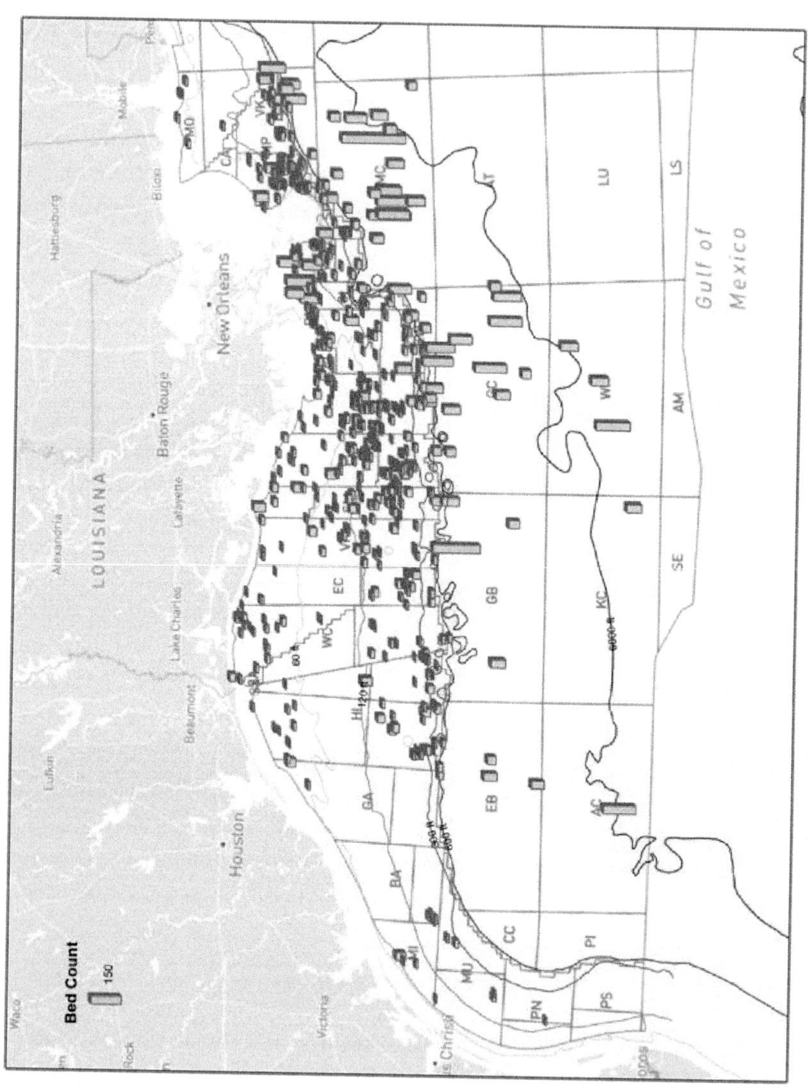

Fig. 23.1 Bed count on manned platforms in the Gulf of Mexico circa 2018. *(Source: BOEM 2018.)*

23.3.1 Marine Vessels

Vessel charters are either the product of direct negotiation or a competitive process that evaluates vessel capability and price. The day rate is the primary bid variable in contract negotiation and selection, but other factors such as safety record, history with the firm, and vessel specifications are also important.

Marine vessels are leased primarily on term or spot charters, although time and bareboat charters have also been used occasionally. Term charters are generally 3 months to 3 years in duration and are typical for drilling or production support. Spot charters are a short-term agreement (from one day up to three months) to provide offshore services. Spot charters are commonly employed for unscheduled or nonrecurring support, as in decommissioning, well work, or incident response. Under a time charter, the operator provides a vessel to a customer and is responsible for all operating expense including crew costs but typically excluding fuel. Under a bareboat charter, the operator provides a vessel to a customer, and the customer assumes responsibility for all operating expenses and associated risks.

Average monthly day rates for crew boats and offshore supply vessels (OSVs) in the GoM are published in WorkBoat Magazine based on contractor surveys and can be found on a consolidated annual (sometimes quarterly) basis in financial statements for public companies. Several consulting firms also provide marine vessel indexes. As a general proxy, market day rates and/or company data can often be used to infer fleet vessel rates because the ships and services are relatively homogeneous and commodity-like (Kaiser, 2015), whilst other activity data are site- and location-specific. High levels of competition mean that day rates are unlikely to deviate significantly among operators unless differentiated by technology or vintage.

Example. Shallow-water vs. deepwater transportation charters

In December 2016, average day rates in the GoM for crew boats <170 ft in length and OSVs <2000 deadweight ton (DWT) were reported in WorkBoat Magazine to be $3230 per day and $7800 per day, respectively (Fig. 23.2). Assuming 1 day per trip for OSVs and 6 h per trip for crew, 2-week on/off schedule for crew and a weekly OSV visit to the platform, an assumed 80% discount to the average OSV spot rates yields the annual cost estimate for crew and material transportation:

$$\text{Crew boat}: \$3230 \text{ per day} \cdot 0.25 \text{ day per trip} \cdot 26 \text{ trips per year} = \$21,000 \text{ per year.}$$
$$\text{OSV}: 0.80 \cdot \$7800 \text{ per day} \cdot 1 \text{ day per trip} \cdot 52 \text{ trips per year} = \$324,000 \text{ per year.}$$

For OSVs >5000 DWT, average day rates reported in WorkBoat Magazine were $30,662 per day, and helicopter spot rates are assumed to be $2500/h. If a round trip to a deepwater facility takes 2 days by boat and 5 h by helicopter, then for a weekly OSV visit and biweekly crew change, the annual crew and logistics cost are estimated as.

$$\text{Helicopter}: \$2500/\text{h} \cdot 5 \text{ h per trip} \cdot 26 \text{ trips per year} = \$325,000 \text{ per year.}$$
$$\text{OSV}: \$30,662 \text{ per day} \cdot 2 \text{ days per trip} \cdot 52 \text{ trips per year} = \$3.19 \text{ million per year.}$$

Structures in deepwater are farther from shore bases and require larger crew than shallow-water structures, which translate into higher labor, catering, transportation, and logistics cost.

Fig. 23.2 Offshore supply vessel *Gloria B. Callais (top)* and fast support vessel *Cougar (bottom)*. *(Source: Seacor.)*

Example. Shallow-water vs. deepwater labor and transportation cost

In Table 23.2, labor, subsistence, and transportation cost for a manned platform in shallow water with five permanent crews and a deepwater facility with 30 permanent crews are estimated circa 2017. Annual labor and transportation cost for a five-man shallow-water platform is estimated at $1.7 million versus $11.6 million per year for a 30-man deepwater facility. Deepwater labor and transportation costs are about an order of magnitude larger than the shallow-water platform.

Table 23.2 Labor and Transportation Cost Comparison—Shallow Water vs Deepwater

Component	Shallow-Water Manned Platform	Cost ($1000/yr)
Labor	5 men · 2 weeks per cycle · $125,000 per man-year	1250
Material transport	0.8 · $7800 per day · 1 day per trip · 52 trips per year	324
Crew transport	$3230 per day · 0.25 day per trip · 26 trips per year	21
Catering	5 men · $50 per man-day · 365 days per year	91
Total		1687

Component	Deepwater Manned Facility	Cost ($Million/yr)
Labor	30 men · 2 weeks per cycle · $125,000 per man-year	7.5
Material transport	$30,662 per day · 2 days per trip · 52 trips per year	3.19
Crew transport	$2500/h · 5 h per trip · 26 trips per year	0.33
Catering	30 men · $50 per man-day · 365 days per year	0.55
Total		11.57

23.3.2 Helicopters

The majority of helicopters in the GoM are chartered through master service agreements, subscription agreements, and day-to-day charter arrangements. Master service agreements and subscription agreement typically require a fixed monthly fee plus incremental payments based on flight hours. These agreements have fixed terms ranging from 1 month to 3 years and contain terms that index fuel costs to market rates so the helicopter operator is not exposed to fuel price variation. Contracts are cancelable by the client with a notice period ranging from 30 to 180 days specified in the contract. In the GoM, short-term contracts for 12 months or less are common. Ad hoc (spot or term charter) services usually entail a shorter notice period and shorter contract duration and are based on an hourly rate or a daily or monthly fixed fee plus additional fees for each hour flown. Generally, ad hoc services have a higher margin than other helicopter contracts due to supply and demand conditions.

Helicopters range in size from small to large. Single-engine and light-twin helicopters perform multiple takeoff/landings to shelf platforms and have a typical passenger capacity of five to nine. Single-engine helicopters are the largest population of helicopters in the GoM, and most are single pilot and use aviation gasoline and reciprocating engines, which not only are cheaper to operate and build than turbine engines but also provide less performance. Medium and heavy helicopters have twin-turbine engines and a typical passenger capacity of 16–19 and can fly in a wider variety of operating conditions using instrument flight rules, travel longer distances, and carry larger payloads than light helicopters. Medium and heavy aircraft are used for crew changes on large production facilities and drilling rigs and all deepwater facilities.

When an aircraft company purchases a helicopter, direct cost is an important feature since it shows the revenue a company must receive to recover its cost and stay in business. The fixed cost and hourly flight cost serve as the primary point of negotiation between operator and customer. Terms depend on the age and type of aircraft, as well as market conditions at the time of negotiation and safety record of the operator. If flight activity is less than anticipated, then the revenue rate may not cover the actual cost per hour, and operators balance this risk through a combination of fixed and variable components.[1] Producers with significant offshore operations may operate their own fleets.

Example. Master Service Agreement for Eurocopter 135/145

Direct cost calculations for a light-twin Eurocopter 135/145 (Fig. 23.3) at a purchase price of $700,000 are considered (Table 23.3). A 5-year depreciation period with a 30% residual value per year is applied. Operators buy hull and liability insurance to protect against damage to aircraft and related liabilities, which is assumed to be 10% of the purchase cost. One pilot with annual salary of $90,000 is assumed. Fuel rates are taken from the EC 145 spec sheet, and the oil and lubricants, maintenance labor, spare parts, and engine overhaul to maintain safe and reliable operations are additional cost terms. Fuel costs are assumed to be $2/gal for Jet A.

[1] Helicopter operators report that they typically receive about half of their revenue from fixed cost, similar to the PHAs described in Chapter 22.

Fig. 23.3 Eurocopter 135 EC. *(Source: Helis.com.)*

Table 23.3 Direct Cost Calculation for a Light-Twin Helicopter

			Cost Per Year ($)		
I	**Fixed Costs**		**500 h**	**1000 h**	**2000 h**
	A Depreciation (5 years, 30% residual value)		98,000	98,000	98,000
	B Insurance				
	1. Liability and property damage	$5000			
	2. Hull insurance (10% of initial cost)	$70,000			
	Total insurance per year	$75,000	75,000	75,000	75,000
	C Pilot ($90,000 per year)		90,000	90,000	90,000
II	**Hourly Costs**	**$**			
	A Fuel (85 gal at $2/gal) per hour	170			
	B Oil and lubricants (10% of fuel) per hour	17			
	C Maintenance labor per hour	20			
	D Spare parts and spares in reserve per hour	25			
	E Engine overhaul per hour	8.50			
	Total cost per flying hour	240.50	120,500	240,500	481,000
	Total cost per year		383,250	503,500	744,000
	Total cost per hour		766.50	503.50	372.00

Note that all values are meant to be illustrative. Purchase cost assumed to be $700,000.

The fixed cost for the 5-year depreciation period is $263,000 per year. The hourly costs for flying are estimated at $240.50/h and are computed for different flight hours. To recover the cost of operations, an operator will need to negotiate a monthly rate of $263,000/12 = $21,900 per month and an hourly rate of $504/h flight time if the flight hours are expected to be 1000 h. If the flight hours are expected to be 500 h, a minimum hourly rate of $767/h is required to breakeven.

23.4 MATERIALS AND SUPPLIES

Materials and supplies are usually not a significant cost in oil and gas operations until late in life since after the equipment is purchased and installed, the primary energy for their operation derives from the reservoir itself and from separated production gas. Facilities that support production from the reservoir, such as gas and water injection, and fields with low-quality crude that exhaust their reservoir energy require more equipment and greater material and energy use relative to young fields with lighter crudes and strong reservoir drives. Gas facilities typically use fewer materials and supplies on a heat-equivalent basis relative to oil facilities, and platforms that support subsea wells require greater utilities and chemical usage relative to dry tree wells.

Chemicals are used for controlling corrosion, emulsions, foaming, scales, paraffins (waxes), asphaltenes, hydrates, hydrogen sulfide, and water quality. The chemistry of produced fluids significantly impacts the design and development of a field, and focused efforts to understand and characterize the fluids are needed during the design stage. Before any treatment is applied, it is important to conduct a thorough investigation of the problems, their root causes, and any implications of the treatment. Chemical cost should be considered from a life-cycle perspective and compared against methods where the problem is managed in a different way. Chemical cost is usually not the overriding driver in chemical selection, but rather, the technical performance must be able to manage the risk.

23.4.1 Chemicals for Water Treatment

Produced oil and water create emulsions that need to be demulsified to satisfy the export pipeline requirements on water (< 1 vol%) and the produced water discharge criteria in the GoM (monthly average of 29 mg/L oil and grease). Therefore, all water–oil emulsions are treated to achieve appropriate separation with mechanical separation (e.g., hydrocyclones) and application of heat and chemicals. Water may also be injected into the reservoir to supplement oil recovery or to dispose produced water. In either case, water will require treatment depending on the source and issues identified (Table 23.4).

Table 23.4 Water Injection Problems and Solutions

Problem	Possible Effect	Solution
Suspended solids	Formation plugging	Filtration or flotation
Precipitates	Scaling and plugging	Scale inhibitors
Bacteria	Loss of injectivity and reservoir souring	Biocides and selection of sour service materials
Dissolved gas	Corrosion and loss of injectivity	Degasification

Source: Jahn, F., Cook, M., Graham, M., 2008. Hydrocarbon Exploration and Production. 2nd ed. Elsevier, Amsterdam, The Netherlands.

Example. FPSO seawater injection cost

For an FPSO with a $100,000 daily operating cost and 16-person crew, water treatment cost per barrel based on seawater injection and produced water treatment at $4/L chemical cost represents about 5% production cost (Table 23.5).

Table 23.5 FPSO Daily Water Treatment Costs per Barrel Based on Seawater Injection and Produced Water Treatment

Chemical Treatment	Dosage (ppm)	Volume (L)	Chemical Cost ($ at $4/L)	General Cost ($/bbl Water)
Seawater injection, 300 Mbbl per day				
Filter aid	2	95	380	0.001
Oxygen scavenger	2	95	380	0.001
Weekly biocide	200	125	500	0.002
Scale inhibitor	2	95	380	0.001
Daily manpower $			4384	0.015
Facility cost $			100,000	0.33
Total cost $			106,024	0.35
Produced water, 10 Mbbl per day				
Water clarifier	5	8	32	0.003
Weekly biocide	200	53	212	0.021
Corrosion inhibitor	20	32	128	0.013
Scale inhibitor	2	95	380	0.001
Daily manpower $			4384	0.004
Facility cost $			100,000	0.614
Total cost $			105,136	

Note: Assumes a 16-person crew and 2–4 h weekly biocide treatment.

Source: Wiggett, A.J., 2014. Water – chemical treatment, management and cost. In: SPE 172191. SPE Saudi Arabia Annual Technical Symposium and Exhibition. Al-Khobar, Saudi Arabia, Apr 21–24.

23.4.2 Chemicals for Corrosion

Highly corrosive fluids (e.g., sour >10,000 ppm H_2S, heavy oil <25° API, gas-oil ratio < 500 cf/bbl, and high water cut >80%) will require greater chemical spending for corrosion control and account for a higher percent of operating expense than non-corrosive fluids. There are many chemical corrosion treatments, including hydrogen sulfide scavenging chemicals, combination treatment chemicals, single-purpose inhibitors, and biocides that can be applied. Pilot studies are often performed to determine the best treatment before implementation (Wiggett, 2014).

Production chemicals are typically injected directly into the wellbore's tubing-casing annulus, wellhead, and flow lines and throughout the separation train to treat for corrosion, emulsions, scale, and H_2S content. Reducing chemical spend can lead to mechanical integrity issues that are more costly to remediate than to prevent (Cavaliaro et al., 2016). Tubing failures are a common impact of corrosion.

Example. Corrosion inhibitors

Natural gas with impurities such as H_2S and CO_2 is highly corrosive and is commonly treated with chemical inhibitors. Typical corrosion inhibitor concentrations are 5–50 ppm for continuous addition and up to 250 ppm for batch dosing. Inhibitor use increases in proportion to flow rate. Gas inhibitor dosage rates are typically in the range of 0.25–0.75 L/MMcf of gas, and at $10/L, chemical cost translates into an annual chemical cost of $1825 and $91,250 for flow rates of 1 and 50 MMcfd (Table 23.6).

Table 23.6 Corrosion Inhibitor Cost for Two Gas Flow Rates

Gas Rate	Inhibitor at 0.5 L/MMcf	Cost at $10/L	Annual Cost
50 MMcfd	25 L per day	$250 per day	$91,250
1 MMcfd	0.5 L per day	$5 per day	$1825

Source: Cavaliaro, B., Clayton, R., Campos, M., 2016. Cost-conscious corrosion control. In: SPE 179949. SPE International Oilfield Corrosion Conference and Exhibition. Aberdeen, Scotland, May 9–10.

In moderately corrosive environments, batch chemical treatments can provide sufficient protection to downhole equipment and flowlines. In highly corrosive environments and larger produced volumes, chemical treatment may be uneconomic. When flow rates are high, turbulence in the pipelines helps ensure that the full interior of the line is inhibited with chemical, but as flow velocities decline and turn laminar, untreated areas are more likely to arise.

23.4.3 Chemicals for Flow Assurance

In general, the greater the change in temperature and pressure that produced fluids experience in traveling from the reservoir to the host, the greater the number of flow assurance problems and chemical treating needs (Fig. 23.4). Flow assurance issues are generally not a concern in shallow water and with dry tree and direct vertical access wells, but are dominate design considerations in subsea development. The list of chemicals utilized in subsea developments may be substantial to avoid problems with solid deposition and to ensure safe and reliable operations. They include methanol, low-dosage hydrate inhibitors (LDHI), asphaltene inhibitor, paraffin inhibitor, pour-point depressant, corrosion inhibitor, and scale inhibitor (Bomba et al., 2018).

Strategies to reduce the magnitude of the changes occurring, especially in temperature that drives wax and hydrate formation and to a lesser extent scale formation, require capital. Flow assurance strategies typically involve a combination of equipment design/selection, operational methodologies, and chemical treatments (Table 23.7). The overall objective of flow assurance is to keep the flow path open.

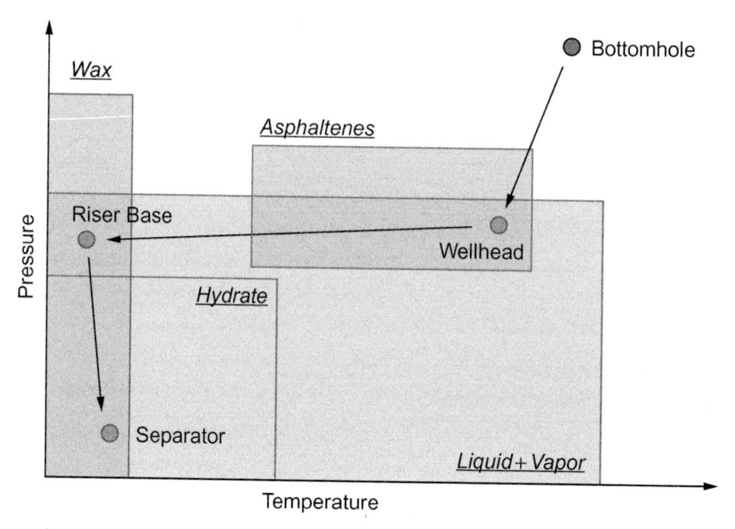

Fig. 23.4 Generalized schematic of pressure-temperature regions that will create asphaltenes, wax, and hydrate production issues.

Table 23.7 Production Chemical Applications and Injection Points

Chemical	Operation Mode	Injection Point(s)	Combinations	Rate (ppm)
Methanol	Continuous (gas) Intermittent (oil)	Tree/downhole	CI	50–100 vol%
Glycol	Continuous (gas) Intermittent (oil)	Tree/downhole	CI	50–100 vol%
PPD	Continuous	Tree/downhole	CI	100–300
PI	Continuous	Tree/downhole	CI or AI	100–300
AD	Continuous	Downhole at packer	CI	100–500
AS	Intermittent	Tree	CI	50–500
CI	Continuous	Tree	MeOH, glycol, AD	10–50
SI	Continuous	Downhole at packer	CI	1–2 vol%
LDHI	Continuous	Tree/downhole	MeOH, MEG	1–2 vol%

Note: AA, antiagglomerant low-dosage hydrate inhibitor; AD, asphaltene dispersant; AS, asphaltene solvent; CI, corrosion inhibitor; MeOH, methanol; MEG, ethylene glycol; LDHI, low-dosage hydrate inhibitor; PI, paraffin inhibitor; PPD, pour-point dispersant; SI, scale inhibitor.

Source: Bomba, J., Chin, D., Kak, A., Meng, W., 2018. Flow assurance engineering in deepwater offshore – past, present, and future. In: OTC 28704. Offshore Technology Conference. Houston, TX, Apr 30–May 3.

Example. Methanol hydrate inhibitor cost

A well produces 1000 bpd water, and 0.5 bbl of methanol per barrel of water is used to prevent hydrate formation. In the GoM, methanol cost varies with market conditions and transportation and historically has ranged from $25 to $75/bbl. For $50/bbl ($1.2/gal) methanol, hydrate inhibitor cost for the well will be $25,000 per day or $9 million per year. For some fields, the methanol delivery system may define the abandonment conditions of the well.

Methanol and LDHI are commonly employed in oil wells and are usually consumed (i.e., not recovered) in the process. Gas-dominated systems typically use MEG rather than methanol because of the lower amounts required and MEG offers the advantage of being recyclable. Selection is normally based on the lowest cost per volume produced, but other factors may also play a role. For example, since methanol is partially soluble in crude oil, excessive use results in the crude oil being contaminated at concentrations ranging from 10 to 10,000 mg/L, which will negatively impact the price received since methanol causes catalyst poisoning in refineries. Most oil contracts limit methanol concentration to below 50 mg/L. Topside storage volume constraints may also play a role in chemical selection.

Example. Methanol vs. LDHI cost comparison

For a subsea well producing a 39° API condensate, application costs for methanol at $1/gal and 35% water flow are compared with LDHI at $30/gal and 0.5% water flow (Swanson et al., 2005). If there is no recovery of the chemical, then the monthly chemical cost to inhibit hydrates for a water flow rate of 1300 bpd is estimated as:

Methanol cost = 0.35(1300 bpd)(42 gal/bbl)($1/gal)(30 day per month) = $573,300 per month.

LDHI cost = 0.005(1300 bpd)(42 gal/bbl)($30/gal)(30 day per month) = $246,300 per month.

Hydrate inhibitors treat the aqueous phase, and therefore, the higher the water rate, the higher the dosage. Hydrate prevention costs vary with the water flow rate, chemical cost, and prevention mechanism.

Example. K2 chemical injection system

The K2 field is a three-well subsea development in 4200 ft water depth tied back 7 mi to its host at the Marco Polo TLP (Brimmer, 2006). Pipe-in-pipe flow lines and insulated risers with a chemical injection system for each well were selected in development. LDHI was selected as the main hydrate inhibitor, and methanol was used as a backup system. At start-up after a shutdown, it is necessary to inject wax inhibitor to prevent wax formation while the flow rates are low and the temperature is below the WAT. Chemical injection to control asphaltenes is used when needed and varies with each well.

The injection points are shown schematically in Fig. 23.5, and the umbilical assembly for chemical delivery is shown in Fig. 23.6. Two deep-set chemical injection mandrels are located above the production packer approximately 18,000 ft below the mudline, and one shallow-set mandrel is located above the

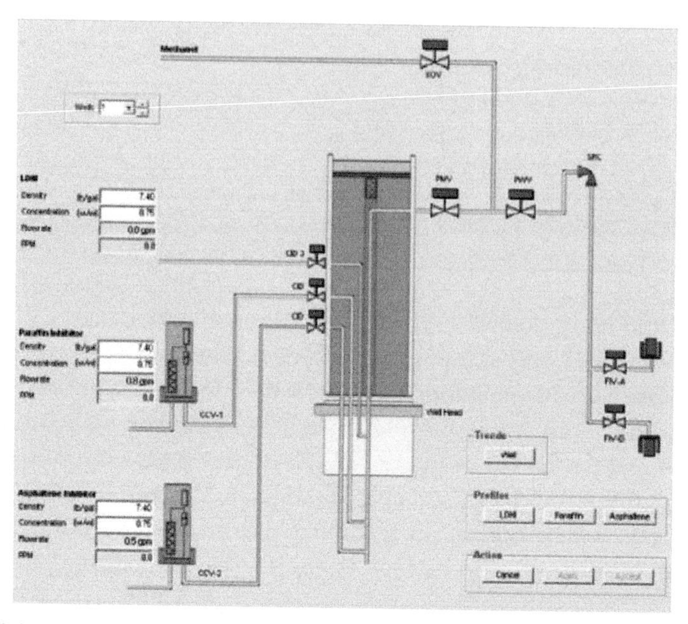

Fig. 23.5 Tree injection system and downhole locations at the K2 field. *(Source: Brimmer, 2006.)*

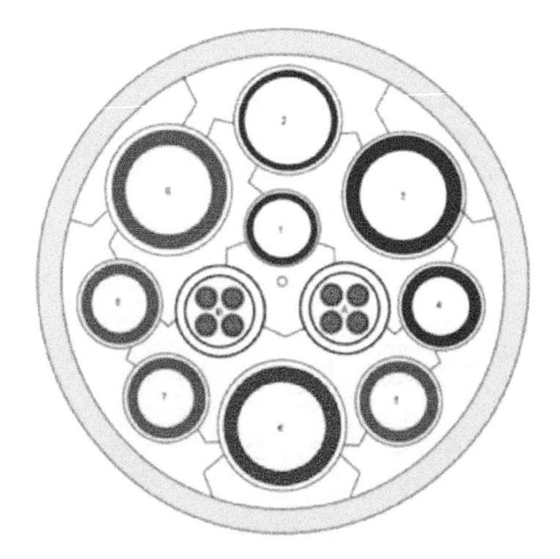

#	Tube Size (ID)	Max Pressure (psi)	Line Function
1	½"	10,000	HP "A"
2	¾"	5,000	LP "A"
3	¾"	15,000	LP "B"/share
4	½"	15,000	HP "B"/share
5	½	15,000	Asphaltenes inhibitor
6	¾"	15,000	Annulus access line
7	½"	15,000	LDHI
8	½"	15,000	Paraffin wax
9	¾"	15,000	Methanol line
A			Power/single cable
B			Power/single cable

Fig. 23.6 Umbilical cross section for K2 subsea development. *(Source: Brimmer, 2006.)*

downhole safety valve that is 3900 ft below the mudline. The deep-set mandrels are used for asphaltenes and wax inhibitor injection to protect the wellbore and pipeline. LDHI is injected downhole for better mixing, and methanol is injected at the tree for start-up.

23.4.4 Fuel, Water and Utilities

"Fuel is free" is a common adage in the oil field, and indeed, as long as wells are not low volume black crude (GOR < 500 cf/bbl), most oil wells produce enough (associated) gas to use at site to satisfy a significant portion of fuel usage and export the surplus. Backup fuels are always required during start-up and after shutdown operations since process systems will be off-line during these times and the gas from processing will not be available.

If a structure produces gas from its lease, it is used free of charge to provide electricity, heat, cooling (e.g., via pumping seawater), and related power requirements. Of course, gas use will reduce sales revenue, but from a cost perspective unless purchased, fuel cost is not a primary design variable. If gas is not available at site, fuel will need to be purchased and delivered to the platform. For start-up operations and as a backup power, diesel is commonly used offshore and stored in bulk or in tanks. Gas supply may be provided by a nearby platform by running a flowline to the facility and installing a metering system.

Utility systems support production operations and include the power plant, seawater and potable water treatment system, chemical and lubricating oils, alarm and shutdown systems, fire protection and firefighting system, and instrumentation and utility air system. Power generation and electric systems are required for large or complex facilities and manned platforms. Water needs to be transported to manned facilities for personnel use.

Example. Power generation at Appomattox

Shell's deepwater Appomattox development required 150 MW power generation equipment for a combined-cycle power plant. The 150 MW power plant features four 27 MW gas turbine-driven generator sets equipped with heat recovery systems and a 40 MW steam turbine generator.

23.5 REPAIRS AND MAINTENANCE—WELLS AND FLOWLINE

23.5.1 Maintenance

Pigging is performed as part of regular operations to facilitate flow and reduce corrosion and buildup in pipelines.

Example. Paraffin cutting at Pompano

In the Pompano development, subsea wells are maintained using a through-flowline (TFL) system to transport (simple) tools and chemicals to the wellbores through the manifold/template (Fig. 23.7). TFL systems were introduced in the late 1960s to provide a low-cost wellbore reentry capability for mechanical removal of paraffin deposits, but many early systems were plagued by operational problems and equipment failures and were not widely adopted (Keprta, 1976). Circulating pressure moves the tool through the looped system. Because of the waxy nature of the crude (3.2 wt% paraffin content, 95°F cloud point, −6°F pour point) and the distance between the subsea wells and host facility, frequent paraffin cutting of the service lines was expected (Kleinhans and Cordner, 1999).

A base-case and a worst-case scenario in the types of TFL operations required and their expected frequencies were estimated by engineers to understand the downtime and operating cost requirements (Table 23.8). Operations require different tools to be employed and contractor cost for equipment and crew

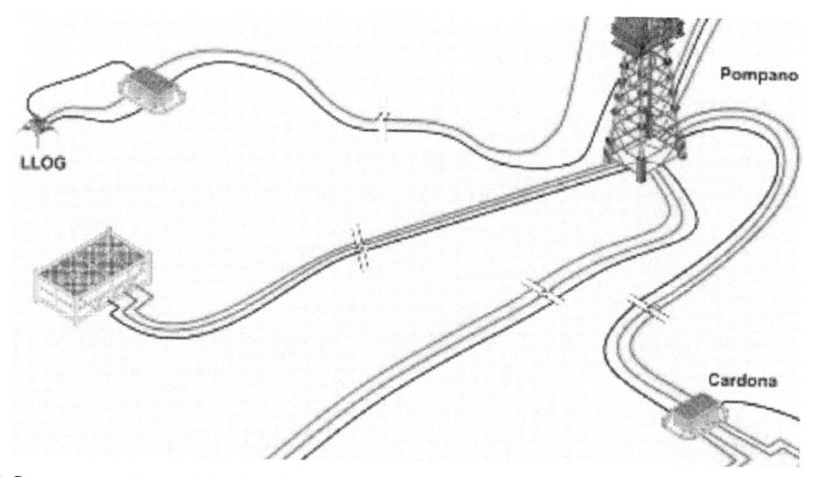

Fig. 23.7 Pompano subsea tieback schematic.

Table 23.8 Pompano Subsea Well Maintenance Operations Requirements (per Well per Year)

Operation	Base Case	Worst Case
Paraffin scraping	6.0	15.0
BHP survey	1.0	1.0
Set and recover plug	0.2	0.5
Tubing caliper survey	0.125	0.25
SCSSV repair	0.05	0.125

Source: Kleinhans, J.W., Cordner, J.P., 1999. Pompano through-flowline system. SPE Prod. Facilit. 14 (1), 37–48.

will depend upon requirements. When wells are producing at high rates (> 2000 bopd), downhole cutting is not expected, but as the well rates diminish, routine paraffin cutting is required. As reservoirs deplete and fluid chemistry changes, the worst-case scenario was considered more likely to arise.

23.5.2 Stimulation

Wells need to be maintained for maximum productivity, and in the event that a well stops producing, a decision has to be made on the merits of attempting to bring the well back online. A useful categorization divides well constraints as completion interval constraints and production tubing constraints (Table 23.9). The type and frequency of activities vary significantly as well as the success of action.

The three primary causes of well impairment include the movement and rearrangement of reservoir fines, deposition of solids from produced crudes, and deposition of mineral scales from produced water (Table 23.10). Scale formation may occur in the reservoir and inside the production tubing and may be removed chemically or mechanically. A well producing at high water cut may be choked back, or a change of perforation interval may be considered to shut off unwanted fluids. Skin problems may be resolved by acidizing or additional perforations.

Organic deposits such as asphaltenes in the oil are usually not a significant concern in operations except when unstable. Deposition can occur in the reservoir, tubing, subsea, and topsides depending on the asphaltene onset pressure and saturation pressure.

Table 23.9 Completion Interval and Tubing Wellbore Constraints

Completion Interval Constraints	Production Tubing Constraints
Damage skin	Tubing string design
Sand production	Artificial lift
Scale formation	Sand production
Emulsion formation	Scale formation
Asphaltene dropout	Choke size

Table 23.10 Primary Causes of Well Impairment

Category	Description	Remedy
Scale	Inorganic minerals deposited from water	Calcium carbonate removed with HCl or organic acids, barium sulfate-treated chelates
Organic deposits	Deposition of solids from the oil phase, usually a combination of asphaltenes and resins	Asphaltenes remediated with aromatic solvents
Fine migration	Flow-induced movement of clay-sized particles from pore bodies to pore throats causing a reduction in permeability	Combination of HCl and/or organic acids with HF acid

Note that remedy represents typical industry guidelines. For example, scale treatment involves injecting treatment solution, soaking (e.g., 12–24 h), and flowing the well back. For fine migration treatments, a 9% HCl/1% HF is commonly used.

Reservoir impairment would require stimulation of the near-wellbore region, whereas tubing or flowline deposition would require a solvent soak. The logistics, expenses, and safety issues are different in each case.

Tubing corrosion requires monitoring, and if a leak develops, the tubing needs to be replaced or the well shut in. As reservoir pressure declines, the tubing may need to be reduced in diameter to maintain maximum flow. When the natural drive energy of the reservoir has reduced, artificial lift may be justified that will increase both capital and operating expense. Sand production from loosely consolidated formations may erode tubulars and valves and cause problems at the surface separators and necessitate recompletion. Paraffin cutting is a common maintenance requirement for high-wax crude.

Personnel separate from the production crew are required for operations and are performed by a service crew that require transportation and logistics support. Because the well is shut in during operations, logistics coordination and efficiency are critical. Stimulation of subsea wells is more challenging and costly than dry tree and direct access wells but may be highly profitable if technology solutions have been developed and outcomes are successful.

23.5.3 Well Failure

When wells stop producing, the probable cause of the shut-in is determined, and the expected cost, benefit and risk of three options are identified:

- Sidetrack the well,
- Perform a workover to remediate problem, or
- Leave the well shut in.

The decision to invest capital in an attempt to bring the well back online depends on many factors. The price of oil/gas and equipment/rig day rates play an important role as well as the uncertainty associated with the problem and its solution. New wells that fail have a much greater chance of getting a workover than an old well that has already drained most of the reservoir. Workover decisions in old wells are risky since the operator may not get their investment back, and so, once an older well stops producing, a workover may not even be contemplated if production rates are low. Decisions may also be based on maintaining a lease position or infrastructure.

Example. Troika well TA-6

The Troika field in Green Canyon 201 is a subsea development tied back to the Bullwinkle platform in GC 65 (Bednar, 1998). Well TA-6 was brought online in November 2000 and produced 3.9 MMbbl of oil and 4.5 Bcf of gas for 14 months before a gravel-pack failure occurred (Gillespie et al., 2005). Sidetracking was not considered the best option due to low oil price forecasts and the uncertainty of the remaining reserves caused by increasing water production. It was decided to clean out the well and run a screen insert inside the failed screen. If this option failed, the well would be shut in.

The workover was completed in one week at a cost of about $8 million. The well was returned to production in January 2003, and the postworkover production was 1.95 MMbbl oil and 2.7 Bcf through June 2005, an obvious success since the postworkover revenue far exceeded the cost of the intervention. Operators seek workovers with strong positive results but cannot always control or predict the outcome of operations.

23.5.4 Flowline and Export Repairs

Hydrates, wax, and asphaltenes in the hydrocarbon streams have the potential to disrupt production due to deposition at many points in the production system.

Example. Stuck pig at Marlin

The Marlin TLP is located in Viosca Knoll 915 in 3250ft water depth and is host to several dry tree and wet tree wells (Fung et al., 2006). Oil export is via a 22 mi noninsulated 10 in line to facilities at Main Pass 225 in 200ft water depth. The oil export management plan used a regular single-trip pigging technique to remove the wax buildup every 14days and was selected over continuous wax inhibition because of the high operating expense of the chemical treatment.

Oil leaves the Marlin TLP at a temperature of approximately 120°F and drops to 40°F over the first 7300ft (1.4 mi) of flowline and then warms to about 65°F at the MP 225 location (Fig. 23.8). Since the pipeline does not have any insulation, higher-flow-rate fluids will retain heat for longer distance until the fluid heat has been lost (at about 20,000ft distance) where slower flow rates will warm up more than fast fluids. The WAT of the comingled oil streams is approximately 95°F. Heavy molecular weight paraffinic

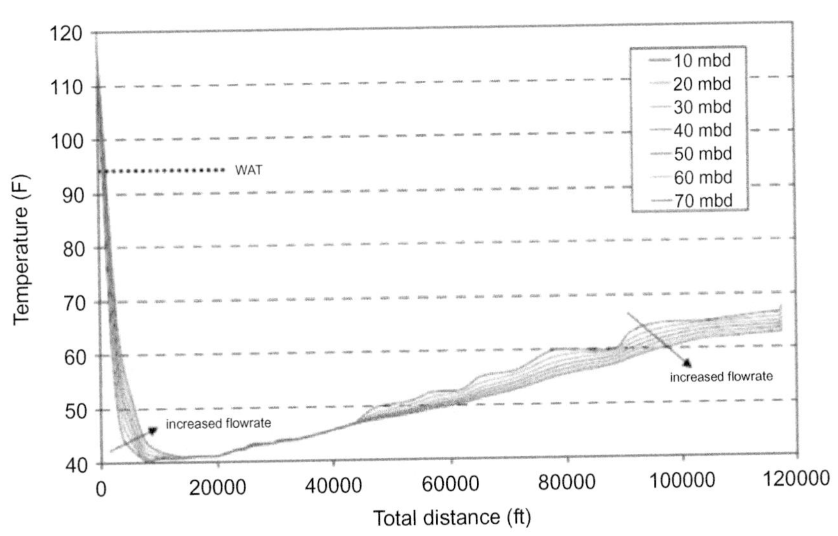

Fig. 23.8 Marlin oil export line temperature profile. *(Source: Fung, G., Backhaus, W.P., McDaniel, S., Erdogmus, M., 2006. To pig or not to pig: the Marlin experience with stuck pig. In: OTC 18387. Offshore Technology Conference. Houston, TX, May 1–4.)*

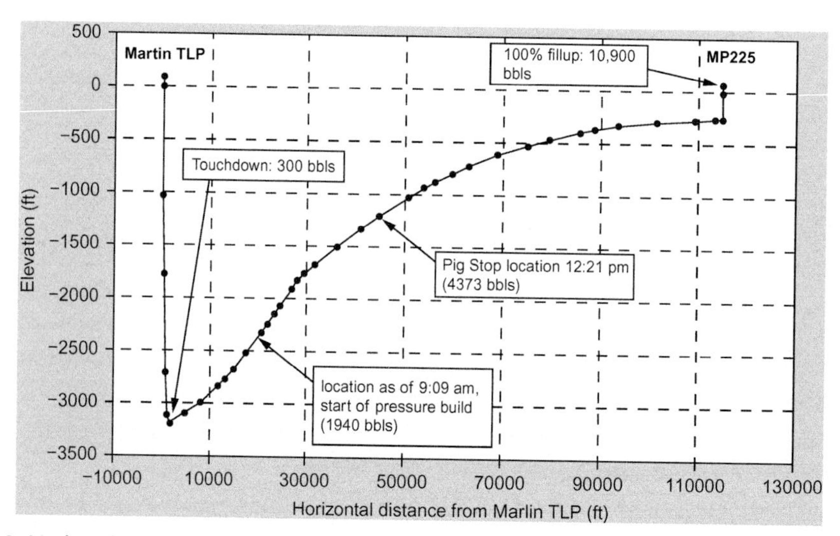

Fig. 23.9 Marlin oil export pipeline elevation and estimated stuck pig location. *(Source: Fung, G., Backhaus, W.P., McDaniel, S., Erdogmus, M., 2006. To pig or not to pig: the Marlin experience with stuck pig. In: OTC 18387. Offshore Technology Conference. Houston, TX, May 1–4.)*

hydrocarbons begin to solidify and deposit on the pipe wall over time and give rise to an increasing pressure drop due to reduction in the flow diameter and increase in the pipe roughness.

Equipment failure stopped the 14-day pigging cycle and resulted in a stuck pig at the next pig run. The stuck pig was estimated to be approximately 9 mi from the MP 225 facility in approximately 1200 ft water depth (Fig. 23.9). Pumping equipment at MP 225 was used to pump the pig back to Marlin using buyback crude with 5000 gal of wax solvent and 300 ppm of wax inhibitor.

23.6 REPAIRS AND MAINTENANCE—EQUIPMENT AND STRUCTURE

Offshore equipment and platforms are inspected on a periodic basis dictated by company practices and regulatory requirements.

23.6.1 Regulatory Requirements

The OCS Lands Act authorizes and requires the BSEE to provide for both an annual scheduled inspection and periodic unscheduled (unannounced) inspection of all oil and gas operations on the OCS. In addition to examining all safety equipment designed to prevent blowouts, fires, spills, or other major accidents, the inspections focus on pollution, drilling operations, completions, workovers, production, and pipeline safety. Inspections also include a review of painting integrity.

Upon detecting a violation, the inspector issues an incident of noncompliance (INC) to the operator and uses one of two main enforcement actions (warning or shut-in),

Table 23.11 In-Service Inspection Intervals for Fixed, Manned, and Unmanned Platforms

	Exposure Category		
Level	**L-1**	**L-2**	**L-3**
I	1 year	1 year	1 year
II	3 years (5 years)	5 years (10 years)	5 years (10 years)
III	6 years (6 years)	11 years (11 years)	

Note: Unmanned platform inspection intervals are denoted in parenthesis.
Source: NTL 2009-G32.

depending on the severity of the violation. If the violation is not severe or threatening, a warning INC is issued. The warning INC must be corrected within a reasonable amount of time specified on the INC. The shut-in INC may be for a single component (a portion of the facility) or the entire facility and must be corrected before the operator is allowed to resume operations.

The BSEE can also assess a civil penalty of up to $40,000 per violation per day if (i) the operator fails to correct the violation in the time specified on the INC or (ii) the violation resulted in a threat of serious harm or damage to human life or the environment. Operators with excessive INCs may be required to cease operations in the Gulf of Mexico.

BSEE regulations require operators perform in-service inspection intervals for fixed platforms according to API Recommended Practice 2A-WSD (NTL 2009-G32, 2009). Section 14 of API RP 2A-WSD describes the inspection program survey levels and frequencies to monitor periodically the adequacy of the corrosion protection system and determine the condition of the platform.

23.6.2 Inspection Schedules

The time interval between platform inspections depends upon structure exposure category (L-1, L-2, or L-3), survey level (Level I, Level II, and Level III), and manned status (Table 23.11).

Exposure Category

Two classes of risk are used to define exposure category, those associated with life safety and those associated with consequences of failure (Ward et al., 2000a, 2000b). The three levels of exposure to life safety are manned and nonevacuated, manned and evacuated, and unmanned (Table 23.12). Consequences of failure are categorized as high, medium, and low and encompass damage to the environment, economic losses to the operator and the government, and public concerns. Economic losses to the operator can include the costs to replace, repair, and/or remove destroyed or damaged facilities; costs to mitigate environmental damages due to released oil; and lost revenue. Economic losses to the government include lost royalty revenues.

Table 23.12 Risk Considered for Consequence-Based Criteria for the Gulf of Mexico

Exposure	Life Safety	Consequences of Failure
L-1	Manned, nonevacuated	High
L-2	Manned, evacuated	Medium
L-3	Unmanned	Low

Source: Ward et al., 2000.

Level Surveys

A Level I survey is required to be conducted for each platform at least annually and a grade assigned to the coating system as A, B, or C. Grade A indicates the coating system is in good condition with no maintenance needed within 3 years. Grades B and C refer to fair and poor coating system conditions requiring maintenance within 3 years or 12 months, respectively.

A Level II survey is required for each platform at the minimum survey interval for each exposure category, at least every 3 years for L-1 platforms and at least every 5 years for L-2 and L-3 platforms.

A Level III survey is required for each platform at the minimum survey interval for each specified exposure category, at least every 6 years for L-1 platforms and at least every 11 years for L-2 platforms.

For unmanned platforms, BSEE may approve an increased interval for Level II and Level III inspections if the operator is in compliance with all structural inspection requirements and the platform is in good structural condition according to previous Level I and Level II surveys.

Inspection Levels

Level I inspections are topside inspections performed annually. Topside maintenance is relatively easy to perform as long as equipment is accessible and paint schedules are followed. Flare towers, crane booms, and lower deck levels present more complicated regions because of access. Level I inspections are normally performed by an operator's maintenance personnel or staff personnel.

Level II inspections are underwater inspections performed every 3–5 years to check for debris and gross damage, measure cathodic potential readings, and verify anode connections. On large shelf structures, anode connections may number in the hundreds and on deepwater structures in the thousands. For example, the cathodic protection system on the Bullwinkle jacket in 1350 ft water depth consists of approximately 6300 aluminum-zinc-mercury anodes totaling about 2400 t of anode material (Wolfson and Kenney, 1989). Level II inspections generally do not involve cleaning or marine growth removal for weld inspections and fractures. One or two days per platform are considered adequate for Level II inspections.

Level III inspections are underwater inspections required to be performed on a 6- to 11-year interval. Typically, Level III inspections include the cleaning and inspection of selected member ends, conductor areas near the bell guides, and connection points of anodes. Member ends with lower fatigue life are selected for inspection, and if damage is detected, the inspection scope is likely to be extended. Level III inspections will usually take several days to a week or longer per platform.

23.6.3 Three Zones

There are three separate zones in platform corrosion control: the immersed zone, the splash zone, and the atmosphere. Corrosion rates for each zone are broadly similar throughout the world, but site-specific factors lead to differences. Tides, water temperature and velocity, salinity, humidity, and wave forces all affect corrosion rates and the design of the corrosion control system.

The bulk of the platform is immersed, and this portion is also the simplest to protect with cathodic protection being the most common method. The process typically involves welding sacrificial nodes throughout the structure before installation to supply protective current to the steel members. The splash zone is the most critical area with the highest corrosion rates due to the alternate submergence and aeration. Splash zone protection in the GoM is often installed a few feet below the waterline to several feet above (e.g., from −3 to 7 ft). Steel members in the splash zone are also normally designed with a greater wall thickness to account for greater corrosion. The atmospheric zone includes structural steel, equipment, piping, vessels, and valves and has the least corrosion rate but the most surface area and is the most expensive to maintain.

Structures and piping in the splash zone are painted to protect the steel from aggressive corrosion due to wave action and continuous wetness. Structures are painted above the splash zone to assure that equipment will remain functional. Bare flanges, valves, and rotating equipment will quickly become inoperable, and pressure equipment may become too thin to hold pressure creating safety hazards if not properly maintained.

23.6.4 Painting

Objective

The primary objective of painting platforms and equipment is to protect structures to ensure integrity and that the structure will last for its intended design life (Choate and Kochanczyk, 1991; Knudsen, 2013). The selection of a coating system is a design choice made by engineers based on the trade-offs between steel thickness and the corrosion allowance that the coating system is engineered to protect (Taekker et al., 2006; Kattan et al., 2013).

During the fabrication of offshore structures, surfaces are painted and treated to the operator's specification, and assuming activities are performed according to requirements, these will provide protection against saltwater corrosion, ultraviolet radiation, rust, splash zones, changing temperatures, and related deterioration for several years. After a period of time, however, depending on the specifications and materials used, environmental conditions, and other factors, the coating will need to be replaced according to the level of deterioration. Coatings are monitored and treatment is performed before major interventions are needed. Regular maintenance extends the structure life integrity and lowers repair frequency.

There are design standards for offshore coating systems (ISO 20340, NORSOK M-501, and NACE TM 0104) and the qualification of products, procedures, companies, and (inspection) personnel. The primary aim of standards is to provide optimum surface protection with a minimum need for maintenance. Each standard describes different coating systems for application on various parts of an offshore structure and is described with requirements on surface preparation, the number of coats, and requirements for prequalification testing.

Inspections are performed to survey the coatings and develop an annual plan and to ensure quality control during application. The blasting crew prepares the surface by hand and power tools and abrasive wet jets to remove contaminants such as oil, grease, and soluble salts, as well as paint chips and metal debris. Garnet and equipment (air compressors, blasting hoses, spray assemblies, and paint-mixing equipment) must be transported to the facility, and blasting generates waste that must be collected, bagged, and transported off the structure and disposed or recycled in dedicated sites associated with industrial (hazardous and nonhazardous) waste.

Maintenance Painting Schedule

There are several methods of conducting maintenance painting and determining which method to use is a philosophical as well as technical decision. Maintenance implies sustaining a particular level of coating integrity and performance, while repair implies restoring a coating to a previously higher level of integrity and performance. Operators with small maintenance budgets or that delay or reduce maintenance activities will often face higher repair costs than operators with a constant painting budget. For low-cost equipment and with a short operation life, corrective maintenance is the common approach (i.e., fix it when it breaks). For high capital cost structures and equipment, preventative and predictive monitoring is standard and painting are year-round activities (Witte and Ribeiro, 2012).

After original painting, spot touch-up and repair are expected to occur at the practical life (or service life) of the coating system shown in Table 23.13 for different coating systems for offshore atmospheric and immersion service.

Structures are often separated by horizontal and vertical zones, and the condition of the coating in each zone is assessed. One example of zones might include below the spider deck, spider deck to the bottom of cellar deck, top of cellar deck to the bottom of

Table 23.13 Estimated Service Life for Selective Offshore Maintenance Coating Systems

Type (Exposure)	Coating System	Preparation	Coats (#)	DFT (Mils)	Life (Year)
Alkyd (Atm)	Alkyd/alkyd/urethane	Blast	3	6	4
Epoxy (Atm)	Surface tolerant epoxy/STE	Power	2	10	9
Epoxy (Atm)	Epoxy/polyurethane	Blast	2	6	8
Epoxy zinc (Atm)	Epoxy zinc/epoxy/epoxy	Blast	3	11	14
Zinc (Atm)	Zinc/epoxy/polyurethane	Blast	3	12	15
Zinc/epoxy (Imm)	Zinc/epoxy/epoxy	Blast	3	10	12
Metallizing (Imm)	Metallizing/epoxy/spoxy	Blast	3	13	18

Note: Atmospheric (Atm) marine exposure is defined as very high corrosion in offshore areas with high salinity. Immersion (Imm) service is saltwater immersion at ambient temperature and pressure. Coating system includes primer/midcoat/topcoat. Service life is the time until 10% coating breakdown occurs and active rusting is present.

Source: Helsel, J.L., Lanterman, R., 2016. Expected service life and cost considerations for maintenance and new construction protective coating work. In: NACE 7422. NACE International Corrosion Conference and Expo. Vancouver, BC, Canada.

production deck, and production deck and above. Component groupings may also be based on where abrasive blasting is not permitted. The structure and decks are usually evaluated separately from equipment and piping.

Paint and Coating Cost

In Table 23.14, typical material costs of paints and protective coatings are depicted based on 2016 survey data from US paint and coating suppliers (Helsel and Lanterman, 2016). Costs are expressed in dollars per square foot at typical dry film thickness (DFT) and assume 30% spray painting losses and 10% brush/roll losses.

There are many factors that impact the volume of paint required to protect steel surfaces. The theoretical spreading rate of paint for a given dry film thickness on a smooth surface is calculated as (Hempel 2016)

$$\frac{\text{Volume solids }(\%) \times 10}{\text{Dry film thickness }(\mu m)} = \frac{10 \text{ VS }(\%)}{\text{DFT }(\mu m)} = \text{m}^2/\text{L}$$

Table 23.14 Typical Offshore Paint and Protective Coating Cost Circa 2016

Coating	DFT (Mils)	Spray ($/ft^2)	Brush/Roll ($/ft^2)
Epoxy, 100% solids	20	1.89	1.47
Polyurethane, aliphatic acrylic	2	0.28	0.22
Siloxane, epoxy	4	1.02	0.79
Zinc rich, inorganic	3	0.40	0.31

Source: Helsel, J.L., Lanterman, R., 2016. Expected service life and cost considerations for maintenance and new construction protective coating work. In: NACE 7422. NACE International Corrosion Conference and Expo. Vancouver, BC, Canada.

Table 23.15 Approximate Conversion of Member Weight in Tons to Footage of Surface

Member	Square Feet per Ton
Typical mix size/shapes	250
Large structural	100
Medium structural	200
Light structural	400
Light trusses	500

where volume solids (VS) express the ratio of dry film thickness (after drying) to wet film thickness (as applied). The practical consumption is estimated by multiplying the theoretical consumption with a relevant consumption factor. Consumption factor is not given in painting specifications since it depends on several conditions such as the size and shape of the surface (small complex area vs square flat area), application method (hand/brush vs spray), surface roughness (rough vs smooth), physical losses, experience of the painters, and atmospheric conditions.

The maintenance and repair cost for an offshore structure depends on the size and location of the project, type and volume of paints used, scope of work, and contracts applied. Work can range from a full blast to repair. Painting may be performed at night (Judice, 2007).

Since total square footage of a structure is rarely reported in public documents, approximations are required to estimate painting areas. Useful conversions based on deck and topside weight are shown in Table 23.15. As an example, deck weight at the Auger TLP is 10,500 t, and total steel surface area was reported as 1.8 million square feet. For medium structural steel and a 200 ft^2/t conversion yields 2.1 million square feet deck surface area, a 15% error from the reported value.

Example. Painting cost estimation at Enchilada, Prince, and Mars

The Enchilada platform was installed in 1997 and sits in 705 ft water depth with a total topside operating weight of 9000 t. The Prince TLP was installed in 2001 in 1490 ft water depth and consists of a three-level deck capable of carrying topside payload of 6100 t. The Mars TLP was installed in 1997 in 2933 ft water depth and has topside operating weight of about 23,000 t.

Assuming 4% of the total area of a deepwater structure requires treatment per year after the service life is reached, and using a topside weight to square foot conversion of 200 ft^2/t, topside painting areas for Enchilada, Prince, and Mars are estimated at 1.8, 1.2, and 4.6 million square feet, respectively.

Normalizing by the unit cost estimated at Shell's Auger TLP ($1.9 million for 75,000 ft^2 treatment per year; see Box 23.1), the annual painting and blasting costs at Enchilada, Prince, and Mars are estimated at $1.8 million, $770,000, and $2.9 million, respectively. The structures are all deepwater facilities and approximately the same age as Auger, serve similar functions, and are of similar complexity. Assuming similar original coat systems and degradation, maintenance cost would likely be similar. Enchilada and Mars are also Shell-operated, which will likely have similar maintenance programs in place.

BOX 23.1 Auger TLP Painting Cost Estimation

Shell's Auger TLP in 2860 ft water depth is composed of four 74 ft diameter by 159 ft high columns connected by four 35 ft wide by 28 ft high pontoons, a drilling rig, production facility, associated power plants, and living accommodations for 142 people (Fig. 23.10). The deck section measures 290 × 330 × 20 ft high, and steel weight is 10,500 t (Bourgeois, 1994). The TLP is held in position with twelve 26-in diameter tendons, three per corner, attached to a foundation template anchored to the seafloor (Fig. 23.11). An eight-point lateral mooring system is also employed. First oil was in April 1994.

Shell reported performing 75,000 ft^2 blasting and painting activities at Auger TLP each year, which is approximately 4% of the total surface area of 1.8 million square feet (Satterlee et al., 2009). Using work decomposition methods and data described in Satterlee et al. (2009), Auger's annual maintenance painting is estimated at $1.9 million per year or about $25 per square foot serviced (Table 23.16).

Shell reported employing a six-man blasting and painting crew with a foreman and inspector. Assuming $150,000 annual salary for foreman and inspector and $100,000 for paint contractors and blasting crew, the total annual labor budget for blasting and painting is $1.2 million per year.

Garnet costs about $500/t and can be reused up to eight times, although there are issues regarding recycle and performance that often limit reuse. In surface preparation, 2–4 lb of garnet is assumed to be used per square foot steel surface[3] and reused once in operations and then disposed onshore. For 75,000 ft^2 steel surface requiring blasting, the annual garnet budget is estimated at $26,000 as follows:

Fig. 23.10 Bird's eye view of the Auger tension leg platform. *(Source: Shell, BOEM.)*

[3] For comparison, the amount of garnet used in ship conversion operations typically requires 10–15 lbs of garnet per square foot steel surface because of the intensive surface preparation requirements (Azevedo, 2011).

Continued

BOX 23.1 Auger TLP Painting Cost Estimation—cont'd

$$0.5 \cdot 75{,}000\,\text{ft}^2 \cdot 3\,\text{lb/ft}^2 \cdot \$500/\text{t} \cdot \text{t}/2200\,\text{lb} = \$26{,}000$$

or \$0.35/ft^2. For high-solid sprayed paint expenditures of \$1.5/ft^2, a similar calculation yields \$112,500 paint cost per year.

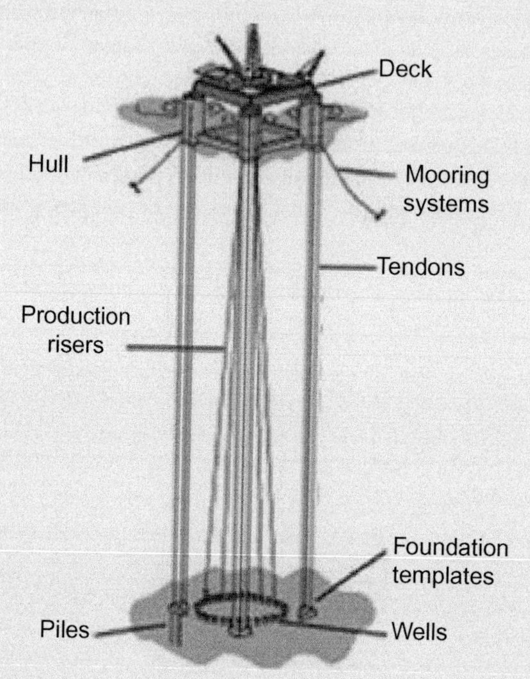

Fig. 23.11 Schematic of the Auger tension leg platform.

Table 23.16 Estimated Annual Maintenance Painting Cost at Auger TLP

	Annual Cost (\$)	Unit Cost (\$/ft^2)
Labor	1,200,000	16.00
Material		
Paint	112,500	1.50
Garnet	26,000	0.35
Disposal		
Nonhazardous	7670	0.10
Hazardous	5100	0.07
Subtotal	1,351,270	18.02
Indirect (40% subtotal)	540,508	7.21
Total	1,891,778	25.22

Note: Assumes six-man crew with garnet reused once. Logistics, transport, and catering costs assumed to be 40% of the direct cost.

Continued

BOX 23.1 Auger TLP Painting Cost Estimation—cont'd

Disposal of spent abrasive material is assumed to cost \$50/t for nonhazardous material and \$300/t for hazardous material. Shell reported about 10% spent abrasive material hazardous due to cadmium-coated bolts that were blasted or older areas containing lead paint. If 5 lb waste per square foot arises from the spent abrasive, corrosion, and paints removed from the steel and 10% is hazardous, annual disposal cost is estimated as

$$\text{Nonhazardous disposal cost} = 0.9 \cdot 170\,\text{t per year} \cdot \$50/\text{t} = \$7670$$
$$\text{Hazardous disposal cost} = 0.1 \cdot 170\,\text{t per year} \cdot \$300/\text{t} = \$5100$$

Indirect costs associated with operations include logistics, catering, travel for the paint crew and equipment, downtime, and weather delays. A 40% indirect rate is assumed to cover these costs.

23.6.5 Underwater Maintenance

Operators deploy underwater inspection schedules according to regulatory requirements and base decisions on the cost-benefit of maintenance. Cleaning is the most time-consuming task, and a good cleaning will remove marine growth down to black oxide 1 to 2 in on each side of welds and cleaned to bright metal. High-pressure water blasters are used for cleaning and need to be set up and properly positioned. Gross damage around the circumference of members (e.g., wide cracks) and fractures are identified, if any, and video and photo documentation performed.

Remotely operated vehicles (ROVs), divers, and supporting equipment may be based on the platform or workboat, but usually, workboats are preferred for economic, safety, and logistic reasons. In shallow water <200 ft, divers are typically used exclusively; in 200–400 ft water depth, divers and/or ROVs are employed; and in >400 ft, ROVs are primarily employed with divers employed in the top water section.

Example. Level II and III inspection cost at Cognac

Shell conducted time studies during Level II and III inspection at Cognac which stands in 1025 ft water depth and documented that it took on average 6 h per member for cleaning and inspection (Miller and Hennegan, 1990). Half of this time was spent on cleaning with the setup time, positioning, and video recording comprising the remainder. On average, 1.5 member ends were cleaned and inspected in a 12 h workday.

Level II and III inspections performed in 1990 were reported to cost \$480,000, about \$293,000 for the Level III inspection and \$175,000 for the Level II inspection (Table 23.17). The top 200 ft of the structure was inspected by divers. The inferred ROV dayrate was \$227,000 per 15 days = \$15,133 per day or \$275/ft of structure. The inferred Level III diver dayrate was \$23,000/200 ft or \$115/ft. The inferred Level II cost was \$161,000/825 ft = \$195/ft. Cost excludes mob/demob fee.

Table 23.17 Level II and III Inspection Costs at Cognac

Activity	Inspection Type	Cost ($1000)
Diver inspection	Level III, top 200 ft	23
ROV-1 inspection	Level II, entire structure	161
ROV-2 inspection	Level III, 200–1025 ft	266
Mooring		28
Total		480

Source: Miller, B.H., Hennegan, N.M., 1990. API Level III inspection of Mississippi Canyon 194 A (Cognac) using an ROV. In: OTC 6355. Offshore Technology Conference. Houston, TX, May 7–10.

Diving

The underwater service industry in the GoM is highly competitive. In deepwater, several companies compete worldwide, while in the shallow water, numerous small companies that operate locally offer services. For services that require less sophisticated equipment, small companies are able to bid for contracts at prices that are uneconomic to larger companies, and this is reflected in the gross margins reported by large contractors for inspection services in the Gulf that are typically the smallest among their business segments. In shallow water, high levels of competition, low day rates, and low technological requirements constrain profits.

Manned diving operations utilize traditional diving techniques of air, mixed gas, and saturation diving, all of which are surface-supplied breathing gas. In water depths >1000 ft, traditional diving techniques are not used, and instead, ROVs are employed (Fig. 23.12). ROVs are tethered submersible vehicles remotely operated from the surface, usually a specially outfitted vessel, either owned or chartered (leased) by the company (Figs. 23.13 and 23.14).

Fig. 23.12 Work class ROV sitting on vessel's deck.

Fig. 23.13 Rendition of ROV manipulating a control valve on a subsea wellhead.

Divers provide high-quality fast cleaning rates and can adapt to difficult positioning and changing conditions, but safety is always a concern with saturation diving as water depths increase. ROVs are expensive and may be difficult to maintain position in high currents or reach difficult areas between conductors, within the structure interior, and at the mudline. Significant technical advancement in ROVs has been made over the years.

Fig. 23.14 Rendition of ROV manipulating subsea wellhead controls under a steel structure.

ROV

Service for ROV contracts is typically awarded on a competitive bid on a dayrate basis for contracts <1 year in duration, although multiyear contracts may also be awarded for significant work campaigns. Under dayrate contracts, the contractor provides the ROV, vessel or equipment, and the required personnel to operate the unit, and compensation is based on a rate per day for each day the unit is used. Lower dayrates often apply when a unit is moving to a new site or a separate mobilization fee is applied or when operations are interrupted or restricted by equipment breakdowns, adverse weather, or water conditions or other conditions beyond the contractor's control. Contracts often specify a 12h workday and an ROV downtime allowance (e.g., 30h downtime per month).

Day rates depend on the market conditions, the nature of the operations to be performed, the duration of the work, the equipment and services to be provided, the geographic areas involved, and other variables. Video inspections typically include wellheads, valve positions, pipeline end terminations and manifolds, flowlines, jumpers, moorings, risers, and associated cabling (Kros, 2011). This equipment is often spaced over many square kilometers requiring the support vessel to maneuver in DP mode for days.

Example. Oceaneering International Inc. ROV dayrates, 2008–16

Oceaneering is one of the largest underwater service contractors in the world and as of 31 December 2016 owned over 300 work-class ROVs, the largest fleet in the world. The average revenue per day on hire from 2008 to 2016 was reported between $8500 and $11,000 per day. Revenue per day on hire is not the same as dayrate but provides an indication of dayrate ranges.

AUV

Autonomous underwater vehicle (AUV) inspection technologies for the offshore oil and gas industry are in its infancy but promise to reduce the cost of inspecting subsea facilities for a range of activities, including pre and post-hurricane inspection, decommissioning structure surveys, and pipeline and riser inspection (McLeod et al., 2012). In diving and ROV operations, support vessels are required with large crews to collect relatively simple visual inspecting records. AUVs use advanced technology to reduce costs and are not proved technology, but the technology is advancing and is likely to play a future role in inspection.

REFERENCES

Azevedo, J., 2011. Coating technology improves project predictability and lowers costs for FPSO conversion projects. In: OTC 21471. Offshore Technology Conference. Houston, TX, May 2–5.

Bednar, J.M., 1998. Troika subsea production system: an overview. In: OTC 8845. Offshore Technology Conference. Houston, TX, May 4–7.

Bomba, J., Chin, D., Kak, A., Meng, W., 2018. Flow assurance engineering in deepwater offshore – past, present, and future. In: OTC 28704. Offshore Technology Conference. Houston, TX, Apr 30–May 3.

Bourgeois, T.M., 1994. Auger tension leg platform: conquering the deepwater Gulf of Mexico. In: SPE 28680. SPE International Petroleum Conference and Exhibition of Mexico. Veracruz, Mexico, Oct 10–13.

Brimmer, A.R., 2006. Deepwater chemical injection systems: the balance between conservatism. In: OTC 18308. Offshore Technology Conference. Houston, TX, May 1–4.

Cavaliaro, B., Clayton, R., Campos, M., 2016. Cost-conscious corrosion control. In: SPE 179949. SPE International Oilfield Corrosion Conference and Exhibition. Aberdeen, Scotland, May 9–10.

Choate, D.L., Kochanczyk, R.W., 1991. Specifications for the selection and application of paint systems for offshore facilities. In: OTC 6596. Offshore Technology Conference. Houston, TX, May 6–9.

Fung, G., Backhaus, W.P., McDaniel, S., Erdogmus, M., 2006. To pig or not to pig: the Marlin experience with stuck pig. In: OTC 18387. Offshore Technology Conference. Houston, TX, May 1–4.

Gillespie, G., Angel, K., Cameron, J., Mullen, M., 2005. Troika field: a well failure, and then a successful workover. In: SPE 97291. SPE Annual Technical Conference and Exhibition. Dallas, TX, Oct 9–12.

Helsel, J.L., Lanterman, R., 2016. Expected service life and cost considerations for maintenance and new construction protective coating work. In: NACE 7422. NACE International Corrosion Conference and Expo. Vancouver, BC, Canada.

Jahn, F., Cook, M., Graham, M., 2008. Hydrocarbon Exploration and Production, 2nd ed. Elsevier, Amsterdam, The Netherlands.

Judice, K., 2007. Painting at night in the Gulf of Mexico. Mater. Perform. 30–32 February.

Kaiser, M.J., 2019. The role of factor and activity-based models in offshore operating cost estimation. J. Petrol. Sci. Eng. 174, 1062–1092.

Kaiser, M.J., 2015. Modeling the Offshore Service Vessel and Logistics Industry in the Gulf of Mexico. Springer-Verlag, London, UK.

Kattan, M.R., Speed, L.A., Broderick, D., 2013. Minimizing the risk of coating failures. In: OTC 23982. Offshore Technology Conference. Houston, TX, May 6–9.

Keprta, D.F., 1976. Seafloor wells and TFL – a review of nine operating years. In: SPE 6072. SPE Annual Technical Conference and Exhibition. New Orleans, LA, Oct 3–6.

Kleinhans, J.W., Cordner, J.P., 1999. Pompano through-flowline system. SPE Prod. Facilit. 14 (1), 37–48.

Knudsen, O., 2013. Review of coating failure incidents on the Norwegian Continental Shelf since the introduction of NORSOK M-501. In: NACE 2500. NACE International Corrosion Conference and Expo. Orlando, FL.

Kros, H., 2011. Performing detailed level I pipeline inspection in deep water with a remotely operated vehicle. In: OTC 21969. Offshore Technology Conference. Houston, TX, May 2–5.

McLeod, D., Jacobson, J.R., Tangirala, S., 2012. Autonomous inspection of subsea facilities – Gulf of Mexico trials. In: OTC 23512. Offshore Technology Conference. Houston, TX, Apr 30–May 3.

Miller, B.H., Hennegan, N.M., 1990. API Level III inspection of Mississippi Canyon 194 A (Cognac) using an ROV. In: OTC 6355. Offshore Technology Conference. Houston, TX, May 7–10.

NTL No. 2009-G32, 2009. Notice to lessees and operators of federal oil and gas leases and pipeline right-of-way holders Outer Continental Shelf, Gulf of Mexico Region. United State Department of the Interior Minerals Management Services, Gulf of Mexico OCS Region.

Satterlee, K., Kuehn, R.B., Guy, C.A., Johnson, C.D., 2009. Best management practices for blasting and painting offshore facilities. In: SPE 121052. SPE Americas E&P Environment and Safety Conference. San Antonia, TX, Mar 23–26.

Swanson, T.A., Petrie, M., Sifferman, T.R., 2005. The successful use of both kinetic hydrate and paraffin inhibitors together in a deepwater pipeline with a high water cut in the Gulf of Mexico. In: SPE 93158. SPE International Symposium on Oilfield Chemistry. Houston, TX, Feb 2–4.

Taekker, N.P., Rasmussen, S.N., Roll, J., 2006. Offshore coating maintenance –cost affect by choice of new building specification and ability of the applicator. In: NACE 06029. NACE International Annual Conference and Exposition, Houston, TX.

Ward, E.G., Lee, G.C., Botelho, D.L., Turner, J.W., Dyhrkopp, F., Hall, R.A., 2000a. Consequence-based criteria for the Gulf of Mexico: philosophy and results. In: OTC 11885. Offshore Technology Conference. Houston, TX, May 1–4.

Ward, E.G., Lee, G.C., Hall, R.A., Turner, J.W., Botelho, D.L., Dyhrkopp, F., 2000b. Consequence-based criteria for the Gulf of Mexico: development and calibration of criteria. In: OTC 11886. Offshore Technology Conference. Houston, TX, May 1–4.

Wiggett, A.J., 2014. Water – chemical treatment, management and cost. In: SPE 172191. SPE Saudi Arabia Annual Technical Symposium and Exhibition. Al-Khobar, Saudi Arabia, Apr 21–24.

Witte, C.C., Ribeiro, D.M., 2012. Structural integrity management: painting predictive control. In: SPE 155857. SPE International Conference and Exhibition on Oilfield Corrosion. Aberdeen, Scotland, May 26–29.

Wolfson, S.L., Kenney, J.J., 1989. Cathodic protection monitoring – Bullwinkle. In: OTC 6053. Offshore Technology Conference. Houston, TX, May 1–4.

INDEX

Note: Page numbers followed by *f* indicate figures, *t* indicate tables, and *b* indicate boxes.

E

F

N

O

S

T

U

V

W

Printed in the United States
By Bookmasters